CELLULAR AND ORGANISMAL BIOLOGY

Readings from
**SCIENTIFIC
AMERICAN**

CELLULAR AND ORGANISMAL BIOLOGY

With Introductions by
Donald Kennedy
Stanford University

W. H. Freeman and Company
San Francisco

All of the SCIENTIFIC AMERICAN articles in
CELLULAR AND ORGANISMAL BIOLOGY are available
as separate Offprints. For a complete list of more than
950 articles now available as Offprints, write to W. H.
Freeman and Company, 660 Market Street, San Francisco,
California 94104.

Library of Congress Cataloging in Publication Data

Kennedy, Donald, 1931–   comp.
  Cellular and organismal biology.

  Includes bibliographies.
  1. Cellular control mechanisms—Addresses, essays,
lectures.   2. Biological control systems—Addresses,
essays, lectures.   I. Scientific American.   II. Title.
QH604.K46   574.8'76   74–775
ISBN 0–7167–0894–9
ISBN 0–7167–0893–0 (pbk.)

Printed in the United States of America

9 8 7 6 5 4 3 2 1

PREFACE

This collection of articles from *Scientific American* has a complex genealogy. It owes something to two earlier anthologies, *The Living Cell* (1965) and *From Cell to Organism* (1967); like this volume, those were intended to serve as readings for introductory courses in cellular and organismal biology. The passage of time, however, made some of the material they contained obsolete; and of course it brought new knowledge that deserved addition. At the same time, curricular reorganization at many universities led to redistribution of the time spent on various topics in introductory biology courses. These changes made it impractical to try to span the levels of biological organization from cell to organism in a single treatment. Instead, molecular biology now has a collection of *Scientific American* articles devoted exclusively to it: *The Chemical Basis of Life*, with introductions by Philip C. Hanawalt and Robert H. Haynes (1973). The basic biochemical functions of the cell are covered in that volume. *Vertebrate Structures and Functions* (1974), an anthology with introductions by Norman K. Wessells, treats a number of the specialized physiological functions in higher animals; and Edward O. Wilson has written introductions for a new collection entitled *Ecology, Evolution, and Population Biology* (1974).

This book, *Cellular and Organismal Biology*, emphasizes, even more strongly than did *From Cell to Organism*, the transition between two levels of organization. It is an inquiry into the *regulatory* feature of multicellularity: how it came about, how it is arranged for in development, and what signals and what responses make it possible in the adult. The process of integration is at the core of this subject matter, and so are the events of intercellular communication that underlie it. We are learning more about how the regulatory processes work, but progress is slower than at the level of molecular biology. For that reason it is still appropriate to remark, as in the Preface for *From Cell to Organism*, that

> many of the answers provided in the following articles are incomplete, and many are necessarily couched in language less precise and less conclusive than that used to define cellular energetics, recombination in bacteriophages, or the structure of a protein molecule. There is a reason for this: the problems are much more complicated. They are also extremely important, and—because their complexity often prohibits a straightforward solution—they have been approached in diverse and ingenious ways. Before we can hope to solve these problems, we must deal with them as they are and obtain dismayingly complex answers in order to ask further questions. For this reason, much of the so-called "physiology" of whole organisms is descriptive because it states the general performance characteristic of a system. Such a description is an essential prelude to an ultimate molecular analysis; the description in itself, moreover, may contain the answers to important questions about the whole organism in its environment.

Yet even in the few years that have passed since that other preface was written, much progress has been made at lifting organismal biology from the merely descriptive. Many of the most encouraging developments are reported in the articles contained here: for example, the discovery and clarification of the function of "second messengers" in chemical intercellular communication, the finding that embryonic cells are joined by bridges that permit the passage of fairly large molecules, and the discovery of particular kinds of nerve circuits that underlie specific behavioral acts.

I could have succumbed to the temptation of including articles that reported only activities along the frontiers of research. I think this collection would be a poorer teaching instrument had I done so. Many of the articles published by *Scientific American* in the decade following World War II were deliberately written to present the entire background of a subject; later articles have often started from that point. For example, the article on vision by George Wald was originally published in 1950; it paints a broad picture of the analogy between the eye and an optical device, and in doing so presents nearly all of the "classical" knowledge about the physiology of vision. Much of the later work on visual systems has built upon just this background. Similarly, the first article in the section on differentiation and development is a historical treatment of the "organizer" concept by George Gray, a great scientific journalist. He surveys, in an article first published in 1957, half a century of embryological research devoted to the mechanism of tissue interactions during development. Although the more recent advances reported in subsequent articles have clarified the problem, the work reported by Gray posed the issues clearly and laid down the basic strategies for attacking them. No amount of current progress can make it less important for students to understand this kind of background; on the contrary, its value increases as we come to know more. In part, too, the longer lifetime of articles on organismal biology reflects the very complexity of the problems, solutions to which depend upon underlying findings from cellular and molecular biology. Progress has thus unfolded at a relatively deliberate pace.

It should be said again that the theme of this book is the business of being an organism. Organisms are integrated populations of cells; and integration depends upon communication. It is thus no accident that the majority of articles in this collection are concerned, directly or indirectly, with mechanisms of intercellular communication. These mechanisms comprise the very core of organismal biology.

The introductions to each section comment on the articles, attempt to relate them to one another and to the field as a whole, and—in some cases—refer explicitly to omissions or to subsequent reinterpretations of an author's finding. In this connection, readers should understand that *Scientific American* articles do not conform to the usual covenants of "scientific publication." They are accounts for the intelligent general reader of the status of a problem, and the author is encouraged to engage in reasonable speculation about its future in a way he would not in a paper reporting only original research. The fact that some interpretations have turned out to be wrong is no criticism of the authors; instead, it is a real indicator of the dynamic state of knowledge, and of the nature of scientific progress. The student who comes to understand that, and along with it to accept (but not become transfixed by) the perishability of the printed word, will have learned something much more important than factual content.

Bibliographic references for all the articles in this volume appear at the end of the book. I am grateful to my Stanford colleagues, Philip C. Hanawalt and Norman K. Wessells, for frequent consultations about how best to divide this subject matter in order to make it most useful to students. Finally, I thank Jeanne Kennedy for help and comfort in so many general ways, and specifically for preparing the index to this book.

December 1973   *Donald Kennedy*

CONTENTS

Note on cross-references: References to articles included in this book are noted by the title of the article and the page on which it begins; references to articles that are available as Offprints, but are not included here, are noted by the article's title and Offprint number; references to articles published by SCIENTIFIC AMERICAN, but which are not available as Offprints, are noted by the title of the article and the month and year of its publication.

CELLULAR AND ORGANISMAL BIOLOGY

I

AN INTRODUCTION TO CELLULAR COMMUNICATION

I AN INTRODUCTION TO CELLULAR COMMUNICATION

INTRODUCTION

As an essay to introduce the scope and purposes of this collection, I have chosen the article "Cellular Communication" by Gunther S. Stent (1972). In explaining its relevance as a theme for the material that will follow, perhaps I can indicate some of my own rationale in constructing this reader.

Stent's opening paragraphs describe the transmission of genetic information, indicating that stability (as opposed to speed, diversity, or some other principle) is of primary importance in the design of the templates. But Stent then points out that for communication over the lifetime of the organism, which is short compared to the time spanned by genetic transmission, other features assume greater importance. He describes the kinds of molecules and secretory mechanisms that are involved in hormonal communication, and then turns to the nervous system for an extended treatment of the way stimuli are abstracted by precisely defined cellular relationships.

In its balance and its selection of material, Stent's article—which was originally published by *Scientific American* in a special issue on natural and man-made communication systems (available as *Communication, A Scientific American Book*, 1972)—reflects the overall design of *Cellular and Organismal Biology*. The basic scheme of communication in biological systems is the transfer of specific chemical messages within the cell, or between cells. In its highest form, in the nervous systems of advanced animals, the specificity comes not so much from the molecular organization of some messenger molecule as from the specific anatomical relationships between the cells engaged in these chemical transactions. We say that in the nervous system, anatomical address is substituted for chemical address—meaning that the *signals* for excitation or inhibition of particular nerve cells derive their meaning not from the identity of some chemical that is broadcast throughout the system, but from the *local* secretion of the appropriate molecule by a connected cell. The specificity of communication resides in the connections rather than in the chemistry. This theme will be sounded repeatedly in later sections. It should be remembered, however, that this merely refers the crucial problem elsewhere. To say that the specificity comes from organized connections leads directly to inquiry about the signals that cause those connections to form during the organism's development.

At the end of his article, Stent refers to a sweeping consequence of the results of this new inquiry. It is significant that we now believe much of the information-processing in the brain to be innate—that is, to depend upon connections established by the nervous system in the course of its development under genetic instructions. Far from being a plastic construct of its immediate environment, the brain and its processes have a history: it is the history of environments past, during which the blueprint for the brain's assembly was shaped and refined by natural selection. The brain is a natural product of that history, an embodiment of the forces that participate in constructing it. Stent shows how this view of brain structure and function moves us toward a philosophical position that recognizes the influence of these innate features upon our own concept of reality.

The breadth of Stent's view, from molecular templates to the structural basis for abstraction in the brain, is a reflection of Stent's personal history. Few scientists have thought in a more penetrating—or literate—way about the history and future of their own discipline. Still fewer have shown Stent's willingness to change fields. Educated first as a physicist, then trained as a microbiologist, he was an early member of the "phage group"—the molecular biologists, largely recruited from the physical sciences, who employed viruses as simple systems for exposing the rules of genetic transmission. Largely out of his conviction that after the unraveling of the genetic code molecular biology would become a less innovative and exciting discipline, Stent became interested in neurobiology in the late 1960's. He wrote generally about the waning excitement of molecular genetics in *The Coming of the Golden Age*, and then made a valedictory address to the field in his book *Molecular Genetics*. He began his transition to neurobiology by spending a year in the laboratory of S. W. Kuffler (whose work is referred to at several points in this book), and then he refinanced and reinstrumented his entire research operation at the University of California at Berkeley. In just a few years, Stent has produced important work on the nervous system: a theoretical paper on how synapses might be selectively preserved or lost during development, and an impressive body of experiments on the neural basis of swimming in leeches.

In a number of more detailed ways, the Stent article foreshadows the organization of what follows. For example, he chooses the visual system as his main example of neural organization; that emphasis rules the selection of material for Section VI. The reason is a simple one: we know more about connectivity and development in the visual system than we do for any other part of the brain, partly because a large number of especially talented investigators have studied the visual system. But the main reason for selecting Stent's article as an introduction is the breadth and thoughtfulness of his approach to integrative processes, which I hope will establish the reader's attitude toward the rest of this book.

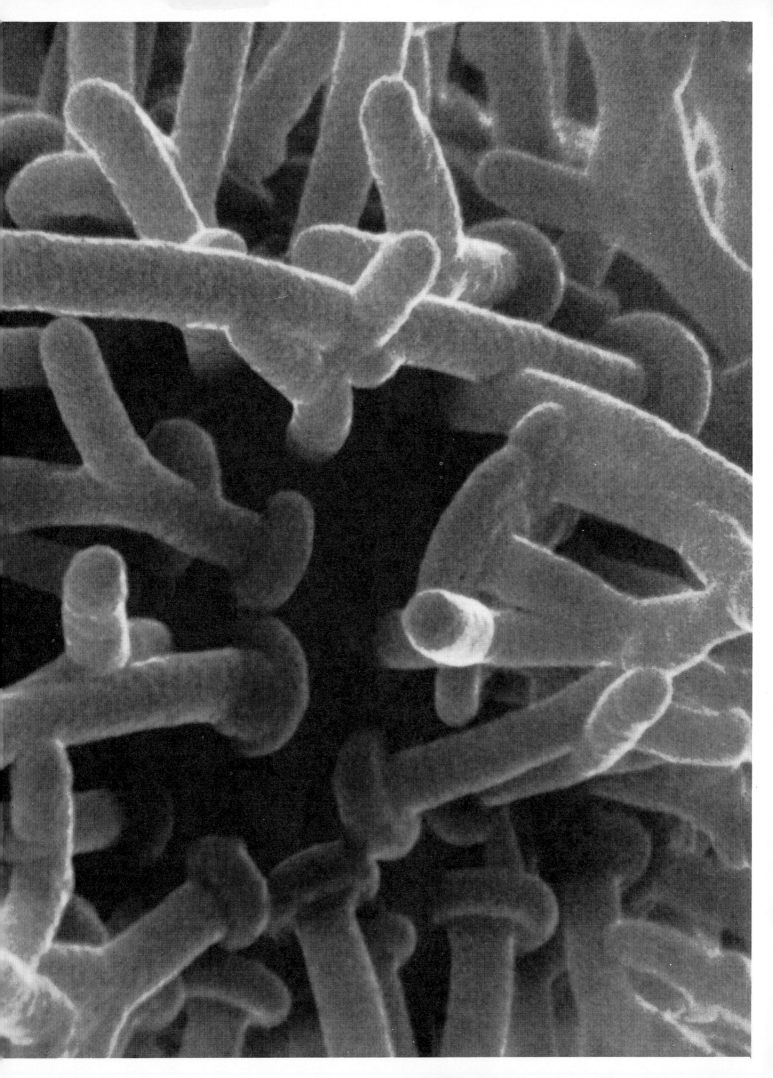

Cellular Communication

by Gunther S. Stent
September 1972

*Cells communicate by means of hormones and nerve
fibers. Such communication and all other forms of
communication are founded on the information
incorporated in the molecules of nucleic acid*

The capacity to communicate is a fundamental feature of living cells. As John R. Pierce notes in "Communication" [Offprint 677], the types of information that are the subject of cellular communication can be grouped into three general classes: genetic, metabolic and nervous. The genetic information of an organism is embodied in the precise sequence of the four kinds of nucleotide base—adenine, guanine, thymine and cytosine—in the DNA molecules of the nuclei of its cellular constituency. The meaning of that information is the specification of the precise sequence of the 20 amino acids in a myriad of different kinds of protein molecule. It is the ensemble of cellular proteins that functions to make the cell what it is: an engine built of highly specific structural members and enzymes that carries out a complex network of catalytically facilitated metabolic reactions. In the course of cell reproduction the parental DNA molecules are replicated and each of the two daughter cells is endowed with a complete store of the genetic information of the mother cell.

In addition to this "vertical" inheritance, in which each cell receives its genetic information from only one parent cell, there is a "horizontal" mode of inheritance, in which cells communicate and thus exchange their genetic information to give rise to offspring of mixed parentage. There are at least four different mechanisms by which such communication of genetic information is realized. The most primitive is genetic transformation, in which a donor cell simply releases some of its DNA molecules into its surroundings. One of the released donor DNA molecules is then taken up by a recipient cell, which incorporates the molecule into its own genetic structures. The second mechanism is genetic transduction. On the infection of a host cell with a virus one or another of the host-cell DNA molecules is incorporated into the shell of one of the progeny virus particles to which the infection gives rise. After its release from the host cell the bastard virus particle infects a recipient cell and thereby transfers to the new host the donor-cell DNA molecule it brought along for the ride. The third mechanism is genetic conjugation. Here two cells meet and establish a conjugal tube, or thin bridge, between them. In the conjugal act the donor member of the cell pair mobilizes a part of its complement of DNA and passes it through the tube to the recipient member.

The natural occurrence of these three mechanisms has so far been demonstrated only for bacteria, the lowest form of cellular life. Higher forms of life, from fungi and protozoa up to man, employ the fourth and most elaborate mechanism of genetic communication, namely sex. Here two gametes, cell types specialized for just such intercourse, meet, fuse and pool their entire complement of DNA. The process gives rise to offspring to which both parents have communicated an equal amount of genetic information. This result is to be contrasted with what happens in the first three processes, where the offspring come into possession of only a small fraction of the total donor-cell DNA and are endowed mainly with the genetic information of the recipient cell.

The biological function of genetic communication is to increase the evolutionary plasticity of the species. The ultimate source of the genetic diversity on which natural selection, the motor that drives evolution, feeds are rare changes, or mutations, in the nucleotide base sequence and hence the genetic information contained in the DNA. Thus in the purely vertical mode of inheritance genetic diversity among the members of a cell population all descended from a single ancestral mother cell could be built up very slowly by the accumulation of mutations in the individual lines of descent. In the horizontal mode of inheritance and its intercellular communication, however, there develops quite quickly a rich individual genetic diversity as mutations that have arisen in different lines of descent are continually combined and recombined among the offspring of the interbreeding population. Thus in anticipation of any environmental changes that may affect its fitness the population presents for natural selection a spectrum of diverse types among which one or another may possess a greater fitness for the future state of the species.

Metabolic information is embodied in the quality and concentration of large and small molecules that participate in the chemical processes by which cells reproduce, develop and maintain their living state. In contrast to the communication of genetic information,

TERMINAL ENDS OF NERVE-CELL AXONS in the nervous system of the "sea hare" *Aplysia* are the button-like objects in the scanning electron micrograph on the opposite page. Information is communicated from cell to cell at the junctions where the axons make synaptic contact. Micrograph, which enlarges structures 30,000 diameters, is by Edwin R. Lewis, Thomas E. Everhart and Yehoshua Y. Zeevi of University of California at Berkeley.

whose utility pertains to time periods that are long compared with the life-span of the members of populations, the communication of metabolic information is of importance to organized societies of cells over much shorter intervals. Here the mechanism of communication gen-erally consists in the release by a secre-tory cell of one of its constituent mole-cules. The released molecule diffuses through the space occupied by the cell society, and on encountering a target cell it intervenes in some highly specific manner in the metabolism of that cell. Such messenger molecules of metabol-ic information are generally called hor-mones.

The biological function of metabolic communication is mainly twofold. In the first instance hormones control the order-ly development of multicellular animals

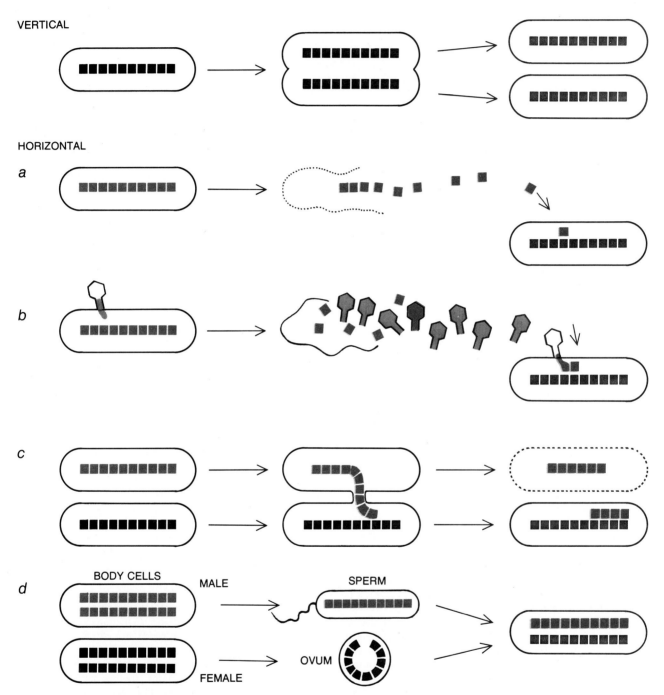

TYPES OF GENETIC COMMUNICATION can be classified as ver-tical or horizontal. In cellular reproduction (*top*) the communi-cation is vertical; the replication of the parental genes before cell division provides both offspring with complete and identical sets of genetic information. The three simpler kinds of horizontal com-munication are found in bacteria. In transformation (*a*) the donor cell releases the DNA molecules embodying its genetic information and some of the molecules are taken up by the recipient cell. In transduction (*b*), which may occur when a virus infects a cell and new viruses are formed, one of the new viruses may incorporate a DNA molecule into its shell; if the new virus now infects another cell, the donor-cell molecule and the information it embodies are transferred to the recipient cell. In conjugation (*c*) donor and re-cipient cells come into physical contact; a bridge is formed be-tween them and some donor DNA molecules are transferred to the recipient. The fourth and most elaborate kind of horizontal genetic communication (*d*) is sex. Parental organisms give rise to male (*color*) and female (*black*) gametes: specialized cells that contain only half of the information an offspring requires. When these fuse, each parent gives the offspring an equal amount of information.

and plants. These organisms represent societies made up of millions or billions of cells that come into being by a series of successive cycles of cell growth and division from a single fused pair of gametes, for example from the sperm-fertilized egg. This process of cell multiplication is accompanied by the process of cell differentiation, in which each cell acquires the particular molecular ensemble that enables it to play its destined specialized role in the life of the organism. In many cases cells receive from hormones their instructions concerning how and when to differentiate in the developmental sequence that leads to the adult organism. The female sex hormone estrogen, which belongs to the steroid class of molecules, is a well-known example of such a developmental messenger. Estrogen is released by secretory cells in the ovary of the female animal, particularly at the onset of puberty. Estrogen reaches target cells in almost all tissues and induces in these cells the metabolic reactions that eventually lead to the development of the secondary sexual characteristics of the body.

In the second instance hormones serve in the homeostatic processes by means of which all organisms minimize for their internal environment the consequences of changes in the external environment. The protein hormone insulin, well known through its connection with diabetes, is an example of such a homeostatic chemical messenger. The rate of release of insulin by secretory cells of the pancreas is accelerated in response to high glucose concentrations in the blood. On reaching target cells in liver and muscles insulin signals to these organs to remove glucose from the blood and either to store it in the form of glycogen or to burn it. Once, thanks to this increase in removal rates, the glucose blood concentration has returned to its normal level, the release of insulin by the pancreas is slowed down and so is the removal of glucose from the blood by liver and muscles. Thus insulin makes possible the maintenance of a relatively constant blood-sugar concentration in the face of great fluctuations of the animal's rate of sugar intake.

Nervous information is embodied in the activity of a special cell type possessed by all multicellular animals, the nerve cell or neuron. Although the physiological time spans over which the communication of metabolic information is relevant are much shorter than the evolutionary periods for which the communication of genetic information is intended, these physiological time spans still extend over hours, days or weeks. In

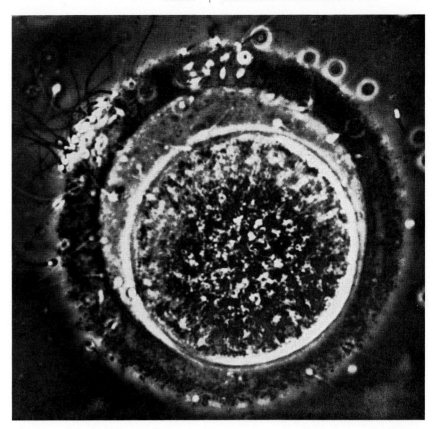

HUMAN OVUM AND SPERM CELLS at the moment of fertilization exemplify the union of gametes and the pooling of an equal amount of genetic information from each parent. Of the numerous sperm cells surrounding the ovum only one will actually deliver the male parental genes. This micrograph, which enlarges the cells some 400 diameters, was made by Landrum B. Shettles of the Columbia University College of Physicians and Surgeons.

order to stay alive, however, most animals must respond to certain events in their environment within time spans of seconds or even milliseconds. And since the diffusion of a molecule such as glucose through the space occupied by the cell society that makes up even so small an animal as a fly requires a few hours, animals must have communication channels that are faster than those provided by hormones. These channels are provided by neurons, and the biological function of the communication of nervous information they perform is to generate the rapid stimulus-response reactions that comprise the animal's behavior.

Neurons are endowed with two singular features that make them particularly suitable for this purpose. First, unlike most other cell types, they possess long and thin extensions: axons. With their axons neurons reach and come into contact with other neurons at distant sites and thereby form an interconnected network extending over the entire animal body. Second, unlike most other cell types, neurons give rise to electrical signals in response to physical or chemical stimuli. They conduct these signals along their axons and transmit them to

other neurons with which they are in contact. The interconnected network of neurons and its traffic of electrical signals forms the nervous system. It is, of course, the nervous system that is both the source and the destination of all the information in communication systems as diverse as those discussed in the September 1972 issue of SCIENTIFIC AMERICAN.

Like Roman Gaul, the nervous system is divisible into three parts: (1) an input, or sensory, part that informs the animal about its condition with respect to the state of its external and internal environment; (2) an output, or effector, part that produces motion by commanding muscle contraction, and (3) an internuncial part (from the Latin *nuncius*, meaning messenger) that connects the sensory and effector parts. The most elaborate portion of the internuncial part, concentrated in the head of those animals that have heads, is the brain.

The processing of data by the internuncial part consists in the main in making an abstraction of the vast amount of data continuously gathered by the sensory part. This abstraction is the result of a selective destruction of portions of

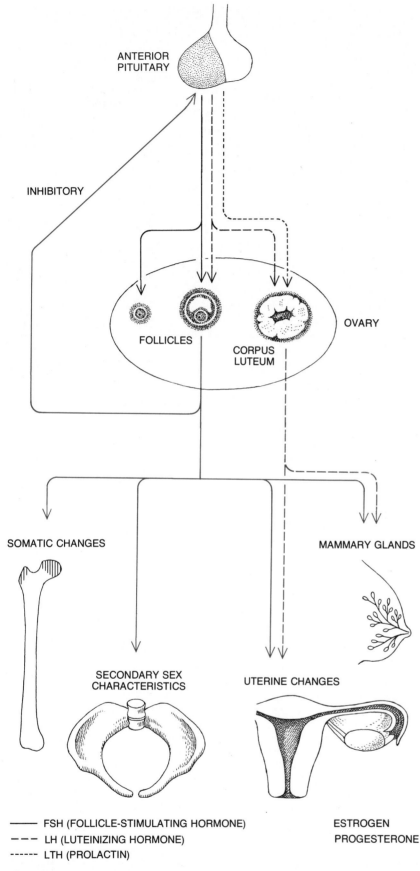

ANTERIOR
PITUITARY

INHIBITORY

OVARY

FOLLICLES

CORPUS
LUTEUM

SOMATIC CHANGES

MAMMARY GLANDS

SECONDARY SEX
CHARACTERISTICS

UTERINE CHANGES

—— FSH (FOLLICLE-STIMULATING HORMONE)
– – – LH (LUTEINIZING HORMONE)
------ LTH (PROLACTIN)

ESTROGEN
PROGESTERONE

METABOLIC COMMUNICATION is largely conducted by the messenger molecules known as hormones. One basic role of these chemical messengers is mediation of cell differentiation. Illustrated here is the action of hormones from the anterior pituitary that induce maturation in the human female. The ovarian hormones that are secreted following stimulus by pituitary hormones affect cells in bone, uterine and mammary tissues.

the input data in order to transform these data into manageable categories that are meaningful to the animal. It should be noted that the particular command pattern issued to the muscles by the internuncial part depends not only on here-and-now sensory inputs but also on the history of past inputs. Stated more plainly, neurons can learn from experience. Until not so long ago attempts to fathom how the nervous system actually manages to abstract sensory data and learn from experience were confined mainly to philosophical speculations, psychological formalisms or biochemical naïvetés. In recent years neurophysiologists, however, have made some important experimental findings that have provided for a beginning of a scientific approach to these deep problems. Here I can do no more than describe briefly one example of these recent advances and sketch some of the insights to which it has led.

Before discussing these advances we must give brief consideration to how electrical signals arise and travel in the nervous system. Neurons, like nearly all other cells, maintain a difference in electric potential of about a tenth of a volt across their cell membranes. This potential difference arises from the unequal distribution of the three most abundant inorganic ions of living tissue, sodium (Na^+), potassium (K^+) and chlorine (Cl^-), between the inside of the cell and the outside, and from the low and unequal specific permeability of the cell membrane to the diffusion of these ions. In response to physical or chemical stimulation the cell membrane of a neuron may increase or decrease one or another of these specific ion permeabilities, which usually results in a shift in the electric potential across the membrane. One of the most important of these changes in ion permeability is responsible for the action potential, or nerve impulse. Here there is a rather large transient change in the membrane potential lasting for only one or two thousandths of a second once a prior shift in the potential has exceeded a certain much lower threshold value. Thanks mainly to its capacity for generating such impulses, the neuron (a very poor conductor of electric current compared with an insulated copper wire) can carry electrical signals throughout the body of an animal whose dimensions are of the order of inches or feet. The transient change in membrane potential set off by the impulse is propagated with undiminished intensity along the thin axons. Thus the basic element of signaling in the nervous system is the nerve impulse, and the information transmitted by an axon is encoded in the frequency with which impulses propagate along it.

Neurophysiologists have developed methods by which it is possible to listen to the impulse traffic in a single neuron of the nervous system. For this purpose a recording electrode with a very fine tip (less than a ten-thousandth of an inch in diameter) is inserted into the nervous tissue and brought very close to the surface of a neuron. A neutral electrode is placed at a remote site on the animal's body. Each impulse that arises in the neuron then gives rise to a transient difference in potential between the recording electrode and the neutral electrode. With suitable electronic hardware this transient potential difference can be displayed as a blip on an oscilloscope screen or made audible as a click in a loudspeaker.

The point at which two neurons come into functional contact is called a synapse. Here the impulse signals arriving at the axon terminal of the presynaptic neuron are transferred to the postsynaptic neuron that is to receive them. The transfer is mediated not by direct electrical conduction but by the diffusion of a chemical molecule, the transmitter, across the narrow gap that separates the presynaptic axon terminal from the membrane of the postsynaptic cell. That is to say, the arrival of each impulse at the presynaptic axon terminal causes the release there of a small quantity of transmitter, which reaches the postsynaptic membrane and induces a transient change in its ion permeability. Depending on the chemical identity of the transmitter and the nature of its interaction with the postsynaptic membrane, the permeability change may have one of two diametrically opposite results. On the one hand it may increase the chance that there will arise an impulse in the postsynaptic cell. In that case the synapse is said to be excitatory. On the other hand it may reduce that chance, in which case the synapse is said to be inhibitory. Most neurons of the internuncial part receive synaptic contacts from not just one but many different presynaptic neurons, some axon terminals providing excitatory inputs and others inhibitory ones. Hence the frequency with which impulses arise in any postsynaptic neuron reflects an ongoing process of summation, more exactly a temporal integration, of the ensemble of its synaptic inputs.

We are now ready to proceed to our example of an important advance in the understanding of the internuncial nervous system, the analysis of the visual pathway in the brain of higher mammals. It is along this pathway that the visual image formed on the retina by light rays entering the eye is transformed into a visual percept, on the basis of which appropriate commands to the muscles are issued. The visual pathway begins at the mosaic of approximately 100 million primary light-receptor cells of the retina. They transform the light image into a spatial pattern of electrical signals, much as a television camera does. Still within the retina, however, the axons of the primary light-receptor cells make synapses with neurons already belonging to the internuncial part of the nervous system. After one or two further synaptic transfers within the retina the signals emanating from the primary light-receptor cells eventually converge on about a million retinal ganglion cells. These ganglion cells send their axons into the optic nerve, which con-

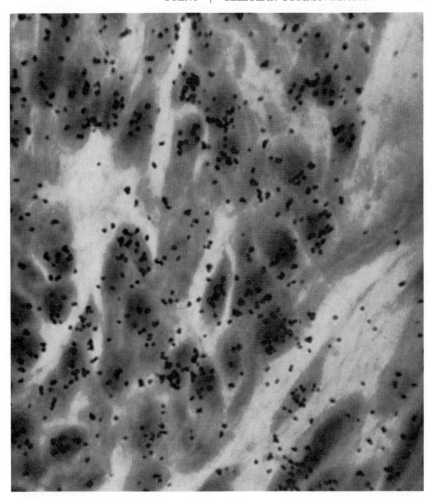

HORMONE AND TARGET CELLS appear in a radioautograph of muscle tissue from a rat uterus. Estradiol, a component of the hormone estrogen that stimulates the growth of uterine muscle, was labeled with tritium; the tissue section was prepared one hour after the hormone was injected. Black dots show concentrations of the hormone. Most are grouped in or near the nuclei of muscle cells. Micrograph was made by W. E. Stumpf, M. Sar and R. N. Prasad of the Laboratories for Reproductive Biology of the University of North Carolina.

nects the eye with the brain. Thus it is as impulse traffic in ganglion-cell axons that the visual input leaves the eye.

In 1953 Stephen W. Kuffler, who was then working at Johns Hopkins University, discovered that what the impulse traffic in ganglion-cell axons carries to the brain is not raw sensory data but an abstracted version of the primary visual input. This discovery emerged from Kuffler's efforts to ascertain the ganglion-cell receptive field, or that territory of the retinal receptor-cell mosaic whose interaction with incident light influences the impulse activity of individual ganglion cells. For this purpose Kuffler inserted a recording electrode into the immediate vicinity of a ganglion cell in a cat's retina. At the very outset of the study Kuffler made a somewhat unexpected finding, namely that even in the dark, retinal ganglion cells produce impulses at a fairly steady rate (20 to 30 times per second) and that illuminating

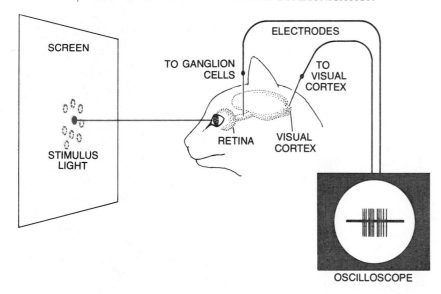

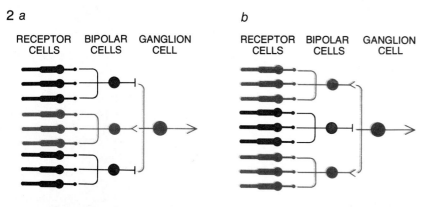

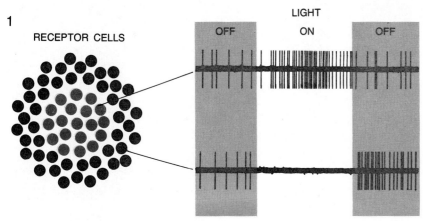

NERVE COMMUNICATION depends for its effectiveness on a process of abstraction that has been demonstrated in studies of visual perception in cats. Use of the experimental situation illustrated at top has revealed the selective destruction of sensory inputs at three successively higher levels of organization within the visual system. At the first of these levels impulses from ganglion cells in the retina, displayed on an oscilloscope, indicate that the cells receive signals from circular arrays of receptor cells. The center and the peripheral cells of the array react differently when stimulated by a spot of light. If the array is an "on center" one (1, *color*), the number of impulses produced by the ganglion cell will increase when light falls on the center but diminish when light falls on the periphery; the reaction is reversed if the array is an "off center" one. An oversimplified diagram of the neuronal circuits involved (2) suggests that in on-center arrays the central receptor cells (*a, color*) feed an excitatory synapse on the ganglion cell while the peripheral cells feed inhibitory synapses. Off-center arrays (*b*) would have the reverse circuitry. Thus the ganglion cell does not report the data on levels of illumination that are collected by the individual receptor cells but sends to the next level of abstraction a summary of the contrast between two concentric regions of the receptive field. The successive stages in this process of selective destruction of sensory inputs are illustrated on the opposite page and on the next two pages.

the entire retina with diffuse light does not have any dramatic effect on that impulse rate. This finding suggested paradoxically that light does not affect the output activity of the retina. Kuffler then, however, projected a tiny spot of light into the cat's eye and moved the image of the spot over various areas of the retina. In this way he found that the impulse activity of an individual ganglion cell does change when the light spot falls on a small circular territory surrounding the retinal position of the ganglion cell. That territory is the receptive field of the cell.

On mapping the receptive fields of many individual ganglion cells Kuffler discovered that every field can be subdivided into two concentric regions: an "on" region, in which incident light increases the impulse rate of the ganglion cell, and an "off" region, in which incident light decreases the impulse rate. Furthermore, he found that the structure of the receptive fields divides retinal ganglion cells into two classes: on-center cells, whose receptive field consists of a circular central "on" region and a surrounding circular "off" region, and off-center cells, whose receptive field consists of a circular central "off" region and a surrounding circular "on" region. In both the on-center and the off-center cells the net impulse activity arising from partial illumination of the receptive field is the result of an algebraic summation: two spots shining on different points of the "on" region give rise to a more vigorous response than either spot alone, whereas one spot shining on the "on" and the other on the "off" region give rise to a weaker response than either spot alone. Uniform illumination of the entire receptive field, the condition that exists under diffuse illumination of the retina, gives rise to virtually no response because of the mutual cancellation of the antagonistic responses from "on" and "off" regions.

It could be concluded, therefore, that the function of retinal ganglion cells is not so much to report to the brain the intensity of light registered by the primary receptor cells of a particular territory of the retina as it is to report the degree of light and dark contrast that exists between the two concentric regions of its receptive field. As can be readily appreciated, such contrast information is essential for the recognition of shapes and forms in the animal's visual field, which is what the eyes are mainly for. Thus we encounter the first example in this discussion of how the nervous system abstracts information by selective destruction of information. The absolute light-intensity data gathered by the primary light-receptor cells are selectively destroyed in the algebraic summation process of "on" and "off" responses and are thereby transformed into the perceptually more meaningful relative-contrast data.

When one thinks about the neuronal circuits that might be responsible for this retinal abstraction process, the first possibility that comes to mind is that they embody the antagonistic function of excitatory and inhibitory synaptic inputs to the same postsynaptic neuron. One might suppose that to produce an on-center receptive field the axon terminals of primary receptor cells from the central "on" territory simply make excitatory synapses with their retinal ganglion cell and primary cells from the peripheral "off" territory make inhibitory synapses [see illustration on opposite page]. Detailed analyses of the anatomy and physiology of retinal neurons conducted in recent years have shown that on the one hand the real situation is much more complicated than this simple picture, but that on the other hand the actual neuronal circuits do involve matching of excitatory and inhibitory synapses in the pathways leading from the primary light receptor of antagonistic receptive-field regions to the ganglion cell.

In the late 1950's David H. Hubel and Torsten N. Wiesel, two associates of Kuffler's (who was by then, and still is, at the Harvard Medical School), began to extend these studies on the structure and character of visual receptive fields to the next-highest stage of information processing [see the article "The Visual Cortex of the Brain," by David H. Hubel, beginning on page 272]. For this purpose they examined the further fate of the impulse signals conducted away from the eye by the million or so retinal ganglion-cell axons in the optic nerve to the brain. The optic nerves from the two eyes meet near the center of the head at the optic chiasm. In animals such as cats and men there is a partial crossover of the optic nerves at the optic chiasm; some retinal ganglion-cell axons cross over to the opposite brain hemisphere and some do not. This partial crossover provides the right hemisphere of the brain with the binocular input that the retinas of both eyes receive from the left half of the animal's visual field, and vice versa. Behind the optic chiasm the output of the retinal ganglion cells passes through a way station in the midbrain that for the purposes of this discussion can be consid-

ered a simple neuron-to-neuron replay and finally reaches the cerebral cortex at the back of the head. The cortical destination of the visual input is designated as the visual cortex. Here the incoming axons make synaptic contact with the nerve cells of the cortex. The first cortical cells with which the axon projecting from the eye comes in contact in turn send their axons to other cells in the visual cortex for further processing of the visual input. From that point one must still find the trail that eventually leads to the motor centers of the brain, where, if the visual stimulus is to elicit a behavioral act, commands must be issued to the muscles.

Hubel and Wiesel's procedure was to insert a recording electrode into the visual cortex of a cat and to observe the impulse activity of individual cortical neurons in response to various light stimuli projected on a screen in front of the cat's eyes. In this way they found that

these higher-order neurons of the visual pathway also respond only to stimuli falling on a limited retinal territory of light-receptor cells. The character of the receptive fields of cortical neurons turned out, however, to be dramatically different from that of the retinal ganglion cells. Instead of having circular receptive fields with concentric "on" and "off" regions, the cortical neurons were found to respond to straight edges of light-dark contrast, such as bright bars on a dark background. Furthermore, for the straight edge to produce its optimum response it must be in a particular orientation in the receptive field. A bright bar projected vertically on the screen that produces a vigorous response in a particular cortical cell will no longer elicit the response as soon as its projection is tilted slightly away from the vertical.

In their first studies Hubel and Wiesel found two different classes of cells in the

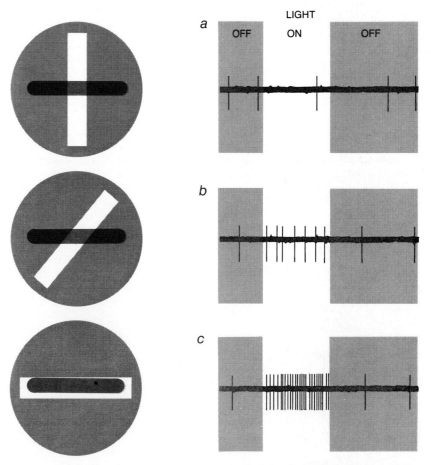

SECOND LEVEL OF ABSTRACTION in the processing of visual inputs occurs among the "simple" cells of the visual cortex, which receive summarized data from many ganglion cells. Cortical-cell impulses increase only when the many retinal receptor cells providing input data are stimulated by light-dark contrast in the form of a straight edge, such as a bar of light shown on a dark background. Moreover, the simple cortical cells fail to react if the edge is not precisely oriented, as the difference in oscilloscope impulse frequencies (a–c) indicates. In effect, by means of selective destruction of input data, the simple cortical cells transform the ganglion-cell information about light-dark contrast at various points in the visual field into information about light-dark contrasts along straight-line sets of points.

visual cortex: simple cells and complex cells. The response of simple cells demands that the straight-edge stimulus must have not only a given orientation but also a precise position in the receptive field. The stimulus requirements of complex cells are less demanding, however, in that their response is sustained on a parallel displacement (but not a tilt) of the straight-edge stimuli within the receptive field. Thus the process of abstraction of the visual input begun in the retina is continued at higher levels in the visual cortex. The simple cells, which are evidently the next abstraction stage, transform the data supplied by the retinal ganglion cells concerning the light-dark contrast at individual points of the visual field into information concerning the contrast present at particular straight-line sets of points. This transformation is achieved by the selective destruction of the information concerning just how much contrast exists at just which point of the straight-line set. The complex cells carry out the next stage of abstraction. They transform the contrast data concerning particular straight-line sets of visual-field points into information concerning the contrast present at parallel sets of straight-line point sets. In other words, here there is a selective destruction of the information concerning just how much contrast exists at each member of a set of parallel straight lines.

The neuronal circuits responsible for these next stages of abstraction of the visual input can now be fathomed. Let us consider first the simple cell of the visual cortex that responds best to a bright bar on a dark background projected in a particular orientation and position on the retinal receptor-cell mosaic. Here we may visualize the simple cell being so connected to the output of

the retina that it receives synaptic inputs from axons reporting the impulse activity of a set of on-center retinal ganglion cells with receptive fields arranged in a straight line. Therefore a bright bar falling on all the central "on" regions of this row of receptive fields but on none of the peripheral "off" regions will activate the entire set of retinal ganglion cells and provide maximal excitation for the simple cortical cell. If the retinal projection of the bar is slightly displaced or tilted, however, some light will also strike the peripheral "off" regions and the excitation provided for the simple cell is diminished.

We next consider the response of a complex cortical cell to a bright bar of a particular orientation in any one of several parallel positions in the receptive field. This response can be easily explained on the basis that the complex cell receives its synaptic inputs from the axons of a set of simple cortical cells. All the simple cells of this set would have receptive fields that respond optimally to a bright bar projected in the same field orientation, but they differ in the field position of their optimal response. A suitably oriented bright bar projected anywhere in the complex receptive field will always activate one of the component simple cells and so also the complex cell.

In their later work Hubel and Wiesel were able to identify cells in the visual cortex whose optimal stimuli reflect even higher levels of abstraction than parallel straight lines, such as straight-line ends and corners. It is not so clear at present how far this process of abstraction by convergence of communication channels ought to be imagined as going. In particular, should one think that there exists for every pattern of whose specific recognition an animal is capable at least

one particular cell in the cerebral cortex that responds with impulse activity when that pattern appears in the visual field? In view of the vast number of such patterns we recognize in a lifetime, that might seem somewhat improbable. So far, however, no other plausible explanation of perception capable of advancing neurophysiological research appears to have been put forward.

Admittedly, ever since the discipline of neurophysiology was founded more than a century ago, there have been adherents of a "holistic" theory of the brain. This theory envisions specific functions of the brain, including perception, depending not on the activity of particular localized cells or centers but flowing instead from general and widely distributed activity patterns. With the discovery of functionally specialized brain loci such as the visual cortex the holistic theory has had to retreat from its original extreme position, yet it may still hold in some more limited way. Quite aside from being hard to fathom, however, the theory seems to be better suited to inspiring experiments that show the defects of the localization concept than to explaining how the brain might actually work.

In any case the findings on the nature of nervous communication described here have some important general implications, in that they lend physiological support to the latter-day philosophical view that has come to be known as "structuralism." In recent years the structuralist view emerged more or less simultaneously, independently and in different guises in diverse fields of study, for example in analytical psychology, cognitive psychology, linguistics and anthropology. The names most often associated with each of these developments are those of Carl Jung, Wolfgang Köhler, Noam Chomsky and Claude Lévi-Strauss. The emergence of structuralism represents the overthrow of "positivism" (and its psychological counterpart "behaviorism") that held sway since the late 19th century and marks a return to Immanuel Kant's late-18th-century critique of pure reason. Structuralism admits, as positivism does not, the existence of innate ideas, or of knowledge without learning. Furthermore, structuralism recognizes that information about the world enters the mind not as raw data but as highly abstract structures that are the result of a preconscious set of step-by-step transformations of the sensory input. Each transformation step involves the selective destruction of information,

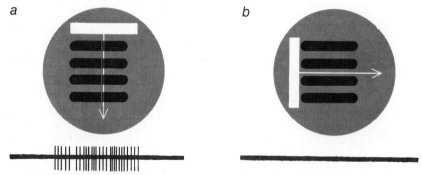

a   b

THIRD LEVEL OF ABSTRACTION occurs when data from the simple cortical cells reach the complex cortical cells. These higher nerve cells increase their impulses only when the retinal stimulus consists of a bar of light in motion across the visual field. Moreover, only the vertical movement of a horizontal bar through the field (*a*) induces the complex cells of the cortex to respond; a vertical bar moved horizontally (*b*) goes unnoticed. A diagram of the circuitry responsible for this type of abstraction appears on the opposite page.

according to a program that preexists in the brain. Any set of primary sense data becomes meaningful only after a series of such operations performed on it has transformed the data set into a pattern that matches a preexisting mental structure. These conclusions of structuralist philosophy were reached entirely from the study of human behavior without recourse to physiological observations. As the experimental work discussed in this article shows, however, the manner in which sensory input into the retina is processed along the visual pathway corresponds exactly to the structuralist tenets.

It should be mentioned in this connection that studies on the visual cortex of monkeys have led to results entirely analogous to those obtained with cats, namely that there are simple and complex cells responding to parallel straight-line patterns in the visual field. It is therefore reasonable to expect that the organization of the human visual cortex follows the same general plan. That is to say, our own visual perception of the outer world is filtered through a stage in which data are processed in terms of straight parallel lines, thanks to the way in which the input channels coming from the primary light-receptors of the retina are hooked up to the brain. This fact cannot fail to have profound psychological consequences; evidently a geometry based on straight parallel lines, and hence by extension on plane surfaces, is most immediately compatible with our mental equipment. It need not have been this way, since (at least from the neurophysiological point of view) the retinal ganglion cells could just as well have been connected to the higher cells in the visual cortex in such a way that their concentric on-center and off-center receptive fields form arcs rather than straight lines. If evolution had given rise to that other circuitry, curved rather than plane surfaces would have been our primary spatial concept. Thus neurobiology has now shown why it is human—and all too human—to hold Euclidean geometry and its nonintersecting coplanar parallel lines to be a self-evident truth. Non-Euclidean geometries of convex or concave surfaces, although our brain is evidently capable of conceiving them, are more alien to our built-in spatial-perception processes. Apparently a beginning has now been made in providing, in terms of cellular communication, an explanation for one of the deepest of all philosophical problems: the relation between reality and the mind.

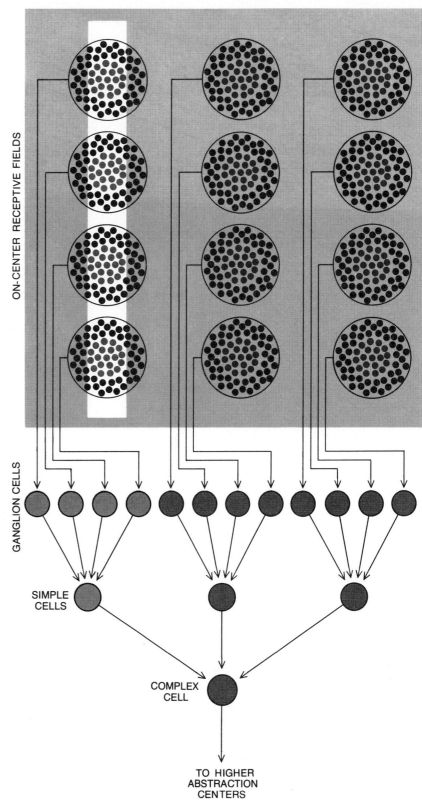

CIRCUITRY OF ABSTRACTION inferred to exist within a cat's visual cortex is illustrated schematically. It begins with a rectangular array of receptive fields. The array is comprised of three parallel vertical rows and each row contains four receptive fields. The axons from the retinal ganglion cells that correspond to each row converge on one of three simple cortical cells; the axons from the simple cells then converge on one complex cortical cell. A bar of light (left) falling on a row of receptive fields excites the four ganglion cells corresponding to that row and also the simple cortical cell connected to them (color).

II

LEVELS OF CELLULAR COMPLEXITY

LEVELS OF CELLULAR COMPLEXITY II

INTRODUCTION

One of the insights that has been most helpful to biologists is the so-called Cell Theory. It is difficult to say precisely what that phrase encompasses. In part, this is because the cellular nature of living systems was revealed not in a dramatic single resolution but rather through a series of clarifications that were developed throughout most of the mid-nineteenth century. The basic postulates of the Cell Theory are, first, that all organisms are composed of subunits resembling one another in the possession of a certain set of organelles and a boundary; second, that these entities arise only through the division of pre-existing cells. Certain organisms, observed a century and a half earlier by Leeuwenhoek and other early lensmakers, are themselves unicellular; a significant new finding in the mid-nineteenth century indicated that the complex structure of large multicellular plants and animals appeared to be composed of microscopic units resembling these solitary microbes. The equally remarkable corollary adds that the cell is the basic unit of reproduction; indeed, we know of no example (though it would not seem to be theoretically unreasonable) in which an element of less-than-cellular complexity serves as the reproductive unit in any organism that has a fully cellular organization itself.

The establishment of the Cell Theory, attended by far less ceremony than the nearly contemporaneous Darwinian Revolution, has made it possible for biologists to deal coherently with what would otherwise be a hopeless welter of diversity. The doctrine of evolution offered explanations for the enormous breadth of the living spectrum; the Cell Theory offered hopeful assurances that these variations, despite their extent, had a theme, that the theme was the cellular organization of living systems, and that one might hope to comprehend basic mechanisms of life without inspecting an infinite series of special cases. That expectation has been amply fulfilled; the articles in this section document the recent knowledge that has resulted from the discovery that all cells show a remarkable similarity of mechanisms for obtaining and using energy and for replicating themselves and their parts.

This section deals with the basic properties of cells, and also with the levels of complexity found in several different types of cells. Despite the usefulness of the Cell Theory as a general explanation of biological structure, nothing is uniform about the components of cells: the "generalized cell" of the textbooks is largely a myth. We recognize several grades of organization in cells and in living units that are below the cellular level. The viruses, clearly not cellular, may in fact correspond to parts of cells. The bacteria and blue-green algae have an organization clearly less complicated than that of protozoan cells or the cells of multicellular organisms.

The first selection, Jean Brachet's "The Living Cell," was written as the introductory article of the September, 1961, special issue of *Scientific American*, also titled "The Living Cell." Brachet's article is a general account of the properties of cells, dealing in particular with the many new revelations made by electron microscopy and correlating their structures with functions assignable to them by biochemical techniques. Brachet discusses those cells found in multicellular organisms or in the higher protists like protozoa and algae—cells characterized by well-developed organelles and a discrete nucleus with visible chromosomes. Such *eucaryotic cells* thus display much more internal complexity and order than the *procaryotic cells* characteristic of the bacteria and blue-green algae. In blue-green algae, defined nuclei with patent chromosomes are missing, and the functions normally performed in mitochondria—especially the chemical reactions of energy conversion—appear to be associated with the membrane instead.

Simpler still, of course, are the viruses. Small in size and missing many of the components of true cells, they are recognizable primarily by the results of their activity: as infections of plant or animal cells or of bacteria. Each of these cell types acts as a host for an assemblage of viruses; some are benign, some destructive. The simplicity of viruses often defeats direct structural analysis, and demands that the investigator use ingenious inferential techniques. For example, in 1961, when the article "Viruses and Genes" was written by François Jacob and Elie L. Wollman, the bacterial chromosome had never been visualized. Now it can be isolated and studied by electron microscopy. Then, however, it was necessary to infer its circular structure indirectly, by experiments involving, in part, the transfer of genetic "markers" from bacterial viruses ("bacteriophages" like the T_2 virus Jacob and Wollman discuss in their article). These genetic manipulations not only contributed to our knowledge of the nature of the bacterial system; they also led to speculation about the origin and nature of viruses. Viral particles, consisting of a protein coat and nucleic acid core, are the simplest living systems known—so simple, in fact, that many biologists would contest the adjective "living." They must intercept the metabolic mechanism of a cellular host in order to reproduce; this requirement arouses speculation about whether viruses are minimal, possibly "primitive" organisms or, as Jacob and Wollman suggest, whether some may be nonintegrated, or stray, parts of the genetic material of a cell.

The bacteria discussed by Jacob and Wollman, of course, differ from the viruses in being complete organisms, capable of carrying out functions such as energy exchange and reproduction without the assistance of other cells. So too are the pleuropneumonia-like organisms discussed by Harold J. Morowitz and Mark E. Tourtellotte in "The Smallest Living Cells" (1962). The smallest organisms in the so-called PPLO group are far smaller than bacteria, nearly as small as the viruses; yet they are able to carry out each of the several dozen enzymatic steps in energy metabolism, for example, and presumably perform all the other biochemical functions performed by self-sufficient organisms, too. Yet their size and performance produce a paradox, because the number of macromolecules needed to encode, synthesize, and regulate a full set of cellular proteins seems almost more than the cell has room for. Since 1962, when the article was first published, it has been discovered that some important large molecules found in bacterial cells—those that assist in regulating the rate of transcription of a particular segment of DNA, for instance—are present in astonishingly low numbers, perhaps only half a dozen molecules per cell. Is there a theoretical lower size limit? We are still left with the riddle that faced the authors: how large must a cell be to house the DNA necessary to encode the structure of enough specific proteins to allow the cell to live?

The next article, Patrick Echlin's "The Blue-Green Algae" (1966), sketches the enormous success of an especially simple group of free-living cells. Possessing the general simplicity of procaryotic cells, blue-green algae are to the bacteria as the more complex green algae are to animal-like protozoans. Like the bacteria, the blue-green algae lack a well-defined nucleus; instead of accomplishing photosynthesis in complex organelles like the chloroplasts of eucaryotic plant cells, they accomplish it with pigments located in peripheral lamellae. Despite this simplicity, as Echlin's article recounts, the blue-green algae have been successful in colonizing unusually diverse habitats.

The last article in this section, that by John Tyler Bonner on "Differentiation in Social Amoebae" (1959), introduces the main theme of this book. For the first time in this consideration of complexity in cells, we confront complex forms of biological organization that we know are multicellular. Nearly all of these are composed of eucaryotic cells: except for a few doubtful examples of colonial bacteria, it appears that multicellularity is a monopoly of the relatively complex form of cellular organization. We do not know the exact point at which complex cellular forms first gave rise to multicellular organization in the course of evolution. It probably originated in several independent instances among the plants, and at least twice in animal evolution. We may never know whether it resulted from the accidental cohesion of prospectively independent cells following division, from some secondary aggregation mechanism, or even from the subdivision of multinucleate single cells. Whatever its origin, multicellularity clearly had advantages; and though a wide range of successful unicellular organisms still exists today, the dominance of complex, well-integrated multicellular plants and animals is easily recognized.

A hallmark of multicellular organization, as will be emphasized in Section III, is the ability of the constituent cells to communicate with one another. This ability is true even of some examples of primitive cellular association, examples so obscurely related to the more familiar world of multicellular organisms that biologists hesitate to put them on the same continuum. Yet they display, in simple and easily analyzed form, the same kind of communication and differentiation that underlies the development of all organisms. Such an example is the one described by Bonner. In social amoebae multicellularity is established by secondary aggregation, which is very different from the embryonic development of most higher plants and animals. The individual amoebae are attracted by a chemical, acrasin, that they themselves produce, and aggregation results after the feeding stage has ended. The cells form a compact mass in which some differentiation between front and rear is evident. The mass responds to surgical separation by re-establishing the original ratio of differentiated cell types. The capacity to regulate this ratio testifies that information is exchanged among individual cells in the mass.

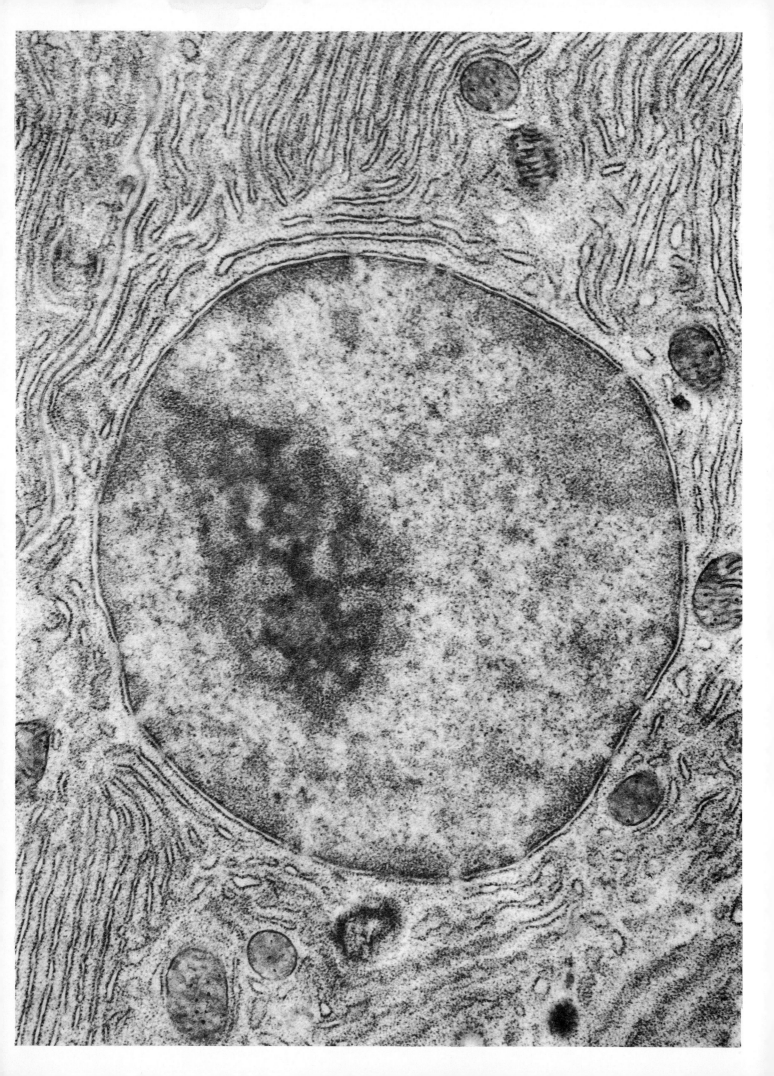

The Living Cell

by Jean Brachet
September 1961

*The living cell is the fundamental particle of life.
Anatomical and chemical views of the cell have now
converged to show that it is not a droplet of
protoplasm but a highly organized molecular factory*

The living cell is the fundamental unit of which all living organisms are made. To a reader who finds this a commonplace, it may come as a surprise that the recognition of the cell dates back only a little more than 100 years. The botanist Matthias Jakob Schleiden and the zoologist Theodor Schwann first propounded the cell theory in 1839 out of their parallel and independent studies of the tissues of plants and animals. Not long after, in 1859, Rudolf Virchow confirmed the cell's unique role as the vessel of "living matter" when he showed that all cells necessarily derive from pre-existing cells: *omnis cellula e cellula*. Since cells are concrete objects and can easily be observed, the experimental investigation of cells thereafter displaced philosophical speculations about the problem of "life" and the uncertain scientific studies that had pursued such vague concepts as "protoplasm."

In the century that followed investiga-

tors of the cell approached their subject from two fundamentally different directions. Cell biologists, equipped with increasingly powerful microscopes, proceeded to develop the microscopic and submicroscopic anatomy of the intact cell. Beginning with a picture of the cell as a structure composed of an external membrane, a jelly-like blob of material called cytoplasm and a central nucleus, they have shown that this structure is richly differentiated into organelles adapted to carry on the diverse processes of life. With the aid of the electron microscope they have begun to discern the molecular working parts of the system. Here, in recent years, their work has converged with that of the biochemists, whose studies begin with the ruthless disruption of the delicate structure of the cell. By observing the chemical activity of materials collected in this way, biochemists have traced some of the pathways by which the cell carries out the biochemical reactions that underlie

the processes of life, including those responsible for manufacturing the substance of the cell itself.

It is the present intersection of the two lines of study that enables an attempt at a synthesis of what is known of the structure and function of the living cell. The cell biologist now seeks to explain in molecular terms what he sees with the aid of his instruments; he has become a molecular biologist. The biochemist has become a biochemical cytologist, interested equally in the structure of the cell and in the biochemical activity in which it is engaged. As the reader will see, the mysteries of cell structure and function cannot be resolved by the exercise of either morphological or biochemical techniques alone. If the research is to be successful, the approach must be made from both sides at once. But the understanding of life phenomena that flows from investigation of the cell has already fully ratified the judgment of the 19th-century biologists who perceived that living matter is divided into cells, just as molecules are made of atoms.

A description of the functional anatomy of the living cell must begin with the statement that there is no such thing as a typical cell. Single-celled organisms of many different kinds abound, and the cells of brain and muscle tissue are as different in morphology as they are in function. But for all their variety they are cells, and so they all have a cell membrane, a cytoplasm containing various

NUCLEUS OF THE LIVING CELL is the large round object in the center of the electron micrograph on the opposite page. The membrane around the nucleus is interrupted by pores through which the nucleus possibly communicates with the surrounding cytoplasm. The smaller round objects in the cytoplasm are mitochondria; the long, thin structures are the endoplasmic reticulum; the dark dots lining the reticulum are ribosomes. Actually the micrograph shows not a living cell but a dead cell: the cell has been fixed with a compound of the heavy metal osmium, immersed in a liquid plastic that is then made to solidify and finally sliced with a glass knife. The electron beam of the microscope mainly detects the atoms of osmium, distributed according to the affinity of the fixing compound for various cell constituents. The micrograph was made by Don W. Fawcett of the Harvard Medical School. The enlargement is 28,400 diameters. The cell itself is from the pancreas of a bat.

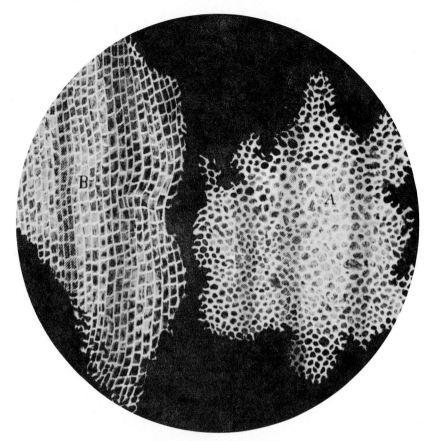

DRAWING OF CELLS in cork was published by Robert Hooke in 1665. Hooke called them cells, but the fact that all organisms are made of cells was not recognized until 19th century.

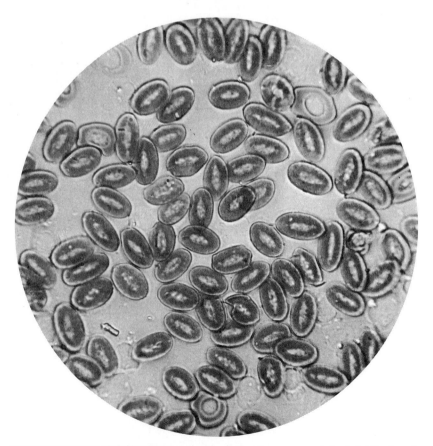

PHOTOMICROGRAPH OF CELLS in the blood of a pigeon was made by J. J. Woodward, a U.S. Army surgeon, in 1871. Woodward had made the first cell photomicrograph in 1866.

organelles and a central nucleus. In addition to having a definite structure, cells have a number of interesting functional capacities in common.

They are able, in the first place, to harness and transform energy, starting with the primary transformation by green-plant cells of the energy of sunlight into the energy of the chemical bond. Various specialized cells can convert chemical-bond energy into electrical and mechanical energy and even into visible light again. But the capacity to transform energy is essential in all cells for maintaining the constancy of their internal environment and the integrity of their structure [see "How Cells Transform Energy," Offprint 91].

The interior of the cell is distinguished from the outer world by the presence of very large and highly complex molecules. In fact, whenever such molecules turn up in the nonliving environment, one can be sure they are the remnants of dead cells. On the primitive earth, life must have had its origin in the spontaneous synthesis of complicated macromolecules at the expense of smaller molecules. Under present-day conditions, the capacity to synthesize large molecules from simpler substances remains one of the supremely distinguishing capacities of cells.

Among these macromolecules are proteins. In addition to making up a major portion of the "solid" substance of cells, many proteins (enzymes) have catalytic properties; that is, they are capable of greatly accelerating the speed of chemical reactions inside the cell, particularly those involved in the transformation of energy. The synthesis of proteins from the simpler units of the 20-odd amino acids goes forward under the regulation of deoxyribonucleic acid (DNA) and ribonucleic acid (RNA), by far the most highly structured of all the macromolecules in the cell [see "How Cells Make Molecules," Offprint 92]. In recent years and months investigators have shown that DNA, localized in the nucleus of the cell, presides at the synthesis of RNA, which is found in both the nucleus and the cytoplasm. The RNA in turn arranges the amino acids in proper sequence for linkage into protein chains. The DNA and the RNA may be compared to the architect and contractor who collaborate on the construction of a nice-looking house from a heap of bricks, stones and tiles.

At one or another stage of life every cell has divided: a mother cell has grown

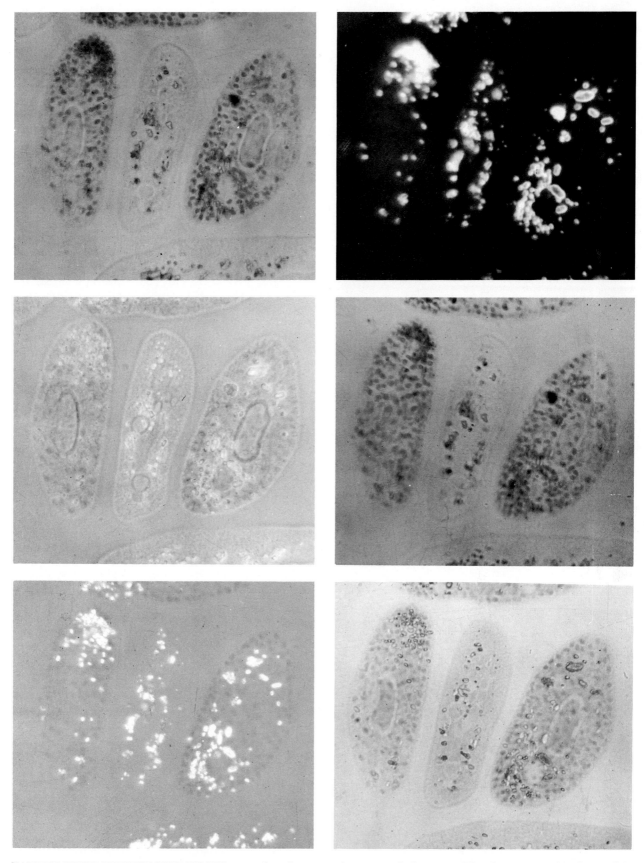

VARIOUS KINDS OF LIGHT MICROSCOPY are used to photograph the same three paramecia. The photomicrograph at top left was made with a conventional light microscope and bright-field illumination; the one at top right, with dark-field illumination. The photomicrograph at middle left was made with a phase microscope at low contrast; the one at middle right, with a phase microscope at high contrast. The photomicrograph at bottom left was made with a polarized-light microscope; the one at bottom right, with an interference microscope of the AO-Baker type. The bright spots that appear in some of the photomicrographs are small crystals that are normally present in paramecia. All the micrographs were made by Oscar W. Richards of the American Optical Company.

and given rise to two daughter cells, according to the delicate process described by Daniel Mazia [see "How Cells Divide," Offprint 93]. Before the turn of the century biologists had observed that the crucial event in this process was the equal division of bodies in the nucleus that accepted a certain colored dye and so were called chromosomes. It was correctly surmised that the chromosomes are the agents of heredity; in their precise self-replication and division they convey to the daughter cells all the capacities of the mother cell. Contemporary biochemistry has now shown that the principal constituent of the chromosomes is DNA, and an important aim of the molecular biologist today is to discover how the genetic information is encoded in the structure of this macromolecule.

The capacity for generative reproduction is not confined exclusively to the living cell. There are in the present world macromolecules called viruses that contain nucleic acids and proteins of great complexity and specificity. When they penetrate into suitable cells, they multiply just as cells do, but at the expense of the cell. They have a heredity, since they breed true when they replicate themselves, and they synthesize their own proteins. But, lacking the full anatomical endowment of the cell, they are unable to generate the energy required for their multiplication. Viruses are thus obligatory parasites of cells and take over the enzyme system of the infected cell in order to supply the energy they need. The cell must, however, furnish exactly the right complement of enzymes. This is why tobacco mosaic virus, for example, will not multiply in human cells and so is harmless to human beings.

Such single-celled organisms as bacteria, having the capacity to make their own enzymes and so to generate the energy required for their growth and multiplication, can live and multiply in a much simpler medium than that provided by the interior of a living cell. They are, therefore, not obligatory parasites. From the viewpoint of anatomy, however, bacteria are much simpler than cells, and the various bacteria are distributed over the range of complexity from the virus upward to the cell.

In addition to the capacity for energy transformation, biosynthesis and reproduction by self-replication and division, the cells of higher organisms possess other capacities that fit them for the concerted community life that is the life of the organism. From the single-celled fertilized egg the multicelled organism arises not only by the division of the daughter cells but also by their concurrent differentiation into the specialized cells that form various tissues. In many cases when a cell has become differentiated and specialized, it does not divide any more; there is a kind of antagonism between differentiation and growth by cell division.

In the adult organism the capacity for reproduction and perpetuation of the species is left to the eggs and spermatozoa. These gametes, like all other cells in the body, have arisen by cell division from the fertilized egg, followed by differentiation. Cell division remains, however, a frequent event in the adult organism wherever cells continuously wear out and degenerate, as they do in the skin, the intestine and in the bone marrow from which the blood cells arise.

During embryonic development the differentiating cells display a capacity for recognition of others of their own kind. Cells that belong to the same family and resemble one another tend to cluster together, forming a tissue from which cells of all other kinds are excluded. In this mutual association and rejection of cells the cell membrane appears to play a decisive role. The membrane is also one of the principal cell components involved in the function of the muscle cells that endow the organism with the power of movement, of the nerve cells that provide communication lines to integrate the activity of the organism and of the sensory cells that receive stimuli from without and within.

Although there is no typical cell, one may usefully put together a composite cell for the purpose of charting the anatomical features that are shared in varying degrees by all cells. Such a cell, based largely upon what is seen in electron micrographs, is presented on the opposite page; comparison of this cell with the corresponding cell drawn from photomicrographs made by Edmund B. Wilson of Columbia University in 1922 suggests the rapid advances that have been brought about by the electron microscope.

Even the cell membrane, which is only 100 angstrom units thick (one angstrom unit is one ten-millionth of a millimeter) and appears as little more than a boundary in the light microscope, is shown by the electron microscope to have a structure. It is true that electron micrographs have not yet revealed much

DIAGRAM OF A TYPICAL CELL (although there is no such thing as a typical cell) is based on what is seen in the conventional light microscope. Diagram is based on one that appears in 1922 edition of Edmund B. Wilson's *The Cell in Development and Inheritance.*

25

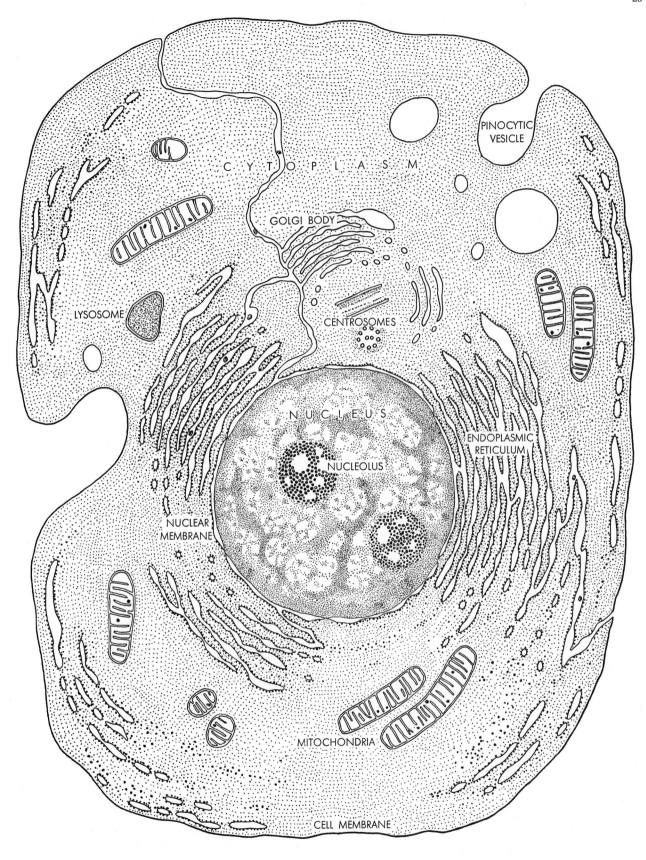

CYTOPLASM

PINOCYTIC
VESICLE

GOLGI BODY

LYSOSOME

CENTROSOMES

NUCLEUS

ENDOPLASMIC
RETICULUM

NUCLEOLUS

NUCLEAR
MEMBRANE

MITOCHONDRIA

CELL MEMBRANE

MODERN DIAGRAM OF A TYPICAL CELL is based on what is seen in electron micrographs such as the one reproduced on page 20. The mitochondria are the sites of the oxidative reactions that provide the cell with energy. The dots that line the endoplasmic reticulum are ribosomes: the sites of protein synthesis. In cell division the pair of centrosomes, one shown in longitudinal section (*rods*), other in cross section (*circles*), part to form poles of apparatus that separates two duplicate sets of chromosomes.

about this structure. On the other hand, such complexity as is shown clearly accords with what is known about the functional properties of the membrane. In red blood cells and nerve cells, for example, the membrane distinguishes between sodium and potassium ions although these ions are alike in size and electrical charge. The membrane helps potassium ions get into the cell and opposes more than a mere permeability barrier to sodium ions; that is, it is capable of "active transport." The membrane also brings large molecules and macroscopic bodies into the interior of the cell by mechanical ingestion [see "How Things Get Into Cells," Offprint 96].

Beyond the membrane, in the cytoplasm, the electron microscope has resolved the fine structure of organelles that appear as mere granules in the light microscope. Principal among them are the chloroplasts of green-plant cells and the mitochondria that appear in both animal and plant cells. These are the "power plants" of all life on earth. Each is adapted to its function by an appro-

priate fine structure, the former to capturing the energy of sunlight by photosynthesis, the latter to extracting energy from the chemical bonds in the nutrients of the cell by oxidation and respiration. From each of these power plants the yield of energy is made available to the energy-consuming processes of the cell, neatly packaged in the phosphate bonds of the compound adenosine triphosphate (ATP).

The electron microscope clearly distinguishes between the mitochondrion, with its highly organized fine structure, and another associated body of about the same size: the lysosome. As Christian de Duve of the Catholic University of Louvain has shown, the lysosome contains the digestive enzymes that break down large molecules, such as those of fats, proteins and nucleic acids, into smaller constituents that can be oxidized by the oxidative enzymes of the mitochondria. De Duve postulates that the lysosome represents a defense mechanism; the lysosomal membrane isolates the digestive enzymes from the rest of the cytoplasm. Rupture of the membrane

and release of the accumulated enzymes lead quickly to the lysis (dissolution) of the cell.

The cytoplasm contains many other visible inclusions of less widespread occurrence among cells. Particularly interesting are the centrosomes and kinetosomes. The centrosomes, or centrioles, become plainly visible under the light microscope only when the cell approaches the hour of division, in which these bodies play a commanding role as the poles of the spindle apparatus that divides the chromosomes. The kinetosomes, on the other hand, are found only in those cells which are equipped with cilia or flagella for motility; at the base of each cilium or flagellum appears a kinetosome. Both of these organelles have the special property of self-replication. Each pair of centrosomes gives rise to another when cells divide; a kinetosome duplicates itself each time a new cilium forms on the cell surface. Long ago certain cytologists advanced the idea that these two organelles have much the same structure, even though their functions are so different. The electron

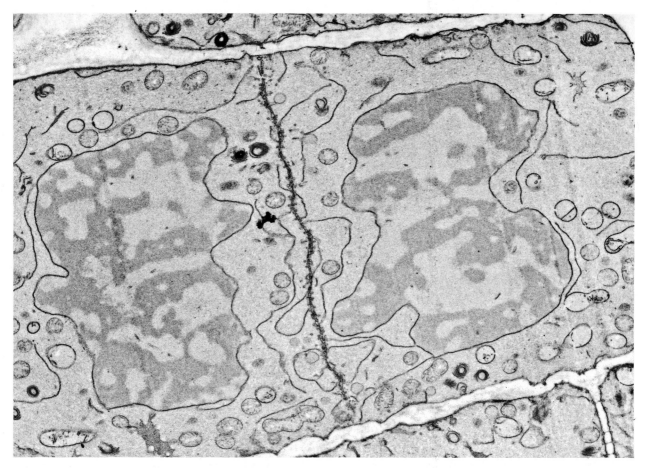

PLANT CELLS (onion root tip) are enlarged 6,700 diameters in this electron micrograph made by K. R. Porter of the Rockefeller Institute. The thin, dark line running from top to bottom of the micrograph shows the membrane between two cells shortly after the cells have divided. The large, irregularly shaped bodies to the left and right of the membrane are the nuclei of the two cells.

microscope has confirmed this suggestion. Each is a cylinder made up of 11 fibers, with two in the center and the other nine on the outside. This is the universal structure of all cilia and of flagella as well. The reason for the structure remains unknown, but it is undoubtedly related to the contractility of the cilia and flagella. It may be that the same "monomolecular muscle" principle underlies the action of the kinetosome and centrosome in their quite diverse functions.

The electron microscope has confirmed another surmise of earlier cytologists: that the cytoplasm has an invisible organization, a "cytoskeleton." Most cells show complicated systems of internal membranes not visible in the ordinary light microscope. Some of these membranes are smooth; others are rough, having tiny granules attached to one surface. The degree to which the membrane systems are developed varies from cell to cell, being rather simple in amoebae and highly articulated and roughened with granules in cells that

specialize in the production of proteins, such as those of the liver and pancreas.

Electron microscopists differ in their interpretation of these images. The generally accepted view is that of K. R. Porter of the Rockefeller Institute, who has given the membrane system its name, the endoplasmic reticulum; through the network of canaliculi formed by the membrane, substances are supposed to move from the outer membrane of the cell to the membrane of the nucleus. Some investigators hold that the internal membrane is continuous with the external membrane, furnishing a vastly increased and deeply invaginated surface area for communication with the fluid in which the cell is bathed. If the membrane does indeed have such vital functions, then it is likely that the cell is equipped with a factory for the continuous production of new membrane. This might be the role, as George E. Palade of the Rockefeller Institute has recently suggested, of the enigmatic Golgi bodies, first noted by the Italian cytologist Camillo Golgi at the end of the last century. The electron micro-

scope reveals that the Golgi bodies are made of smooth membrane, often continuous with that of the endoplasmic reticulum.

There is no doubt about the nature of the granules, which appear consistently on the "inner" surface of the membrane. They appear particularly in cells that produce large amounts of protein. As Torbjörn O. Caspersson and I showed some 20 years ago, such cells possess a high RNA content. Recent studies have revealed that the granules are exceedingly rich in RNA and correspondingly active in protein synthesis. For this reason the granules are now called ribosomes.

The membrane that surrounds the cell nucleus forms the interior boundary of the cytoplasm. There is still much speculation about what the electron microscope shows of this membrane. It appears as a double membrane with annuli, or holes, in the outer layer, open to the cytoplasm. To some investigators these annuli represent pores through which large molecules may move in either direction. Since the outer layer is often in close contact with the endoplasmic reticulum, it is also argued that the nuclear membrane participates in the formation of the reticulum membrane. Another possibility is that fluids percolating through the canaliculi of the endoplasmic reticulum are allowed to accumulate between the two layers of nuclear membrane.

Inside the nucleus are the all-important filaments of chromatin, in which the cell's complement of DNA is entirely localized. When the cell is in the "resting" state, that is, engaged in the processes of growth between divisions, the chromatin is diffusely distributed in the nucleus. The DNA thus makes maximum surface contact with other material in the nucleus from which it presumably pieces together the molecules of RNA and replicates itself. In preparation for division the chromatin coils up tightly to form the chromosomes, always a fixed number in each cell, to be distributed equally to each daughter cell.

Much less elusive than the chromatin are the nucleoli; these spherical bodies are easily resolved inside the nucleus with an ordinary light microscope. Under the electron microscope they are seen to be packed with tiny granules similar to the ribosomes of the cytoplasm. In fact, the nucleoli are rich in RNA and appear to be active centers of protein and RNA synthesis. Finally, to complete this functional anatomy of the

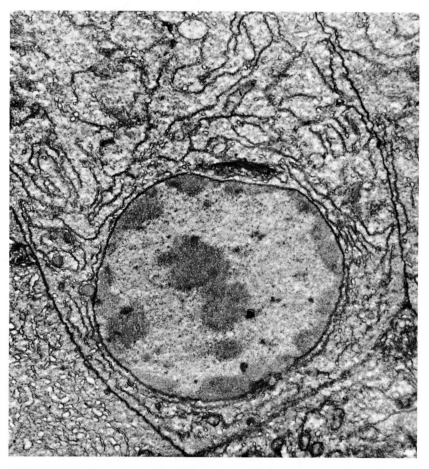

ANIMAL CELL (hepato-pancreatic gland of the crayfish) is enlarged 12,500 diameters in electron micrograph by George B. Chapman of the Cornell University Medical College. The large round object is the nucleus; the smaller dark region just above it is the Golgi body.

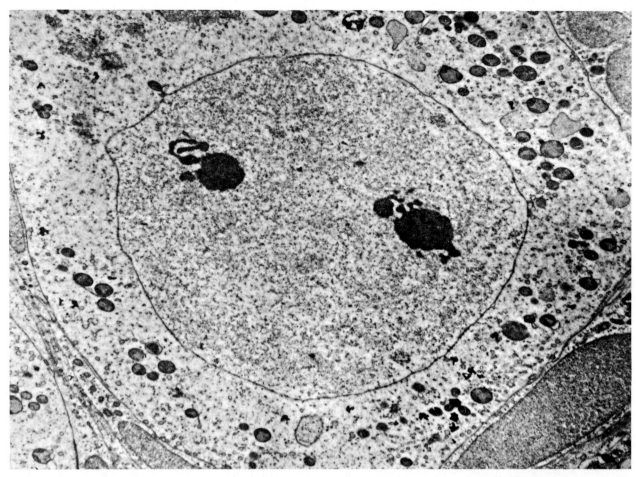

EGG CELL of rabbit is enlarged 7,500 diameters. The large round object is the nucleus; the two prominent dark bodies within it are nucleoli. Electron micrograph was made by Joan Blanchette of Columbia University College of Physicians and Surgeons.

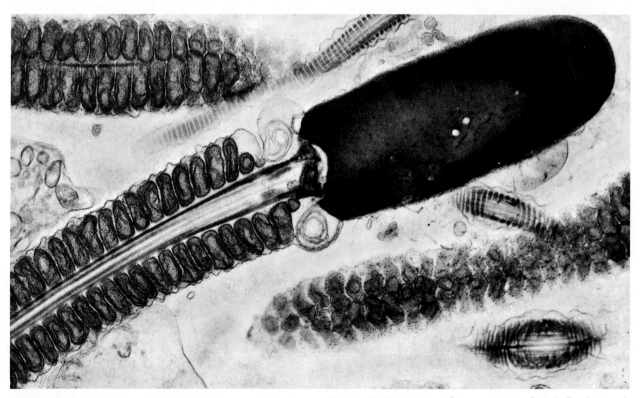

SPERM CELL of a bat is enlarged 21,500 diameters in electron micrograph by Fawcett and Susumu Ito of the Harvard Medical School. Nucleus (*top right*) constitutes almost all of sperm's head; arranged behind head are numerous mitochondria (*left*).

cell, it should be added that the chromatin and nucleoli are bathed together in the amorphous, proteinaceous matrix of the nuclear sap.

A remarkable history in the development of instruments and technique has gone into the drawing of the present portrait of the cell. The ordinary light microscope remains an essential tool. But its use in exploring the interior of the cell usually requires killing the cell and staining it with various dyes that selectively show the cell's major structures. To see these structures in action in the living cell, microscopists have developed a range of instruments—including phase, interference, polarizing and fluorescence microscopes—that manipulate light in various ways. In recent years the electron microscope, as the reader has gathered from this article, has become the major tool of the cytologist. But this instrument has a serious limitation in that it requires elaborate preparation and fixation of the specimen, which must inevitably confuse the true picture with distortions and artifacts. Progress is being made, however, toward the goal of resolving under the same high magnification the structure of the living cell.

Biochemistry has had an equally remarkable history of technical development. Centrifuges of ever higher rotation speed have made it possible to separate finer fractions of the cell's contents. These are divided and subdivided in turn by chromatography and electrophoresis. The classical techniques have been variously adapted to the analysis of quantities and volumes 1,000 times smaller than the standard of older micromethods; investigators can now measure the respiration or the enzyme content of a few amoebae or sea-urchin eggs. Finally, autoradiography, employing radioactive tracer elements, allows the worker to observe at subcellular dimensions the dynamic processes in the intact living cell.

The achievements and prospects that have been generated by the convergence of these two major movements in the life sciences furnish the subject of the articles that follow. To conclude this discussion it will be useful to consider how the two approaches have been employed to illuminate a single question: the role of the nucleus in the economy of the cell.

A simple experiment shows, first of all, that removal of the nucleus in a unicellular organism does not bring about the immediate death of the cytoplasm. The nucleate and enucleate

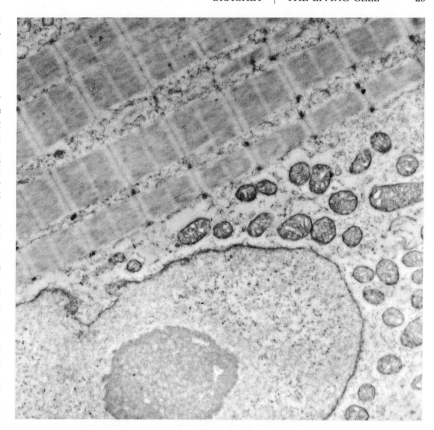

PART OF MUSCLE CELL of a salamander is enlarged 19,500 diameters in electron micrograph by George D. Pappas and Philip W. Brandt of Columbia College of Physicians and Surgeons. The nucleus is at bottom; around it are mitochondria. At top are muscle fibers.

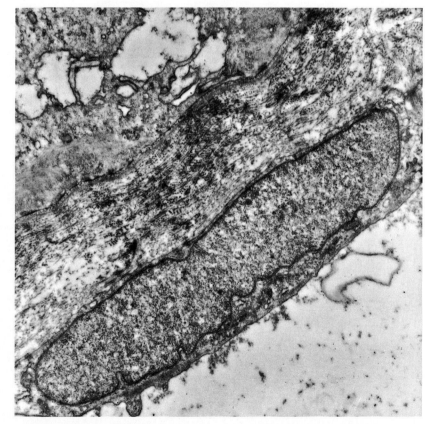

PART OF CONNECTIVE-TISSUE CELL of a tadpole is enlarged 14,500 diameters in electron micrograph by Chapman. Nucleus is oblong object; above it are fibrils of collagen.

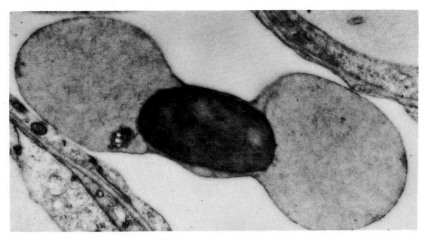

RED BLOOD CELL of a fish is enlarged 8,000 diameters in electron micrograph by Fawcett. Large dark body in center is nucleus. The mature red cells of mammals have no nuclei.

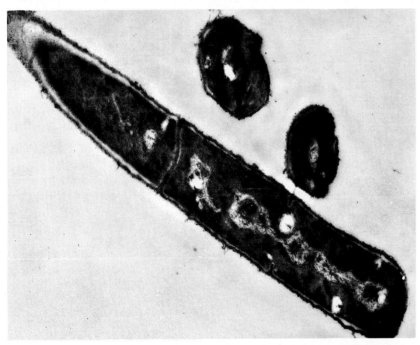

BACTERIUM *Bacillus cereus* (*long object*) is enlarged 30,000 diameters in electron micrograph by Chapman and by James Hillier of RCA Laboratories. Bacillus has several nuclei.

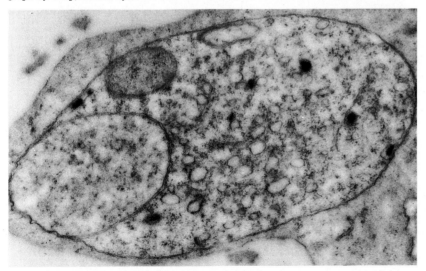

PROTOZOON *Plasmodium berghei* is enlarged 21,000 diameters in electron micrograph by Maria A. Rudzinska of the Rockefeller Institute. The nucleus is the large body at lower left.

halves of amoebae, if kept fasting, attain the same survival time of about two weeks; the cilia of an enucleate protozoon such as the paramecium continue to beat for a few days; the enucleate fragments of the unicellular giant alga *Acetabularia* may survive several months and are even capable of an appreciable amount of regeneration. Many of the basic activities of the cell, including growth and differentiation in the case of *Acetabularia,* can therefore proceed in the total absence of the genes and DNA. In fact, the enucleate pieces of *Acetabularia* are perfectly capable of making proteins, including specific enzymes, although enzyme synthesis is known to be genetically controlled. These synthetic activities, however, die out after a time. One must conclude that the nucleus produces something that is not DNA but which is formed under the influence of DNA and is transferred from the nucleus to the cytoplasm, where it is slowly used up. From such experiments—employing the combined techniques of cell biology and biochemistry—a number of fundamental conclusions emerge.

First, the nucleus is to be considered as the main center for the synthesis of nucleic acid (both DNA and RNA). Second, this nuclear RNA (or part thereof) goes over to the cytoplasm, playing the role of a messenger and transferring genetic information from DNA to the cytoplasm. Finally, the experiments show that the cytoplasm and in particular the ribosomes are the main site for the synthesis of specific proteins such as the enzymes. It should be added that the possibility of independent RNA synthesis in the cytoplasm is not ruled out and that such synthesis can, under suitable conditions, be demonstrated in enucleate fragments of *Acetabularia.* From this brief description of recently observed facts it is clear that the cell is not only a morphological but also a physiological unit.

Perhaps the reader will wonder how such knowledge of this unit helps to answer questions under the more general headings of "life" and "living." All one can venture to say is that the results of investigation invariably point in the same direction: Life, in the case of the cell and its constituents, is more a quantitative than an "all or none" concept. This dissection of cells into their constituents does not, therefore, throw much light on the questions posed by philosophy. But without this dissection, without experimentation, we would know next to nothing about the cell. And, after all, the cell is the fundamental unit of life.

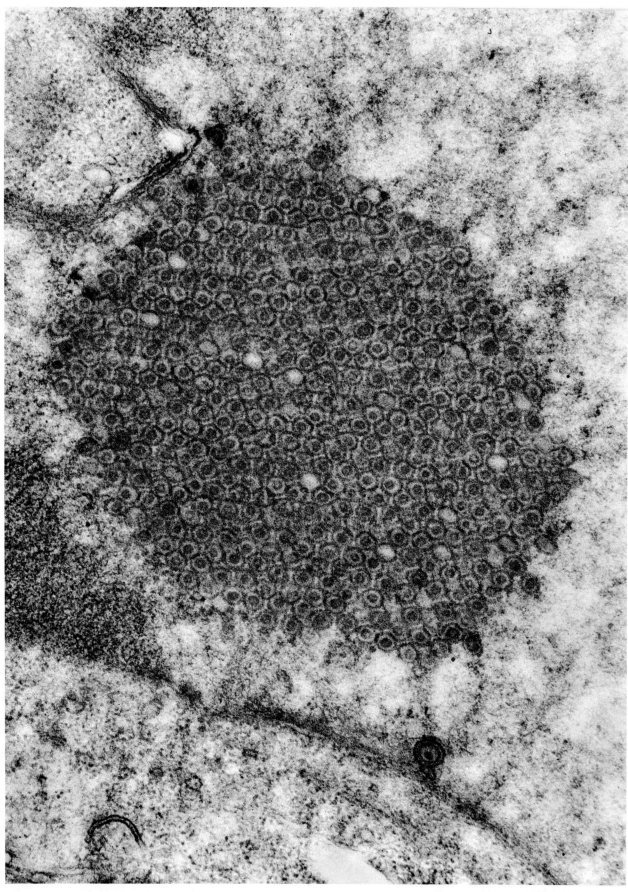

PARTICLES OF VIRUS *Herpes simplex* form a crystal within the nucleus of a cell. This electron micrograph, which enlarges the particles 73,000 diameters, was made by Councilman Morgan of the Columbia College of Physicians and Surgeons. Although viruses are exceptions to the rule that all living things are cells or are made of cells, they can reproduce only when they are inside cells.

Viruses and Genes

3

by François Jacob and Elie L. Wollman
June 1961

*When a virus infects a bacterium, the genes of the
virus sometimes act as genes of their host. The
phenomenon has illuminated the mechanism of both
heredity and infection*

Almost everyone now accepts the unity of the inanimate physical world. Physicists do not hesitate to extrapolate laboratory results obtained with a small number of atoms to explain the source of the energy produced by stars. In the world of living things a comparable unity is more difficult to demonstrate; in fact, it is not altogether conceded by biologists. Nevertheless, most students of bacteria and viruses are inclined to believe that what is true for a simple bacillus is probably true for larger organisms, be they mice, men or elephants.

Accordingly we shall be concerned here with seeking lessons in the genetic behavior of the colon bacillus (*Escherichia coli*) and of the still simpler viruses that are able to infect the bacillus and destroy it. Viruses are the simplest things that exhibit the fundamental properties of living systems. They have the capacity to produce copies of themselves (although they require the help of a living cell) and they are able to undergo changes in their hereditary properties. Heredity and variation are the subject matter of genetics. Viruses, therefore, possess for biologists the elemental qualities that atoms possess for

SCORES OF VIRUSES of the strain designated T₂ are attached to the wall of a colon bacillus in this electron micrograph. The viruses are fastened to the bacterial wall by their tails, through which they inject their infectious genetic material. (Walls of the cell collapsed when the specimen was dried by freezing. "Shadowing" with uranium oxide makes objects stand out in relief.) The electron micrograph was made by Edouard Kellenberger of the University of Geneva. The magnification is 70,000 diameters.

physicists. When a virus penetrates a cell, it introduces into the cell a new genetic structure that interferes with the genetic information already contained within the cell. The study of viruses has thus become a branch of cellular genetics, a view that has upset many old notions, including the traditional distinction between heredity and infection.

For a long time geneticists have worked with such organisms as maize and the fruit fly *Drosophila*. They have learned how hereditary traits are transmitted from parents to progeny, they have discovered the role of the chromosomes as carriers of heredity and they have charted the results of mutations—the events that modify genes. Complex organisms, however, multiply too slowly and in insufficient numbers for the high-resolution analyses needed to clarify such problems as the chemical nature of genes and the processes by which a gene makes an exact copy of itself and influences cellular activity. These detailed problems are most readily studied in bacteria and in viruses. Within the space of a day or two the student of bacteria or bacterial viruses can grow and study more specimens than the fruit-fly geneticist could study in a lifetime. An operation as simple as the mixing of two bacterial cultures on a few agar plates can provide information on a billion or more genetic interactions in which genes recombine to form those of a new generation.

It is the events of recombination, together with mutations, that model and remodel the chromosomes, the structures that contain in some kind of code the entire pattern of every organism. In recent years geneticists and biologists have clarified the nature of the hereditary message and have gained some clues as to what the letters of the code

are. The primary, and perhaps the unique, bearers of genetic information in all forms of life appear to be molecules of nucleic acid. In living organisms, with the exception of some of the viruses, these long-chain molecules are composed of deoxyribonucleic acid (DNA). In all plant viruses and in some animal viruses the genetic substance is not DNA but its close chemical relative ribonucleic acid (RNA). DNA molecules are built up of hundreds of thousands or even millions of simple molecular subunits: the nucleotides of the four bases adenine, thymine, guanine and cytosine. These subunits, in an almost infinite variety of combinations, seem capable of encoding all the characteristics that all organisms transmit from one generation to the next. RNA molecules, which are somewhat shorter in length and not so well understood, act similarly for the viruses in which RNA is the genetic material.

Ultimately the role of the genes—the words of the hereditary message—is to specify the molecular organization of proteins. Proteins are long-chain molecules built up of hundreds of molecular subunits: the 20 amino acids. The sequence of nucleotides in the nucleic acid that contains the hereditary message is thought to determine the sequence of amino acids in the protein it manufactures. This process involves a "translation" from the nucleic-acid code into the protein code through a mechanism that is not yet understood.

The Bacterial Chromosome

Before considering viruses as cellular genetic elements, we shall summarize the present knowledge of the genetics of the bacterial cell. In bacteria the hereditary message appears to be written in a single linear structure, the bacterial chromosome. For the study of this chromosome an excellent tool was discovered in 1946 by Joshua Lederberg and Edward L. Tatum, who were then working at Yale University. They used the colon bacillus, which is able to synthesize all the building blocks required for the manufacture of its nucleic acids and proteins and therefore to grow on a minimal nutrient medium containing glucose and inorganic salts. Mutant strains, with defective or altered genes, can be produced that lack the ability to synthesize one or more of the building blocks and therefore cannot grow in the absence of the building block they cannot make. If, however, two different mutant strains are mixed, bacteria like the original strain reappear and are able to grow on a minimal medium.

Lederberg and Tatum were able to demonstrate that such bacteria are the result of genetic recombination occurring when a bacterium of one mutant strain conjugates with a bacterium of another mutant strain. Further work by Lederberg, and by William Hayes in London, has shown that the colon bacillus also has sex: some individuals act as males and transmit genetic material by direct contact to other individuals that act as recipients, or females. The difference between the two mating types may be ascribed to the fertility factor (or sex factor) F, present only in males. Curiously, females can easily be converted into males; during conjugation certain types of male, called F$^+$, transmit their sex factor to the females, which then become males.

The Chromosome "Essay"

Our own work at the Pasteur Institute in Paris has shed light on the different steps involved in bacterial conjugation and on the mechanism ensuring the transfer of the chromosome from certain strains of male, called Hfr, to females. When cultures of such males and of females are mixed, pairings take place between male and female cells through random collisions. A bridge forms between the two mating bacteria; one of the chromosomes of the male (bacteria have generally two to four identical chromosomes during growth) begins to migrate across the bridge and to enter the female. In the female, portions of the male chromosome have the ability to recombine with suitable portions of one of the female chromosomes. The chromosomes may be compared to written essays that differ only by a few letters, or a few words, corresponding to the

mutations. Portions of the two essays may become paired, word for word and letter for letter. Through the process known as genetic recombination, which is still very mysterious and challenging, fragments of the male chromosome, which can be anything from a word or a phrase up to several sentences, may be exactly substituted for the corresponding part of the female chromosome. This process gives rise to a complete new chromosome that contains a full bacterial essay in which some words from the male have replaced corresponding words from the female. The new chromosome is then replicated and transmitted to the daughter cell.

Perhaps the most remarkable feature of bacterial conjugation is the way in which the male chromosome migrates across the conjugation bridge. For a given type of male the migration always starts at the same end of the chromosome, which, if we represent the bacterial chromosome by the letters of the alphabet, we can call A. Then, with the chromosome proceeding at constant speed, it takes two hours before the other end, Z, has penetrated the female. After the mating has begun, conjugation can be interrupted at will by violently stirring the mating mixture for a minute or so in a blender. The mechanical agitation does not kill the cells but it disrupts the bridge and breaks the male chromosome during its migration. The fragment of the male chromosome that has entered the female before the interruption is still functional and has the ability to provide words or sentences for a chromosome [*see illustration on pages 36 and 37*]. If conjugation is mechanically interrupted at various intervals after the onset of mating, it is found that any gene carried by the male chromosome, from A to Z, enters the female at a precise time. We have therefore been able to draw two kinds of detailed chromosome map showing the location of genes. One map, the conventional kind, is based on the observed frequency of different sorts of genetic recombination; the second is a new kind of map reflecting the time at which any gene penetrates the female cell. The latter can be compared to a road map drawn by measuring the times at which a car proceeding at a constant speed passes through various cities.

Finally, the mode of the male chromosome's migration has provided a unique opportunity for correlating genetic measurements with chemical measurements of the chromosome. In collaboration with Clarence Fuerst, who is now working at the University of Toronto, we have grown male bacteria in a medium con-

taining the radioactive isotope phosphorus 32, which is incorporated into the DNA of the bacterial chromosome. The labeled bacteria are then frozen and kept in liquid nitrogen to allow some of the radioactive atoms to disintegrate. At various times samples are thawed and the labeled males are then mated with unlabeled females. The experiments show that the radioactive disintegrations sometimes break the chromosomes. If the break occurs between two markers, say E and F, the head part, ABCDE, is transferred to the female, but the tail part, FGHIJKLMNOPQRSTUVWXYZ, is not. Therefore the greater the number of phosphorus atoms between the A extremity of the chromosome and a given gene, the greater the chance that a break will prevent this gene from being transferred to the female. It is thus possible to draw a chromosomal map showing the location of the genes in terms of numbers of phosphorus atoms contained in the chromosome between the known genes. When we compare this map with those obtained by genetic analysis or by mechanical interruption, we find that for a given type of male all three maps are consistent.

In some types of male mutant the genetic characters have the same sequence along the chromosome but the character injected first differs from one mutant to another. The characters can also be injected either in the forward direction or in the backward direction, that is, from A to Z or from Z to A, with the alphabet capable of being broken at any point. These observations can be explained most simply by assuming that all the genetic "letters" of the colon bacillus are arranged linearly in a ring and that the ring can be opened at various points by mutation. It seems, furthermore, that the opening of the ring is a consequence of the attachment of the sex factor to the chromosome. The ring opens at precisely the point where the factor F, which is free to move, happens to affix itself. A cell with the F factor affixed to the chromosome is called an Hfr male, or "supermale," because it enhances the transmission of chromosomal markers. Hfr stands for "high frequency of recombination." When the chromosome is opened by the F factor, one of the free ends initiates the penetration of the chromosome into the female, carrying the sequence of characters after it. The other end carries the sex factor itself and is the last to enter the female. The sex factor has other remarkable properties and we shall bring it back into our story later.

The long-range objective of such stud-

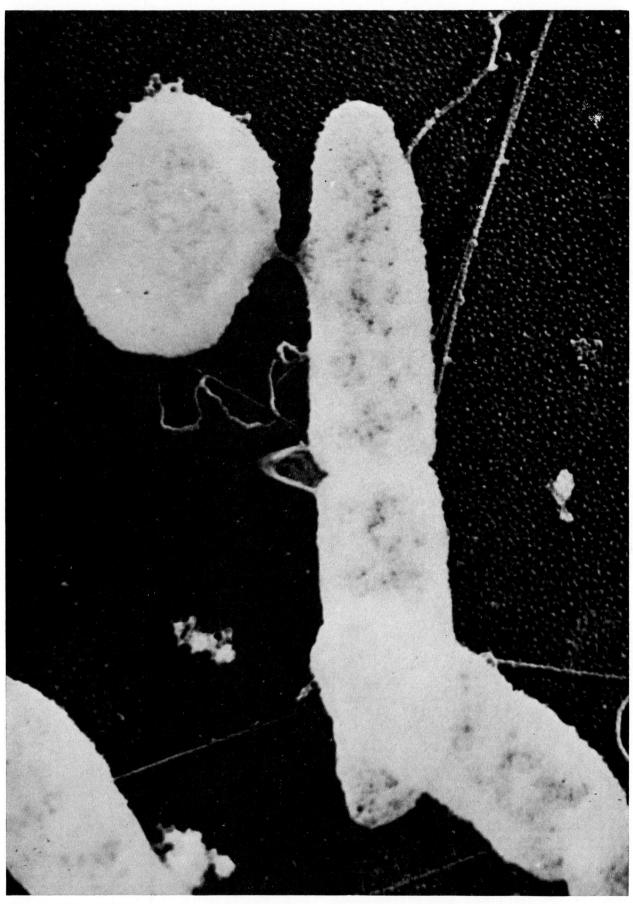

CONJUGATING BACTERIA conduct a transfer of genetic material. Long cell (*right*) is an *Hfr* "supermale" colon bacillus, which is attached by a short temporary bridge to a female colon bacillus (*see illustration on next two pages*). This electron micrograph, shown at a magnification of 100,000 diameters, was made by Thomas F. Anderson of the Institute for Cancer Research in Philadelphia.

ies is to learn how the thousands of genes strung along the chromosome control the molecular pattern of the bacterial cell: its metabolism, growth and division. These processes imply precise regulatory mechanisms that maintain a harmonious equilibrium between the cellular constituents. At any time the bacterial cell "knows" which components to make and how much of each is needed for it to grow in the most economical way. It is able to recognize which kind of food is available in a culture medium and to manufacture only those protein enzymes that are required to get energy and suitable building blocks from the available food.

At the Pasteur Institute, in collaboration with Jacques Monod, we have recently found new types of gene that determine specific systems of regulation. Mutants have been isolated that have become "unintelligent" in the sense that they cannot adjust their syntheses to their actual requirements. They make, for example, a certain protein in large amounts when they need only a little of it or even none at all. This waste of energy decreases the cells' growth rate. It seems that the production of a particular protein is controlled by two kinds of gene. One, which may be called the structural gene, contains the blueprint for determining the molecular organiza-

tion of the protein—its particular sequence of amino acid subunits. Other genes, which may be called control genes, determine the rate at which the information contained in the structural gene is decoded and translated into protein. This control is exercised by a signal embodied in a repressor molecule, probably a nucleic acid, that migrates from the chromosome to the cytoplasm of the cell. One of the control genes, called the regulator gene, manufactures the repressor molecule; thus it acts as a transmitter of signals. These are picked up by the operator gene, a specific receiver able to switch on or off the activity of the adjacent structural genes. Metabolic

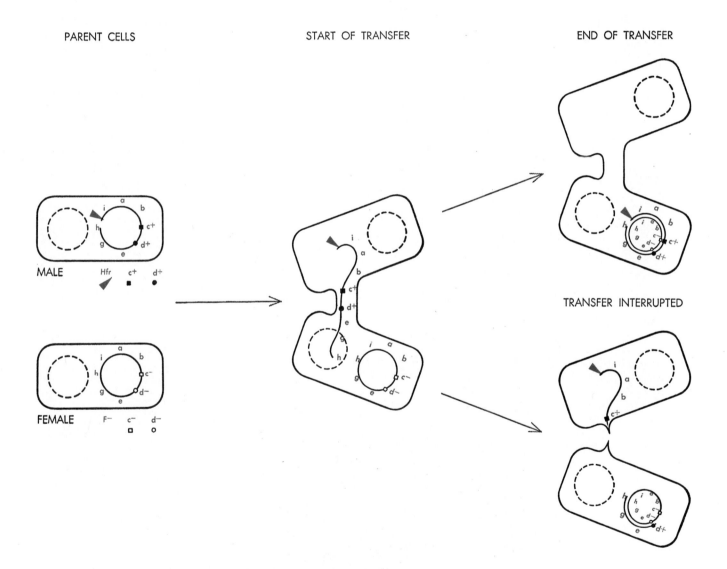

PARENT CELLS   START OF TRANSFER   END OF TRANSFER

TRANSFER INTERRUPTED

CHROMOSOMAL TRANSFER provides a primitive sexuality for colon bacillus. The bacterial chromosome, which appears to be ring-shaped, carries genetic markers (*designated by letters*), the presence or absence of which can be determined by studying cell's nutritional requirements. When the sex, or *F*, agent is attached to the chromosome, opening the ring, the cell is called an *Hfr* supermale. Two markers, labeled *c+* and *d+* when present and *c-* and *d-* when absent, can be traced from parents to daughter cells. When male and female cells conjugate, one of the male chromosomes (there are usually several, all identical) travels through the bridge.

products can interfere with the signals, either activating or inactivating the proper repressor molecules and thereby initiating or inhibiting the production of proteins.

Within the bacterial cell, then, there exists a complex system of transmitters and receivers of specific signals, by means of which the cell is kept informed of its metabolic requirements and enabled to regulate its syntheses. The bacterial chromosome contains not only a series of blueprints for the manufacture of individual molecular components but also a plan for the co-ordinated production of these components.

Let us now turn to the events that

DAUGHTER CELLS

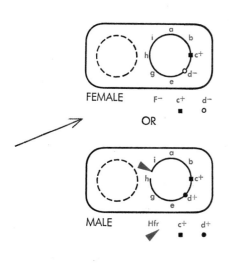

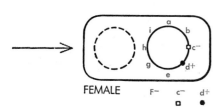

If transfer is complete, daughter cells may be male or female and carry any marker of the male. If transfer is interrupted, daughters are all female and can carry only those markers passed before bridge was broken.

take place when a bacterial virus of the strain designated T_2 infects the colon bacillus. A T_2 virus is a structure shaped like a tadpole; by weight it is about half protein and half DNA. The DNA is enclosed in the head, the outside of which is protein; the tail is also composed of protein. The roles of the DNA and the protein in the infective process were clarified in 1952 by the beautiful experiments of Alfred D. Hershey and Martha Chase of the Carnegie Institution of Washington's Department of Genetics in Cold Spring Harbor, N. Y. By labeling the DNA fraction of the virus with one radioactive isotope and the protein fraction with another, Hershey and Chase were able to follow the fate of the two fractions. They found that the DNA is injected into the bacterium, whereas the protein head and tail parts of the virus remain outside and play no further role. Electron micrographs reveal that the tail provides the method of attachment to the bacterium and that the DNA is injected through the tail. The Hershey-Chase experiment was a landmark in virology because it demonstrated that the nucleic acid carries into the cell all the information necessary for the production of complete virus particles.

How Viruses Destroy Bacteria

A bacterium that has been infected by virus DNA will break open, or lyse, within about 20 minutes and release a new crop of perhaps 100 particles of infectious virus, complete with protein head and tail parts. In this brief period the virus DNA subverts the cell's chemical facilities for its own purposes. It brings into the cell a plan for the synthesis of new molecular patterns and the cell faithfully carries it out. The infected cell creates new protein subunits needed for the virus head and tail, and filaments of nucleic acid identical to the DNA of the invading particle. These pools of building blocks pile up more or less at random, and in excess amounts, inside the cell. Then the long filaments of virus DNA suddenly condense and the protein subunits assemble around them, creating the complete virus particle. The whole process can be compared to the occupation of one country by another; the genetic material of the virus overthrows the lawful rule of the cell's own genetic material and establishes itself in power.

A virus can therefore be considered a genetic element enclosed in a protein coat. The protein coat protects the genetic material, gives it rigidity and stability

and ensures the specific attachment of the virus to the surface of the cell. As André Lwoff of the Pasteur Institute has pointed out, viruses can be uniquely defined as entities that reproduce from their own genetic material and that possess an apparatus specialized for the process of infection. The definition excludes both the cell and the specialized particles within the cell that serve its normal functions.

Another important criterion of viral growth is that of unrestricted synthesis. Infection with a virus is a sort of molecular cancer. The replication of the genetic material of the virus and the synthesis of the viral building blocks do not appear to be subject to any control system at all.

Lysogenic Bacteria

When a T_2 virus infects a bacterium, it forces the host to make copies of it and ultimately to destroy itself. Such a virus is said to be virulent, and when it is inside the cell, reproducing itself, it is said to be in the vegetative state.

There are, however, other bacterial viruses, called temperate viruses, which behave differently. After entering a cell the genetic material of a temperate virus can take two distinct paths, depending on the conditions of infection. It can enter the vegetative state, replicate itself and kill the host, just as a virulent virus does. Under other circumstances it does not replicate freely and does not kill the host. Instead it finds its way to the bacterial chromosome, anchors itself there and behaves like an integrated constituent of the host cell. Thereafter it will be transmitted for years to the progeny of the bacterium like a bacterial gene. We know that the bacterial host has not destroyed the invading particle, because from time to time one of the daughter cells in the infected line will break open and yield a crop of virus particles, as it would if it had been freshly attacked by a virulent virus. When the virus is in the subdued and integrated state, it is called a provirus. Bacteria carrying a provirus are called lysogenic, meaning that they carry a property that can lead to lysis and death.

Lysogeny was discovered in the early 1920's, soon after the discovery of the bacterial virus itself, and it remained a profound mystery for some 25 years. The mystery was explained by the fine detective work of Lwoff and his colleagues [see "The Life Cycle of a Virus," by André Lwoff; SCIENTIFIC AMERICAN, March, 1954]. Lwoff found that when he exposed certain types of lysogenic bac-

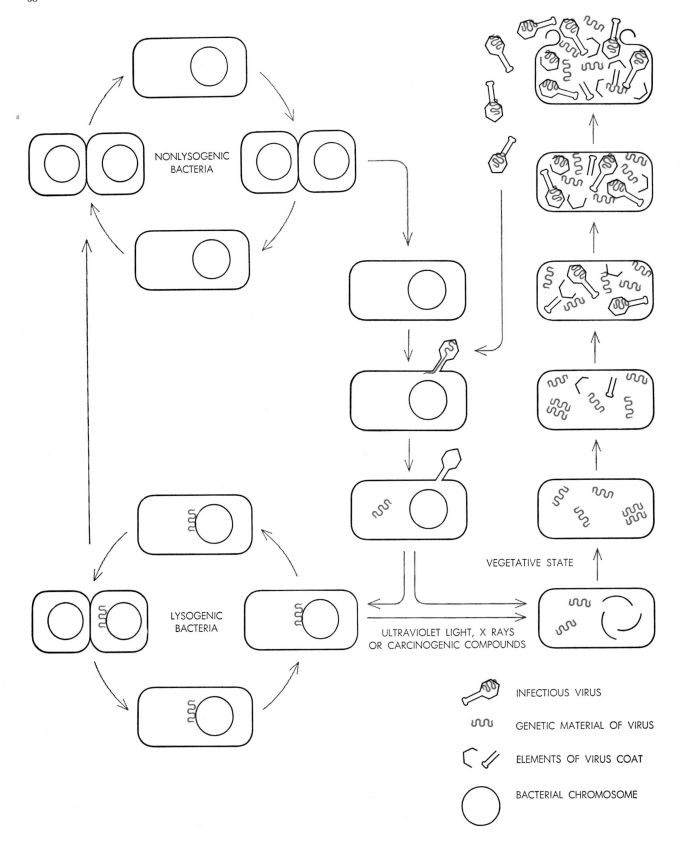

NONLYSOGENIC
BACTERIA

LYSOGENIC
BACTERIA

ULTRAVIOLET LIGHT, X RAYS
OR CARCINOGENIC COMPOUNDS

VEGETATIVE STATE

INFECTIOUS VIRUS

GENETIC MATERIAL OF VIRUS

ELEMENTS OF VIRUS COAT

BACTERIAL CHROMOSOME

LIFE CYCLE OF BACTERIAL VIRUS shows that, for the bacterium attacked, infection and death are not inevitable. After the genes of the virus (*color*) enter a cell descended from a completely healthy line (*top left*), the cell may take either of two paths. One (*far right*) leads to destruction as the virus enters the vegetative state, makes complete copies of its infective self and bursts open the cell, a process called lysis. The other path leads to the so-called lysogenic state, in which the viral genes attach themselves to the bacterial chromosome and become a provirus; the cell lives. Exposure to ultraviolet light, however, can dislodge the provirus and induce the vegetative state. The provirus is sometimes lost during cell division, returning the cell to the nonlysogenic state.

teria to ultraviolet light, X rays or active chemicals such as nitrogen mustard or organic peroxides, the whole bacterial population would lyse within an hour, releasing a multitude of infectious virus particles. When a provirus is thus activated, or "induced," it leaves the integrated state and enters the vegetative state, eventually destroying the cell [*see illustration on opposite page*].

To determine the position of the provirus inside the host cell, we can apply the method of interrupting the sexual conjugation of bacteria that carry a provirus and are therefore lysogenic. In this way we can correlate the location of the provirus with that of known characters on the bacterial chromosome. Each of 15 different types of provirus takes a particular position at a specific site on the bacterial chromosome. Only one is an exception; it seems free to take a position anywhere. In the proviral state the genetic material of the virus has not become an integral part of the bacterial chromosome; instead it appears to be added to the chromosome in an unknown but specific way. However it may be hooked on, the genetic material of the virus is replicated together with the genetic material of the host. It behaves like a gene, or rather as a group of genes, of the host.

Nonviral Effects of Provirus

The presence of this apparently innocuous genetic element, the provirus, can confer on the lysogenic bacteria that harbor it some new and striking properties. It is not at all obvious why some of these properties should be related to the presence of a provirus. As one example, diphtheria bacilli are able to produce diphtheria toxin only if the bacilli carry certain specific types of provirus. The disease diphtheria is caused solely by this toxin.

In other instances the presence of a provirus is responsible for a particular type of substance coating the surface of a bacterium. The substance can be identified by various immunological tests (typically by noting if a precipitate forms when a certain serum is added). The nonlysogenic strain, carrying no provirus, will bear a different substance. In such cases the genes of the virus are responsible for hereditary properties of the host. They can scarcely be distinguished from the genes of the bacterium.

The most striking property the provirus confers on its bacterial host is immunity from infection by external viruses of the same type as the provirus. When

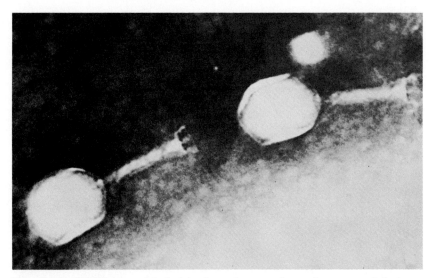

INTACT T₂ VIRUS has polyhedral head membrane and a curious pronged device at the end of its tail. The magnification is 200,000 diameters. This electron micrograph and the two below were made by S. Brenner and R. W. Horne at the University of Cambridge.

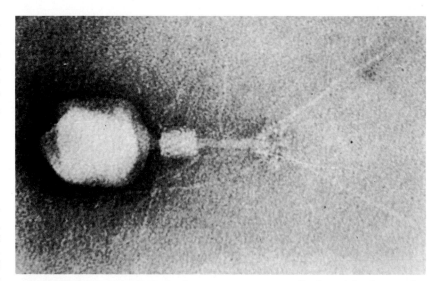

"TRIGGERED" T₂ VIRUS results from exposure to a specific bacterial substance that causes contraction of the tail sheath (*stubby cylinder*) and discharge of viral genes.

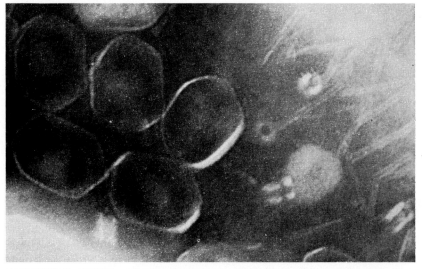

ISOLATED T₂ PARTS can be found still unassembled if host cell is forced to burst open before synthesis of virus particles is complete. Parts include head membranes and tails.

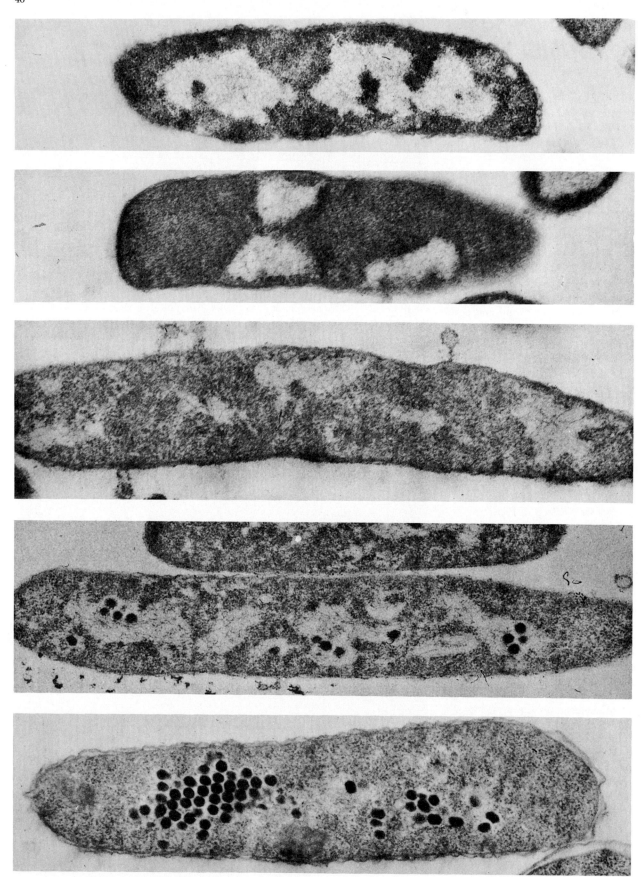

GROWTH OF T$_2$ VIRUS inside bacterial host is revealed in a striking series of electron micrographs by Kellenberger. Top picture shows the colon bacillus before infection. Four minutes after infection (*second from top*) characteristic vacuoles form along the cell wall. Ten minutes after infection (*third from top*) the virus has reorganized the entire cell interior and has created pools of new viral components. Twelve minutes after infection (*fourth from top*) new virus particles have started to condense. Thirty minutes after infection (*bottom*) more than 50 fully developed T$_2$ viruses have been produced and the cell is about ready to burst open.

lysogenic cells are mixed with such viruses, the virus particles adsorb on the cell and inject their genetic material into the cell, but the cell survives. The injected material is somehow prevented from multiplying vegetatively and is diluted out in the course of normal bacterial multiplication.

In the past two years we have attempted to learn more about the mechanism of this immunity. It seems clear that the mere attachment of the provirus to the host chromosome cannot account for the immunity of the host. The provirus must do something or produce something. We have evidence that the immunity is expressed by a substance or factor not tied to the chromosome. Remarkably enough, the system of immunity appears to be similar to the cellular systems already described that regulate the synthesis of protein in growing bacteria. It seems that the provirus produces a chemical repressor capable of inhibiting one or several reactions leading to the vegetative state. Thus immunity can be visualized as a specific system of regulation, involving the transmission of signals (repressors), which are received by an invading virus particle carrying the appropriate receptor.

Transduction

The close association that may take place between the genetic material of the virus and that of the host becomes even more striking in the phenomenon of transduction, discovered in 1952 by Norton D. Zinder and Lederberg at the University of Wisconsin [see " 'Transduction' in Bacteria," by Norton D. Zinder; SCIENTIFIC AMERICAN Offprint 106]. They found that when certain proviruses turn into infective viruses, thereby killing their hosts, they may carry away with them pieces of genetic material from their dead hosts. When the viruses infect a host that is genetically different, the genes from the old host— the transduced genes—may be recombined with the genes of the new host. The transduction process seems able to move any sort of gene from one bacterial host to another.

Lysogeny and transduction therefore represent two complementary processes. In lysogeny the genes of the virus become an integral part of the genetic apparatus of the host and replicate at the pace of the host's chromosome. In transduction genes of the host become linked to the genes of the virus and can replicate at the unrestricted viral pace when the virus enters the vegetative state.

Viruses, like all other genetic ele-

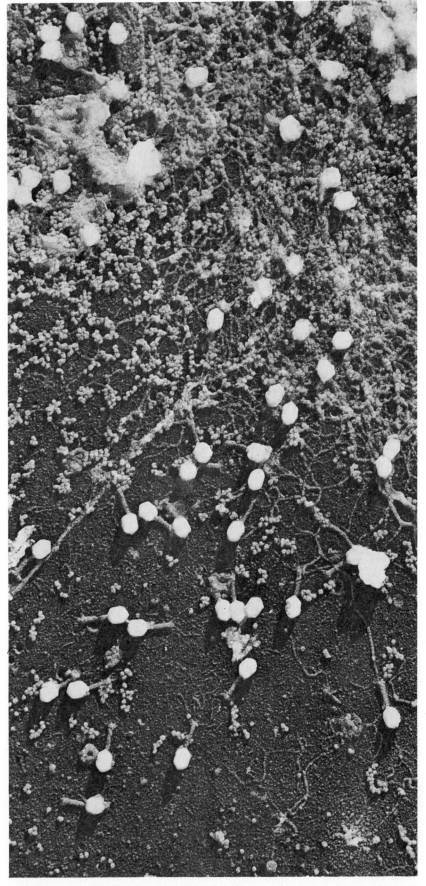

DEATH OF A BACTERIUM occurs when T_2 virus particles, having multiplied inside their host (*see sequence on opposite page*), dissolve the walls of the bacterial cell and spill out— a phenomenon called lysis. Viruses are the large white objects; the other matter is cellular debris. The electron micrograph (magnification: 50,000 diameters) is by Kellenberger.

ments, can undergo mutations, and these produce a variety of stable, heritable changes. The mutations of particular interest are those that prevent the formation of mature, infectious virus particles. Lysogenic bacteria in which such mutations have taken place are called defec-tive lysogenic bacteria. These bacteria hereditarily perpetuate a mutated pro-virus, which is perfectly able to replicate together with the host's chromosome. If these cells are exposed to ultraviolet radiation, which activates the provirus, we observe that the defective lysogenic cells die without releasing any infectious viruses. Examination of such bacteria usu-ally shows that virus subunits have start-ed to appear inside the cell but have failed to reach maturity [*see illustra-tions on pages 44 and 45*]. Evidently some essential step in the formation of

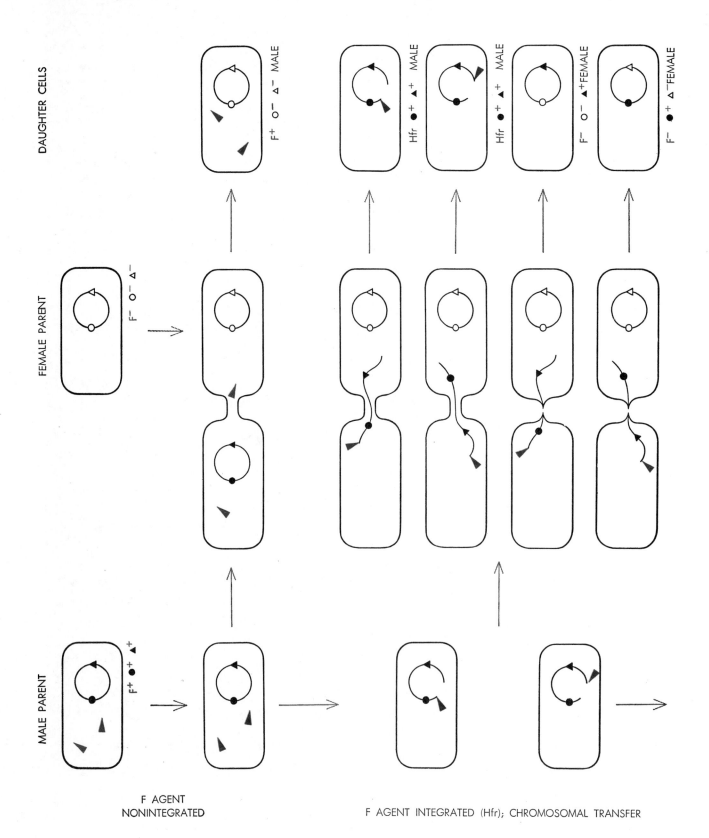

an infectious virus has been blocked by the mutation.

Just as we can study how other kinds of mutation block biochemical pathways associated with cell nutrition, we can try to identify the biochemical blockages that keep the provirus from multi-plying normally. When a defective pro-virus turns to the vegetative state, some viral components begin to appear but the process halts. By using various biological tests, together with electron microscopy, we try to establish how far the process has gone. We have been able to identify two ways in which the process is halted and to relate the blockage to two main groups of viral genes.

One group of genes is concerned with the autonomous reproduction of the genetic material of the virus. The DNA of the provirus, which was able to repli-

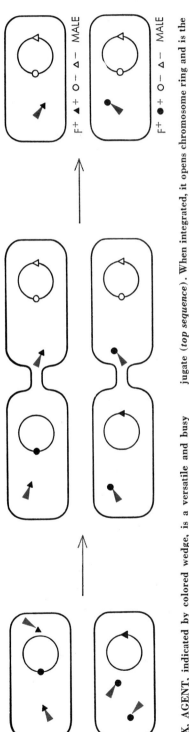

SEX-DUCTION
F AGENT NONINTEGRATED

F, OR SEX, AGENT, indicated by colored wedge, is a versatile and busy "broker" in genes. It can be attached to the bacterial chromosomes (*inte-grated*) or unattached (*nonintegrated*) and can alternate between the two states. When nonintegrated, it usually transmits only itself when bacteria con-jugate (*top sequence*). When integrated, it opens chromosome ring and is the last marker transferred in conjugation (*middle sequence*). Daughters may inherit markers in combinations other than those shown. When *F* agent leaves integrated state (*bottom*), it may remove a marker and transfer it (*sex-duction*).

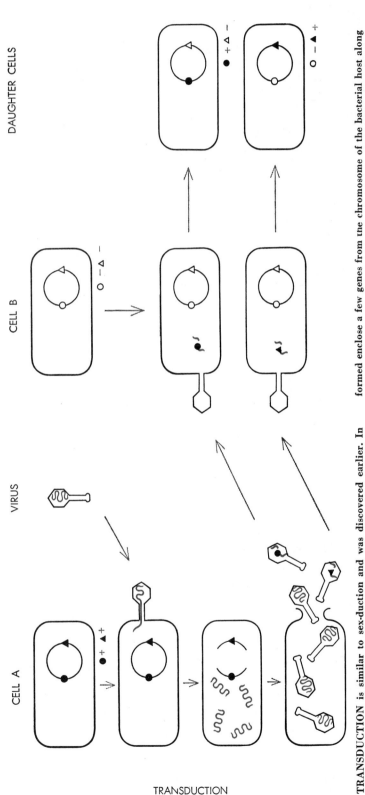

TRANSDUCTION

TRANSDUCTION is similar to sex-duction and was discovered earlier. In transduction the agent for transferring bacterial genes is a virus particle rather than an *F* agent. The virus injects its genes (*color*) into bacterial cell *A* and the genes create new copies of the virus. Occasionally the new virus particles so formed enclose a few genes from the chromosome of the bacterial host along with a few viral genes. These imperfect viruses are able to inject their contents into another cell (*cell "B"*) but are unable to destroy it. In this way genes (*solid black shapes*) can be transferred from cell *A* to the daughters of cell *B.*

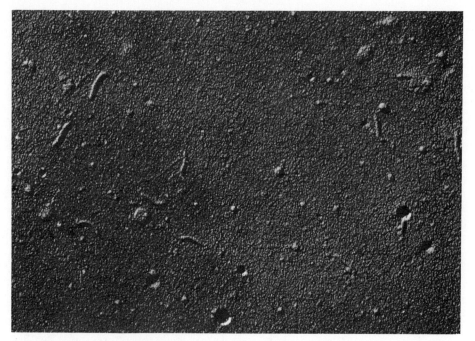

INCOMPLETE VIRUS PARTICLES are created by defective proviruses (*see illustration below*). The electron micrograph at left shows virus heads and tails that remain unassembled because of some defect. Occasionally (*right*) only heads can be found. Electron

cate when attached to the host chromosome, becomes unable to replicate on its own. A second group of genes is involved in the manufacture of the protein molecules that provide the coat and infectious apparatus of a normal virus. We have examples in which there is plenty of viral DNA, and many components of the coat material, but one or another essential protein is missing.

This study leads us to conclude that what distinguishes the genetic material of a virus from genetic elements of other types is that the virus carries two sets of information, one of which is necessary for the unrestricted multiplication of the viral genes and the other for the manufacture of an infectious envelope and traveling case.

The concept of a virus as it has emerged from the study of bacterial vi-

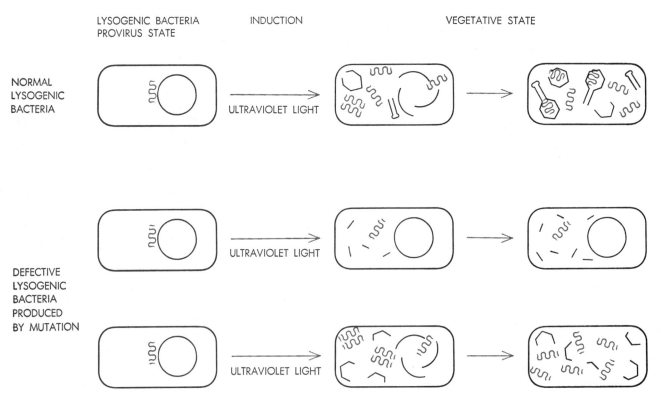

DEFECTIVE LYSOGENIC BACTERIA appear as mutations among normal lysogenic bacteria. Upon induction with ultraviolet light a normal provirus (*color, top left*) leaves the bacterial chromosome, replicates, produces infectious virus particles and kills its host. When defective proviruses are induced, the host cell may also be killed, but no infectious viruses appear at lysis. In

micrographs (magnification: 57,000) were made by Kellenberger and W. Arber.

ruses is far more complex and more fascinating than the concept that prevailed only a decade ago. As we have seen, a virus may exist in three states; the only thing common to the virus in the three states is that it carries at all times much the same genetic information encoded in

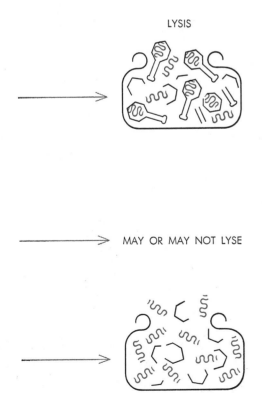

LYSIS

MAY OR MAY NOT LYSE

some cases (*middle*) the viral genes fail to replicate. In others (*bottom*) they replicate but the jacketing components are defective.

DNA. In the extracellular infectious state the nucleic acid is enclosed in a protective, resistant shell. The virus then remains inert like the spore of a bacterium, the seed of a plant or the pupa of an insect. In the vegetative state of autonomous replication the genetic material is free of its shell, overrides the regulatory mechanism of the host and imposes its own commands on the synthetic machinery of the cell. The viral genes are fully active. Finally, in the proviral state the genetic material of the virus has become subject to the regulatory system of the host and replicates as if it were part of the bacterial chromosome. A specific system of signals prevents the genes of the virus from expressing themselves; complete virus particles are therefore not manufactured.

The Concept of the "Episome"

Less than a decade ago there was no reason to doubt that virus genetics and cell genetics were two different subjects and could be kept cleanly apart. Now we see that the distinction between viral and nonviral genetics is extremely difficult to draw, to the point where even the meaning of such a distinction may be questionable.

As a matter of fact there appear to be all kinds of intermediates between the "normal" genetic structure of a bacterium and that of typical bacterial viruses. Recent findings in our laboratory have shown that phenomena that once seemed unrelated may share a deep identity. We note, for example, that certain genetic elements of bacteria, which we have no reason to class as viral, actually behave very much like the genetic material of temperate viruses. One of these is the fertility, or *F*, factor in colon bacilli; in the so-called *Hfr* strains of males the *F* agent is attached to one of various possible sites on the host chromosome. In the males bearing the *F* agent designated F^+ the agent is not fixed to the chromosome and so it replicates as an autonomous unit. It bears one other striking resemblance to provirus. The integrated state of the *F* factor excludes the nonintegrated replicating state, just as a provirus immunizes against the vegetative replication of a like virus.

Another genetic agent resembling provirus is the factor that controls the production of colicines. These are extremely potent protein substances that are released by some strains of colon bacillus; the proteins are able to kill bacteria of other strains of the same or related species. The colicinogenic factors also seem

to exist in two alternative states: integrated and nonintegrated. In the latter state they seem able to replicate freely and eventually at a faster rate than does the bacterial chromosome. Bacteria that lack these genetic elements—F agents and colicinogenic factors—cannot, so far as we know, gain them by mutation but can only receive them (by sexual conjugation, for example) from an organism that already possesses them. They may replicate either along with the chromosome or autonomously. Such genetic elements, which may be present or absent, integrated or autonomous, we have proposed to call "episomes," meaning "added bodies" [*see illustration on page 46*].

The concept of episomes brings together a variety of genetic elements that differ in their origin and in their behavior. Some are viruses; others are not. Some are harmful to the host cell; others are not. The important lesson, learned from the study of mutant temperate bacterial viruses, is that the transition from viral to nonviral, or from pathogenic to nonpathogenic, can be brought about by single mutations. We also have impressive evidence that any chromosomal gene of the host may be incorporated in an episome through some process of genetic recombination. During the past year, in collaboration with Edward A. Adelberg of the University of California, we have shown that the sex factor, when integrated, is able to pick up the adjacent genes of the bacterial chromosome. Then this new unit formed by the sex factor and a few bacterial genes is able to return to the autonomous state and to be transmitted by conjugation as a single unit. This process, in many respects similar to transduction, has been called sex-duction [*see illustration at left on pages 42 and 43*].

Do episomes exist in organisms higher than bacteria? We do not know; but if we accept the basic unity of all cellular biology, we should be confident that the answer is yes and that mice, men and elephants must harbor episomes. So far the great precision and resolution that can be achieved in the study of bacterial viruses cannot be duplicated for more complex organisms. There is, nevertheless, evidence for episome-like factors in the fruit fly and in maize. There have been reports of two viruses in the fruit fly, transmitted through the egg to the offspring, which may exist either as nonintegrated or as integrated elements. Although it does not seem that the virus is actually located on the chromosome in the latter state, the resemblance to provirus is striking. Barbara McClintock, of

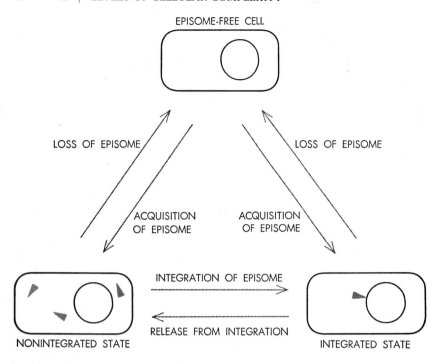

CONCEPT OF THE "EPISOME," as put forward by the authors, describes a genetic element, such as the *F* agent, that may be either attached to the chromosome or unattached. When integrated, it replicates at host's pace; nonintegrated, it replicates autonomously.

the Carnegie Institution of Washington's laboratory at Cold Spring Harbor, has discovered in maize "controlling elements" that are able to switch a gene off or on. (A gene responsible for a reddish color in corn may be switched on and off so fast that a single kernel may turn out speckled.) The controlling elements in maize are not always present, but when they are, they are added to specific chromosomal sites and can move from one site to another or even from one chromosome to another. These elements, therefore, act like episomes.

The discovery of proviruses and episomes has brought to light a phenomenon that biologists would scarcely have considered possible a few years ago: the addition to the cell's chromosome of pieces of genetic material arising outside the cell. The bacterial episomes provide new models to explain how two cells that otherwise possess an identical heredity can differ from each other. The episome brings into the cell a supplementary set of instructions governing additional biochemical reactions that can be superimposed on the basic metabolism of the cell.

The episome concept has implications for many problems in biology. For example, two main hypotheses have been advanced for the origin of cancer. One assumes that a mutation occurs in some cell of the body, enabling the cell to escape the normal growth-regulating mechanism of the organism. The other suggests that cancers are due to the presence in the environment of viruses that can invade healthy cells and make them malignant [see "The Polyoma Virus," by Sarah E. Stewart; SCIENTIFIC AMERICAN Offprint 77]. In the light of the episome concept the two hypotheses no longer appear mutually exclusive. We have seen that proviruses, living peacefully with their hosts, can be induced to turn to the vegetative, replicating state by radiation or by certain strong chemicals—the very agents that can be used to produce cancer experimentally in mice. If defective, the provirus will not even make viral particles. Malignant transformation involves a heritable change that allows a cell to escape the growth control of the organism of which it is a part. We can easily conceive that such a heritable change may result from a mutation of the cell, from an infection with some external virus or from the action of an episome, viral or not. Thus in the no man's land between heredity and infection, between physiology and pathology at the cellular level, episomes provide a new link and a new way of thinking about cellular genetics in bacteria and perhaps in mice, men and elephants.

The Smallest Living Cells

4

by Harold J. Morowitz and Mark E. Tourtellotte
March 1962

A microbe known as the pleuropneumonia-like organism gives rise to free-living cells smaller than some viruses. They suggest the question: What are the smallest dimensions compatible with life?

What is the smallest free-living organism? The most likely candidate for this niche in the order of nature was discovered by Louis Pasteur when he recognized that bovine pleuropneumonia, a highly contagious disease of cattle, must be caused by a microbial agent. But Pasteur was unable to isolate the microbe: he could not grow it in nutrient broth nor could he see it under the microscope. Apparently it was too small to be seen.

Then, in 1892, the Russian investigator D. Iwanowsky succeeded in demonstrating that certain infectious agents were so small that they could pass easily through the porcelain filters used to trap bacteria. The size of the microbes postulated by Pasteur was comparable to that of Iwanowsky's organisms, which were subsequently named viruses. All viruses, however, are parasites of the living cell. The pleuropneumonia agent, on the other hand, is not. In 1898 Pasteur's successors E. I. E. Nocard and P. P. E. Roux were able to grow the pleuropneumonia agent in a complex, but cell-free, medium. In this respect the agent seemed more like a bacterium than a virus. In 1931 W. J. Elford of the National Institute for Medical Research in London, who developed the first filters in which pore size could be precisely determined, showed that cultures of the pleuropneumonia agent contained viable particles only .125 to .150 micron (.0000125 to .000015 centimeter) in diameter. Thus the particles were smaller than many viruses. Yet, as subsequent investigations have shown, the particles fully satisfy the definition "free-living": they have the ability to take molecules out of a nonliving medium and to give rise to two or more replicas of themselves.

More than 30 strains of this tiny organism have now been isolated from soil and sewage, as contaminants from tissue cultures and from a number of animals,

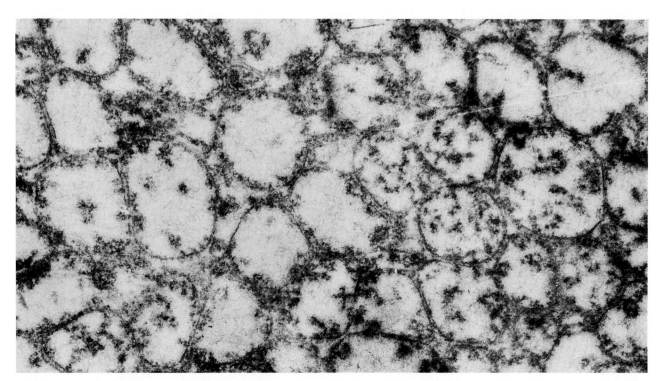

CELLS OF PLEUROPNEUMONIA-LIKE ORGANISM, abbreviated PPLO, are seen in cross section in this electron micrograph made by Woutera van Iterson of the University of Amsterdam. The cells, which are enlarged 72,000 diameters, are not the smallest PPLO's that have been observed. Nevertheless, they are only about 50 per cent larger than the vaccinia virus. Unlike the virus, however, these cells and smaller PPLO's meet a biologist's criterion for life: they are able to grow and reproduce in a medium free of other cells.

including man. In veterinary medicine one or another of them has been identified as the cause of a respiratory disease in poultry, of a type of arthritis in swine and of an udder infection in sheep. Although a pleuropneumonia organism was implicated in cases of human urethritis (inflammation of the urethra), it was not until January of this year that one of them was positively identified as an agent of disease in man. Robert M. Chanock and Michael F. Barile of the National Institutes of Health and Leonard Hayflick of the Wistar Institute of Anatomy and Biology then published their finding that an organism called the Eaton agent, first isolated in 1944, is actually a member of the pleuropneumonia group and is the cause of a common type of pneumonia. Because these organisms pass through filters (like viruses) and grow in nonliving media (like bacteria) they are considered by some workers to be a bridge between these two large classes of organism, and because they show obvious differences from both bacteria and viruses they have been accorded the status of a separate and distinct order: *Mycoplasmatales.* Because of their similarity to the original pleuropneumonia

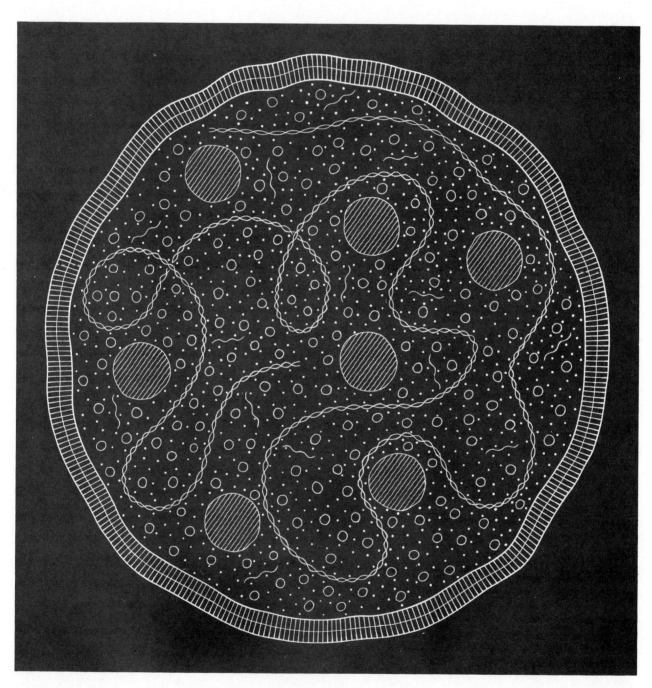

 LIPOPROTEIN MEMBRANE

○ SOLUBLE PROTEIN

RIBOSOME

• METABOLITE

SOLUBLE RNA

DNA

SCHEMATIC REPRESENTATION of a single cell of a PPLO is based on the authors' chemical analysis of *Mycoplasma gallisepticum*, which causes a respiratory disease in poultry. Deoxyribonucleic acid (DNA) and ribonucleic acid (RNA), found both in the ribosomes and in soluble particles, constitute 12 per cent of the total weight of the cell. The soluble proteins are similar to those in larger cells. The delicate cell membrane is composed of successive layers of protein, lipid, lipid and protein.

organism they are usually referred to as pleuropneumonia-like organisms, abbreviated to PPLO.

Although some very small bacteria are smaller than the larger PPLO, none is as small as the smaller PPLO: .1 micron (.00001 centimeter) in diameter. This is a tenth the size of the average bacterium; it is only a hundredth the size of a mammalian tissue cell and a thousandth the size of a protozoon such as an amoeba. But as the British mathematical biologist D'Arcy Wentworth Thompson observed some years ago, a major factor in any comparison of living things is mass, and mass varies as the cube of linear dimension. By such reckoning a protozoon is a billion times heavier than a PPLO. This vast gap in size gains vividness in the mind's eye from the reckoning that a laboratory rat is about a billion times heavier than a protozoon. A protozoon weighs .0000005 gram; a PPLO weighs a billionth as much: 5×10^{-16} gram.

In terms of linear dimensions the smallest PPLO is as close in size to an atom as it is to a 100-micron protozoon. A hydrogen atom measures one angstrom unit (.0001 micron) in diameter; a PPLO cell .1 micron in diameter is only 1,000 times larger. The existence of such a small cell raises intimate questions about the relationship of molecular physics to biology. Does a living system only a few orders of magnitude larger than atomic dimensions possess sufficient molecular equipment to carry on the full range of biochemical activity found in the life processes of larger cells? Or does the minuscule amount of molecular information it can carry compel it to operate in a simpler way? What biological or physical factors place a lower limit on the size of living cells?

In our laboratory at Yale University we have cultured 10 distinct strains of PPLO, clearly distinguished from one another by their metabolic behavior and by the antibody responses they produce in rabbits. Our work so far has been concentrated primarily on two of these strains: *Mycoplasma laidlawii*, a strain that is normally free-living in nature, and *Mycoplasma gallisepticum*, which causes chronic respiratory disease in poultry. In the first, which contains the smallest cells we have thus far studied, we have been able to follow the life cycle. In the second we have been able to determine details of chemical composition and structure.

At many stages in its life cycle the individual PPLO cell is too small to be

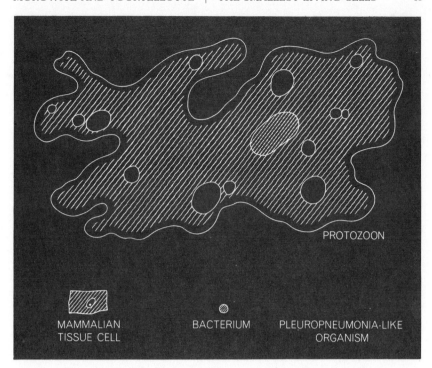

SIZES OF VARIOUS CELLS are compared. A protozoon, with a diameter of .01 centimeter, is 10 times bigger than a tissue cell, 100 times bigger than a bacterium and 1,000 times bigger than the smallest PPLO, with a diameter of .1 micron, or .00001 centimeter.

seen in the light microscope. In the electron microscope, however, we have been able to examine at least four different types of cell in *M. laidlawii*. One, called an elementary body, is a small sphere between .1 and .2 micron in diameter. A second is somewhat larger than this. A third is still larger: up to a full micron in diameter, about the size of a bacterium. A fourth type, which is of similar size, contains inclusions that are about the size of elementary bodies. In addition to observing the cell sizes directly, we measured them by forcing the cultures through filters with pores of various sizes and then examining in the electron microscope the material that had gone through the filters [see *illustration on page 54*]. To determine the size of the smallest PPLO cells we calibrated our filters by performing filtrations on two viruses of known size: the influenza virus, which is .08 to .1 micron in diameter, and the vaccinia virus, which is .22 by .26 micron in size. The smallest PPLO cells lie between these two; they are larger than the influenza virus but smaller than the vaccinia virus.

To separate the smallest cells of the strain from the others we had to employ the method of density-gradient centrifugation [see *bottom illustration on next two pages*]. This technique derives its effectiveness from the fact that cells as

small as the PPLO vary in density as well as in size as they go through their life cycle. The density at each phase depends on the changing chemical composition of the cell and closely approximates the mean of the densities of its constituents. Salt solutions of different concentration are layered in a centrifuge tube, and the cell culture is added at the top. When the tube is inserted in the centrifuge and spun at high speed, cells of various sizes settle in the layer of salt solution that has a density equal to their own. Centrifuging a 72-hour culture of *M. laidlawii* in solutions that varied in density from 1.2 to 1.4 (the density of water is 1.0) revealed three bands. Examination of these bands in the electron microscope showed the bottom band contained large cells; the top band, elementary bodies; and the middle band, cells of intermediate size and large cells with inclusions.

Starting with elementary bodies thus isolated from a culture, we have been able to follow a culture of *M. laidlawii* through its life cycle. Our method is to sample the culture at periodic intervals and inspect the samples in the electron microscope. Young cultures—about 24 hours old—are primarily composed of large cells. Cultures about six days old, on the other hand, are predominantly elementary bodies. Samples

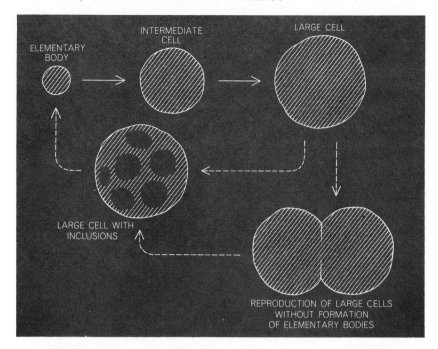

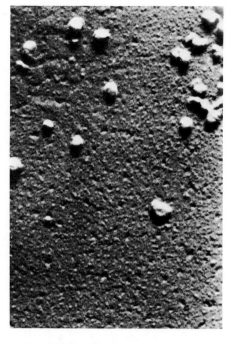

LIFE CYCLE of the PPLO *Mycoplasma laidlawii* is outlined. Elementary bodies grow to intermediate cells and then large ones. The large cells may divide, some developing inclusions released as elementary bodies, or may develop and release inclusions directly.

FOUR TYPES OF CELL in the PPLO *M. laidlawii* are seen in these electron micrographs. First micrograph (*far left*)

taken over the course of the five-day interval suggest that this strain has two methods of reproduction. In both cases the organism goes through a cycle in which elementary bodies are transformed first into intermediate cells, then into large cells and then back into elementary bodies again. Differences in composition between young and old cultures show, however, that the organism can probably adopt one of two courses once it has reached the large cell stage. In one cycle the large cells develop inclusions, which are apparently released as elementary bodies. In the second the large cells seem to reproduce by binary fission. Thereafter it appears that some of them form inclusions from which new elementary bodies are liberated. In either case the new elementary bodies begin the life cycle all over again.

We have not so far been able to establish the mode of reproduction in *M. gallisepticum*. None of our cultures has

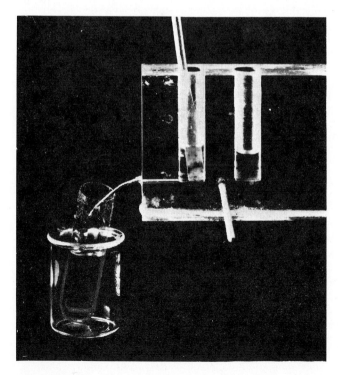

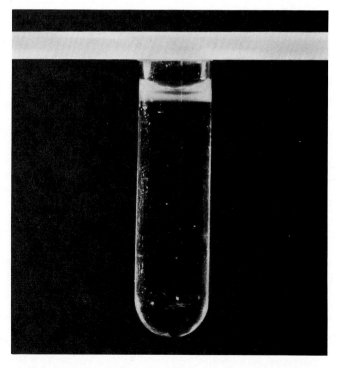

SEPARATION OF PPLO CELLS BY SIZE AND TYPE is achieved by density-gradient centrifugation. This method can be used because small cells have different densities at different times in their life cycle.

In photograph at far left two densities of salt solution are layered in a centrifuge tube. In the second photograph a PPLO culture has been added at the top of the tube. In the third photograph the

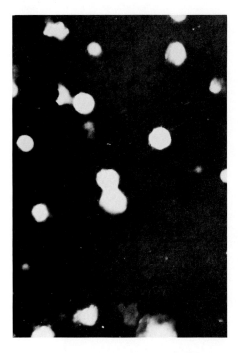
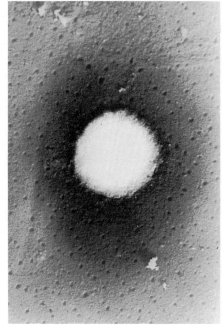

shows elementary bodies about .1 micron in diameter. The second shows intermediate cells. The third shows large cells about 1 micron in diameter, and fourth shows large cells that have developed inclusions. The inclusions may be released as elementary bodies to begin the life cycle again. All the micrographs, which enlarge the cells 17,750 diameters, were made by the authors.

revealed either elementary bodies or large cells. All the cells we have seen appear uniformly spherical and all appear to be about .25 micron in diameter [*see illustration on page 48*]. Our work with *M. gallisepticum* has helped, however, to settle the question of whether or not these tiny organisms conduct the same biochemical processes as larger cells.

Chemical analysis shows that the *M. gallisepticum* cell has the full complement of molecular machinery. In the first place, the nonaqueous substance of the cell contains 4 per cent deoxyribonucleic acid (DNA) and 8 per cent ribonucleic acid (RNA). These large molecules have been identified in larger cells as the bearers of the genetic information that governs the synthesis of the other components of a cell. Moreover, we find that in the tiny cells the composition of these molecules, the so-called base ratios, falls within the normal range. The DNA ap-

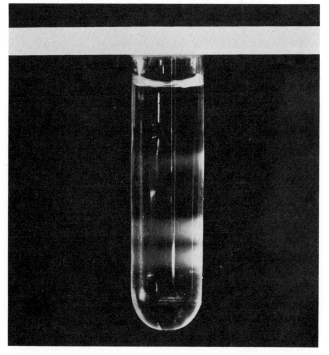

tube is ready to be placed in a container (*left*) and fastened to a rotor, part of which is seen at right. The rotor is then placed in a centrifuge. The last photograph shows the tube after centrifuga-tion. The PPLO's have settled in three bands. The bottom band contains large cells; the middle band, intermediate cells and large cells with inclusions; the top band, elementary bodies.

COLONIES OF THE EATON AGENT, now known to be a PPLO, are seen at a magnification of 600 diameters in this light micro- graph. Discovery that the Eaton agent causes a type of pneumonia in man is the first proof that a PPLO can produce human disease.

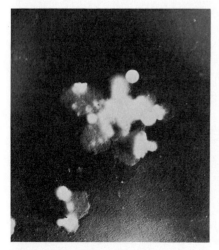

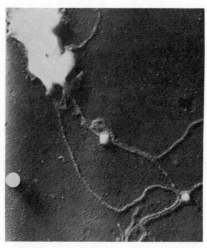

PPLO CELLS from a rat strain are contrast- ed in size with a .26-micron sphere. The large cell seen at the center contains inclusions.

FILAMENTS are observed in many strains of PPLO. The cells shown here are the same strain as those in micrograph at left.

HUMAN PPLO is seen in this electron mi- crograph. All three micrographs, magnified 16,000 diameters, were made by the authors.

pears as the familiar double-stranded helix found in the chromosomes of larger cells, and most of the RNA appears to be in the form of particles resembling ribosomes, the organelles that are believed to conduct protein synthesis in larger cells. The soluble proteins in the cell seem to have the usual range of size and variety, and the amino acid units of which they are composed occur in the expected ratios.

In several respects this PPLO cell appears to resemble animal cells more than it does plant cells or bacteria. The composition of its fatty substances, including cholesterol and cholesterol esters as an essential element, is characteristic of animal cells. More important, it has no rigid cell wall but has instead a flexible membrane that, in other strains of PPLO, permits the cells to assume a great variety of shapes. In spite of its delicacy the PPLO membrane is able to fulfill the functions of a cell membrane. It effectively distinguishes the cell from its environment and it is firm enough to contain the cell's internal structures in a coherent way. Indirect measurement of its electrical properties shows that they fall within the normal range. At a sufficiently high magnification the membrane can be seen in the electron microscope. It measures about 100 angstrom units (.01 micron) in thickness, which is typical of many animal cells.

Thus far we have been able to demonstrate more than 40 different enzymatic functions in *M. gallisepticum.* These include the entire system of enzymes necessary for the metabolism of glucose to pyruvic acid, one of the processes by which cells extract energy from their nutrients. Therefore the evidence points to a considerable biochemical complexity in these organisms. In spite of their size they seem to compare in structure and function with other known cells.

The demonstration that these tiny cells are indeed free-living compels a further question: Can there be other cells, even smaller than the PPLO and as yet undiscovered, that possess the capabilities for growth and reproduction in a cell-free medium? The mere detection of such cells presents a challenge to the ingenuity and technical resources of the biologist. If the cells happen to be pathogens, they might be discovered by the diseases they produced. If they are harmless to other forms of life, they might put in a visible appearance by causing turbidity in a culture medium through mass growth, or they might

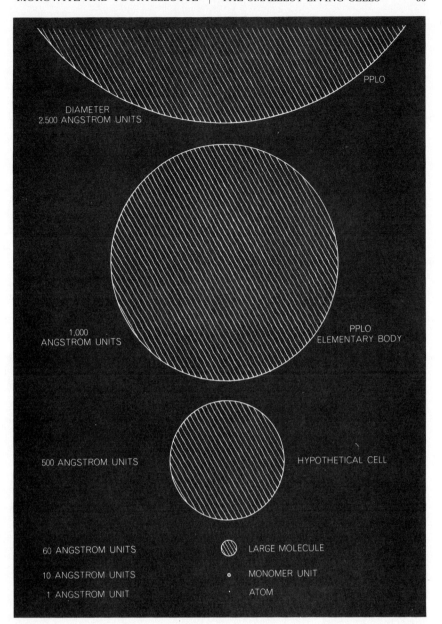

	NUMBER OF ATOMS IN DRY PORTION OF CELL	MOLECULAR WEIGHT OF DNA (DNA = 4 PER CENT OF CELL CONTENT)	NUMBER OF MONOMER UNITS (AMINO ACIDS AND NUCLEOTIDES)	NUMBER OF LARGE MOLECULES
PLEUROPNEUMONIA-LIKE ORGANISM (DIAMETER: 2,500 ANGSTROM UNITS)	187,500,000	45,000,000	9,375,000	18,750
PLEUROPNEUMONIA-LIKE ORGANISM ELEMENTARY BODY (DIAMETER: 1,000 ANGSTROM UNITS)	12,000,000	2,880,000	600,000	1,200
HYPOTHETICAL CELL (DIAMETER: 500 ANGSTROM UNITS)	1,500,000	360,000	75,000	150

PPLO CELLS AND ATOMS can be shown on the same scale; a PPLO elementary body is only 1,000 times larger than a hydrogen atom. Table shows number of atoms, molecular weight of DNA, number of monomer units (the repeating units of a large molecule) and number of large molecules anticipated in PPLO cells and in smallest theoretical cell.

show up in electron microscope preparations. A very small cell, however, might escape detection by any of these methods. If the population it formed grew to a concentration of only 100,000 cells per cubic centimeter of the growth medium, the chances of finding it would be slight. There is no assurance that the proper growth medium could be found to culture such cells. It may well be, therefore, that the PPLO is not the smallest living cell.

Yet there are lower limits, in theory at least, to the size of a living organism. Biological considerations suggest one such limit. A cell must have a membrane, if only to provide coherence for its structure. Since all cell membranes so far studied appear to be on the order of 100 angstrom units (.01 micron) in thickness, it would seem that no cell could exist that had a diameter less than 200 to 300 angstrom units (.02 to .03 micron), or about a tenth the diameter of the *M. gallisepticum* cell. Biochemistry suggests that the smallest cell would have to be somewhat larger in size. The complexity of function necessary to growth and reproduction indicates that the minimal organism must be equipped to conduct at least 100 enzymatically catalyzed reactions. If each reaction were mediated by a single enzyme molecule, the molecules would require a sphere 400 angstrom units in diameter to encompass them and the raw materials on which they operate.

Biophysics suggests another limit to the smallest size. In his little book *What Is Life?* the physicist Erwin Schrödinger pointed out that a cell has to survive against the ceaseless internal deterioration caused by the random thermal motion of its constituent molecules. In a very small cell even small motions are large in proportion to the entire system, and small motions are statistically more likely to occur than large ones. With only one or at most a few molecules of each essential kind present in the smallest conceivable cell, the most minute dislocation might be enough to disable the cell.

The foregoing reasoning seems to set 500 angstrom units as the minimum diameter of a living cell. A cell of this size would have, in its nonaqueous substance, about 1.5 million atoms. Combined in groups of about 20 each, these atoms would form 75,000 amino acids and nucleotides, the building blocks from which the large molecules of the cell's metabolic and reproductive apparatus would be composed. Since these large molecules each incorporate about

FILTRATION of a PPLO culture requires filters calibrated for viruses. At left is a tube holding an unfiltered culture of PPLO's. In middle, PPLO culture is forced through a filter with .22-micron pores. Tube at right contains filtered PPLO elementary bodies.

500 building blocks, the cell would have a complement of 150 large molecules. This purely theoretical cell would be delicate in the extreme, its ability to reproduce successfully always threatened by the random thermal motion of its constituents.

The smallest living organism actually observed—the .1-micron, or 1,000-angstrom, elementary body of *M. laidlawii*—is only twice this diameter. Its mass, of course, is eight times larger, and calculation from its observed density shows that it may contain 1,200 large molecules. This is a quite finite number, and since the organism grows to considerably larger size in the course of its reproductive cycle it cannot be said that 1,200 large molecules constitute its complete biochemical equipment. In the case of *M. gallisepticum*, however, we know that a diameter of .25 micron—only

five times the theoretical lower limit—does encompass an autonomous metabolic and reproductive system. Our chemical analysis shows that the entire system is embodied in something less than 20,000 large molecules [*see chart on page 53*].

This is still an exceedingly small amount of material to sustain the complexity of biochemical function necessary to life. In fact, the portion of it allotted to the genetic function seems inadequate to the task. The 4 per cent of its dry substance that is DNA has a total molecular weight of 45 million. Since, according to current views of genetic coding, it takes an amount of DNA with a molecular weight of one million to encode the information for the synthesis of one enzyme molecule, *M. gallisepticum* would seem to contain enough genetic material to encode only

a few enzymes beyond the 40 we have identified so far. That is far short of the 100 enzymes thought to be the minimum for cellular functions. It may be that the enzymes of very small cells are less specific and hence more versatile in their action than the enzymes of larger ones. On the other hand, it may prove necessary to re-examine prevailing ideas about the way information is encoded in the genetic material.

Questions of this kind suggest the principal challenge of very small cells. If they are indeed simpler than other cells, they can tell much about the basic mechanisms of cell function. If, on the other hand, they are functionally as complicated as other cells, they pose the fundamental question of how such functional complexity can be carried in such tiny pieces of genetic material.

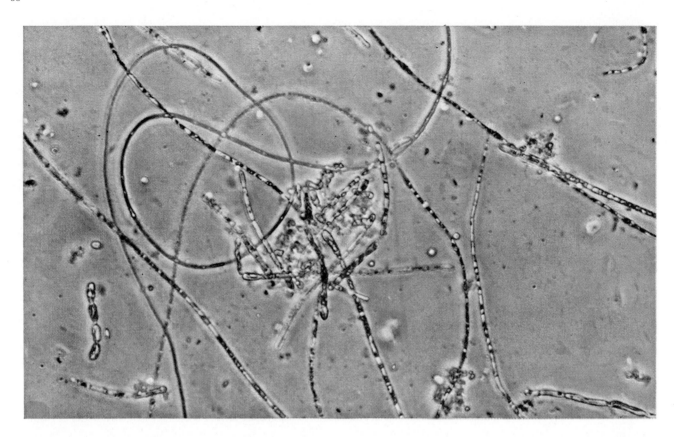

BLUE-GREEN ALGAE of the species *Gloeotrichia echinulata* consist of numbers of individual cells strung together in attenuated filaments radiating from a central point. The enlargement in this photomicrograph is about 150 diameters. Blue-green algae actually appear in a variety of colors. They were named blue-green because the first specimens to be classified were of that color.

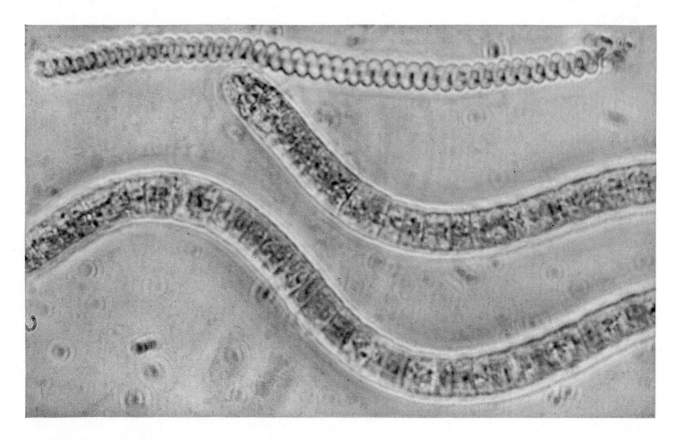

FILAMENTOUS FORMS of blue-green algae also include the species *Spirulina versicolor* (top) and two filaments of the genus *Arthrospira*; the enlargement is approximately 1,500 diameters. *Arthrospira* is one of the blue-green algae responsible for the pink color often seen in flamingos; the birds acquire the color from carotenes in the blue-green algae that are part of their diet.

5

The Blue-Green Algae

by Patrick Echlin
June 1966

*These primitive plants more closely resemble bacteria
than they do other algae. They live in an
extraordinary range of environments, and they have
both beneficial and harmful effects in human affairs*

The algae are the simplest members of the plant kingdom, and the blue-green algae are the simplest of the algae. Indeed, the blue-green algae resemble bacteria more closely than they do other forms of algae. By this token they occupy a distinctive niche in the evolutionary order of things. They provide insights into the evolution of bacteria and algae, and also, since they are among the most primitive living cells, into the beginnings of the cell itself. Today, moreover, the blue-green algae have a considerable and increasing economic importance: they have both beneficial and harmful effects on human life.

Most of the blue-green algae are blue-green, but not all; they are found in a wide range of colors. Their name stems from the fact that the first species to be recognized as members of the group were blue-green. A few of the 2,000 species now known live as single cells of microscopic size. The cells of other species gather in colonies but still live essentially as individuals. Most species of blue-green alga, however, are filamentous: their cells are strung together in a hairlike structure. This is the form in which blue-green algae are most likely to be visible to the unaided eye, either as a mosslike growth on land or as a soft mass in water.

Whatever the form, the individual cells of blue-green algae are much alike. Each is surrounded by a gelatinous sheath. Inside the sheath is a thin membrane that encloses the cell's cytoplasm. The cell of a blue-green alga lacks a well-defined nucleus and the elaborate intracellular membranes and separate organelles of cells in advanced plants and animals. In the peripheral parts of the cytoplasm there are, however, complex lamellae, or thin sheets, that apparently form from the cell membrane [*see illustrations on page 60*].

Reproduction in most blue-green algae is a simple and asexual process. The cell merely divides. Some of the filamentous forms reproduce by a breaking of the filament, the two parts of which then grow by cell division. A few species of blue-green alga are able to reproduce by forming spores.

Like most other algae, the blue-greens manufacture their food by photosynthesis. They are distinctive, however, in that their photosynthetic pigments are distributed throughout the peripheral lamellae rather than in discrete bodies such as chloroplasts. A unique feature of certain filamentous species is the tendency to form heterocysts: large colorless cells at irregular intervals along a filament. The function of these cells is uncertain. According to various hypotheses they represent a vestigial reproductive cell, a store of food or a structure associated with either cell division or the formation of internal spores.

The blue-green algae are widely distributed over land and water, often in environments where no other vegetation can exist. They live in water that is salt, brackish or fresh; in hot springs and cold springs, both pure and mineralized; in salt lakes; in moist soils, and in symbiotic or parasitic association with other plants and animals. Most marine forms grow along the shore, fixed to the bottom in the narrow zone between the high and low tidemarks; a few float about as plankton. The largest number of species live in fresh water. Some can be found in fast-moving or turbulent waters, such as the water falling on rocks under a waterfall; others flourish in quiet waters—even in bodies of water that appear only temporarily.

On land in places of high humidity, such as gullies on the lower slopes of mountains, both tree trunks and rocks may be covered with gelatinous mats of many hues; the mats consist of single-celled blue-green algae. The filamentous algae form a feltlike growth over extensive land areas. Where high temperatures are combined with high humidity, as in the Tropics, the growth can be quite luxurious.

The wide variety of colors among blue-green algae has two main sources: the pigmentation of the gelatinous sheath and the pigmentation within the cell. The sheath, particularly in species that grow on land, is often deeply pigmented. Yellow and brown tints predominate, although shades of red and violet are also seen. The sheath coloration appears to be related to the environment—chiefly to the amount of sunlight received by the algae and to the acidity of their medium.

Within the cell are three pigments that are found in all plants capable of photosynthesis: chlorophyll, carotene and xanthophyll. In addition there are two pigments that are found only in this group of organisms: the blue pigment c-phycocyanin and the red pigment c-phycoerythrin. It is these two pigments that are principally responsible for the group's diverse coloration. The colors range from the red of the species *Oscillatoria cortiana* and *Phormidium persicinum* through the emerald green of the genus *Anacystis* to the near black of some algae that live on rocks. The color of the algae is also dependent on the age and physiological state of the organism. Healthy cultures of *Anacystis montana* are a bright emerald green, whereas an old culture is a dirty yellow. There are even a few members of the blue-green group that are colorless; they live in such diverse habitats as the bottom of lakes, the intestines of animals and the human mouth.

Certain forms of blue-green algae have the capacity for changing color in relation to the color of the light that falls on them. The filamentous alga *Oscillatoria sancta* is green in red light, blue-green in yellow-brown light, red in green light and brownish yellow in blue light. Some deep-water algae are red, apparently for the following reason. As sunlight penetrates deeper into water its longer wavelengths, starting with the wavelengths of red light, are progressively absorbed. The last part of the visible spectrum to be absorbed is the blue. Such light could best be utilized for photosynthesis by algae with a preponderance of red pigments.

The capacity of the blue-green algae for adjusting to light of different intensities and colors means they are better adapted than other photosynthetic organisms to utilize the available light. This characteristic probably has much to do with the occurrence of these algae in so many terrestrial and aquatic habitats. Blue-green algae can grow in full

sunlight and in almost complete darkness. In the Jenolan caves of Australia, for example, they grow on moist limestone near incandescent lamps that are lighted for only six hours a week while tourists pass through.

The majority of blue-green algae are aerobic photoautotrophs: their life processes require only oxygen, light and inorganic substances. The process of photosynthesis uses the energy of light to build carbohydrates (and some fats) out of carbon dioxide in the air or the water. The process of respiration then uses oxygen to "burn" these products and supply the energy needed for the rest of the alga's activities. Unlike more advanced organisms, these algae need no substances that have been preformed by other organisms.

A few forms of blue-green algae, such as a species of *Oscillatoria* that is found in mud at the bottom of the Thames, are able to live anaerobically: their life processes do not require free oxygen. They obtain their energy from inorganic compounds such as hydrogen sulfide. Other species, notably some that live in the soil, can grow in the dark if they are

supplied with suitable organic nutrients. Some forms are able to fix atmospheric nitrogen in soluble salts that can then be utilized by the alga itself.

A remarkable feature of blue-green algae in water is their ability to move even though they possess no recognizable locomotory parts such as flagella and cilia. The filamentous forms can move fairly rapidly; the unicellular forms move much more slowly. All the blue-green algae exhibit a gliding movement parallel to the alga's long axis. The movement can be either forward or backward. Sometimes it is accompanied by rotation.

Several mechanisms have been put forward to explain these movements. One currently receiving attention is that the excretion of a mucilaginous substance propels the alga. Other proposed mechanisms involve osmosis, surface tension, the streaming of cytoplasm within the cell and the propagation of rhythmic waves of contraction through the cell. In connection with this last hypothesis a scalloped edge visible in electron micrographs of some species of blue-green algae may be significant. Such an edge could be associated with contractile movement.

Another remarkable feature of the blue-green algae is their ability to withstand environmental extremes such as high and low temperatures and high concentrations of salt. Indeed, blue-green algae are perhaps best known to biologists as inhabitants of hot springs. Most of these "thermal" species live at temperatures of 50 to 60 degrees centigrade (122 to 140 degrees Fahrenheit); a few have been known to exist at temperatures as high as 85 degrees C. (185 degrees F.). Yet most of the thermal blue-green algae can also thrive at normal temperatures. Several species that flourish in my laboratory at the University of Cambridge at a temperature of 15 degrees C. (59 degrees F.) were collected from a hot spring in Yellowstone National Park.

As for the other end of the environmental temperature scale, blue-green algae are found in mountain streams and in Antarctic lakes. Some species live in association with lichens on bare mountain rocks in polar regions, where the surface temperature may vary in a few hours from −60 to +15 degrees C. (−76 to +59 degrees F.). In fact, the only cold environment in which blue-green algae are rare is snow.

About a fifth of all the blue-green algae live in saline environments. A few

ORDER AND FAMILY	GENERA	CHARACTERISTICS
ORDER CHROÖCOCCALES	ANACYSTIS CHROÖCOCCUS GLOEOCAPSA GLOEOTHECE MERISMOPEDIA MICROCYSTIS SYNECHOCOCCUS SYNECHOCYSTIS	Unicellular and free-living, solitary or colonial.
ORDER CHAMAESIPHONALES	CHAMAESIPHON DERMOCARPA	Unicellular and free-living, solitary or colonial. Spore-producing.
ORDER PLEUROCAPSALES	PLEUROCAPSA	Filamentous. Growing on or in rocks.
ORDER NOSTOCALES		Filamentous, simple or branched.
FAMILY OSCILLATORIACEAE	LYNGBYA OSCILLATORIA	Simple filaments. No heterocysts.
FAMILY SCYTONEMATACEAE	PLECTONEMA SCYTONEMA TOLYPOTHRIX	False branching. Generally with heterocysts.
FAMILY RIVULARIACEAE	GLOEOTRICHIA RIVULARIA	False branching. Colonial.
FAMILY NOSTOCACEAE	ANABAENA CYLINDROSPERMUM NOSTOC	Simple filaments.
ORDER STIGONEMATALES	FISCHERELLA STIGONEMA	True branching. Heterocysts.

CLASSIFICATION of blue-green algae according to a system used by many botanists lists them in the division Schizophyta, class Myxophyceae (Cyanophyceae) of the plant kingdom. About 2,000 species of blue-green alga are grouped in the orders and families shown here.

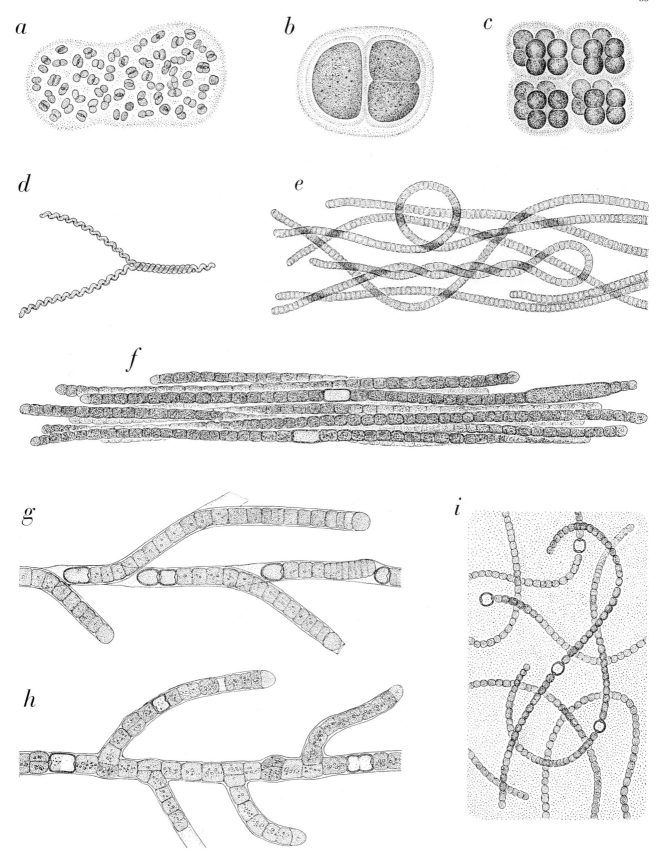

TYPICAL SPECIMENS of blue-green algae include, at top, three of the unicellular and free-living variety: *a, Coccochloris* in a loosely formed colony; *b, Anacystis dimidiata* dividing; *c, Anacystis thermalis* in a compact colony. The remaining examples are of genera in which several cells exist together in a filament: *d, Spirulina,* with two intertwined filaments; *e, Oscillatoria; f, Aphanizomenon,* with two of the large, colorless cells called heterocysts; *g, Tolypothrix,* which is a genus with false branching; *h, Hapalosiphon,* which has true branching; *i, Nostoc* in a gelatinous mass. Each species is represented at an enlargement of about 500 diameters.

60

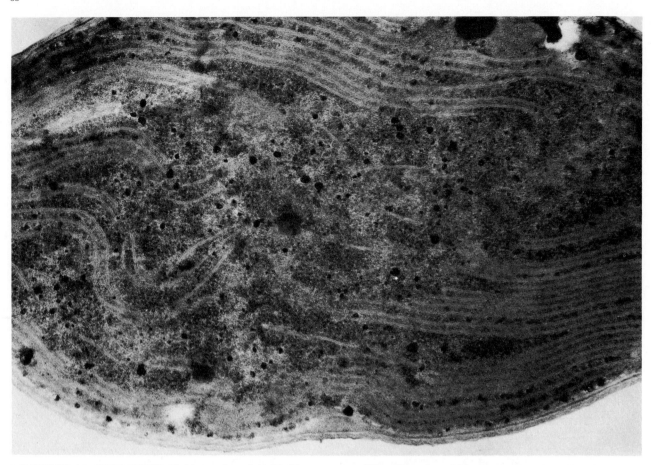

UNICELLULAR BLUE-GREEN ALGA is enlarged 90,000 diameters in this electron micrograph of a specimen of *Anacystis mon-* *tana*. Like all species of blue-green alga, the cell lacks a well-defined nucleus. A cell of the same species is shown in detail below.

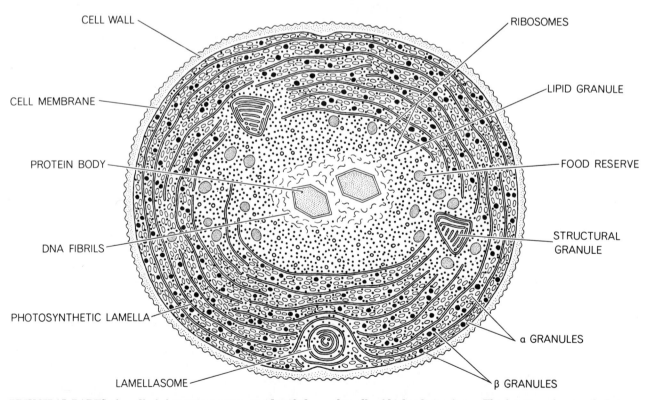

CELL WALL

RIBOSOMES

CELL MEMBRANE

LIPID GRANULE

PROTEIN BODY

FOOD RESERVE

DNA FIBRILS

STRUCTURAL GRANULE

PHOTOSYNTHETIC LAMELLA

α GRANULES

LAMELLASOME

β GRANULES

PRINCIPAL PARTS of a cell of *Anacystis montana* are identified. This genus, like other blue-green algae, lacks the highly specialized structures such as mitochondria and chloroplasts that are found in the cells of higher living forms. The function of many algal structures, including the larger ones, is unknown. Not shown is a gelatinous outer sheath characteristic of the blue-green algae.

species are truly halophilic, or salt-tolerant; such species are found in southern France and in California in brines with a salt concentration as high as 27 percent. (The concentration of salt in seawater is about 3 percent.) An extreme example of high salt tolerance is the single-celled species *Gomphosphaeria*, which grows abundantly in the spring called Bitter Waters in Death Valley. The concentration of magnesium salts there is so high that the mud around the spring is encrusted with them.

From the evolutionary standpoint the algae in general are of interest because of the trends they reveal in the internal organization of the cell and the mechanism of cellular reproduction. The blue-green algae in particular, with their simple cell structure and asexual reproduction, appear to stand in close relation to the first organisms on earth. It is in this context that the similarities between the blue-green algae and the bacteria have intrigued a number of investigators.

Three features that set the blue-green algae and the bacteria apart from all other cellular organisms have been known for some time. These features are the absence of nuclei, the absence of specialized organelles and the absence of sexual reproduction. Citing the absence of certain features from two groups of organisms is, however, a rather negative way of establishing similarity between the groups. With the advances of electron microscopy in recent years it has become possible to examine the fine structure of blue-green algae and bacteria and so to find some positive similarities between them.

One such similarity is a fairly close resemblance between the cell wall of blue-green algae (not the gelatinous sheath) and the cell wall of bacteria [*see illustration on this page*]. In both cases an important component of the cell wall belongs to the class of molecules known as mucopeptides. Another similarity is that in both blue-green algae and the photosynthetic bacteria (as in other photosynthetic organisms) the essential feature of the photosynthetic apparatus is a set of membranes enclosing a space. In the blue-green algae these are the lamellae; in the photosynthetic bacteria they are called the chromatophores.

These similarities and others argue strongly for isolating blue-green algae and bacteria in a group distinct from all other organisms. I would also argue

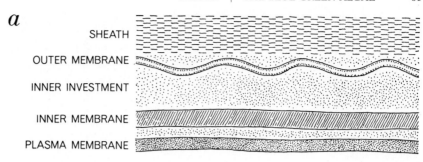

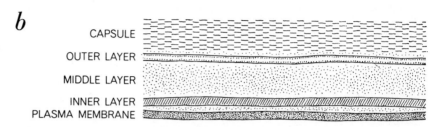

STRUCTURAL RESEMBLANCE between blue-green algae and bacteria is typified by the similarities of their exteriors. At *a* are the sheath and cell wall of a cell of blue-green alga; at *b*, the capsule and wall of a typical bacterium. The particular similarities are in the structure of the walls. In each organism the distance between the outside of the outer membrane or layer and the inside of the inner one is between 200 and 400 angstrom units.

that the blue-green algae and the bacteria have probably descended from the same type of ancestral cell, although here I must rest my case more on speculation than on firm evidence. It is true that there are differences between the two types of organism. Blue-green algae are generally aerobic and photosynthetic and are more complex in form than the bacteria. The bacteria are either aerobic or anaerobic, but they are usually not photosynthetic; they require for their existence substances preformed by other organisms. In cases where bacteria are photosynthetic they differ from blue-green algae in photosynthetic mechanism, particularly in the chemical nature of the pigments.

Still, the similarities are too marked to be overlooked. It is unlikely that so many kindred features would arise independently. This is the root of the argument for a common ancestral cell. The conspicuous differences between the blue-green algae and the bacteria presumably arose at later stages in their evolution.

Perhaps the most important contribution made by the blue-green algae to human affairs lies in the fact that the nitrogen-fixing species increase the fertility of the soil. It is more than likely that this contribution can be enhanced. R. N. Singh of the Banares Hindu University in India has shown that the introduction of blue-green algae to saline and alkaline soils in the province of Uttar Pradesh increases the soils' content of nitrogen and organic matter and also their capacity for holding water. The treatment has so much improved these formerly barren soils that they can now be used to grow crops.

Other studies have shown that blue-green algae can fix enough nitrogen to support a good crop of rice and leave about 70 pounds of nitrogen in each acre after the harvesting of the rice. Astushi Watanabe of the University of Tokyo tested several species of blue-green alga in rice paddies and found that the filamentous form *Tolypothrix tenuis* yielded 1¼ tons of nitrogen per acre per year. Watanabe grew the algae in his laboratory on fine gravel soaked with a culture medium and aerated with air rich in carbon dioxide. The cultures were transferred to the experimental site in plastic bags and then inoculated into it. This kind of inoculation over a period of years increased the nitrogen in the soil by 30 percent and the yield of the rice crop by 20 percent.

Experiments conducted some years ago in Kansas, Oklahoma and Texas by W. E. Booth of the University of Kansas showed that a coating of blue-green algae on prairie soil binds the particles of the soil together, maintains a high water content and reduces erosion. Singh has envisioned the agricultural application of algal "blooms": huge shoals of algae that sometimes form in

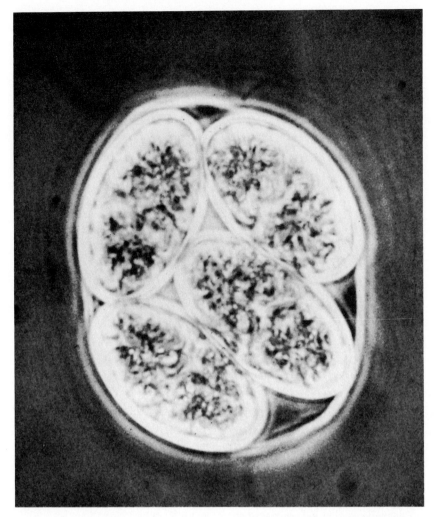

SYMBIOTIC RELATION of blue-green algae and another organism is represented by four cells of *Glaucocystis nostochinearum*, which are enlarged some 2,200 diameters in this photomicrograph. The host cells are a green alga that has lost its chloroplasts. Inside the host cells are filaments of a blue-green alga that apparently function as chloroplasts.

gus receives carbohydrates and perhaps nitrogen compounds from the alga, and the alga is able to survive in otherwise inhospitable environments because of the moist medium for growth provided by the fungus.

Several examples have come to light in which blue-green algae live parasitically within bacteria or other algae (including some blue-green species). I am currently investigating such an organism, *Glaucocystis nostochinearum*. This is a complex of two organisms: the host cell is a green alga that lacks the chlorophyll-containing chloroplasts, and within it are several filaments of a blue-green alga. Apparently the blue-green algae act as chloroplasts for the host cell.

There are many associations between blue-green algae and animals. The exact physiological relation between the partners is not known; presumably the partnership produces an essential substance that neither partner could produce by itself. In associations of this kind it seems likely that the blue-green algae originally entered the host as food but resisted digestion and stayed permanently—even, in the case of simple algae that reproduce by dividing, unto succeeding generations.

Whereas some blue-green algae act to break down rock, the species that live in hot springs actually build rock. This they accomplish by depositing salts of calcium and possibly silica within the gelatinous sheath of the algal cell wall. At Mammoth Hot Springs in Yellowstone National Park, for example, the algae deposit travertine at the rate of two feet a year. The bright colors of the basins and terraces around such hot springs are caused by the algae living in the outer layers.

As one might expect of so large and widely distributed a group of organisms, the blue-green algae have some effects inimical to the interests of man. Today their most harmful effect is undoubtedly the formation of blooms in bodies of water. When these epidemic growths occur in fresh water, they can be hazardous to human health. They choke the intakes of water-supply systems and give the water a disagreeable odor. To make matters worse, the increasing pollution of fresh waters in heavily populated areas favors the growth of algae and not of other organisms. Lake Erie, which once had white beaches and supported a prosperous fishing industry, is now seriously infested with blue-green algae. Sewage,

bodies of water. (The Red Sea, for example, owes its occasional red color to blooms of the blue-green alga *Trichodesmium*.) Singh's proposal is to use blooms as a manure because of their high content of nitrogen and phosphorus. He found that adding dried blooms to soil in which sugarcane was growing substantially increased the crop yield. Periodic applications of fertilizers sufficient to maintain a continuous bloom of algae in fishing lakes in Alabama and Mississippi nourished the large numbers of small animals on which fish feed.

Blue-green algae are often the first plants to colonize bare areas of rock and soil. A dramatic example of such colonization is provided by the island of Krakatoa in Indonesia, which was denuded of all visible plant life by its cataclysmic volcanic explosion of 1883. Filamentous blue-green algae were the first plants to appear on the pumice and

volcanic ash; within a few years they had formed a dark green gelatinous growth. The layer of blue-green algae formed in such circumstances eventually becomes thick enough to provide a soil rich in organic matter for the growth of higher plants. The algae further contribute to soil formation by acting to break down the surface of the rock.

Some of the blue-green algae that live symbiotically with other organisms undoubtedly have an economic impact, but it is difficult to measure. The best-known example of such symbiosis is provided by lichens, which are a combination of a fungus and a blue-green alga. Usually the lichen fungus can grow only if the appropriate alga is present. Lichens, like blue-green algae alone, play an important role in pioneering plant growth on bare rock. The relation between the fungus and the alga is not clearly understood. Presumably the fun-

industrial wastes and an estimated 80 tons a day of phosphates in water running off from farmlands have turned parts of the lake into a vast tank for the culture of algae. The algae rob the water of its oxygen, and the lake becomes incapable of supporting fish life. They also wash onto the beaches and cover them with a malodorous green slime.

The measures needed to ameliorate this problem may cost the states that adjoin the lake up to a billion dollars apiece. The states have to expand sewage plants and establish tighter controls over industrial wastes. They may also have to treat the water in areas where blooms occur with such algicides as copper sulfate, which in concentrations as low as two parts per million prevent all species of blue-green alga from growing.

Under certain circumstances freshwater algae of such genera as *Microcystis, Aphanizomenon* and *Anabaena* can cause death or injury to animals. Cases of human illness associated with blue-green algae are on record, even though forms such as *Oscillospira* and *Anabaenolium* normally reside in the human gastrointestinal tract. Various gastrointestinal, respiratory and skin disorders have been traced to the ingestion or inhalation of blue-green algae or to contact with them. In South Africa in 1943 thousands of cattle and sheep were killed along a dam in the Transvaal, where the reservoir developed a poisonous bloom of *Microcystis*. The toxic substance, which was ingested by the animals when they were watered, was later identified as an alkaloid that affects the liver and central nervous system.

In general the harmful effects of blue-green algae can be controlled. The beneficial effects are open to further development. The time may come, for example, when the pressure of population on food supplies will justify mass cultivation of the blue-green algae, perhaps in conjunction with other algae. At present blue-green algae are seldom eaten by humans. There is, however, a species of *Nostoc* that forms into balls called "water plums," which have a high content of protein and oil and are eaten in parts of China and South America. Most blue-green algae are rather unpalatable, but that difficulty could be overcome by suitable flavorings or by feeding the algae to fish, poultry and cattle destined for human consumption.

6

Differentiation in Social Amoebae

by John Tyler Bonner
December 1959

*Certain amoebae gather to form a mass of spores and a
stalk. The way in which spore cells and stalk cells
segregate may shed light on how the cells of many-celled
organisms differentiate into various types*

Recently I was asked to talk to two visiting Russian university rectors (both biologists) about the curious organisms known as slime molds. Communication through the interpreter was somewhat difficult, but my visitors obviously neither knew nor really cared what slime molds were. Then, without anticipating the effect, I wrote on the blackboard the words "social amoebae," a title I had used for an article about these same organisms some years ago [see "The Social Amoebae," by John Tyler Bonner; Scientific American, June, 1949]. The Russians were electrified with delight and curiosity. I described how individual amoebae can come together under certain conditions to form a multicellular organism, the cells moving into their appropriate places in the organism and differentiating to divide the labor of reproduction. Soon both of my guests were beaming, evidently pleased that even one-celled animals could be so sophisticated as to form collectives.

Of course there are other reasons why slime molds hold the interest of biologists. The transformation of free-living, apparently identical amoebae into differentiated cells, members of a larger organism, presents some of the same questions as the differentiation of embryonic cells into specialized tissues. In the budding embryo, moreover, cells go through "morphogenetic movements" which seemingly parcel them out to their assigned positions in the emergent organism. The only difference is that the simplicity of the slime molds provides excellent material for experiments.

The slime-mold amoebae, inhabitants of the soil, do their feeding as separate, independent individuals. Flowing about on their irregular courses they engulf bacteria, in the manner of our own amoeboid white blood cells. At this stage they reproduce simply by dividing in two. Once they have cleared the food away, wherever they are fairly dense, the amoebae suddenly flow together to central collection points. There the cells, numbering anywhere from 10 to 500,000, heap upward in a little tower which, at least in the species *Dictyostelium discoideum*, settles over on its side and crawls about as a tiny, glistening, bullet-shaped slug, .1 to two millimeters long. This slug has a distinct front and hind end (the pointed end is at the front) and leaves a trail of slime as it moves. It is remarkably sensitive to light and heat; it will move toward a weak source of heat or a light as faint as the dial of a luminous wrist watch. As the slug migrates, the cells in the front third begin to look different from the cells in the two thirds at the rear. The changes are the early signs of differentiation; eventually all the hind cells turn into spores—the seeds for the next generation—and all the front cells cooperate to make a slender, tapering stalk that thrusts the mass of spores up into the air.

To accomplish this transformation the slug first points its tip upward and stands on end. The uppermost front cells swell with water like a bit of froth and become encased in a cellulose cylinder which is to form the stalk. As new front cells arrive at the frothy tip of the stalk they add themselves to its lengthening structure and push it downward through the mass of hind-end cells below. When this process, like a fountain in reverse, has brought the stalk into contact with the surface, the continued upward migration of pre-stalk cells heightens the stalk lifting the presumptive spore cells up into the air. Each amoeba in the spore mass now encases itself in cellulose and becomes a spore. The end result is a delicate tapering shaft capped by a spherical mass of spores. When the spores are dispersed (by water or by contact with some passing creature such as an insect or a worm), each can split open to liberate a tiny new amoeba.

What mechanism brings the independent slime-mold amoebae together in a mass? More than a decade ago we found that they are attracted by the gradient of a substance which they themselves produce. In our early experiments we were unable to obtain cell-free preparations of this substance (which we named acrasin); cells actively secreting it were always necessary to start an aggregation. Later B. M. Shaffer of the University of Cambridge got around this barrier in an ingenious experiment. He took water that had been near acrasin-producing cells (but was itself free of cells) and applied it to the side of a small agar block placed on top of some amoebae. The amoebae momentarily streamed toward the side where the concentration of acrasin was higher. Shaffer found that the water must be used immediately after it is collected in order to achieve this effect, and that it must be applied repeatedly. He therefore concluded that acrasin loses its potency rapidly at room temperature. The loss of potency, he showed, is caused by enzymes that are secreted by the amoebae along with acrasin; when he filtered the fluid through a cellophane membrane to hold back the large enzyme molecules, he was able to secure a stable preparation of acrasin. Presumably the enzymes serve to clear the environment of the substance and so enhance the establishment of a gradient in the concentration of acrasin when it is next secreted. Maurice Sussman and his

SPHERICAL MASSES OF SPORES of the social amoeba *Dicty-ostelium discoideum* are held aloft by stalks composed of other amoebae of the same species. When the spores are dispersed, each can liberate a new amoeba. The stalks are about half an inch high.

AGGREGATING AMOEBAE of *Dictyostelium discoideum* move in thin streams toward central collection points. Each of the centers comprises thousands of cells. This photograph and facing one were made by Kenneth B. Raper of the University of Wisconsin.

co-workers at Brandeis University in Waltham, Mass., have confirmed Shaffer's work and are now attempting the difficult task of fractionating and purifying acrasin, steps leading toward its identification.

Meanwhile Barbara Wright of the National Institutes of Health in Bethesda dropped a bombshell. She discovered that urine from a pregnant woman could attract the amoebae under an agar block just as acrasin does. The active components of the urine turned out to be steroid sex hormones. This does not necessarily mean that acrasin is such a steroid. Animal embryologists were thrown off the track for years when they found that locally applied steroids induce the further development of early embryos. Only after much painful confusion did it become clear that steroids do not act directly on the embryo, but stimulate the normal induction substance. We must therefore consider the possibility that the steroids act in a similarly indirect manner on the amoebae. The purification of acrasin will, we hope, soon settle the question.

From observations of the cells during aggregation, Shaffer has come to the interesting conclusion that the many incoming amoebae are not responding to one large gradient of acrasin but to relays of gradients. That is, a central cell will release a puff of acrasin that produces a small gradient in its immediate

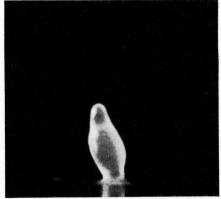

DEVELOPMENT OF THE FRUITING BODY of a slime mold is shown in this series of photographs made at half-hour intervals. At far left the tip cells are starting to form a stalk. In the next two pictures the stalk has pushed down through the mass to the

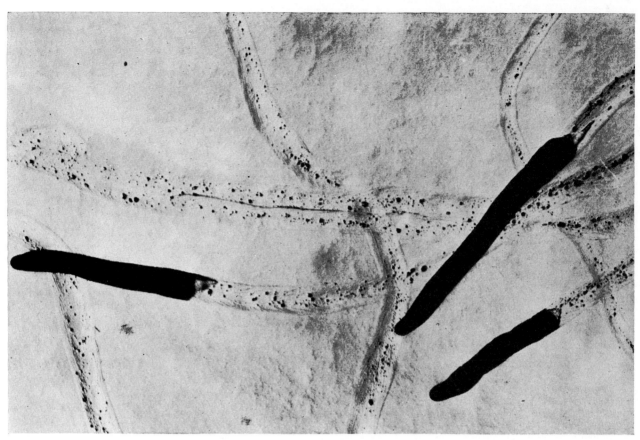

MIGRATING SLUGS of *Dictyostelium discoideum* leave trails of slime behind them as they move. The photographs in this article appear in *The Cellular Slime Molds*, by John Tyler Bonner, and are reproduced with permission of Princeton University Press.

vicinity. The surrounding cells become oriented, and now produce a puff of their own. This new puff orients the cells lying just beyond, and in this way a wave of orientation passes outward. Time-lapse motion pictures show the amoebae moving inward in waves, which could well represent the relay system. If this interpretation is sound, then the rapid breakdown of acrasin by an enzyme plainly serves to clear the slate after each puff in preparation for

the next. The cells do not depend entirely on acrasin for orientation; once they are in contact they tend to stick to one another and the pull-tension of one guides the cells that follow. This is a special case of contact guidance, a phenomenon well known in the movements of embryonic cells of higher animals.

After the amoebae have gathered together, what determines their position within the bullet-shaped slug?

One might assume that the cells that arrive at the center of the heap automatically become the tip of the slug, and that the last cells to come in from the periphery make up the hind end. If this were the case, chance alone would determine whether a cell is to become a front-end cell and enter into the formation of the stalk, or a hind-end cell and become a spore. If, on the other hand, the cells rearrange themselves as they organize into a slug, then it is conceiv-

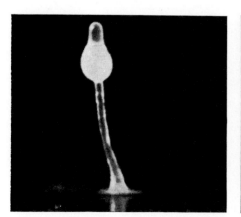
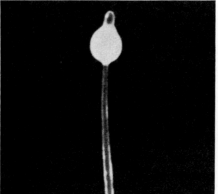

surface and is starting to lift the cell mass. In the fourth picture the spores have formed their cellulose coats, making the ball more opaque. In the last two pictures the spore mass moves up to the very top of the stalk, as the stalk itself becomes still longer.

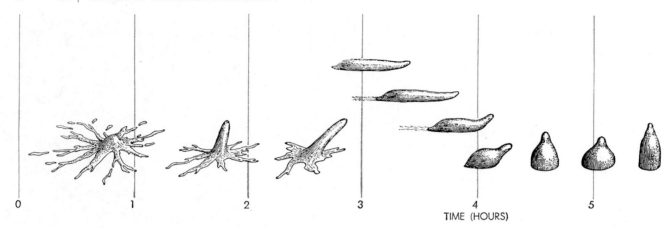

TIME (HOURS)

LIFE CYCLE OF A SLIME MOLD, typified by *Dictyostelium discoideum*, involves the aggregation of free-living amoebae into a unified mass (*first three drawings*), then the formation of a slug which moves about for a time (*next four drawings*) and finally

able that the front end might contain selected cells, differing in particular ways from those in the hind end. I am embarrassed to say that in 1944 I presented some evidence to support the idea that their chance position was the determining factor—evidence that, as will soon be clear, was inadequate. It is some comfort, however, that I was able to rectify the error myself.

The first faint hint that the cells do redistribute themselves in the slug stage came when we repeated some experiments first done by Kenneth B. Raper of the University of Wisconsin. We stained some slugs with harmless dyes and then grafted the hind half of a colored slug onto the front half of an unstained slug. The division line remained sharp for a

number of hours, just as Raper had previously observed. But later we noticed that a few stained cells were moving forward into the uncolored part of the slug. In the reverse graft, with the front end stained, a similar small group of colored cells gradually migrated toward the rear end of the slug. Still, the number of cells involved was so small that it could hardly be considered the sign of a major redistribution. Next we tried putting some colored front-end cells in the hind end of an intact slug. The result was a total surprise: now the colored cells rapidly moved to the front end, traveling as a band of color up the length of the slug.

Here was a clear demonstration that the cells do rearrange themselves in the

slug and that there is a difference between the cells at the front and hind ends. The difference between front-end and hind-end cells—whatever its nature—was confirmed in control experiments in which we grafted front-end cells to the front ends of the slugs and hind-end cells to the hind ends; in each case the cells maintained their positions.

It looked as if front-end cells were selected by their speed; the colored cells simply raced from the rear end to the front. When we placed hind-end cells in the front end, they traveled to the rear, outpaced by the faster-moving cells, which again assumed their forward positions. We tried to select fast cells and slow cells over a series of genera-

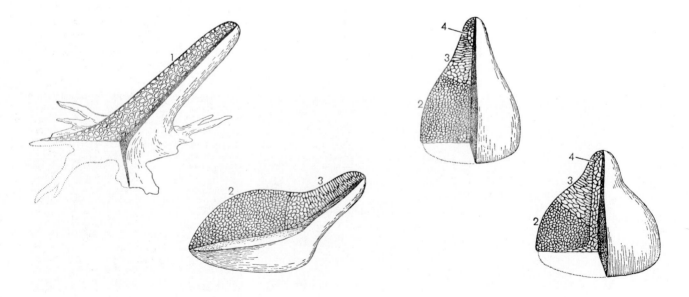

CUTAWAY DRAWINGS of five stages show how the cells change. At the end of aggregation all cells appear the same (1), but in the slug they are of two types (2 and 3). The cells near the tip (3) gradually turn into stalk cells (4) and move down inside the

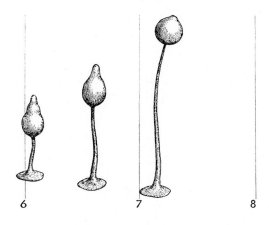

6 7 8

the development of a fruiting body (*last six drawings*). Times are only approximate.

tions to see if speed was a hereditary trait, but after selection the cultures showed no differences from one another or from the parent stock.

Quite by accident a new bit of evidence turned up in an experiment designed for totally independent reasons. Instead of using the fully formed slug we stained amoebae colonies in the process of aggregation and made grafts at this stage by removing the center of the stained group and replacing it with a colorless center, or vice versa. In either case the resulting slug was always uniformly colored, indicating a rapid reassortment of the cells during the formation of the slug.

The evidence for a rearrangement of

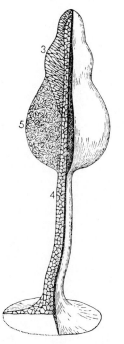

mass. The others (2) become spores (5) as the growing stalk lifts them into the air.

cells was becoming impressive, but I felt uneasy about the reliability of tests with dyes because such tests had led me into my earlier error. We needed to confirm our results by a different method.

At about this time M. F. Filosa, who was working in our laboratory on his doctoral dissertation, discovered that many of our amoeba cultures contained more than one genetic type. By isolating and cultivating single cells of each type he was able to obtain pure strains that displayed various recognizable abnormalities—in the way they aggregated, in the shape of their slugs or in the form of their spore masses [*see illustration on page 71*]. The discovery of these strains furnished natural "markers" for identifying and following cells.

Of course there remained one technical problem: How could the individual cells be identified? Fortunately Raper had shown some time earlier that each fragment of a slug that has been cut into pieces will form a midget fruiting body. Spores derived from the several fragments can then be cultured individually. The amoeba from each spore will give rise to many daughter amoebae which can be scored for mutant or normal characteristics as they proceed to form slugs and fruiting bodies.

In one experiment we started with a culture of cells in the free-living feeding stage, into which was mixed 10 to 15 per cent of mutant cells. If we were to find a higher concentration of one type of cell in one part of the resulting slug, then we could conclude that there had been a rearrangement. We allowed the cells to form a slug and cut it up into three parts. Upon culturing the individual spores produced by each part, we found that the hind third had 36 per cent mutant cells, the middle third 6 per cent and the front third 1 per cent. Nothing could be more clear-cut; obviously the cells sort themselves out in a way that brings the normal cells to the front end of the slug. In another experiment, with a larger percentage of mutant cells in the mixture, hind and middle fractions contained 91 per cent mutant cells, and the front end only 66 per cent. Further experiments, including some with other species of slime mold, all led to the same conclusion. During the process of slug formation some cells are more likely to reach the front end than others, and the position of a cell in the slug does not merely depend upon its chance position before aggregation.

One must assume that certain cells move to the front because they travel the fastest, while the other, slower cells are

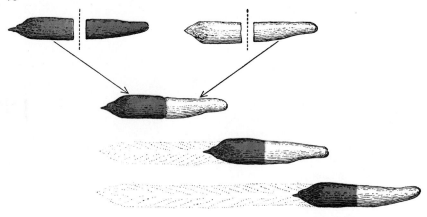

GRAFTED SLUG composed of the hind end of a stained slug and the front end of an unstained slug retains a sharp line of demarcation between the parts even after several hours.

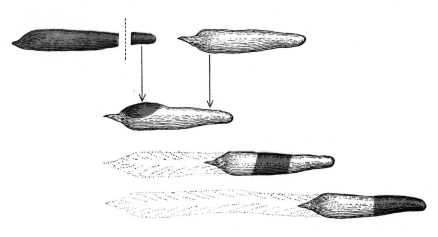

COLORED TIP taken from a stained slug can be inserted into the hind part of an intact slug. The colored cells then move forward as a band until they again are at the front tip.

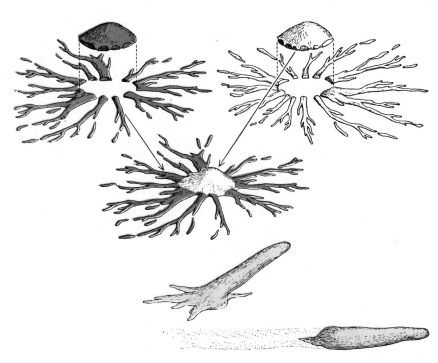

COLORED AGGREGATE in which the center has been replaced by a colorless center produces a uniformly colored slug, indicating that the cells are rearranged as the slug forms. The experiment illustrated in these drawings was originally performed by Kenneth B. Raper.

fastest, while the other, slower cells are left behind in the rear end of the slug. Considering the different fates of the front and rear cells, however, it is natural to wonder whether there are any other discernible differences between the front and hind cells. Size is one of the easiest qualities to measure, and comparison of spores from the front and rear portions showed that cells of the front segment are larger. From this it might be concluded that the fastest cells are the largest. But size is related to many other factors; some evidence indicates that cells in the front end divide less frequently than those in the hind regions, and this could affect their size. The possibility of a correlation between size and speed can only be settled by further experiment and observation.

But one fact is inescapable. The cells that tend to go forward are not identical with those that lag behind. Do the differences ultimately determine which cells become stalk cells and which will be spores? The most obvious deduction is that among feeding amoebae roughly a third are presumptive stalk cells, and the rest are predestined to be spores. This interpretation is clearly false, however, because then it would be impossible to explain how a single fragment of a cut-up slug can produce a perfect miniature fruiting body. The cells in the hind piece, which would normally yield spores, recover from the surgery that isolates them from the large slug, and one third of these presumptive spore cells proceed to form the midget stalk. This remarkable accommodation to a new situation is also exhibited by many types of cells in embryos and in animals capable of regenerating limbs and organs.

A more reasonable way to explain the relation between sorting-out and differentiation is to visualize the aggregating amoebae as having all shades of variation in characteristics between the extremes found at the ends of the slug. As they form a slug the cells place themselves in such an order that from the rear to the front they display a gradual increase in speed, in size and perhaps in other properties not yet measured. Thus each fragment of a cut-up slug retains a small gradient of these properties. It is conceivable that the gradient, set up in the process of cell rearrangement, actually controls the chain of events that leads the front cells to form a stalk and the hind cells to become spores. For the present, however, this is only conjecture.

At this point let me emphasize that the sorting-out process is not unique to slime molds. Recently A. A. Moscona of the University of Chicago and others have found that if the tissues of various embryos or simple animals are separated into individual cells, the cells can come together and sort themselves out [see "Tissues from Dissociated Cells," by A. A. Moscona; SCIENTIFIC AMERICAN, May, 1959]. For instance, if separate single pre-cartilage cells are mixed with pre-muscle cells, the cartilage cells will aggregate into a ball and ultimately form a central mass of cartilage surrounded by a layer of muscle. By marking the cells in a most ingenious way Moscona showed that there was no transformation of pre-cartilage cells into muscle cells or vice versa; each cell retained its original identity but moved to a characteristic location. In animals, then, sorting-out appears to be a general phenomenon when the cells are artificially dissociated. Since the movement in slime molds is part of their normal development, this raises the challenging question whether such sorting-out occurs in the normal development of animal embryos as well.

One must concede that slime-mold amoebae do profit by collectivization: the aggregate can do things the individuals cannot accomplish alone. In the amoebae's society, however, all are not created equal; some rise to the top and others lag behind. And then there is this distressing moral: Those that go forward with such zest to reach the fore are rewarded with sacrifice and destruction as stalk cells. It is the laggards that they lift into the air which survive to propagate the next generation.

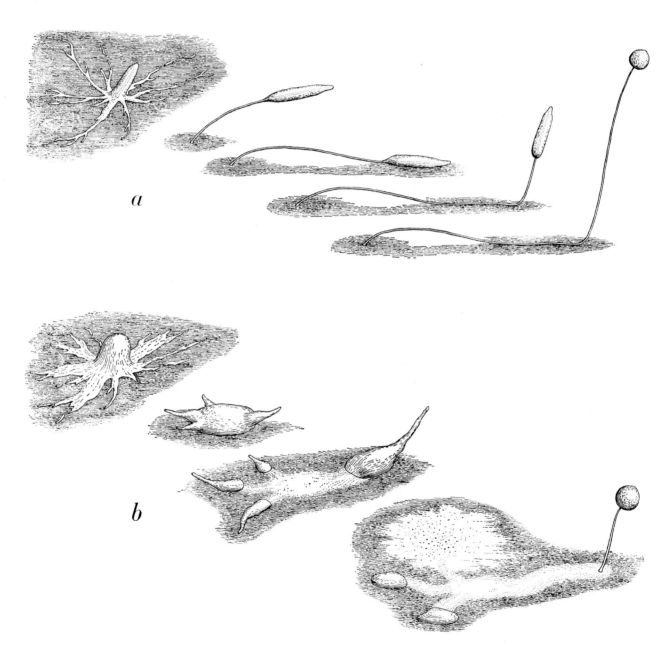

NORMAL AND MUTANT STRAINS of *Dictyostelium mucoroides* are contrasted in these drawings. The normal form (*a*) aggregates in thin streams, and its slug remains anchored by a thin stalk. The "MV" mutant (*b*) aggregates in broad streams and produces a starfish-like slug which then breaks up into smaller slugs. The stalk of the mutant is usually shorter than that of normal strain.

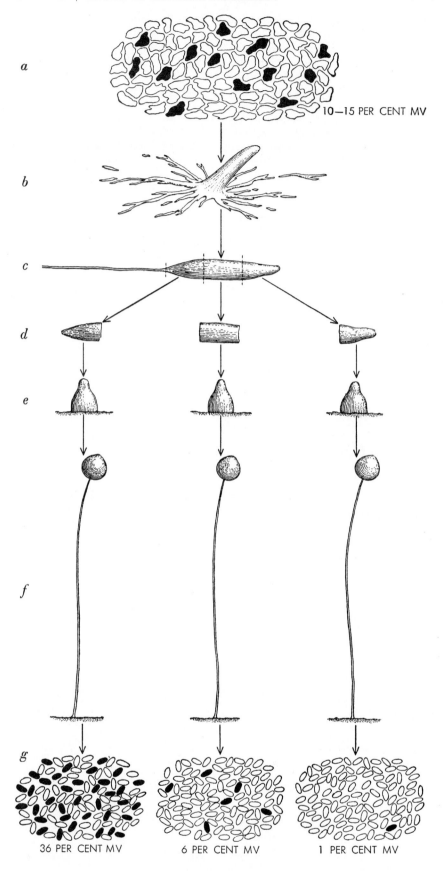

a

10—15 PER CENT MV

b

c

d

e

f

g

36 PER CENT MV   6 PER CENT MV   1 PER CENT MV

REDISTRIBUTION OF CELLS was proved in an experiment in which MV mutant cells (*black*) were randomly mixed with normal cells at feeding stage (*a*). The cells aggregated (*b*), and the resulting slug (*c*) was cut into three parts (*d*). Each part produced a fruiting body (*e* and *f*). Spores of each were then identified (*g*) by culturing them separately. The concentration of mutant cells was markedly higher in spores from the hind part of the slug.

III

CELLULAR DIFFERENTIATION AND DEVELOPMENT

CELLULAR DIFFERENTIATION AND DEVELOPMENT

III

INTRODUCTION

The article on the cellular slime molds that closed the preceding section served also to introduce the subject of development. All multicellular organisms, whether aberrant ones like the slime molds in which aggregation is a secondary process, or more familiar ones like higher plants and animals in which cellular aggregation results from the adherence of the products of division following fertilization, make a critical transition from one cell to many. In higher organisms this transition begins with a succession of divisions of a fertilized egg, followed by processes of individual growth, migration, association, and differentiation of the resulting cells to construct the organism designated or its genetic "blueprint." This is a demanding project for the individual cell: at various times it may be called upon to grow to a certain length and then to stop, to move among other cells and then to take up association with a specific group, or to specialize according to the dictates of messages from its immediate environment. All these feats require of cells a common competence—the ability to communicate with other cells.

All the articles in this section deal with cellular differentiation, in different ways and with different experimental systems. Surely the most distinctive event in the development of multicellular organisms is the way in which the specialization of component cells unfolds. We can measure the extent of that specialization in several different ways. First of all, the cells begin to *look* different: epithelial cells lining or covering organs become flattened and sheetlike; fibroblasts that will become the main cellular constituents of connective tissue become elongated and spindle-shaped; still other cells that will form muscle develop a tubular configuration. A second aspect of the differentiated state is measured in the relationship of one cell to other cells: migration of entire cells, or movement of their parts, often establishes specific arrangements that are crucial to cell functions. The lining up of presumptive skeletal muscle cells to form multicellular fibers, or the formation of connections between developing nerve cells, are examples of this important measure of specialization. But perhaps the most sensitive—and experimentally the most useful—measure of differentiation is the emerging biochemical uniqueness of cell types during their development. The specific pattern of protein synthesis performed by a cell is perhaps the most distinctive signature of its special personality. A rod cell in the retina manufactures huge amounts of the light-absorbing protein rhodopsin; a red blood cell makes none, but instead specializes in the synthesis of the oxygen-carrying protein hemoglobin. Both kinds of cells, of course, have some shared protein-synthesis requirements: each one must, for example, manufacture all of the enzymes required for the basic chemical reactions of energy conversion, and for the synthesis of basic constituents such as membrane lipids, structural proteins, and the like. No cell can evade these universal synthetic obligations; but beyond them, each has a repertoire of its own.

How does the differentiating cell become established in its own particular channel? This question has been a central one in developmental biology for three-quarters of a century. Several possibilities can be recognized, including: (1) a progressive restriction in the genetic material of the differentiating cell; (2) a segregation of cytoplasmic materials soon after cleavage, as a result of being unequally apportioned among the products of cleavage; (3) a response to chemical signals issued by neighboring cells that have already begun their own differentiation. In principle, these possibilities are not mutually exclusive: a combination of all three could control cell specialization.

The article by George Gray, " 'The Organizer' " (1957), provides an important historical perspective on differentiation. Gray shows how the classical debate between *preformation* (the view that the results of differentiation are fully determined before differentiation begins) and *epigenesis* (the view that differentiation unfolds continuously from undifferentiated starting materials) ultimately resolved into a debate over whether differentiation did or did not involve a restriction in the potency of the nucleus. This issue can be couched in modern genetic terms, even though it antedates them historically. Is there, during the program of cell divisions that is a feature of early development, a progressive loss of genetic material, thereby restricting the cell's ability to organize the synthesis of particular proteins? A contemporary statement of the preformationist position would hold that there is a programmed loss of genetic instructions in cells, ultimately leading to a limited, therefore specialized, array of capabilities. Skin cells, on this view, act like skin cells because they have lost the capacity to supervise the synthesis of muscle, blood, bone, and other kinds of proteins.

It was this question, Gray points out, that Roux, Driesch, and others were testing just before the turn of the century. Their strategy was simple and entirely sound: determine whether a whole embryo can be grown from a single cell isolated during a relatively early stage of cleavage; if it can, then surely the nucleus cannot have lost any of its genetic competence; if it cannot, then perhaps the potency of the nucleus has become restricted.

All the experiments discussed by Gray, except the first one by Wilhelm Roux, gave the same answer: entire embryos could be grown from isolated cells and thus there appeared to be *no* restriction in nuclear potency. We will see that the most advanced experiment conceptually and technically—that of Spemann in which he delayed the nuclear supply to dividing cells—gave an interesting and complex extra result: the ability of a nucleus to organize development in a normal way depended upon whether it inhabited a particular kind of cytoplasm. Here Gray could have given some more history. At the same time Spemann was working on amphibians in Germany, a group of American embryologists was concentrating on development in various marine organisms, including molluscs and the primitive vertebrate ancestors, the tunicates. E. G. Conklin of Princeton University and others, working in the summers at the Marine Biological Laboratory in Woods Hole, Massachusetts— even then Woods Hole was a world center for biological research—were showing that the eggs they studied, unlike those of the frog and the sea urchin, appeared to undergo a very early restriction in nuclear potency. A single cell that was separated from a four-celled molluscan embryo, for example, typically could *not* develop into a normal larva; instead, the embryo was deficient in certain structures. But in a series of experiments, Conklin showed that this result could be explained by the distribution of cytoplasmic substances in the egg cell. Even before fertilization, he noticed that various granules and pigmented substances were very unevenly arranged, so much so that the very first division would apportion almost all of a particular substance to one daughter cell, and little if any to the other. He was able to interfere with development in predictable ways experimentally by rearranging the contents of the eggs through such techniques as centrifugation of all the dense, granular material to one side.

Thus it became possible to distinguish two sorts of eggs: *mosaic* eggs, in which the distribution of cytoplasm is so critical that products of the very earliest division are already specialized by virtue of which cytoplasmic substances they get, or lack; and *regulative* eggs, in which all products of early division receive a share of the essentials. The term *regulative* reflects an important property of the developing embryo, shown in the very earliest experiments of Driesch and others. When a single cell is separated from others at, say, a four-celled stage, and subsequently develops into a full embryo, that cell has changed its behavior in an important way. If left unseparated, it would have made only some of the ultimate adult structures; but in isolation, it makes all of them. The difference, clearly, is in the absence of signals from the other cells—signals that accomplish the same kind of functional restriction achieved by the unequal division of cytoplasmic products.

It is sometimes entertaining to play the game of rewriting scientific history. Suppose, for example, that Mendel had selected only characters that displayed incomplete dominance for his early experiments. Or suppose that for the two-factor crosses in which he established the independence of assortment he had selected characters whose controlling genes had been located on the same chromosome. Certainly the development of the understanding of genetics that began with the rediscovery of Mendel's work would have been very different. Similarly, we might have a radically different history of embryology had Driesch and his successors in Europe been devoted to molluscs as experimental material. All the experiments would have resulted in incomplete embryos; the conclusion, given the assumptions most biologists were willing to make at the time, would likely have been that the restriction in potency was due to the loss of "nuclear determinants."

Fortunately, the experiment by Spemann, in which he deliberately restricted material of the "gray crescent" to one or the other half-embryo, brought both approaches together. The gray crescent marks the distribution of an essential cytoplasmic material. It happens that in frogs the first cleavage apportions this material relatively evenly among the early blastomeres, so that each blastomere retains all of its potential. (Later cleavages divide it unequally.) By restricting access of part of the early embryo to the material of the gray crescent, Spemann showed that the gray crescent is just as essential as the various substances in molluscan eggs, and that its distribution has a profound—if delayed—effect on the topography of the developing organism. We shall return to that point later.

The confusion that would have resulted if early experiments had been done on mollusc embryos and failed to demonstrate regulative development, as in our rewritten history, is relevant to the design of such experiments. If an early blastomere (from whatever kind of embryo) developed into a full adult, there could be no ambiguity in the result: full nuclear potency had been retained. But the *failure* to develop cannot prove the converse, inasmuch as several explanations can be offered as alternatives to a restriction in nuclear potency. For one thing, surgical intervention can interfere with subtle, sensitive, developmental processes and stop further growth for reasons having nothing to do with nuclear determinants. (That objection, Gray points out, was raised against the early experiment of Roux.) Or, as we have just discussed, the restriction of potency could be cytoplasmic rather than nuclear in origin. This last possibility profoundly altered the further strategy of experiments on the restriction of potency—an alteration foreshadowed by Spemann's experiments on delayed nuclear supply. The modern approach has been to test the potency of nuclei, not the potency of cells; in that way the experimenter can select a cytoplasm of his own choosing, and use the nucleus as the variable. The technique of putting a nucleus into a foreign cytoplasm is called *nuclear transplantation*.

John Gurdon of Oxford University describes the technique in his article "Transplanted Nuclei and Cell Differentiation" (1968). Although the experiments of Spemann and the other pioneers had shown that the first few cell divisions following fertilization indicated no restriction in nuclear potency, these findings did not lay to rest the notion that loss of genetic material could be a factor in differentiation. In the first place, the really impressive *visible* indications of differentiation appear much later than in the stages of cleavage with which Spemann and the others worked. The so-called primary germ layers of ectoderm, mesoderm, and endoderm are not easily discernible in an animal embryo until the occurrence of gastrulation—that complex, revolutionary set of movements that converts the hollow blastula into the three-layered gastrule. Only then does the formation of organs occur—and only during the formation of organs do most cell types approach full differentiation. It would be perfectly logical to propose that only at this stage in development, with the specialization of the cells already apparent, should one expect to see a restriction in the activity of the nucleus. Indeed, the pioneers of the nuclear transplantation technique, Briggs and King, found evidence suggesting that just such irreversible changes in nuclear potency had occurred.

Gurdon's article summarizes convincing experimental evidence that the genetic potential of the nucleus, even of cells that are rather advanced in their own specialization, is called forth by the cytoplasmic environment in which that nucleus is placed. Of particular interest, in the historical context supplied by Gray's article, is the careful approach Gurdon makes to the interpretation of his own results. Because *some* intestinal cell nuclei are totipotent, it does not follow that some have not lost genetic potency; on the other hand, a failure of cells to develop may have explanations other than genetic restriction. By demonstrating that such alternative explanations are quite likely for those nuclei that fail to support development, Gurdon makes a convincing case that the interaction between nucleus and cytoplasm determines cell specialization, rather than any genetic deprivation of the nucleus.

But how does the cytoplasm acquire its special character? We have already described one way: by receiving a special allocation of cytoplasmic products from the egg during early cleavage. At the end of Gray's article, " 'The Organizer'," another mechanism emerged: populations of cells exchange signals that influence the course of their differentiation. When one early-segregating population controls a number of other populations, one might conclude that it is critical in establishing the overall plan of development: hence the name "organizer." We now know, however, that no single tissue has a monopoly on *embryonic induction*—the term applied to designate control of differentiation by signals from other cells. Most often proximity is a requirement for the exchange of such signals; it is in part the function of gastrulation to bring previously distant populations of cells close together, so that they can engage cooperatively in the formation of organs. Experiments on the interaction itself, however, quickly yielded confusion, as Gray recounted. No single chemical is central to interaction; on the contrary, an astonishingly large number seem to work. Since Gray wrote his article, the approach of identifying diffusible substances led to inconclusive ends, and the search for an induction mechanism has taken different lines. It has been shown, for example, that in tissue culture one tissue can influence the differentiation of another across thin pieces of Millipore filter, but not across thicker ones. This finding suggests that materials associated with the cell surface are responsible for mediating the inductive message. In other experiments, particularly those studying the action of hormones on late-differentiating tissues, induction obviously requires that chemicals act over a long distance. Readers of Gray's treatment of induction are given a basic message that is still valid, however:

the specificity of the response resides in the target cell and not in the chemicals that pass from cell to cell. That is why it may sometimes seem that any stimulus will cause induction. This returns us to our theme, which focuses on the *process* of differentiation.

Norman K. Wessells and William J. Rutter, in "Phases in Cell Differentiation" (1969), emphasize again that differentiation depends on many things: a pattern of morphogenetic movements, a shape, or a biochemical specialty. To this list, however, they make a crucial addition: differentiation is dependent on an early commitment by cells, made long before their ultimate state is visible or even detectable by chemical assays of the conventional sort. Yet the commitment can be tested: once cells have made it, they can no longer be kept from differentiating by blocking genetic transcription, and small but significantly increased concentrations of specific protein can be detected by appropriately sensitive tests. Wessells and Rutter suggest that an altered cytoplasm has already turned on, selectively, those parts of the genome that will fill the future requirements of differentiation, but that a realization of the cell's commitment may await additional signals.

Intercellular messages must be important in other ways as well. For example, a striking feature of early embryonic development in animals is the degree to which changes in form are generated by the migration of individual cells or groups of cells from place to place. In order to account for the precision of individual cell movements, we must assume that the cells have means of "recognizing" one another and of thereby selecting the place in which they ultimately settle down in the embryo. A. A. Moscona describes the experimental analysis of such mechanisms in "How Cells Associate" (1961). It has been known for some time that dissociated sponge cells or tissue-cultured mammalian cells will "sort" themselves out from mixtures. Moscona describes experiments showing that whereas in sponges dissociated cells seem able to recognize other cells from the same species, in vertebrates recognition is of other cells from the same tissue; indeed, cells from species as diverse as chicken and mouse will cooperate in the formation of mosaic kidneys or other tissues. Such experiments have exposed some of the ground rules that govern cell associations, and they have led to the hypothesis that surface-associated glycoproteins may be the essential materials that bind compatible neighbors together or provide the signals that enable them to recognize one another.

The workings of cellular affinity are nowhere more critical than in the construction of the nervous system. There, every aspect of function depends upon which cells form functional connections with which others. The article by Roger Sperry entitled "The Growth of Nerve Circuits" (1959) might appear, at first thought, to belong in a later section of this book. But the article, about the development of the nervous system, focuses upon a different aspect of cell specialization. Essentially, it enquires about the way a nerve cell acquires the specific relationships with others that allow the proper networks to form in the brain, and about the extent to which these are permanent. Sperry recounts a significant revolution in our thinking about such connections. At one time, it was thought that nervous connections were formed largely under the influence of sensory experience, and that they could be broken or redirected whenever the need for a changed configuration became apparent. Sperry's own experimental work has emphasized that, on the contrary, nerve cell associations are relatively stable once they are formed.

"The Growth of Nerve Circuits" also introduces a property of the nervous system that will frequently be referred to later in the book. In the brain, various regions of the body are "represented" by clusters of nerve cells whose discharge is associated with activity—either sensory or motor—in those regions of the body. Thus when a particular region of the motor area of the cerebral cortex is stimulated with a weak electric shock, a muscle contraction is produced at a particular location in the body; similarly, mechanical stimulation of a local area of skin produces impulses in a concentrated location in the sensory cortex. The relation of the cortex to peripheral nerve endings is said to be *topographic*—the points on the body surface are mapped out in the nervous system systematically, so that adjacent points on the body map are also adjacent on the brain map. This has to mean that nerve cells correlate their peripheral and central connections—in other words, by virtue of the location of its axon in the periphery of the body, a nerve cell will make a particular set of central connections. Sperry proposes a biochemical mechanism whereby a neuron "recognizes" a region of the body surface, and adjusts the affinities of its central portion.

We still do not know how the central and peripheral regions of the nervous system are connected together to produce this precise correspondence. One idea is that a genetically endowed specificity is imposed on neurons in the periphery, and that this information then establishes their central affinities. This view has been supported by recent experiments on the inward growth of optic nerve fibers in larval amphibians, performed by Marcus Jacobsen at The Johns Hopkins University. Jacobsen has shown that very early in development the optic nerve fibers from a particular region of the retina will form connections with different points in the brain if the orientation of the retina is changed experimentally. After a certain critical period, however, the neurons have developed a specific affinity for a specific region of the brain, toward which they will grow even if the pathway is made very long or tortuous. It appears that this commitment is entirely analogous to that for specific protein synthesis discussed by Wessells and Rutter: it arises at a specific time in the differentiation of the nerve cells, even though its origin is not indicated by any change in the appearance of the cells. At the time specificity comes about in the optic nerve cells, the axons have not yet started to grow inward toward the brain, so the specificity must reside within the sheet of retinal cells themselves.

Can a subsequent change in peripheral neural connections dictate a rearrangement of the central connections? Sperry's results suggest that it cannot. At one time it was thought that if a nerve were redirected in a young amphibian so that it leads to a different, inappropriate muscle, the central connections of the motor fibers would also become "reprogrammed" so that the nerve would be discharged at the time the new muscle would be expected to act. Careful tests of this idea in several different systems suggested, on the contrary, that when motoneurons form new peripheral connections they do not make a corresponding alteration of their central connections. The one exception to this generalization about the stability of relationships between the center and the periphery is the experiment by Nancy Miner on transplanted and reinnervated frog skin, which Sperry describes. In this experiment new peripheral nerve connections seemed to result in changed central connections. Jacobsen has since performed the experiment with new controls. It appears that the interpretation given by Miner is correct: the transplanted skin is reinnervated by sensory cells near the new graft, and these cells make whatever central connections will result in nerve impulses appropriate for the transplanted skin. It is difficult to interpret the changes in terms of foreign sensory neurons searching out, from a vast area, their old patch of skin in a new place. If we conclude, as it seems we must, that the sensory nerves can match changes in peripheral connections by changing their central connections, then we must also ask how many other nervous systems may exist, beside that of amphibians, in which the plasticity shown by the nervous system before its commitment to differentiation is extended into adulthood.

7

"The Organizer"

by George W. Gray
November 1957

How are the unspecialized cells of the dividing egg organized into the specialized cells of a plant or animal? For 70 years biologists have been searching for the answer by experiments

It's a very odd thing—
As odd as can be—
That whatever Miss T. eats
Turns into Miss T.
—Walter de la Mare, *Peacock Pie*

Individuality is the hallmark of life. In the realm of physics and chemistry an investigator must deal with crowds; he can rarely if ever single out one atom or one molecule for study. But a biologist can focus on a single cell, on the nucleus of the cell, on the individual strands of material that make up the nucleus, even, indirectly, on the activity of a single gene. And so he learns that not only is Miss T. unfailingly able to convert steak and potatoes into the unique pattern of the tall, angular, blond woman that is Miss T., but every one of the billions of cells that make up her body carries the individual design that marks it as exclusively her own. The cells are not a crowd but members of an organized community, each serving a special function according to a pre-established plan.

How this organization is brought about is the central problem of biology. If man is ever to understand what life is, he must solve the mystery of how a living thing takes inanimate material and builds it into a germ cell, and how this one cell, after fertilization by merger with another cell, divides into two, and then each into two more, and so on

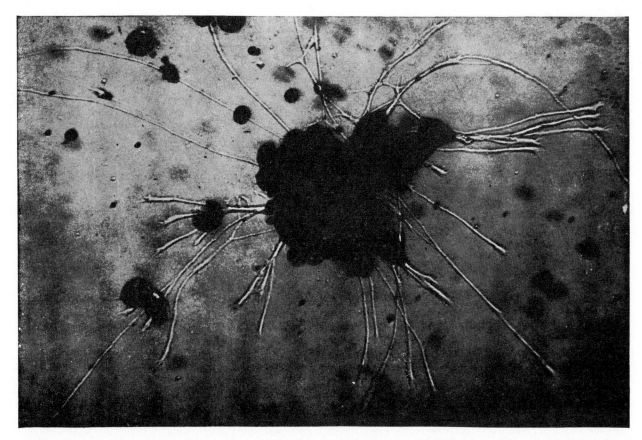

NERVE CELLS with their typical long fibers were unorganized ectoderm eight days before this photomicrograph was made by M. C. Niu of the Rockefeller Institute. The change was induced by fluid taken from a culture of embryonic "organizer" cells.

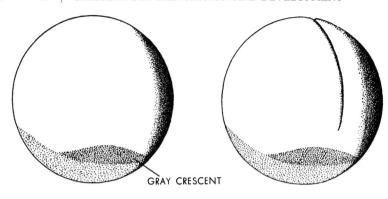

GRAY CRESCENT

DIVISION OF THE EGG of the salamander is depicted in this series of somewhat schematic drawings. The first drawing shows the fertilized egg, with its characteristic "gray crescent." The second drawing shows the first cleavage of the egg, which is perpendicular

through a succession of 40 to 50 cell generations until a human being is born. Everything that is now expressed in the 25 million million cells of the newborn baby was precisely blueprinted in the original germ cell. Not only the architect's plan, but the machinery for building according to the plan was carried in that seed of life not much bigger than the point of a pin.

But how? By the operation of what laws is a single cell able to multiply into such different structures as skin cells, bone cells, muscle cells, blood cells, brain cells and all the rest—and at the same time marshal this wide diversity into a closely coordinated and smoothly working whole?

Epigenesis v. Preformation

Embryology dates back to the shadowy dawn of Greek medicine. Two thousand years before there was any knowledge of the biological cell, physi-

cians observed the differences between organs and began to speculate on how the organs were formed. The question was raised by one of the Hippocratic writers, and quite early in history two concepts arose.

Aristotle argued that the mother contributes the substance and the father the structure of their offspring. He pictured the male's semen as the moving element which organized the substance provided by the female, just as an artist "imparts shape and form to his material." From observations of animals, but mainly of the developing chick in the egg, Aristotle deduced that the first organ to emerge was the heart. He said:

"Either [the organs] are formed simultaneously—heart, lung, liver, eye, and the rest of them—or successively, as we read in the poems ascribed to Orpheus, where he says that the process by which an animal is formed resembles the plaiting of a net. As for the simultaneous formation of the parts, our senses plainly

tell us that this does not occur; some of the parts are clearly to be seen in the embryo while others are not."

Epicurus held another view. Believing that matter is everything ("there are only atoms and the void"), he contended that both parents contribute material and that a child must be completely formed from conception, though in miniature. The Roman rhetorician Seneca later epitomized the idea in these words:

"In the seed are enclosed all the parts of the body of the man that shall be formed. The infant in his mother's womb hath the roots of the beard and hair that he shall wear some day. In the little mass, likewise, are all the lineaments of the body and all that which posterity shall discover in him."

Here are two strikingly contrasting ideas. In Aristotle's view there was a gradual emergence of form from undifferentiated material—a process which has come to be called epigenesis (from

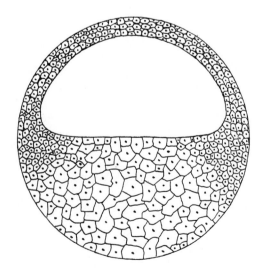

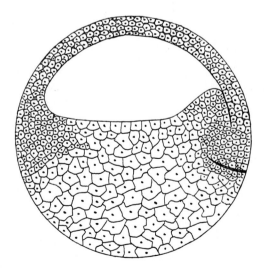

FORMATION OF THE GASTRULA of the frog, which differs somewhat from that of the salamander, is depicted in cross section. The first drawing shows the blastula, a hollow ball partly filled with yolk cells. The second drawing shows the cells beginning to fold

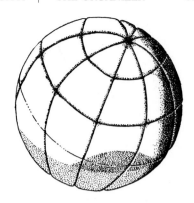

to the gray crescent. The third drawing shows the egg divided into four cells; the fourth, into eight cells; the fifth, into 16; the sixth, into 32. The region just below the gray crescent later gives rise to the blastopore (*see drawings at the bottom of these two pages*).

the Greek, meaning "ensue upon"). The other view, called preformation, asserted the presence of a full-structured organization from the beginning, so that the development of the embryo was simply an enlargement of what already existed in small scale.

It is interesting that in Renaissance Europe it was the scholastics who accepted Aristotle's theory, while medical men and other biologists turned increasingly to the embryology of the Epicureans. By the 17th century the sway of the preformationists was almost unchallenged, and the concept was pushed to the most absurd extremes. If the embryonic germ is a complete body, it was argued, then the germ must contain all the organs and parts, including the seed of the next generation, and that seed in turn the seed of the next, and on and on, like a series of Chinese puzzle boxes. Mother Eve, it was said, carried in her body the forms of all the people to be born.

Meanwhile the power of the glass lens had been discovered. The pioneering Dutch microscopist Anton van Leeuwenhoek focused his "optik glass" on a drop of human semen and saw the spermatozoa, which he named "animalcules." Others took up the new instrument, and among them was an ardent preformationist who thought he saw in each animalcule the form of a tiny human being, complete with head, body, hands and feet! This observation led to a great schism among preformationists. For it suggested that Adam, rather than Eve, contained all mankind. Many forsook Eve to espouse the new dogma, and they were called animalculists; those who remained loyal to Mother Eve were ovists.

Observations and Experiments

The first breath of fresh air came from Germany. At the University of Halle, Kaspar Friedrich Wolff watched the development of the tip of a growing plant through his microscope and made careful drawings of what he saw. Shoot after shoot showed only homogeneous tissue. There was no sign in this tissue of the leaves, flowers and other organs which later emerged from the shoots. Wolff noticed, moreover, that when the specialized parts did begin to form, each appeared first as an almost imperceptible prominence or swelling in the undifferentiated tissue.

Wolff next trained his microscope on the developing chick in the egg to see what he could learn of animal tissue. He found that the intestine gradually formed from tissue which at the beginning showed no rudiment of the organ that was to come. And so with other organs. Neither in a plant nor in an animal could he see any trace of preformation, and he concluded that in both the developmental process was epigenesis.

Wolff's 18th-century observations inaugurated a rational approach to embry-

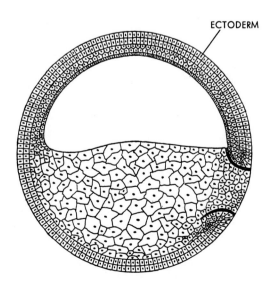
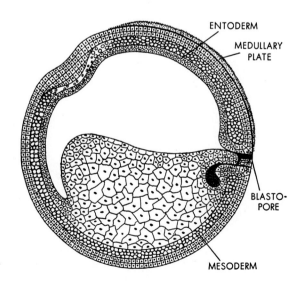

inward to form the gastrula. The third and fourth drawings show the formation of the blastopore and the three layers of the gastrula: the ectoderm, the mesoderm and the entoderm. The medullary plate, from which springs the nervous system, grows out of the ectoderm.

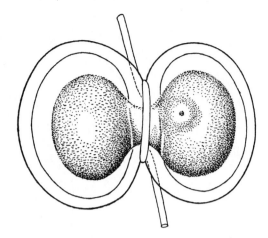

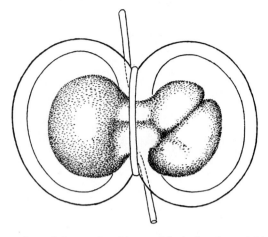

CLASSIC EXPERIMENT performed by Hans Spemann involved tying the salamander egg into two halves across the gray crescent.

The nucleus of the egg (*marking at right in drawing at left*) was confined to one half. At first only the half containing the nucleus

ology, and other advances paved the way: improvement of the microscope, the firm establishment of the cellular theory, the invention of the microtome, the discovery of evolution, of the gene, of proteins and of nucleic acids. Up to the latter part of the 19th century, however, the epigeneticists and the preformationists continued their war of words without direct experimental test, and embryology remained almost entirely a descriptive science. "Take 20 or more eggs," an experimenter of the fourth century B.C. had instructed, "and let them be incubated by two or more hens. Then, each day from the second to that of hatching remove an egg, break it and examine it." This preoccupation with what we may call the natural history of the embryo, paying little attention to causal relationships, continued to domi-

nate embryological research until almost the turn of the century.

In the 1880s the speculations of a German zoologist, August Weismann, precipitated a significant investigation. He developed a germ-plasm theory which pictured the nucleus of the fertilized egg as a mosaic in which "primordia" (starting points of the organs and tissues) "stand side by side, separate from each other like the stones of a mosaic, and develop independently, although in perfect harmony with one another, into the finished organism." If this were the case, then an embryo at the two-cell stage would have one half of the individual in each cell. Wilhelm Roux, an anatomist at the University of Breslau, decided to test Weismann's hypothesis. He figured that if he removed one of the two cells at this stage,

the incubation of the remaining cell would provide the needed test. And so in 1888 Roux performed a historic experiment. Taking the two-cell embryo of a frog, he killed one of the cells with a hot needle and let the other develop in its natural water medium. The result was a half-tadpole. This seemed to demonstrate that each cell carried half of the machinery for constructing the frog, and the experiment was hailed as proof of the mosaic theory of preformation.

But there were doubting Thomases. They pointed out that the killed cell had remained attached to the living one and might conceivably have influenced its development. Various attempts were then made to separate embryonic cells, and in 1891 this was accomplished by Hans Driesch, working in a laboratory in Naples. Driesch shook two-cell em-

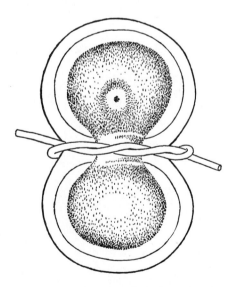

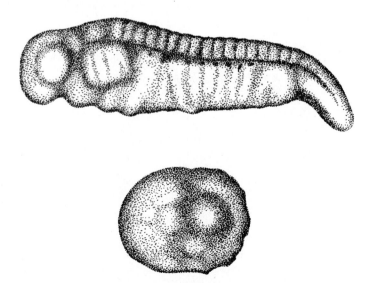

SECOND EXPERIMENT by Spemann was to tie the egg parallel to the gray crescent. The half of the egg with the gray crescent de-

veloped into a normal embryo (*upper right*), but the other half produced only an unorganized "belly-piece" (*lower right*).

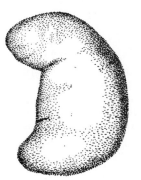 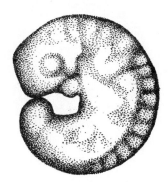

divided (*drawing second from left*). Later the other half began to divide. Eventually both halves gave rise to normal embryos, one younger than the other (*drawings at right*).

bryos of sea urchins in a vial of sea water and succeeded in disjoining the two cells without injury. Under incubation each cell developed into a whole sea-urchin larva. In later experiments he separated a four-cell embryo into its components, then an eight-cell, and finally one that had reached the 16-cell stage—and from each he obtained on incubation a complete animal.

This spectacular emergence of the whole from a fragment was so contrary to the popular dogma of the day that it aroused embryologists. There was a rush to the laboratories, and investigators who had been content merely to watch life unfold now became ardent experimenters. By the turn of the century a whole new school of laboratory workers were engaged on problems which the conflicting results of Roux and Driesch had posed. Among them were Hans Spemann, a 31-year-old zoologist at the University of Würzburg, and Ross G. Harrison, a 30-year-old associate professor of anatomy at the Johns Hopkins University. These two became the leaders, builders and teachers of the modern science of developmental biology.

Spemann and Harrison

Both men were superb experimenters. They knew how to ask the right questions of nature. Employing techniques which in some cases had been pioneered by others without definitive results, they had the imagination to strip the problem to its simplest terms, bend the experimental procedures to the new approach and thereby frame a question in such a way that the subject could respond. Spemann used to say that he regarded the embryo as "a conversational partner who must be permitted

to answer in his own language"; Harrison had the same attitude. As one reads the papers of these two masters of research, Harrison appears to be more matter-of-fact, more objective, a greater realist, while Spemann seems more philosophically minded, more concerned with symbols, a searcher for wholeness. They never collaborated in investigation but were warm personal friends, frequently consulting each other on speculations and experimental results. Harrison often spent part of his summer vacation visiting Freiburg, where Spemann was professor from 1919 to his retirement in 1935.

Harrison is most widely known, perhaps, for his invention of the tissue-culture technique. It was devised to tackle a problem in embryology, namely, the origin of the peripheral nerves that connect the brain with the end-organs of touch, taste, smell and the rest. The connection is through the spinal column, of course, but how are the lines laid down during development? Some investigators believed that the nerves grew out from the neural tube, the primitive spinal cord of the embryo. Others argued that the fibers originated in the organs and grew toward the cord.

Harrison conceived the idea of trying to culture a minute fleck of tissue from an embryonic neural tube in the nutrient fluid to which it was accustomed, and watching to see whether nerves could grow out of it. He devised the "hanging drop" technique, putting the bit of tissue in a drop of fluid which was then suspended from a glass cover slip laid over a hollowed-out well in the center of a glass slide [*see drawing on page 87*]. The tissue was a particle of neural tube cut out of a frog embryo, and the fluid was a drop of the animal's lymph.

Harrison watched the bit of tissue almost continuously under a microscope, and soon results began to show. After a few hours delicate protuberances began to bud out from the tissue. They grew into filaments which extended into the clotted texture of the lymph. Then, in the same way, Harrison cultured bits of non-nervous tissue—a piece of embryonic muscle, a particle of intestinal wall, other fragments of early organs. All of them grew, but none sent out any nerve fibers. Thus he demonstrated that nerve development is a growth outward from the neural tube, not inward from the organs.

The hanging drop became one of the most powerful tools of experimental biology. In 1917 the Nobel prize committee of the Swedish Karolinska Institute chose Harrison for the Nobel award in physiology and medicine, but for some reason the Institute decided not to give the prize that year, and Harrison never received this honor. But he has had many others, and only last year the Academy of the Lincei in Italy, the world's oldest learned society, sought him out, at the age of 86, to give him its Antonio Feltrinelli prize of $8,000.

Spemann, too, was the recipient of numerous decorations. In 1935, six years before his death, the committee selected him for the Nobel prize and this time the Karolinska Institute decided to give it. The citation said that the award was for "his discovery of the organizer effect in embryonic development."

The Organization of a Newt

Spemann's discovery was made in experiments with newts, a variety of salamander. These small, lizard-like amphibians originate from eggs whose habitat is water, so it is a simple matter to keep the temperature and other conditions favorable for incubation and nurture. Comparative studies have shown that the development of the salamander egg parallels closely that of other backboned creatures, including man.

A peculiarity of the salamander's egg is that about half of its surface is dark-colored, the other half light or colorless. Immediately after fertilization a small, crescent-shaped segment of the boundary region between the light and dark areas takes a grayish hue. This so-called "gray crescent" is the first visible manifestation of profound changes occurring within the egg. It appears just before the self-duplication of the fertilized cell.

Spemann, like many other biologists of his day, was fascinated by the prob-

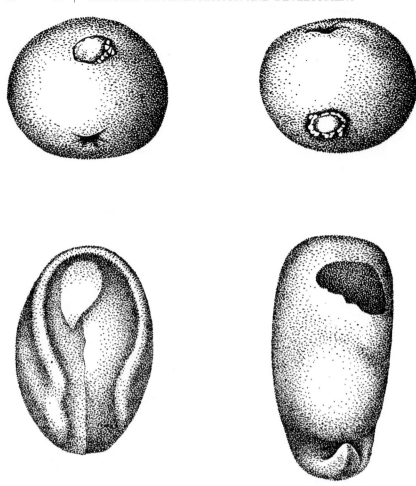

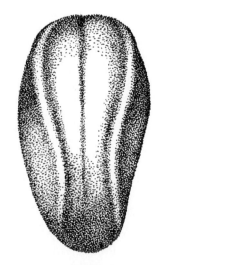

TISSUE WAS TRANSPLANTED by Spemann from a region above the blastopore of a salamander egg (where it would normally grow into belly skin) to a region below the blastopore of another egg (where it would normally grow into nerve tissue), and vice versa (*drawings at top*). Later the cells which would have become skin tissue became nerve tissue (*lower left*), and the cells which would have become nerve tissue became skin (*lower right*).

TISSUE WAS REMOVED from the lip of the blastopore of a colorless salamander gastrula and implanted in the ectoderm of a pigmented gastrula. The blastopore tissue of the colorless gastrula induced the pigmented gastrula to make a second medullary plate. At left is the medullary plate of a pigmented embryo; at right, the induced medullary plate on the other side of the embryo. The light area in the induced plate is the implanted tissue.

lem of testing Weismann's mosaic theory of preformation. Suppose, instead of separating the two-cell embryo into its halves as Driesch had done, the egg were simply constricted across its middle just before it cleaved? Spemann procured a strand of baby hair, made a slip noose, looped it over the cleaving egg and drew the strand tight, leaving only a slender bridge of protoplasm between the two halves. The nucleus of the cell was sequestered in one half [*see drawing at top of page 84*]. It began to split and draw apart in two spindles, duplicating itself in the process known as mitosis. The cell on its side of the constricting noose thus divided into two cells, then into four. Meanwhile the cell on the other side of the noose remained just a single cell. Soon, however, it too began to divide, after mitosis had occurred on the other side in a cell close enough to the pinched waist to send nuclear material through the narrow bridge. The net result was two embryos, the one on the side with the original nucleus developing first, but the other finally catching up. Eventually two complete salamander larvae developed. Spemann repeated this experiment many times, once getting 32 cells on one side before nuclear material slipped through and started development on the other.

Suppose the cleaving egg were pinched in a direction at right angles to the first—that is, parallel to the gray crescent instead of across it [*see drawing at the bottom of page 84*]. Would this make a difference? Spemann made the experiment and found the result indeed quite different. The half without the nucleus eventually divided and subdivided, but its final product this time was only an unorganized mass which Spemann called *Bauchstück*—bellypiece. This belly-piece contained liver cells, lung cells, intestinal cells and other abdominal material—but it had no axial skeleton, no nervous system, no unifying pattern.

Why? Why should the unnucleated half of the egg now produce only a crowd of miscellaneous cells? Spemann decided that the distinguishing difference between the experiments must be the fact that in the first, each half of the egg had part of the gray crescent, whereas in the second, one half had most or all of the gray crescent, while the other—the half that failed to develop—had little or none of it.

What was so significant about the gray crescent? Spemann considered its "geographical" relationship to the known facts about development of an embryo.

After the embryo has developed to the

form of a hollow sphere, a dimple begins to form on its surface, and this deepens into a crater called the blastopore. The cells around the lip of the crater suddenly begin to slide in over the brink, as if pushed by an invisible hand. At the end of this process, known as gastrulation, the salamander embryo looks like a rubber ball with half of it pushed in. Now the cells on the concave outer surface, called ectoderm, will become skin, brain, nerves, ears, eye lenses and other sensory organs. A layer of the cells that have migrated to the inside, called entoderm, will develop into lungs, liver, stomach, intestines and certain other abdominal organs. Another infolded layer of cells, the mesoderm, is destined to produce bones, cartilage, connective tissue, muscle, the blood and its vessels, and organs of the urogenital system.

The Organizing Lip

How are all these fates fulfilled? Spemann knew that the lens of the eye is formed by a process of "induction." An eye begins as a tiny protuberance budding off from the embryonic brain; the protuberance grows into a vesicle, and then the vesicle apparently induces the skin overlying it to become a transparent lens. Reflecting on his experiments with pinched eggs, Spemann began to suspect that cells associated with the gray crescent might be the primary inducers of the whole chain of development that produces an organized individual. The gray crescent is merely a surface feature of the egg which soon disappears as the cells multiply. But the blastopore forms just below the area the crescent occupied. For other good and sufficient reasons, Spemann focused his attention on the lip of the blastopore and addressed a series of questions to it.

How would it be if one took a bit of tissue from the specific area of the embryo that is destined to form belly skin and transplanted it to the area behind the blastoporal lip that is to form brain? Spemann did that, using two specimens. From one embryo he cut a microscopic patch from the presumptive flank; from the other he sliced a fragment of equal size from the area that he knew would form brain if left undisturbed. Then he exchanged the two bits, transplanting the presumptive flank tissue into the wound left in the brain area and the presumptive brain into the cut surface of the flank. The grafts grew, the two embryos developed, and lo!—the transplanted flank turned into brain and the transplanted brain into belly skin.

After repeating this and similar experiments many times, and always getting the same answer, Spemann took up a new line of questioning. Suppose, he said, the lip itself were cut out and implanted in another embryo?

The two-story zoology building at the University of Freiburg where Spemann worked was teeming with students, and among them was Hilde Proescholdt. It was her good fortune to be looking for a research project to fulfill the requirements for her Ph.D. Spemann outlined his proposed experiment to the young woman, appointed her his assistant to carry it through, and thus she shared in the great discovery published two years later under their joint authorship.

For this experiment Spemann used two varieties of salamander—one dark-hued, the other colorless. The blastoporal lip was cut from the colorless embryo and grafted into the belly of the pigmented individual. The latter now had two blastoporal lips—its own on its topside, and the implanted lip from the colorless salamander on its underside. Thus there were two centers of organization in a single embryo. And what happened?

Gastrulation occurred at each place: that is, ectodermal cells migrated over each lip into the mesodermal layer beneath. Eventually there developed two axial systems, each complete with a spinal column, head, trunk, legs and tail —two baby salamanders joined like Siamese twins! The extra salamander had whole sheets of tissue made of dark

cells, as well as organs and parts built of colorless cells. This showed that the lip transplanted from the colorless embryo had extended its organizing power to multiplying cells of the host embryo.

Here, said Spemann, in the cells that flow around the blastoporal lip, is the primary center of induction. He named this region "the organizer." He did not, however, suppose that it controlled the whole process of development: he saw a succession of organizers at work, one taking up where another left off, each having its part in the sequence which began with the migration of a sheet of epidermal cells over the lip of the blastopore and ended with the birth of a coherent, sentient being.

The New Questions

Spemann's work is one of the great landmarks of biological research. It exemplifies, in the simplicity and directness of its approach, how the mind of a master investigator works. And his experiments and interpretations brought a sense of unity to the science of embryology, leading it sufficiently out of the trees to see the forest and to ask more intelligent questions.

How the organizer and suborganizers work is still a mystery. At first it was thought that the movement of the cells over the lip generated dynamic effects which induced differentiation. But a simple experiment by A. Marx, one of Spemann's students, demonstrated that the primary induction occurred when

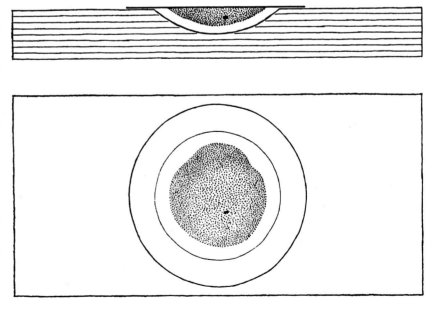

"HANGING-DROP" TECHNIQUE was devised by Ross G. Harrison to grow bits of embryonic tissue *in vitro*. He grew frog tissue in a drop of lymph on the bottom of a microscope cover slip. At the top, the drop and tissue are shown from the side; at the bottom, from above.

there was no movement. Next followed tests in which organizer tissue was subjected to crushing, freezing, heating and other injuries, and then implanted into gastrulating embryos—whereupon it induced the emergence of a central nervous system just as undamaged organizer tissue had done. So the conclusion was that induction must be a chemical effect, and a search began for the potent chemical or chemicals released by the organizer cells.

Johannes Holtfreter, who served his apprenticeship under Spemann, found that even after the lip tissue was killed in alcohol it was able to induce development. The same discovery was independently made by C. H. Waddington at the University of Cambridge. Holtfreter went on to try other salamander tissues and then tested both living and dead tissues of other animals—mouse kidney, liver and brain and extracts from chick embryos. Each induced the embryo to form a neural tube. Various efforts were made to analyze these alien substances and blastoporal lip tissue chemically. But all the analyses have been inconclusive, and the situation was further confused when Waddington, Joseph Needham of Cambridge and Jean Brachet of Brussels discovered the astonishing fact that even the synthetic dye methylene blue induced formation of nerve tissue when injected into a gastrulating embryo.

With this medley of causes and effects in the record, and more seeming paradoxes being added almost every month, it would seem that the era of experiment has changed into an era of perplexity. Perhaps that is another way of saying that the embryo is more versatile and complex than anybody dreamed. The situation can be likened to that which arose in biology when it was discovered that an unfertilized egg could be made to develop into an animal by pricking its membrane with a needle; indeed, that virgin birth ("parthenogenesis") could also be induced by heat, electric shock, ether and many other agents. Apparently all that is needed is some impulse to activate the egg. Similarly, while blastoporal lip tissue normally supplies the agent which induces cells to differentiate, the inductive force can also be supplied by various other materials, both living and dead.

Recently Victor C. Twitty and M. C. Niu performed an experiment at Stanford University which demonstrated that the action of the primary organizer is mediated through some diffusible substance which it exudes. They cut a minute piece of blastoporal lip from the early embryo of a salamander and cultured it in a hanging drop. After several days they introduced a bit of ectoderm, consisting of about 15 cells, into the drop. Although the ectoderm fragment was not in contact with the lip tissue, within 10 days it developed into nerve and pigment tissue. They then changed the experiment, this time using not the lip tissue itself but only the fluid in which it had been cultured for several days. As before, the implanted ectodermal cells gave rise to pigment and nerve tissue. Thus it would appear that the organizer is some secreted substance, and Niu (who last year joined the staff of the Rockefeller Institute for Medical Research) is now trying to isolate the active material. He thinks it may turn out to be not just one substance but a group of nucleoproteins, each specific to the induction of a particular tissue. Indeed, embryologists are fairly sure that "the organizer" is not a single substance but a complex of agents and reactions.

The Organizer and the Genes

All that the organizer does, apparently, is to release capabilities already present in the cell but dormant. You cannot force a cell into an alien pattern. You can change the direction of development, but each cell has only a limited "repertory." Its repertory becomes more and more restricted as development proceeds, and this restriction of potentialities is the very essence of embryonic development. It is a restriction under the influence of the genes which reside in the cell's nucleus.

Oscar E. Schotté, now professor of biology at Amherst College, who studied under Spemann at Freiburg and later at Yale University under Harrison, conducted an experiment which brings out beautifully the influence of the genes on development. His project was to transplant an embryonic bit from a frog to a salamander. These two animals belong to different orders and have striking contrasts in structure. The salamander larva has teeth, and on each side of its head are balancers which aid it in swimming. The tadpole of the frog, on the other hand, has no teeth but horny jaws, and the protuberances on the sides of its head are suckers. From the underside of a frog embryo Schotté took a slice of cells which normally would become flank skin and transplanted it to the prospective head area of a salamander embryo. The salamander developed horny jaws instead of teeth, and the head suckers of a tadpole instead of balancers!

From this we conclude that while the organizer determines in general what organs are to be formed, the genes control the details of those organs. It is as if the genes in the frog cells said to the salamander: "You tell us to form a mouth, but we don't know how to make your mouth; we can make only a frog mouth." It is a case of the genes of one animal confronting the alien organizer of a different animal. The fact that they are still able to team up to produce an animal rather than a confused collection of cells is an unsolved puzzle.

This much is sure: development is a business of *both* preformation and epigenesis. The blueprint of the individual is carried in the fertilized egg, but the pattern takes form, organ by organ, as it is called into being by the organizer and is shaped in detail by the genes.

Medical Implications

All these studies, with their changing tactics, have meaning for us. The plastic embryonic cells of sea urchins, salamanders, frogs and chicks are, if not brothers to our own cells, at least cousins. What is learned of them applies to all. And there are many practical problems whose solution may hang on the scientist's fuller understanding of what the organizer is. The cruel malformations that occasionally arise during the development of the embryo—such as Siamese twins or babies born without arms or legs—are failures of organization. Cancer is a lawless crowd of unorganized cells. As we gain in knowledge of the laws of organized growth we may get new clues to the nature of the wild growth that produces malignancy.

Recently, at the Rockefeller Institute, Paul A. Weiss obtained striking evidence of the capacity of cells to organize themselves. From the embryo of a chicken he cut bits of skin tissue, of limb-bud cartilage and of tissue destined to become the coating of the eyeball. He treated all these with an enzyme which dissolves or loosens the "glue" holding the cells together, and so got a mixture of completely dissociated cells of three kinds of tissue. Yet in a tissue culture the cells reassembled themselves according to their kind, and the limb-bud cartilage proceeded to form bone, the eye cells to form eyeball coating and the skin cells to form feathers. "These experiments imply," said Weiss, "that a random assortment of cells which have never been part of any adult tissue can set up conditions—a 'field,' I call it—which will cause members of the cell group to move and grow in concert, following the pattern of a feather in one case, of an eye in another and of a bone in still another."

In a second set of experiments Weiss was able to watch organization at the subcellular level. He cut a salamander and then observed the healing of the wound. The wound cavity filled with a mucus-like liquid. While new skin grew over the cut, connective tissue beneath sent fibrils into the mucus. The tiny fibrils at first were a jumble, like a log jam in a river, but presently they began to assume an orderly arrangement, forming alternate layers like cordwood being stacked crisscross. "Two processes were at work," said Weiss. "The underlayer of connective-tissue cells produced the fibrils, organizing them out of mole-cules, while the overlying layer of skin cells organized them into a subcellular construction."

Down to the Molecules

Basically, of course, cell differentiation depends on the operation of molecule-forming processes. The molecular building blocks of one tissue (*e.g.*, muscle) differ radically from those of another (*e.g.*, brain). Heinz Herrmann, head of the laboratory of chemical embryology at the University of Colorado Medical School, has called my attention to the fact that investigators of developmental biology have begun to focus on the protein-forming systems of the embryonic cell in their search for the "organizer." Recent experiments in bacteriology are highly suggestive. They show that in a bacterium a new enzyme may suddenly emerge in response to an environmental change. In other words, a new differentiation suddenly appears in one of the molecular building blocks, and as a result the cell acquires a new property, for every feature of its molecular construction of course is controlled by the catalytic action of enzymes.

Herrmann points out that embryology, having arrived at the protein-forming systems as the center of its search, now joins forces with other branches of biology—physiology, immunology, microbiology—which likewise are seeking ultimate answers to their basic questions in the hidden mechanisms by which cells make proteins. Thus the mystery of the development of an organism emerges from its long isolation as a separate study and becomes an integral part of the many-sided inquiry into the nature of life itself.

8

Transplanted Nuclei and Cell Differentiation

by J. B. Gurdon
December 1968

The nucleus of a cell from a frog's intestine is transplanted into a frog's egg and gives rise to a normal frog. Such experiments aid the study of how genes are controlled during embryonic development

The means by which cells first come to differ from one another during animal development has interested humans for nearly 2,000 years, and it still constitutes one of the major unsolved problems of biology. Much of the experimental work designed to investigate the problem has been done with amphibians such as frogs and salamanders because their eggs and embryos are comparatively large and are remarkably resistant to microsurgery. As with most animal eggs, the early events of amphibian development are largely independent of the environment, and the processes leading to cell differentiation must involve a redistribution and interaction of constituents already present in the fertilized egg.

Several different kinds of experiment have revealed the dependence of cell differentiation on the activity of the genes in the cell's nucleus. This is clearly shown by the nonsurvival of hybrid embryos produced by fertilizing the egg of one species (after removal of the egg's nucleus) with the sperm of another species. Such hybrids typically die before they reach the gastrula stage, the point in embryonic development at which major cell differences first become obvious. Yet the hybrids differ from nonhybrid embryos only by the substitution of some of the nuclear genes. If gene activity were not required for gastrulation and further development, the hybrids should survive as well as nonhybrids. The importance of the egg's non-nuclear material—the cytoplasm—in early development is apparent in the consistent relation that is seen to exist between certain regions in the cytoplasm of a fertilized egg and certain kinds or directions of cell differ-

entiation. It is also evident in the effect of egg cytoplasm on the behavior of chromosomes [see "How Cells Specialize," by Michail Fischberg and Antonie W. Blackler; Scientific American Offprint 94]. Such facts have justified the belief that the early events in cell differentiation depend on an interaction between the nucleus and the cytoplasm.

Nuclear transplantation is a technique that has enormously facilitated the analysis of these interactions between nucleus and cytoplasm. It allows the nucleus from one of several different cell types to be combined with egg cytoplasm in such a way that normal embryonic development can take place. Until this technique was developed the only kind of nucleus that could be made to penetrate an egg was the nucleus of a sperm cell, and this was obviously of limited use for an analysis of those interactions between nucleus and cytoplasm that lead to the majority of cell differences in an individual.

The technique was first applied to the question primarily responsible for its development. The question is whether or not the progressive specialization of cells during development is accompanied by the loss of genes no longer required in each cell type. For example, does an intestine-cell nucleus retain the genes needed for the synthesis of hemoglobin, the protein characteristic of red blood cells, and a nerve-cell nucleus the genes needed for making myosin, a protein characteristic of muscle cells? If unwanted genes are lost, the possibility exists that it is the progressive loss of different genes that itself determines the specialization of cells, as August Weismann originally proposed in 1892. The

clearest alternative is that all genes are retained in all cells and that the genes are inactive in those cells in which they are not required. Before describing the nuclear-transplant experiments that distinguish between these two possibilities, we must outline the methods used to transplant living cell nuclei into eggs.

The aim of a nuclear-transplant experiment is to insert the nucleus of a specialized cell into an unfertilized egg whose nucleus has been removed. Ingenious attempts in this direction were made many years ago by constricting an egg just after fertilization and then letting one of the early-division nuclei that appeared in the nucleated half of the egg enter the non-nucleated half. This method, however, is applicable only to the nuclei of early embryos whose cells are not normally regarded as being specialized. The first real success in transplanting living cell nuclei into animal eggs was achieved in 1952 by Robert W. Briggs and Thomas J. King, both of whom were working at the Institute for Cancer Research in Philadelphia. Their method, which has been generally adopted in subsequent work, involves three steps [see *illustration on page 93*]. Owing to the fortunate circumstance that the unfertilized egg of an amphibian has its nucleus (in the form of chromosomes) located just under the surface of the egg at a point visible through the microscope, it is not difficult to obtain an egg with no nucleus. This can be done by removing the region of the egg that contains chromosomes with a needle or by killing the nuclear material with ultraviolet radiation. The second step is to dissociate a tissue into separate cells,

NORMAL FROG was raised in the author's laboratory at the University of Oxford from an egg that had been fertilized in the usual way by a sperm. The frogs used in the experiments described in this article were the South African clawed species *Xenopus laevis*.

TRANSPLANT FROG was raised from an egg cell from which the nucleus had been removed and into which the nucleus of an intestine cell had been transplanted. The frog is normal in all respects, indicating that intestine-cell nucleus has full range of genes.

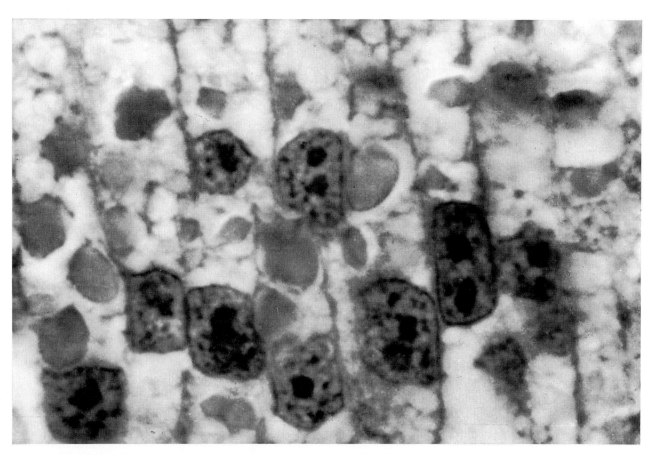

DONOR CELLS used in the author's transplant experiments were taken from the epithelial layer of intestine. The micrograph shows the cells' characteristic columnar shape, central nucleus and yolk contents. Cells will be dissociated in preparation for transplant.

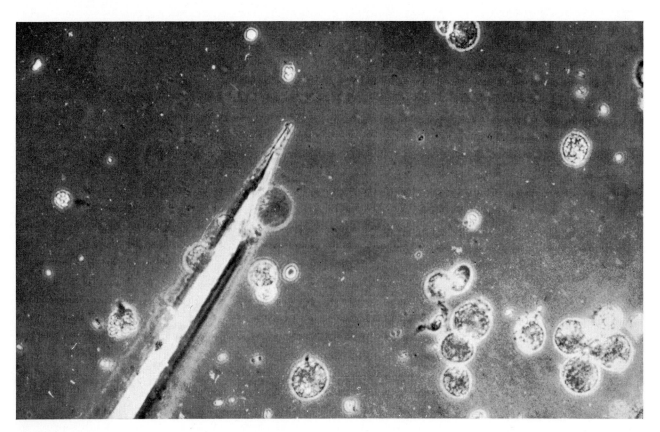

SINGLE NUCLEI from intestine epithelium cells are obtained for transplanting by sucking the whole cell into a micropipette (left). Smaller in diameter than the cell, the pipette breaks the cell wall; only the nucleus and a coating of cytoplasm enter the host egg.

each of which can be used to provide a donor nucleus for transplantation. The cells separate from one another in a medium lacking calcium and magnesium ions, which are removed from the embryo more quickly by adding to the medium a chelating substance such as Versene.

The third and most difficult stage in the procedure involves the insertion of the donor-cell nucleus into the enucleated egg. Briggs and King found that this can be done by sucking an isolated cell into a micropipette that is small enough to break the cell wall but large enough to leave the nucleus still surrounded by cytoplasm. This compromise is required because the nucleus in an unbroken cell does not make the necessary response to egg cytoplasm, and conversely a bare nucleus without surrounding cytoplasm is readily damaged by exposure to any artificial medium. The broken cell with its cytoplasm-protected nucleus is injected into the recipient egg. The amount of donor-cell cytoplasm injected is very small and does not have any effect.

A useful extension of the basic nuclear-transplant technique is called serial nuclear transplantation. It involves the same procedure as the one just described except that instead of the donor nuclei being taken from the cells of an embryo or larva reared from a fertilized egg, they are taken from a young embryo that is itself the result of a nuclear-transplant experiment. The effect is the same as in the vegetative propagation of plants, namely the production of a clone: a population consisting of many individuals all having an identical set of genes in their nuclei.

One other feature of nuclear-transplant experiments is of the greatest importance for their interpretation. It is the use of a nuclear marker whereby the division products of a transplanted nucleus can be distinguished from those of the host egg nucleus. A nuclear marker is virtually indispensable where attention is to be paid to the development of a very small percentage of eggs that have received transplanted nuclei, since one cannot otherwise be sure that an occasional error in enucleation by hand or by ultraviolet irradiation has not occurred. Only by the presence of a marker in the nuclei of a transplant embryo does one have proof of its origin.

A nuclear marker must be replicated and therefore be genetic. One of the most useful for nuclear transplantation is found in a mutant line of the South African clawed frog *Xenopus laevis*, discovered at the University of Oxford by Mi-

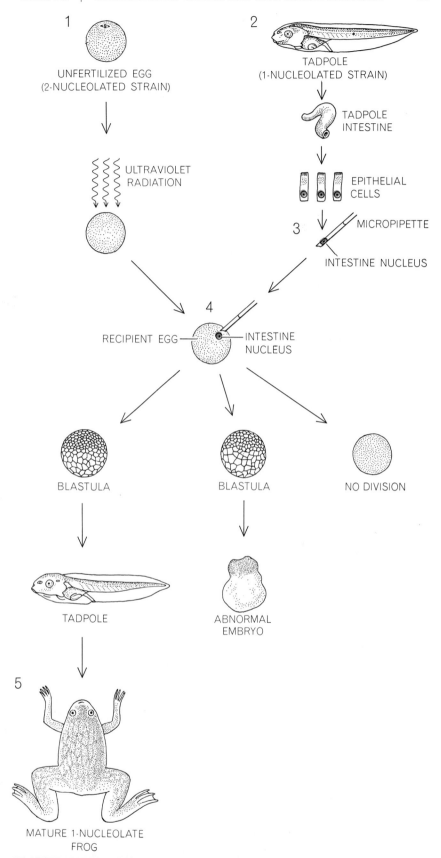

TRANSPLANT PROCEDURE starts with preparation of a frog's unfertilized egg (*1*) for receipt of a cell nucleus by destroying its own nucleus through exposure to ultraviolet radiation. Next, intestine is taken from a tadpole that has begun to feed (*2*) and cells are taken from its epithelial layer. A single epithelial cell is then drawn into a micropipette; the cell walls break (*3*), leaving the nucleus free. The intestine-cell nucleus is transplanted into the prepared egg (*4*), which is allowed to develop. In some 1 percent of transplants the egg develops into a frog that has one nucleolus in its nucleus instead of the usual two (*5*).

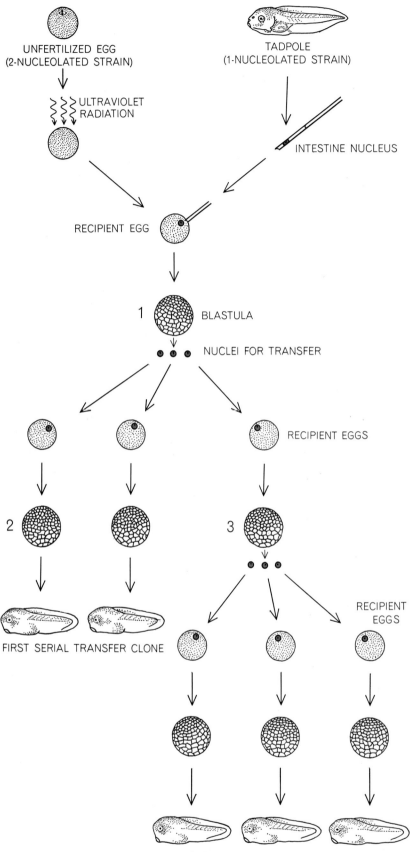

UNFERTILIZED EGG
(2-NUCLEOLATED STRAIN)

ULTRAVIOLET
RADIATION

TADPOLE
(1-NUCLEOLATED STRAIN)

INTESTINE NUCLEUS

RECIPIENT EGG

1   BLASTULA

NUCLEI FOR TRANSFER

RECIPIENT EGGS

2

3

FIRST SERIAL TRANSFER CLONE

RECIPIENT
EGGS

SECOND SERIAL TRANSFER CLONE

SERIAL TRANSPLANTS involve the same first steps as the transplantation procedure illustrated on the preceding page. At the blastula stage (1) the cells of a transplant-embryo are dissociated. The genetically identical nuclei from these cells are then transplanted into enucleate eggs, giving rise to a clone: a population comprised of genetically identical individuals (2). The procedure can be continued indefinitely (3).

chail Fischberg. The nuclei of most normal frog cells contain two of the bodies called nucleoli; the nuclei of cells carrying the mutation never have more than one. This mutation is almost ideal as a nuclear marker because a sample of cells taken from any tissue at any developmental stage beyond the blastula (the hollow sphere from which the gastrula arises) can be readily classified as being mutant or not.

We can now return to the question of whether or not genes are lost in the course of normal cell differentiation. Nuclear-transfer experiments are performed to answer this question on the assumption that if the combination of egg cytoplasm with a transplanted nucleus can develop into a normal embryo possessing all cell types, then the transplanted nucleus cannot have lost the genes essential for pathways of cell differentiation other than its own. For example, if a normal embryo containing a specialized cell type such as blood cells can be obtained by transplanting an intestine-cell nucleus into an enucleated egg, then the genes responsible for the synthesis of hemoglobin cannot have been lost from the intestine-cell nucleus in the course of cell differentiation. The only assumption here is that a gene, once lost, cannot be regained in the course of a few cell generations. It happens that the best evidence for the retention of genes in fully differentiated cells comes from two series of experiments carried out at Oxford on eggs of the frog *Xenopus*.

The fully differentiated cells used for these experiments were taken from the epithelial layer of the intestine of mutant tadpoles that had begun to feed. Intestine epithelium cells have a "brush border," a structure that is present only in cells specialized for absorption and that is assumed to have arisen as a result of the activity of certain intestine-cell genes. Not all the cells of the intestine are epithelial, but when the epithelial cells are dissociated, they can be distinguished from the other cell types by their large content of yolk, by the ease with which they dissociate in a medium that contains Versene and sometimes by their retention of the brush border.

The first experiments with intestine-cell nuclei were designed to show that at least *some* of these nuclei possess *all* the genes necessary for the differentiation of all cell types, and therefore that some of the transplant embryos derived from intestine nuclei could be reared into normal adult frogs. Both male and female adult frogs, fertile and normal in every respect, have in fact been obtained from transplanted intestine nuclei [*see bottom illustration on page 91*]. Although only about 1.5 percent of the eggs with transplanted intestine nuclei developed into adult frogs, all of these frogs carried the mutant nuclear marker in their cells; their existence therefore proves that at least some intestine cells possess as many different kinds of nuclear genes as are present in a fertilized egg.

Subsequent experiments with intestine nuclei were designed to show that *many* of these nuclei have retained genes required for the differentiation of at least *some* quite different cell types. In these experiments the criterion for gene retention was the differentiation of functional muscle and nerve cells by nuclei whose mitotic ancestors had already promoted the differentiation of intestine cells. Functional muscle and nerve cells are present in any nuclear-transplant embryo that shows the small twitching movements, or muscular responses, characteristic of developing tadpoles just be-

fore they swim. Out of several hundred intestine nuclear transfers, about 2.5 percent of the injected eggs developed as far as the muscular-response stage or further. The reason why the remainder did not reach this stage is not necessarily because that proportion of intestine nuclei lack the necessary genes. In some cases it is known to be the inability of certain recipient eggs to withstand injection; in others it is the incomplete replication of some of the transplanted nuclei or their daughter nuclei during cleavage. In either case a nuclear-transplant embryo should contain some cells with normal nuclei as well as some abnormal cells responsible for the early death of the embryo.

Serial nuclear transplantation offered a means of overcoming both difficulties. A sample of nuclear-transplant embryos whose development was so abnormal they would have died before reaching the muscular-response stage provided nuclei for serial transplant clones. Many of the serial transplant clones included embryos which developed as far as

the muscular-response stage or beyond it. By adding the proportion of nuclei shown by first transplants to be able to support muscular-response differentiation to the proportion shown by serial transplantation to possess this capacity, we can conclude that at least 20 percent of the intestine epithelium cells must have retained the genes necessary for muscle-cell and nerve-cell differentiation [*see illustration below*].

There is no reason to believe that muscle-cell or nerve-cell genes have been lost or permanently inactivated in the remaining 80 percent of transplanted intestine nuclei. There are many reasons why it might not have been possible to demonstrate their presence. For example, about 50 percent of all the eggs that received intestine nuclei failed to divide even once. When a sample of these eggs was sectioned, they were found to contain either no nucleus at all or else a nucleus that was still inside an intact intestine cell. In the first instance the nucleus presumably stuck to the injection pipette and was never deposited in the egg; in

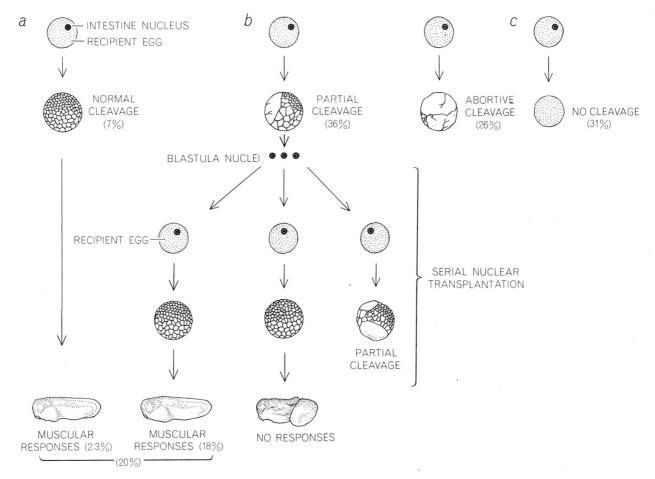

NERVE AND MUSCLE TISSUE in nuclear-transplant embryos shows that 20 percent of the nuclei of differentiated cells still possess the genes necessary for other kinds of function. The criterion of success is development of an embryo to the point at which small twitching motions are observed; this was demonstrated for more than 2 percent of all intestine nuclei using first transfers (*a*), and in 18 percent using serial transfers (*b*). Many eggs that failed to develop (*c*) contained either no transplant nuclei or still unruptured cells.

the second, the donor cell was never broken so as to liberate its nucleus, a technical error that is easy to make with very small cells. In both cases the developmental capacity of the intestine nuclei was not tested, and the recipient eggs that failed to divide should not be counted in the results.

It is clear from these experiments that the loss or permanent inactivation of genes does not necessarily accompany the normal differentiation of animal cells. This conclusion is not inconsistent with a recent finding: the "amplification" of the genes responsible for synthesizing the ribonucleic acid (RNA) of the subcellular particles called ribosomes. This phenomenon was demonstrated in amphibian oöcytes, the cells that give rise to mature eggs. The nuclear-transfer experiments just described do not exclude such amplification, which simply alters the number of copies of one kind of gene in a nucleus. Instead they show that specialized cells always have at least one copy of every different gene.

The inability of some transplanted nuclei to support normal development has attracted considerable interest because it is always found that the propor-

tion of nuclei showing a restricted developmental capacity increases as the cells from which they are taken become differentiated. Furthermore, serial nuclear-transplant experiments conducted by Briggs and King (and subsequently by others) have shown that all the embryos in a clone derived from one original nuclear transplant often suffer from the same abnormality, whereas the embryos in a clone derived from another original transplant may suffer from a different abnormality. Some of the abnormalities of nuclear-transplant embryos can therefore be attributed to nuclear changes that can be inherited.

The discovery that these changes arise as a result of nuclear transplantation, and not in the course of normal cell differentiation, was an important one. This was first established by Marie A. DiBerardino of the Institute for Cancer Research, who made a detailed analysis of the number and shape of chromosomes in nuclear-transplant embryos. Abnormal embryos were usually found to suffer from chromosome abnormalities that were not present in the donor embryos, a finding that at once explains why the factors causing many of the developmental abnormalities of nuclear-transplant

embryos are inherited. The fact that chromosome abnormalities arise after nuclear transplantation does not necessarily mean that they are of no interest; there could be a connection between the kind of chromosome abnormality encountered and the cell type of the donor nucleus concerned. In spite of an intensive search, however, no such relationship has yet been found.

The origin of these chromosome abnormalities is probably to be understood as an incompatibility between the very slow rate of division of differentiating cells—only one division every two days or more—and the rapid rate of division in an egg, which starts to divide (and causes any injected nucleus to try to divide) about an hour after injection. Unless an injected nucleus can complete the replication of its chromosomes within this brief period, they will be torn apart and broken at division. This concept is supported by the observation, made at Oxford in collaboration with my colleagues C. F. Graham and K. Arms, that many transplanted nuclei continue to synthesize the genetic material DNA right up to the time of the first nuclear division, whereas sperm and egg nuclei always complete this synthesis well be-

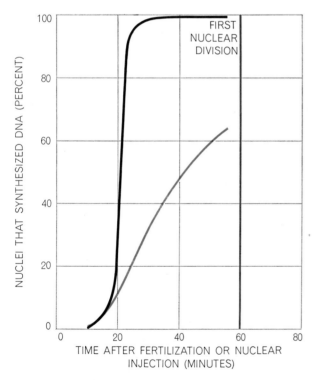

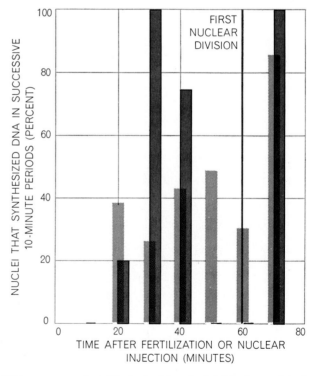

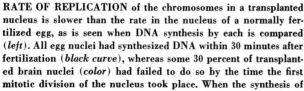

RATE OF REPLICATION of the chromosomes in a transplanted nucleus is slower than the rate in the nucleus of a normally fertilized egg, as is seen when DNA synthesis by each is compared (*left*). All egg nuclei had synthesized DNA within 30 minutes after fertilization (*black curve*), whereas some 30 percent of transplanted brain nuclei (*color*) had failed to do so by the time the first mitotic division of the nucleus took place. When the synthesis of

DNA was assessed at 10-minute intervals (*right*), the nuclei of fertilized eggs (*black*) were found to have ceased synthesis, an indication of completed chromosome replication, before time for mitotic division. Some 30 percent of transplanted gastrula nuclei, however, were still making DNA (*color*). Failure of transplant nuclei to complete replication before division results in damage to the chromosomes as they are torn apart, producing abnormalities.

fore division. Presumably molecules associated with the DNA of specialized cells prevent the chromosomes of such cells from undergoing replication as rapidly as those of sperm nuclei, thereby leading to the chromosome abnormalities commonly observed in nuclear-transplant embryos.

Having concluded that the specialization of cells involves the differential activity of genes present in all cells, rather than the selective elimination of unwanted genes, we can now consider how genes are activated or repressed during early embryonic development. Nuclear transplantation has been used to demonstrate that the signals to which genes or chromosomes respond are normal constituents of cell cytoplasm. This information has come from experiments in which the nucleus of a cell carrying out one kind of activity is combined with the enucleated cytoplasm of a cell whose nucleus would normally be active in quite another way. One of two results is to be expected: either the transplanted nucleus should continue its previous activity or it should change function so as to conform to that of the host cell to whose cytoplasm it has been exposed. For the purposes of these experiments changes in nuclear activity have to be recognized by the appearance of direct gene products and not by the much less direct criterion of the normality of nuclear-transplant embryo development. Many of these experiments have been carried out in collaboration with Donald D. Brown of the Carnegie Institution of Washington or with another of my Oxford colleagues, H. R. Woodland.

The first experiments were designed to find out if the different functions performed by any one gene—the synthesis of DNA, the synthesis of RNA and chromosome condensation in preparation for cell division—are determined by cytoplasmic constituents. Three kinds of host cell were used: unfertilized but activated eggs whose nucleus would normally synthesize DNA but no RNA; growing oöcytes in which the nucleus synthesized RNA but not DNA, and oöcytes maturing into eggs, in which situation the nucleus consists of condensed chromosomes arranged in the "spindle" of cell division, and synthesizes neither RNA nor DNA. Two kinds of test nuclei were used: nuclei from adult brain tissue, which synthesize RNA but almost never synthesize DNA or divide, and nuclei from embryonic tissue at the mid-blastula stage of development; mid-blastula nuclei do not synthesize RNA but

FERTILIZED EGGS

FERTILIZED EGG

NO RNA
SYNTHESIS

MID-BLASTULA
(10TH DIVISION)

LARGE RNA
MOLECULES

LATE BLASTULA

TRANSFER
RNA

GASTRULA

RIBOSOMAL
RNA

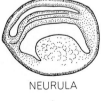

NEURULA

NUCLEAR TRANSPLANT EMBRYOS

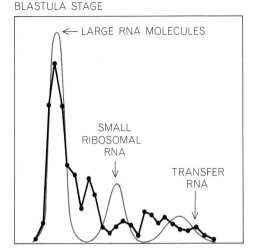

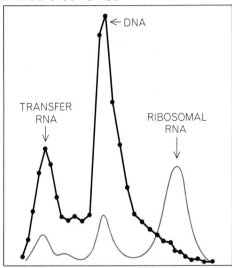

CYTOPLASMIC CONTROL of the activities of genes is demonstrated in a nuclear-transplant experiment. As a normal embryo develops (*left*), early growth stages are marked by the nuclear synthesis of different varieties of RNA. The nucleus of a cell from an embryo at the neurula stage, which synthesizes all varieties of RNA, is transplanted into an egg. At first the transplanted nucleus halts RNA synthesis. On reaching the blastula stage (*top right*) the nucleus starts to synthesize large RNA molecules. In this graph and the other ones the black curve indicates the extent to which radioactive precursors of various nucleic acids are incorporated. The quantity of preexisting nucleic acids is in red. At the late-blastula stage (*middle right*) transfer RNA, but not ribosomal RNA, is synthesized. By the neurula stage (*bottom right*) nucleic acid synthesis has gone on to include the making of ribosomal RNA.

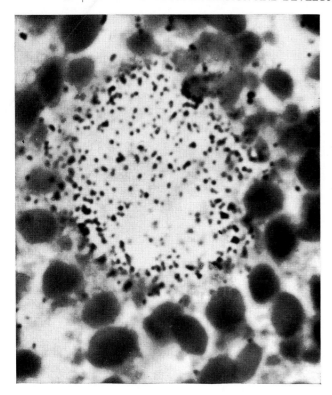

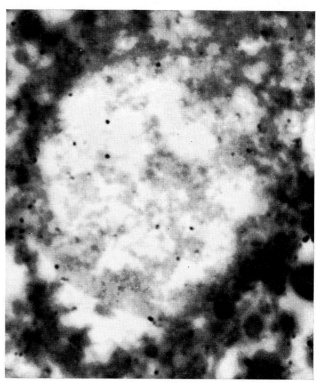

CHANGE IN ACTIVITY of a transplanted nucleus to conform to the normal activity of the host cell's missing nucleus is shown in two photomicrographs. Nuclei from embryos at the mid-blastula stage of development, which synthesize DNA but not RNA, have been injected into an oöcyte (*left*) and an egg (*right*). Oöcytes, the cells that grow into eggs, synthesize RNA but not DNA; eggs do the opposite. A substance that is a precursor of RNA, labeled with a radioactive isotope, was injected simultaneously. The many black dots, formed by the radioactive molecules, show that the substance was incorporated in newly synthesized RNA in the oöcyte, a factor in the host cell's cytoplasm having altered the injected nuclei's activity. The nuclei in the egg, however, make no RNA.

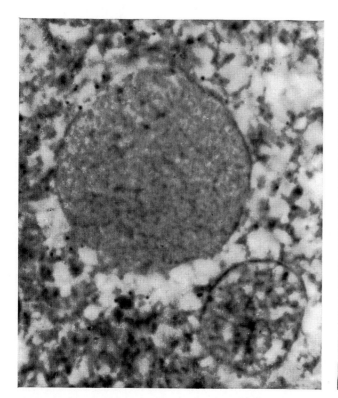

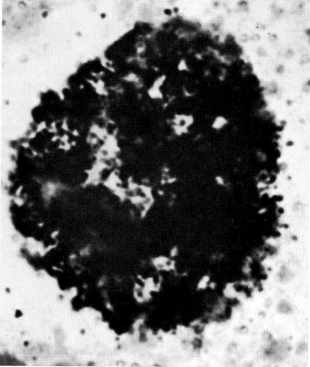

REVERSE RESULTS are shown in two other photomicrographs. The nuclei to be injected are from frog brain cells; they synthesize RNA but almost never DNA. The radioactively labeled precursor is one that is taken up only in DNA synthesis. In the oöcyte (*left*), where the synthesis of RNA is progressing, the nucleus contains no radioactive DNA. The intense radioactivity in the egg (*right*) shows that the injected brain nuclei have switched from RNA to DNA synthesis in response to a factor in the host cell's cytoplasm.

synthesize DNA and divide about every 20 minutes. For technical reasons it was desirable to inject each host cell with many nuclei, even though this can prevent the subsequent division of the injected cell. The results were clear: In all respects tested the transplanted nuclei changed their function within one or two hours so as to conform to the function characteristic of the normal host-cell nucleus [*see illustrations on opposite page*]. Mid-blastula nuclei injected into growing oöcytes stopped synthesizing DNA and dividing and entered a continuous phase of RNA synthesis that lasted for as long as the injected oöcytes survived in culture (about three days). Adult brain nuclei injected into eggs stopped RNA synthesis and began DNA synthesis. When the same nuclei were injected into maturing oöcytes, they synthesized neither RNA nor DNA but were rapidly converted into groups of chromosomes on spindles.

The next set of experiments was designed to find out if cytoplasmic components can repress or activate genes, that is, if they can select which genes in a nucleus will be active at any one time. Advantage was taken of the natural dissociation that exists in the time of synthesis of different classes of RNA during the early embryonic development of *Xenopus*. The work of several investigators has established the following sequence of events in *Xenopus* embryos. For the first 10 divisions after fertilization no nuclear RNA synthesis can be detected. Just after this—at the mid-late-blastula stage—the cells synthesize large RNA molecules, which are believed not to include ribosomal RNA but which are likely to include "messenger" RNA. Toward the end of the blastula stage "transfer" RNA synthesis is first detected; this is followed a few hours later, during the formation of the gastrula, by the synthesis of ribosomal RNA.

The extent to which these events are under cytoplasmic control has been investigated by transplanting into enucleated eggs single nuclei from embryonic tissue at the neurula stage of development, the one that follows the gastrula stage. As the nuclear-transplant embryos develop, RNA precursor substances that have been labeled with radioactive atoms (for example uridine triphosphate labeled with tritium, the radioactive form of hydrogen) are used to determine the classes of RNA being synthesized at each stage. Autoradiography has shown that a neurula nucleus, which synthesizes each main kind of RNA, stops all detectable RNA synthe-

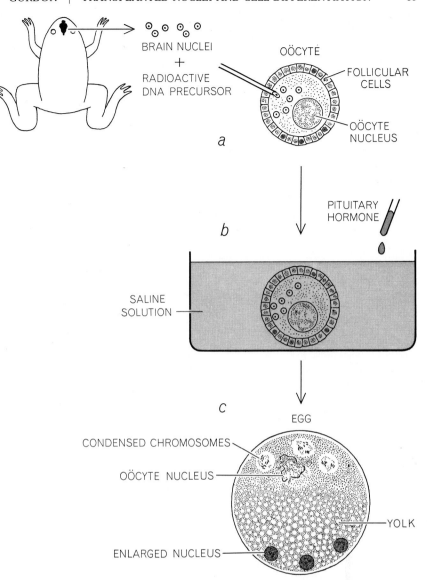

INDUCING FACTOR in egg cytoplasm that alters the activity of the nucleus was found to be absent at the oöcyte level of egg-cell development; when brain nuclei and labeled DNA precursor were injected into oöcytes (*a*), the nuclei did not synthesize DNA. After the oöcyte was brought to maturity (*b*), however, the inducing factor made its appearance (*c*). Brain nuclei near the oöcyte's ruptured nucleus underwent chromosomal condensation like that of the oöcyte nucleus. Brain nuclei in the yolky region began to synthesize DNA.

sis, that is, it no longer incorporates labeled RNA precursors, within an hour of transplantation into egg cytoplasm. Furthermore, chromatography and other kinds of analysis show that, when the transplant embryos are reared through the blastula and gastrula stages, they synthesize heterogeneous RNA, transfer RNA and ribosomal RNA in turn and in the same sequence as do embryos reared from fertilized eggs.

Taken together, these experiments have shown that changes in the type of gene product (for example the synthesis of RNA or DNA), as well as changes in the selection of genes that are active (for example the synthesis of different types of RNA), can be experimentally induced.

Since a high proportion of transplanted neurula nuclei support entirely normal development, the results show that egg cytoplasm must contain constituents responsible for independently controlling the activity of different classes of genes in normal living nuclei.

We can now consider what is perhaps the most interesting question of all: What is the mechanism by which cytoplasmic components bring about changes in gene activity? Of the various changes in chromosome and gene activity that can be experimentally induced in transplanted nuclei, special attention has been devoted to the induction of DNA synthesis by egg cytoplasm. It is

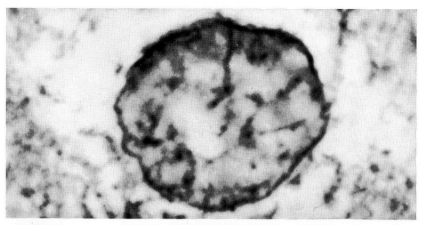

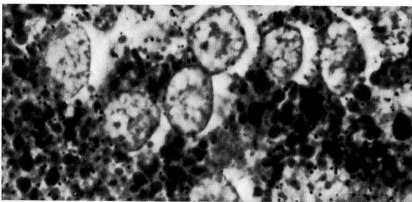

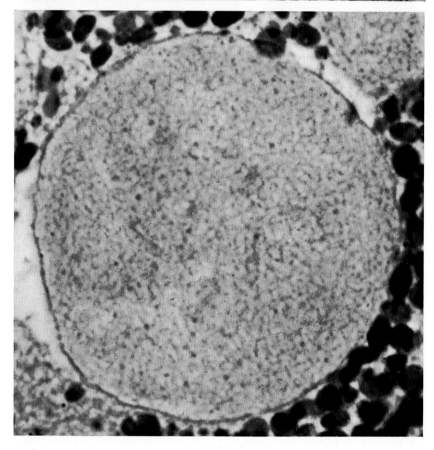

BLASTULA NUCLEI swell up following their injection into an egg (*top*) or into an oöcyte (*bottom*). Within the oöcyte the enlargement may be as much as 200 times. The center micrograph shows blastula nuclei of normal dimensions; the bottom micrograph, a nucleus 48 hours after injection. Genetic material in the nucleus is dispersed during swelling.

easier to analyze than other changes, and it seems likely to exemplify certain general principles of cytoplasmic regulation in early embryonic development.

The origin of the cytoplasmic condition that induces DNA synthesis has been investigated by injecting adult brain nuclei, together with a radioactive labeling substance, into growing and maturing oöcytes. The inducing factor appears just after an increase in the level of pituitary hormone has caused an oöcyte to mature into an egg, an event that is accompanied by intensive RNA and protein synthesis [*see illustration on preceding page*].

Concerning the identity of the inducing factor, the first candidate to be considered was simply the presence of an adequate supply of DNA precursor substances. Woodland, however, has injected growing oöcytes with 10 times the amount of all four common DNA precursors believed to be present in the mature egg. One of the precursors, thymidine triphosphate, had been labeled with tritium. In spite of the availability of these precursors, the brain nuclei did not incorporate the labeled thymidine into DNA. Although this experiment requires further analysis before DNA precursors can be excluded as inducers of DNA synthesis, it encourages a search in other directions.

The next candidate to be considered was DNA polymerase, an enzyme that promotes the incorporation of precursor substances into new DNA in a way that is specified by the composition of the preexisting "template" DNA. DNA polymerase activity in living cells has been tested by introducing purified DNA and tritium-labeled thymidine into eggs. In collaboration with Max Birnstiel of the University of Edinburgh we have established that the injected DNA serves as a template for synthesis of the same kind of DNA. When DNA and labeled thymidine are introduced into oöcytes, no DNA replication can be detected. This means that the cytoplasmic factor inducing DNA synthesis in eggs includes DNA polymerase or something that activates this enzyme. It is doubtful, however, that this is the *only* constituent of the inducer. If it were, the injection of egg cytoplasm (which contains DNA polymerase) into oöcytes might be expected to induce DNA synthesis, a result that is not in fact obtained. This experiment, in which purified DNA is replicated in the cytoplasm of unfertilized eggs, also serves to demonstrate that constituents of injected brain nuclei other than their DNA are not required in order to initiate the particular reaction being discussed here.

The last aspect of this reaction on which some information is available concerns the mechanism by which the inducing factors in the cytoplasm interact with the DNA in the nucleus. It was noticed several years ago by Stephen Subtelny, now at Rice University (and subsequently by others), that transplanted nuclei increase in volume soon after they have been injected into eggs. A pronounced swelling is also observed in nuclei injected into oöcytes; the swelling is therefore not directly related to a particular type of nuclear response. During this nuclear enlargement chromatin (which contains the genetic material in the nucleus) becomes dispersed and, as Arms has demonstrated, cytoplasmic protein also enters the swelling nuclei. While working at Oxford, Robert W. Merriam of the State University of New York at Stony Brook found a close temporal relation between the passage of cytoplasmic protein into enlarging nuclei and the initiation of DNA synthesis. The interpretation of these events currently favored by those of us involved in the experiments is that the nuclear swelling and chromatin dispersion facilitate the association of cytoplasmic regulatory molecules with chromosomal genes, thereby leading to a change in gene activity of a kind determined by the nature of the molecules that enter the nucleus.

The experiments described here have established two general conclusions. First, nuclear genes are not necessarily lost or permanently inactivated in the course of cell differentiation. Second, major changes in chromosome function as well as in different kinds of gene activity can be experimentally induced by normal constituents of living cell cytoplasm. The same type of experiment is now proving useful in attempts to determine the identity of the cytoplasmic components and their mode of action.

We have had to restrict our attention to what can be described as sequential changes in gene activity, that is, differences between one developmental stage and the next. These may be compared with regional variations in nuclear activity, that is, differences between one part of an embryo and another at the same developmental stage. The latter are hard to study biochemically because of the difficulty in obtaining enough material. There is no obvious reason, however, why the processes leading to the two types of differentiation should be fundamentally different.

Experiments analogous to those described here have been conducted with bacteria infected with viruses, with nuclear transplantation in protozoans and with fusion in mammalian cells. Each kind of material is well suited for certain problems; nuclear transplantation utilizing amphibian eggs and cell nuclei is especially suited to the analysis of processes that lead to the first major differences between cells. Only after these differences have been established by constituents of egg cytoplasm are cells able to respond differentially to other important agents that guide development, such as inducer substances and hormones. Finally, the technique of nuclear transplantation may be used to introduce cell components other than the nucleus into the cytoplasm of different living cells; this is likely to be of great value for the more detailed analysis of early development and cell differentiation.

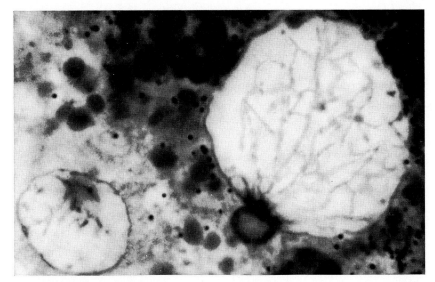

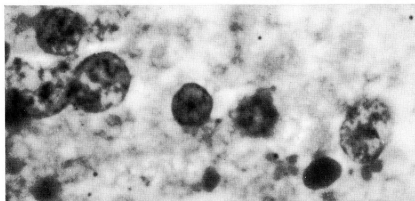

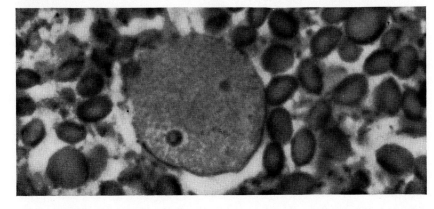

BRAIN NUCLEI also swell up when injected into an egg (*top*) or an oöcyte (*bottom*), but they do not enlarge as much as blastula nuclei. The center micrograph shows brain nuclei of normal dimensions. During enlargement, dispersal of genetic material and entry of cytoplasmic protein into the nucleus facilitate contact of cytoplasmic regulators with the genes.

9 Phases in Cell Differentiation

by Norman K. Wessells and William J. Rutter
March 1969

By cultivating embryonic tissue in the laboratory one can study the specialization of cells in the mammalian pancreas. There seem to be three regulatory phases, each leading to a new stage of differentiation

How does a single cell—the fertilized egg—give rise to the many different cell types of a multicellular organism? After fertilization of an egg by a spermatozoon the number of cells in the developing embryo increases dramatically. Soon three classes of cells can be distinguished: ectodermal and endodermal cells respectively make up an outer and an inner embryonic layer, and mesodermal cells constitute an intermediate layer. These cells come to be arranged in groups that develop into recognizable tissues and organs. Ultimately several hundred kinds of cells can be distinguished in an adult mammal. The process of functional and structural specialization of cells is called differentiation, and the mechanisms controlling differentiation remain major mysteries of biology.

Experimental work done by embryologists since 1900 has demonstrated that interactions between cells play an important role in differentiation. The development of some organs, including the pancreas, the liver and the lungs, depends on discrete sets of cells derived from the embryonic gut endoderm and also on adjacent mesodermal cells; the brain, the mammary glands and the limbs involve combinations of certain ectodermal and mesodermal cells. Experiments with intact embryos and with laboratory cultures of combinations of tissues indicate that for normal development the two interacting tissues must be adjacent to each other. A fundamental and still unanswered question is: How does one cell influence the development of another?

What happens within cells during development to make one cell type different from another? Since nearly all cells have some physiological activities in common, all must contain some of the same enzymes and structures, and mere quantitative differences among cells in this common metabolic machinery would not in themselves confer unique characteristics on a given cell. For that a cell has a group of specific proteins responsible for its specialized functions: muscle cells contain contractile proteins, plasma cells make antibodies, red blood cells form hemoglobin and so on. What controls these qualitative differences? This is a second fundamental and still unanswered question.

In this article we shall approach these two major questions by discussing the differentiation of the mammalian pancreas. First we shall outline the development of the pancreas and then describe a detailed investigation of interactions between cells and the regulation of specific protein synthesis in this organ. The view of differentiation gleaned from this work, carried out in our laboratories at Stanford University and the University of Washington, may be generally applicable to other organ systems.

The mammalian pancreas manufactures enzymes that digest foodstuffs and also secretes two hormones that regulate the metabolism of carbohydrates. Different populations of cells are involved in the two functions. The exocrine cells make about a dozen specific proteins that are involved in the breakdown of proteins, carbohydrates, fats and nucleic acids. Some are synthesized as active enzymes and others as inactive precursors called zymogens; all of them are stored within the cells in granules. The exocrine cells are arranged in clusters called acini and secrete the contents of their granules into a duct system that leads into the small intestine. There the zymogens are converted to active enzymes, and digestion takes place.

There are two kinds of endocrine cell. The *B* cells synthesize insulin, the hormone that regulates the uptake of blood glucose and its storage as glycogen in muscle and fat. Insulin is stored in beta granules. Donald F. Steiner of the University of Chicago has demonstrated that the synthesis of insulin (like the synthesis of the protein-digesting enzymes) has two steps: first the formation of a precursor called proinsulin and then its conversion to insulin by an enzyme. The site of this conversion and the specific enzymes involved are not known. The *A* cells produce the hormone glucagon, a protein that controls the degradation of glycogen in the liver. Glucagon is stored in alpha granules. Both *B* cells and *A* cells are in the islets of Langerhans, spherical aggregates of cells that lie near blood capillaries, so that the hormones pass easily into the blood for transport to their sites of action elsewhere in the body.

In the embryo of a mouse or a rat the formation of the pancreas begins about midway through the gestation period. A group of cells in the upper wall of the primitive gut begins to bulge upward on the ninth day of gestation in the mouse, and about 36 hours later in the rat. The base of this evagination constricts, giving rise to the future pancreatic duct; the upper part, the pancreatic epithelium, expands into the surrounding mesodermal tissue. (Note the proximity of endoderm and mesoderm.) In the next three or four days rapid cell division continues and the cells become arranged in typical acini and primitive islets. During the later stages of gestation cell division tends to be restricted to the peripheral regions of the pancreas; near the center one can detect the signs of advanced differentiation: zymogen granules in the acinar cells and alpha

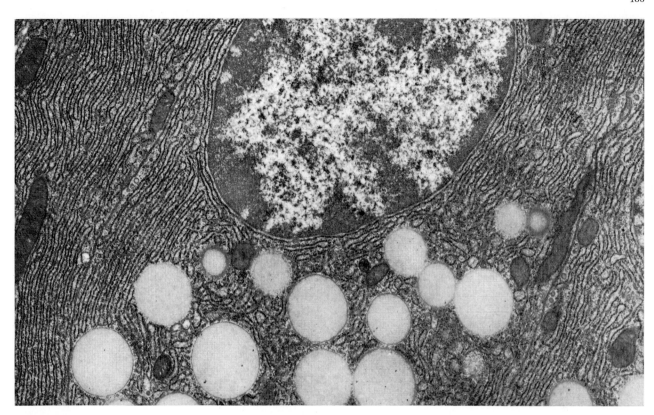

EXOCRINE CELLS of the pancreas secrete zymogens, proteins that are precursors of digestive enzymes. In this electron micrograph made in the laboratory of one of the authors (Wessells) a thin section of exocrine cells from the adult mouse pancreas is enlarged 14,000 diameters. The extensive endoplasmic reticulum, a folded membrane studded with ribosomes (*small black dots*), is the site of protein synthesis. Zymogens are stored in zymogen granules (*spheroidal vesicles at bottom*) until they are secreted into the intestine.

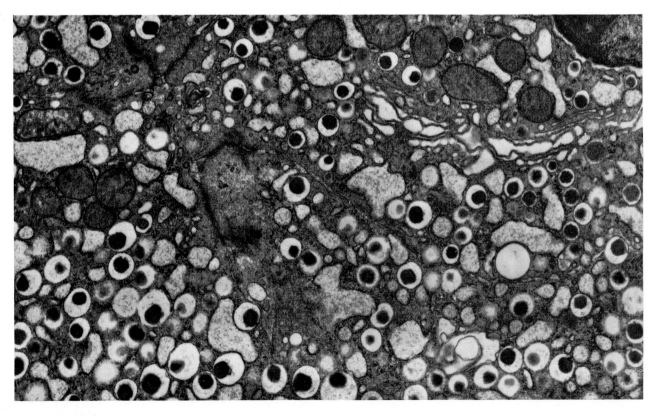

ENDOCRINE CELLS of the pancreas secrete the hormones glucagon and insulin. In this electron micrograph several adult mouse B cells, which secrete insulin, are enlarged 17,000 diameters. The light spheroidal vesicles present in large numbers are beta granules and the dense material in many of the granules is thought to be insulin or proinsulin, the precursor protein of the hormone.

and beta granules in the endocrine cells. As birth approaches, cell division stops in the peripheral acini and they too differentiate.

To establish the course of differentiation at the molecular level one can monitor the specific functional attributes of the various cell types. For the exocrine cells the digestive enzymes produced in the largest quantities are the best indexes of differentiation; for the endocrine cells insulin and glucagon serve the same purpose. The problem is to measure these proteins specifically and sensitively in the mixture of several hundred different proteins that are present in the cells. We developed microassays based on specific catalytic activity (for the individual digestive enzymes), the ability of specific antibodies to recognize single species of proteins (for insulin and glucagon) and such distinctive physical characteristics of the protein molecules as their size and their mobility in an electric field.

With these assay procedures we discovered that both the exocrine enzymes and glucagon and insulin are present at concentrations many thousands of times higher in functional pancreatic cells than in other cells of the adult organism; these proteins are indeed cell-specific. This finding gave us a considerable experimental advantage, since it meant we could attribute the proteins to pancreas cells even when we measured them in extracts of several cell types, such as pancreas plus gut. We set out to determine the pattern of accumulation of the specific products and relate it to the development of acini and islets and the secretory granules.

By measuring the activity of the major exocrine enzymes present in extracts of tissues we established their developmental profiles during gestation [*see top illustration on page 106*]. Two significant and rather unexpected features of their accumulation patterns became apparent.

First, the concentration of proteins (expressed as enzyme molecules per cell) does not change in a simple way, rising from zero to finite high levels in one step. Instead there are three discrete stages in the developmental process. In what we call State I the enzymes are present at a relatively low concentration; then the concentration is 1,000 times or more higher (State II), and finally the concentrations of individual enzymes are adjusted a little with respect to one another (State III). Even the low State I level was much higher than the level in other cell types, and the relative proportions of the individual enzymes approximate those found in State II, indicating that State I is a stage in development unique to the pancreas rather than simply a reflection of lack of development.

The second unexpected feature of the developmental profiles is that the concentrations of the exocrine enzymes do not change as a coordinated set during

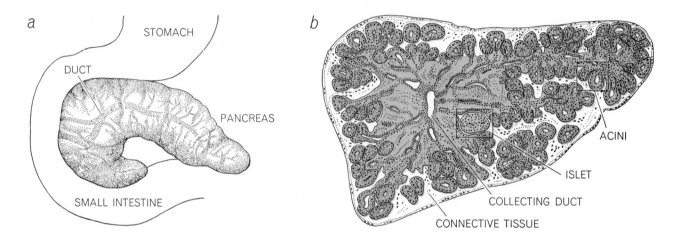

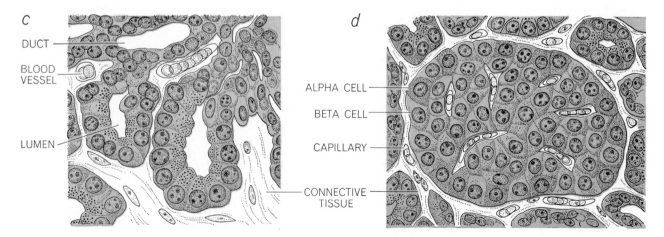

PANCREAS is a secretory organ that lies behind the stomach, as indicated in the schematic drawing of mouse viscera (*a*). A cross section of a small portion of the pancreas (*b*) contains cell clusters called acini, ranged along a duct system, and distinctive cell groups called islets of Langerhans. The acini are composed of exocrine cells with zymogen granules (*c*); islets contain endocrine cells (*d*).

the transition from State I to State II; the pattern of accumulation between 15 and 19 days differs significantly for the various proteins. For example, trypsin and carboxypeptidase B lag significantly behind the other proteins; lipase B activity is first detected in the rat at 18 days, when the other enzymes have already reached high levels. The exocrine proteins, then, are not all regulated as a unit. On the other hand, the relative concentrations of certain groups of enzymes—for instance lipase A and carboxypeptidase A, trypsin and carboxypeptidase B—appear to change in concert; the curves for the former pair are similar, for example, and their midway points between low and high levels occur at the same time. Perhaps the simplest way to explain the developmental pattern of these proteins is to assume that small sets of them are regulated together and that these sets are synchronized to rise together over a certain period of embryonic development.

With the aid of the electron microscope we correlated these changes in protein content with the appearance of intracellular structures. During the period of great amplification of enzyme content between State I and State II we observed a dramatic formation of the rough endoplasmic reticulum, the site of enzyme synthesis, and saw the first zymogen granules appear. The experiments of George E. Palade and Philip Siekevitz and their collaborators at Rockefeller University had suggested that newly synthesized enzymes first appear within the endoplasmic reticulum and then proceed to the larger cavities of the Golgi apparatus, where they are packaged in zymogen granules. In none of our experiments have we ever observed high levels of specific enzyme formation without the development of these intracellular organelles; the accelerated synthesis of specific proteins appears always to be coupled with the formation of new structures involved in their synthesis, storage and secretion.

Turning to the endocrine cells, we first established that the insulin content of the adult mouse or rat pancreas is remarkably high: more than a billion molecules per B cell. (Very low but significant levels were found in other adult tissues, presumably reflecting the normal dissemination of the hormone from the pancreas.) Our findings in the early embryo were more unexpected: we were unable to detect any insulin before the ninth day, although our assay could have detected one molecule of insulin in 10 cells. Clearly there is little, if any, transfer of

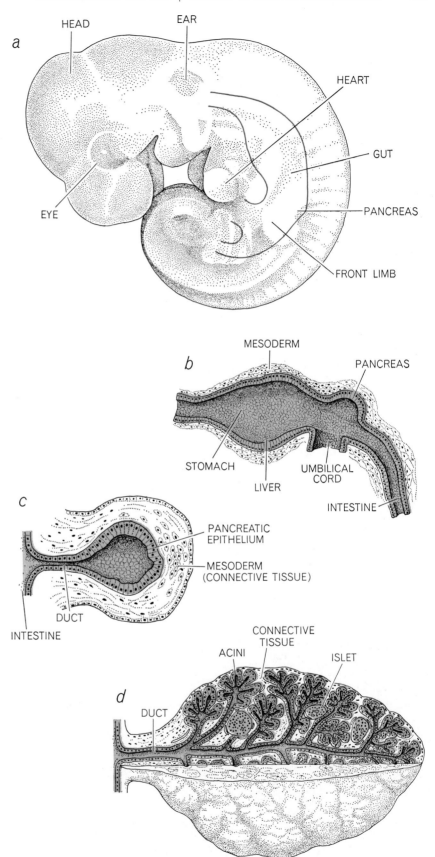

DEVELOPMENT of the mouse pancreas begins when the embryo is in its ninth day of gestation (*a*). The organ forms from the central part of the gut by a process of evagination (*b*): a group of endodermal cells (*color*) bulge into adjacent mesoderm (*black*). By the 11th day the pancreas forms a pouchlike diverticulum (*c*). By the 15th day some pancreas cells have stopped dividing and begun to synthesize large quantities of specific proteins (*d*).

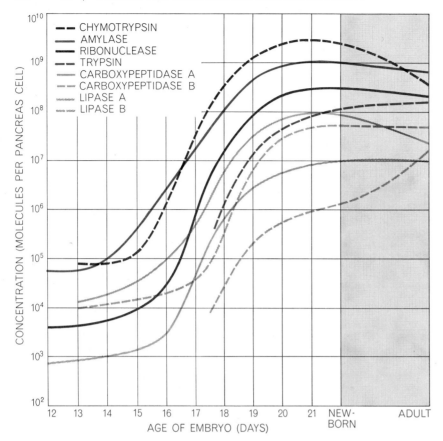

PATTERN OF ACCUMULATION of the digestive enzymes is an index of exocrine-cell development. Extracts of pancreatic rudiments of various ages were prepared and assayed for enzyme content on the basis of specific catalytic activity. The midpoints between low and high levels differ for most of the proteins. Clearly they are not regulated as a single group.

insulin from the mother to the embryo and the early embryo itself must synthesize insignificant quantities. Early embryonic development apparently proceeds in the absence of this hormone. There must be remarkably stringent regulation of specific protein synthesis in embryonic cells. Until a given time in development, proteins such as insulin simply are not made.

Late on the ninth day insulin is detected in the gut region, where it is presumably present in primitive pancreas cells [see top illustration on opposite page]. The concentration then increases until, during the period of State I, there may be as many as a million insulin molecules per B cell. Later, at the time the exocrine cells make the transition from State I to State II, the insulin content of the endocrine cells increases another several hundredfold. By applying an antigen-antibody assay for glucagon we have shown that glucagon too is present at the earliest stages of pancreatic development. Electron micrographs show that there are secretory granules in the endocrine cells at this early stage of differentiation. In the exocrine cells, on the other hand, zymogen granules only appear later, during the large rise in specific activity from State I to State II. It seems that the details of the differentiation processes in these two cell types are different.

These correlated studies of pancreatic development suggest a model for the differentiation of pancreatic exocrine and endocrine cells [see illustration on page 108]. The model recognizes several distinct stages of development. The protodifferentiated state corresponds to the State I we described earlier. The differentiative state (State II) is found in the late embryo, and the modulated levels found in the adult characterize State III. This implies that there are three regulatory transitions, each introducing a new stage of differentiation.

It is difficult, if not impossible, to identify the factors that control these transitions in the intact embryo, because one cannot effectively manipulate the cellular environment or the physical relations among cells. The cultivation of cells or tissues in laboratory vessels seemed to offer a way to obtain further information about the differentiative process, particularly the nature of the regulatory events. We found that a pancreas isolated from a 10-day or 11-day mouse embryo would develop normally in culture, and that both the pattern of increase in enzyme levels and the appearance of zymogen granules closely mimic

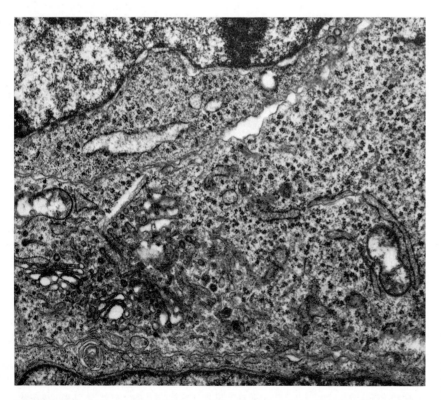

CYTOPLASM of an exocrine cell from a 10-day embryo reflects the cell's differentiation. Some ribosomes are organized into endoplasmic reticulum but there are no zymogen granules yet, perhaps because there is not enough specific protein to require such structures.

the developmental sequence in intact embryos. This developmental competence in tissue culture opened-up attractive experimental vistas. We could investigate relations among different cell types and also monitor developmental events within cells much more effectively.

First we tested the ability of progressively younger pancreatic tissues and more primitive gut tissues to develop into normal pancreas. We found that even the youngest pancreatic rudiment differentiates normally, provided that both mesodermal and epithelial elements are present. Then we found that gut tissue from the eight-day mouse embryo, excised 18 hours before the pancreas would become visible, develops into normal pancreatic tissue (plus some other tissues) in culture; gut material taken from earlier embryos forms liver-like and stomach-like tissues in culture but no pancreatic cells. It seems, therefore, that some significant developmental event must occur on the eighth embryonic day that confers pancreatic potential on a group of precursor cells. This event presumably requires an input from some part of the embryonic system that is absent in the laboratory culture. After this event, however, the tissue is somehow "determined," so that when it is removed from the embryo, it will develop just as it would in the intact embryo.

The simplest way to test for an interaction between two tissues in an organ is to separate the tissues and see whether each is capable of differentiating normally alone or whether recombination is necessary for normal development. We found we could carry out such an experiment with the pancreatic rudiment in the protodifferentiated state (the 11-day mouse embryo or the 13-day rat embryo). At this point the epithelial cells exist as a bulbar structure encased in mesoderm. We separated the two tissues by first incubating the intact rudiment for a short time in a mixture of protein-digesting enzymes and then stripping off the mesoderm by drawing the rudiment up into a micropipette. We then tested for an interaction between the two components with a technique, perfected by Clifford Grobstein of the University of California at San Diego, in which a thin, porous membrane serves as a platform to support the individual tissues or as a barrier to separate them [see bottom illustration on page 109]. We found that the epithelium would not develop normally alone but that it did differentiate if it was cultured directly across the filter from mesoderm. To our surprise,

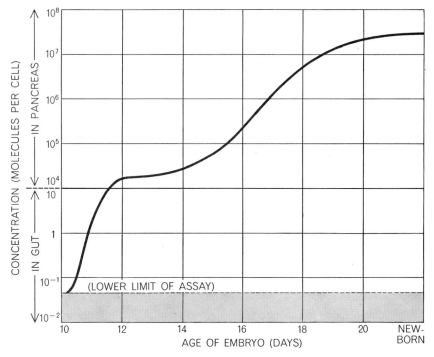

INSULIN CONCENTRATION in tissue extracts was measured by an antigen-antibody assay and is given here in molecules per gut cell and (when the embryo is large enough for dissection) molecules per pancreas cell. The sensitivity of the assay makes it possible to detect insulin at an earlier developmental stage than one can detect any of the exocrine proteins.

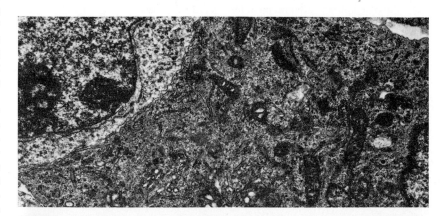

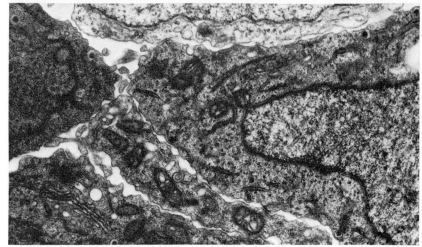

SECRETORY GRANULES appear in endocrine cells as soon as hormones begin to be synthesized. There are no granules in the cytoplasm of future endocrine cells when the pancreas begins to take shape (top), but in pancreas cells from an early 10-day embryo, when insulin is first detected (see illustration at top of page), a number of granules can be seen (bottom).

the mesoderm did not have to be from the pancreas; we learned that mesodermal tissues from salivary gland, kidney, stomach, lung or spleen were just as effective in promoting epithelial development.

Apparently the mesoderm contributed some factor that is required for epithelial differentiation and that is small enough to pass through the filter. To get enough material from which to attempt to isolate the factor, we first prepared a crude homogenate of an entire embryo, which turned out to contain enough of the mesodermal factor to promote pancreatic differentiation. We were able to remove most of the active material from the homogenate by low-speed centrifugation, indicating that the factor was part of (or was bound to) some rather large particulate structure. A number of experiments suggest that it is a protein, or at least that a protein is involved in the activity. The active material has now been solubilized and partially purified.

Is the mesodermal factor required at all times during development of the pancreas or just at a certain time? We have noted an epithelial requirement for the presence of mesoderm as early as the ninth day, when a restricted region of the gut responds to mesodermal tissue by forming pancreatic acini. Nearby epithelium, treated similarly, forms oth-

er gut derivatives such as stomach or intestine but does not form pancreatic tissue. Several hours before the pancreas can be seen, then, a discrete group of gut cells has the capacity to form the organ, provided that mesodermal tissue is present. Presumably this ability reflects the determination event that occurred some eight hours earlier. Similar experiments with older pancreatic epithelium indicate that the mesodermal requirement ceases just before the accelerated synthesis of the specific proteins; the mesoderm can be removed from the epithelium late in the protodifferentiated state without any effect on subsequent development. In summary, the mesodermal factor must be present from a time before the pancreas can be seen until about five days later, when cell division ceases and enzymes begin to accumulate at high levels.

How does the mesodermal factor act? It could promote growth, cause the formation of acini or actually cause differentiation in the epithelial cells. Recent results suggest that a major action of the mesodermal factor is to stimulate DNA synthesis and thus the proliferation of epithelial cells. (As we shall see, such DNA synthesis may be required for differentiation to occur.) The eventual availability of pure preparations of the factor may give us clear answers to these questions.

Our efforts to describe and analyze some of the changes that take place at the cellular and molecular level during development have been focused on the secondary transition leading to the differentiated state. First we sought to trace changes in the pattern of protein synthesis during the transition. (What we had measured in intact embryos was enzyme content, not the rate at which enzymes were being synthesized.) We incubated 14-day and 19-day pancreatic rudiments in a medium containing radioactively labeled amino acids—the building blocks of proteins—and then sorted out the resulting radioactively labeled proteins by electrophoresis in polyacrylamide gel. In this technique specific proteins are identified by the rate at which they move through the gel in the presence of an electric field, and the amount of radioactivity at each point shows how much of each protein has been synthesized.

There was a remarkable difference between the two patterns of protein synthesis [see illustration on page 110]. Similar experiments with rudiments at various stages of the secondary transition confirmed that there is about a fiftyfold increase in the rate of synthesis of the enzymes during the transition. This increase in rate accounts quantitatively for the increase in enzyme content between State I and State II noted in the earlier assays. There must therefore be little, if any, secretion of these proteins by exocrine cells during this period of development; the proteins that are synthesized simply accumulate within the cells.

Protein synthesis depends on the activity of a number of species of RNA. We decided to see if the increase in specific enzyme synthesis in pancreas depends on the synthesis of new RNA. For these studies we employed actinomycin D, a compound that is bound to the DNA template and thus blocks RNA synthesis. Very low levels of actinomycin applied to dividing cells in the protodifferentiated state inhibited only part (about 70 percent) of the RNA synthesis in those cells but completely prevented the accelerated synthesis of the differentiative proteins. This was a specific effect, since the cells remained healthy and continued to synthesize cellular (as opposed to secretory) proteins at nearly the normal rate. If the tissue was treated with actinomycin at a later time, when the central group of cells in the culture had stopped dividing, those cells differentiated more or less normally and accumulated zymogen granules; cells located on the periphery of the same

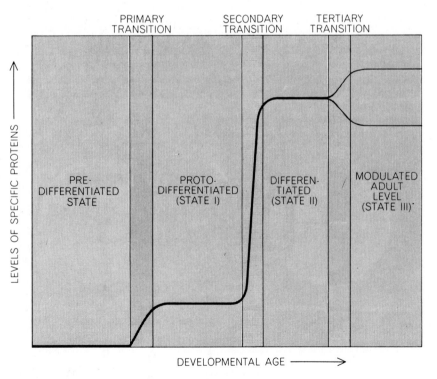

POSSIBLE REGULATORY PHASES in cell differentiation are indicated by this schematic curve. The three stages of differentiation correspond to the numbered states associated with the exocrine enzyme levels. Each of them is introduced by a regulatory transition phase.

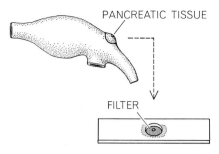
PANCREATIC TISSUE

FILTER

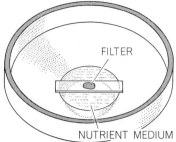
FILTER

NUTRIENT MEDIUM

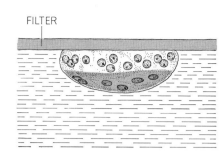
FILTER

PANCREATIC TISSUE is grown in the laboratory. Nine-day gut tissue is excised (*left*) and placed in a culture dish (*middle*). The **tissue, mesoderm (*color*) as well as endoderm (*black*), is suspended under a plastic support and incubated in nutrient (*right*).**

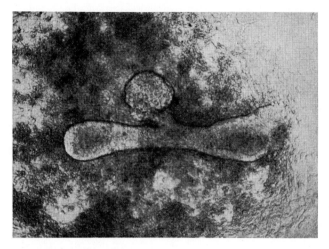
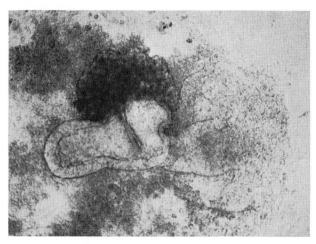

PANCREAS DEVELOPS in culture. After three days in culture the pancreas bulges upward (*left*). After another four days it has **grown, acini have formed and the acinar tissue has been darkened by zymogen granules (*right*). The enlargement is 50 diameters.**

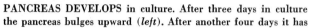
FILTER   ENDODERM

MESODERM

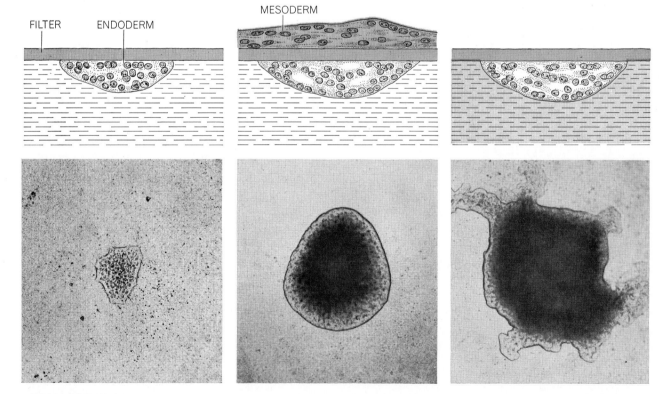

TRANSFILTER CULTURE tests for mesoderm-endoderm interaction. When pancreatic epithelium (endoderm) is cultured alone (*top left*), the cells fail to differentiate in five days (*bottom left*). When epithelium and mesoderm (*color*) are separated by a thin **filter and cultured together, there is substantial growth in five days, acini are visible and zymogen granules darken the cells (*center*). Epithelium that is cultured without mesoderm, but with an embryo extract included in the nutrient medium, does equally well (*right*).**

rudiments were still dividing and they (like all cells treated in the protodifferentiated state) never made the secondary transition. If cultures were treated still later, when most of the cells had ceased to divide, there was no significant effect of actinomycin on the synthesis or accumulation of the specific proteins. In collaboration with Fred H. Wilt of the University of California at Berkeley we showed that actinomycin has the same inhibitory effect on RNA synthesis in early pancreatic epithelial cells as it does in older cells. The results were therefore not due to a difference in the susceptibility of the two populations of cells to actinomycin but to the fact that the synthesis of new RNA is required for the accelerated synthesis of the pancreatic proteins.

One further experiment with actinomycin is of special significance in relation to our earlier observations about different enzymes being regulated together in groups. When actinomycin D is added to cultures early in the secondary transition, the synthesis of the various specific proteins is inhibited to different degrees. This sequential inhibition suggests that the "messenger" RNA's that encode the various proteins are synthesized at slightly different times, supporting our earlier contention that the exocrine proteins are not all regulated as a single set.

Inhibition of DNA synthesis also blocks exocrine cell differentiation. A culture treated with fluorodeoxyuridine (a strong inhibitor of DNA synthesis and hence of cell division) during the protodifferentiated state does not differentiate, although it appears otherwise healthy; cells treated after they have ceased to divide, however, develop normally. Similar results are obtained with bromodeoxyuridine, a structural analogue of thymidine, one of the four nucleosides of DNA. Bromodeoxyuridine is actually incorporated into the DNA during the synthesis period; by forming a faulty DNA that does not allow the synthesis of functional RNA, bromodeoxyuridine blocks the differentiative process. In contrast to the action of actinomycin D (which has a sequential inhibitory effect on the synthesis of certain proteins), bromodeoxyuridine exhibits an "all or none" effect, depending on just when it is administered: differentiation is either blocked completely or is unaffected. These experiments suggest that some event that occurs only in dividing cells or is facilitated by DNA synthesis must precede the subsequent changes in RNA synthesis that bring about the new pattern of protein synthesis.

Our present studies have provided some insight into the basic mechanisms underlying the formation of organs. They demonstrate the importance of cell interaction in the developmental process. Specifically, after the determination event, a mesodermal-epithelial interaction is required to maintain the protodifferentiated state and initiate the secondary transition. It may be that interactions between the exocrine and endocrine cells are also involved in the developmental process.

Perhaps the most striking thing about the molecular events within maturing pancreatic cells is the fact that many complex regulatory events are apparently coordinated at a few distinct times. This is most obvious in the exocrine cells. There some early initiating event in a population of cells starts the synthesis of specific proteins and the morphogenesis of the pancreas. The ensuing protodifferentiated state is primarily a phase of growth and acinus formation. Then there is a transition to the differentiated state, when cells stop dividing and undergo intracellular differentiation, which is marked by the accelerated synthesis and the accumulation of specific proteins in zymogen granules.

The two major developmental phases must involve hundreds of genes. We believe sets of genes are activated and other sets are inactivated simultaneously, rather than each of the hundreds of individual genes' being activated in sequence. One possible mechanism to accomplish such a concerted transition is an alteration of chromosomal structure so that some sets of genes are chemically unmasked, and thus activated, while others are chemically sequestered and thus repressed. These experiments have not yielded a precise description of the molecular mechanisms involved, but they have simplified our view of the problem and suggested a hypothesis that will be the basis for continuing explorations.

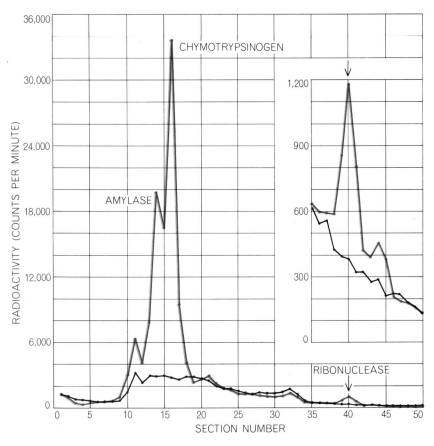

PROTEIN SYNTHESIS in 14-day and 19-day rat pancreatic tissue is compared. The tissues were incubated with a radioactive amino acid, which was incorporated in the proteins being synthesized. When the proteins were subjected to electrophoresis in a gel, they migrated to different points in the gel. The gel was cut into sections and the amount of radioactivity in each section was measured. In the 19-day tissue three new peaks corresponding to specific exocrine proteins are identified. An expanded scale is used (*inset*) to visualize the ribonuclease peak. If 14-day rudiments are cultured for five days, they exhibit a synthetic pattern like that of the 19-day pancreas, unless an inhibiting agent such as actinomycin D or bromodeoxyuridine (*see text*) is in the medium during the early part of the culture period.

How Cells Associate

by A. A. Moscona
September 1961

The cells of many-celled organisms are marshaled and held together by specific physical and chemical factors. These factors are studied by dispersing the cells of a tissue and allowing them to recombine

To explain how cells join with one another to form the tissues and organs of multicellular organisms, the biologist must answer questions that are as basic and pressing in their way as those that surround the nature of the chemical bond. In the absence of the intercellular bonds that hold cells together, the human body would collapse in a heap of disconnected, individual cells, many of them quite indistinguishable from certain free-living protozoa. Were it not for the high specificity of these bonds and the selectivity with which cells interact with one another, there would be no tissues or organs, only nondescript clumps of cells. To devise an approach to these questions—to submit masses of cells to experimental test as they proceed to associate, interact and synthesize tissues—challenges the ingenuity of the investigator.

The study of cell association proceeds along the parallel paths of analysis and synthesis. Since the turn of the century workers in this field have been developing techniques of tissue culture that make it possible to study tissue cells in the simplified environment of laboratory glassware and, by one means or another, to cause the tissues to dissociate into cells. Biochemical analysis has sought to identify the substances involved in the bonding and interaction of cells; morphological analysis, facilitated by the electron microscope, has concentrated on the connection between function and structure. But it is the relatively novel and direct method of synthesis—the experimental synthesis of tissues from free cells under controlled conditions—that offers particular promise in this field. Only by such frontal approach can one put hypothesis to the test and find out how cells actually associate.

In nature, as the fertilized egg pro- liferates into a mass of rapidly dividing cells, the cells first bunch together in no clearly apparent order. But the lack of order is only superficial. The cells have fundamentally identical genetic endowments. Their initial diversification must arise, therefore, in large measure from their different positions in the embryo. There is an impressive body of evidence for this. In the early embryo, for example, one can graft cells from a skin-forming area to the eye-forming one. The grafted cells develop in harmony with their new site, acquiring their neighbors' "eyeness" as their persisting identity, recognized as such by their kind and by other cells. If they are thereafter transferred to other sites, they remain unchanged.

While the embryonic cell may thus "learn" a specific functional identity in response to influences in its environment, it also retains an intrinsic identity established by its genetic endowment. Oscar E. Schotté and Hans Spemann performed an experiment many years ago that strikingly demonstrates this principle. In amphibians the ectodermal tissue (the outer of the three primary embryonic layers) of the mouth forms the teeth. But it does so only if it is in contact with the mouth endoderm (the inner embryonic layer). A "signal" from the endodermal cells triggers a sequence of events in the ectoderm that leads to the formation of teeth. Actually the matter is more complex; the endodermal signal apparently reciprocates a prior stimulus from the ectoderm. So before any noticeable appearance of teeth several "messages" may have been exchanged by the cells in this region. In the early embryo it is possible to transfer ectoderm from any part of the body to the mouth region and make it form teeth by placing it in proper association with the endoderm. Schotté and Spemann took advantage of the fact that newts have bony teeth and frogs have horny teeth to see what would happen if they transplanted frog ectoderm cells to the mouth endoderm of the newt. It turned out that the frog cells get the "message" to form teeth but, being frog cells, they form horny teeth. The learned identity acquired in this experimental association is interpreted by the cells in accordance with their genetic endowment.

The movement of cells from one place to another in the embryo constitutes an essential and conspicuous feature of normal development. Singly and in groups, cells move to new sites where, in association with new neighbors, they form new structures. The mammalian kidney, for example, arises from two separate and initially distant components. A little pocket of cells on each side of the cloaca elongates into a finger-like process, destined to form the ureters, and extends into the body cavity toward a mass of mesodermal (middle layer) cells that at this stage shows no definite structure. As soon as the two groups of cells come into contact, however, they begin to change rapidly. The ureter branches and sends out secondary processes; the mesodermal cells with which these processes make contact are organized into kidney tubules. These changes come quite promptly, as if by an exchange of signals between the two groups of cells. Proximity and association are necessary to the interaction. If the cell groups are kept separate, they do not produce their typical responses. In one strain of mice a genetic defect keeps the two kidney components from making contact, and the kidney does not form.

Next to nothing is known about the

signals that are supposed to be involved in such "inductive" interactions. Jean Brachet and H. de Scoeux, working at the Catholic University of Louvain, found many years ago that the messages did not get across when they interposed a strip of cellophane between two prospectively reactant masses of cells. Cellophane allows the passage of only very small molecules. On the other hand, L. W. McKeehan of the University of Chicago used thin strips of agar, through which larger molecules can diffuse, and observed interaction between two tissues. Clifford Grobstein of Stanford University has performed similar experiments with the two components of mouse kidney isolated in tissue culture; he has found that cellophane blocks their interaction, whereas a filter that passes larger molecules permits the interaction to proceed.

The simplest and perhaps likeliest deduction from these experiments is that the tissues, as they associate, react toward one another through the medium of certain metabolic products. These products may provide both the signals and the means of linking the cells in a specific manner. It must be emphasized, however, that at present, with one possible exception, no such products have been isolated from the cells of any higher organism; moreover, there are acceptable alternative explanations for the experimental results.

But it would seem that some means of intercommunication between cells in a developing system must exist. The cells act as if they were capable of mutual recognition and of specific responses to messages conveyed by their neighbors. There is support, on general biological grounds, for the idea that the messenger is chemical in work on slime molds initiated by Kenneth B. Raper of the University of Wisconsin and continued by John Tyler Bonner of Princeton University, by Maurice and Raquel Sussman at Brandeis University and others. The slime molds live part of their life cycle as free amoebae; under certain conditions they come together and form aggregates that differentiate into "fruiting bodies." Their aggregation is directed by a substance (named acrasin) that has been isolated by Brian Shaffer of the University of Cambridge and that is being investigated in a number of laboratories. It emanates from the initial cluster of amoebae and attracts other cells to them. Here is an established case of chemical communication and guidance in the interaction of cells.

It is not too farfetched to assume that all cell contact implies interaction through the production of specific reaction products. The Australian biologist Sir Macfarlane Burnet suggested recently that production of antibodies by cells in adult organisms might present a

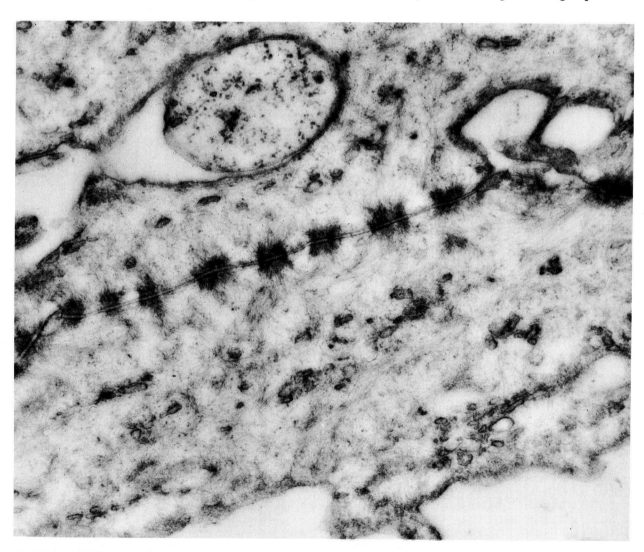

BRIDGES BETWEEN CELLS (desmosomes) are apparently special devices for mutual attachment of cells across their membranes. In this electron micrograph by K. R. Porter of the Rockefeller Institute more than a dozen such bridges (*dark, squarish areas*) connect two skin cells from a salamander larva. The cell membranes run horizontally across picture. Magnification is 35,000 diameters.

model, and perhaps an extreme case, of specific cellular response to chemical signals. The interactions among embryonic cells are, of course, different in detail from the true antibody reaction, and the subtlety and intricacy of these processes are probably of a different order. But it is precisely such subtle chemistry that could provide embryonic cells with the means of mutual communication and integration.

As for the intercellular bond, the term must not be taken as implying that the cells are firmly stuck together or even in direct contact with one another. Electron micrographs made by K. R. Porter of the Rockefeller Institute, by Don W. Fawcett of the Harvard Medical School and by others have suggested that cells may have special devices for mutual attachment on the outer surface of their membranes [*see illustration on opposite page*]. Furthermore, there is always some distance between cells in contact; this space may be extremely narrow or quite wide, and it seems to be filled with a cementing substance. Unlike brick-binding mortars, these intercellular cements have remarkably flexible and dynamic properties. Although they bind the cells, they permit them to move about and regroup without actual dissociation or loss of contiguity.

This dynamic linking is a cardinal feature of cell contact at all levels of multicellular organization. Consider the case of the everted hydra, described by R. L. Roudabush of Iowa State College in 1933. This tiny, vase-shaped animal can be made to turn itself inside out like the finger of a glove. Its internal digestive cells are then on the outside and the skin cells inside. The cells sense this change, and the hydra promptly proceeds to revert to normal. With the intercellular bonds destabilized, the cells migrate, gliding past each other from wrong side to right side. Throughout the process the hydra retains its over-all configuration, keeping its identity as an organization despite the flux of its constituent parts.

It is in terms of such flexibility of contact and such perception of position by the cells that one must visualize the nature of the intercellular bonds. Variations and changes in the stability of cell contacts are part and parcel of any organism—embryonic or adult. Pigment cells begin their embryonic development in the so-called neural crest; they soon lose their contact with this tissue and move out, singly and in groups, to find positions throughout the integument. Their migrations are clearly not random: they reach specific destinations and form

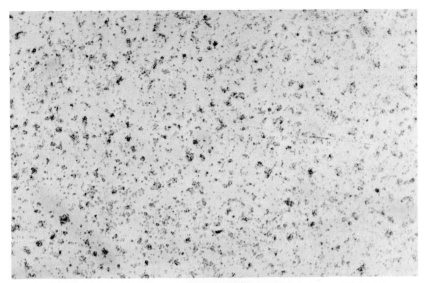

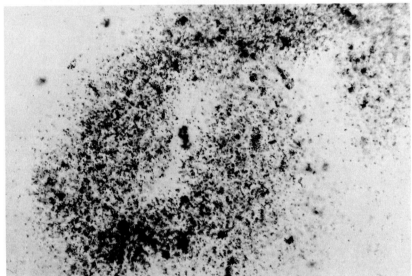

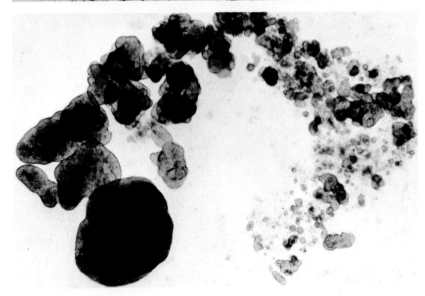

AGGREGATION IN ROTATING FLASK is illustrated in this series of photographs by the author. At top is a suspension of cells from the retina of a seven-day-old chick embryo. In middle is the initial stage of aggregation in a gyrating flask with cells and intercellular material accumulating in the vortex of the liquid culture medium. At bottom a later stage shows compact aggregations at the "head" of the spiral, with continuing aggregation toward the "tail." Magnification in these photographs is approximately 30 diameters.

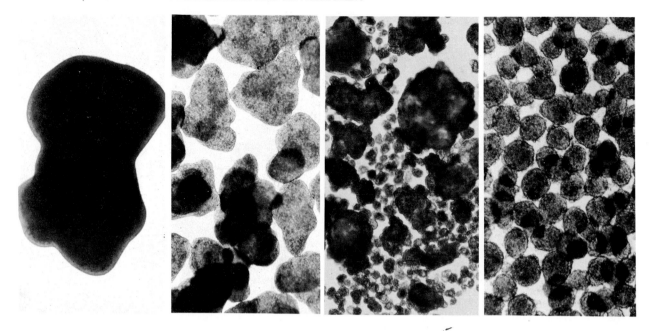

SPECIFIC AGGREGATION PATTERNS characterize each type of cell population. These aggregations were made by (*left to right*) liver cells, retina cells, kidney cells and limb-bud cells, all rotated for 24 hours at 70 revolutions per minute. The first three types came from seven-day-old chick embryos, the last from a four-day chick embryo. Enlargement is approximately 30 diameters.

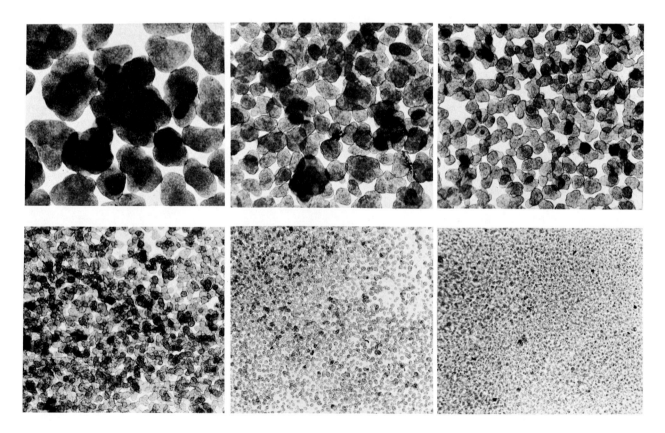

AGGREGATIONS ARE SMALLER when older cells are used. In these photomicrographs the concentration of dissociated retina cells was the same in every case; all were rotated for 24 hours at 70 r.p.m. The cells, however, were taken from chick embryos aged 7, 9, 11, 14, 17 and 19 days respectively. At 19 days the cells simply do not form aggregations. Lowering the temperature or increasing the rotation rate, while all other experimental conditions remain the same, has a similar effect on the size of the aggregations.

typical pigmentation patterns. Other cells leave the neural crest in loose swarms, "homing" toward certain sites in the head of the embryo, where, in conjunction with the cells of that region, they form the lower jaw.

Changes in cell-contact stability continue to play an important part in the life of the organism past the embryonic stage. The steady supply of blood cells involves the continuous disconnection of precursor cells from the bone marrow and their entry as free cells into the circulatory system. Similarly, sperm and egg cells free themselves, as they mature, from their tissues of origin. Elsewhere stability is greater, but definitely relative. Living cells cannot be disengaged from their places in the skin by mere pinching. But when the skin is cut, cells rapidly dissociate from the periphery of the wound, move into the gap, fill it and reestablish stable contacts.

Few questions about cell association yield to fruitful study in the intact organism. It is necessary to separate the cells and tissues from the complexity of the organism in order to control the conditions of observation and experiment. The first steps in this direction necessarily involved the tissues of lower organisms. At the turn of the century Curt Herbst, working at the Zoological Station in Naples, found that young sea-urchin embryos would fall apart and dissociate into single cells when placed in sea water from which he had removed the calcium. He then made the even more interesting discovery that the cells would coalesce and re-form into an embryo when calcium was restored to the water. Calcium has since proved to be an important element in the binding of cells, but not always so dramatically as in the sea-urchin embryo. In general calcium acts more directly as a cell binder in early embryos; later on it seems to operate in conjunction with organic materials to which the primary role seems to shift. There are, however, many invertebrates whose tissues fall apart in the adult state when deprived of calcium. In 1927 James Gray of the University of Cambridge isolated living ciliated cells from the mantle tissue of mussels by placing fragments of the mantle in calcium- and magnesium-free sea water.

An experiment by H. V. Wilson of the University of North Carolina in 1907 pointed to even deeper questions. By gently pressing a marine sponge through a fine sieve he found that he could dissociate it into free cells. He then noticed that as soon as the dispersed cells settled through the sea water onto the dish they started to coalesce. The resulting clumps, when suitably cultured, grew into small but complete sponges. At first it was thought that the sponges regenerated from cells called archeocytes, which, along with skin and digestive cells, make up the loosely associated tissues of the sponge. But further observation showed that all three types of cell persisted following dissociation and that they reassociated in the new aggregations.

Work by later investigators, particularly by Paul S. Galtsoff at the Marine Biological Laboratory in Woods Hole, Mass., and by Tom Humphreys of our laboratory at the University of Chicago, has added new dimensions to these early findings. When cells of different sponge species, preferably of different color for easy recognition, are dispersed and then mixed together, they separate and reaggregate by species, forming separate clusters [see illustration on page 119]. The cells, in other words, are able to identify one another, to give out and register some kind of signal and so associate preferentially with their kin.

A sponge is in some respects a differentiated colony of cells rather than a true multicellular organism. One might question whether the capacity of sponge cells for mutual recognition and sorting out represents a phenomenon of general significance, found in other cellular systems and particularly in higher organisms. Certainly in the case of mammalian tissues it would be difficult to answer the question one way or another in the absence of techniques for dissociating them into individual cells. Some years ago, however, I found that the cementing substances in these tissues will yield to digestion by trypsin and certain other enzymes that break down proteins without serious injury to the cells. Practically any tissue of embryonic origin can now be dissociated and reduced to a suspension of its constituent cellular units. The cells may then be maintained in suitable nutrient media in a germ-free, temperature-controlled environment. It was now possible to conduct studies of the bonding and interaction of the tissue cells of mammals, birds and other higher organisms.

The next step—the resynthesis of complete systems from individual cells—also proved to be feasible. We found that, like sponge cells, the dispersed cells of mammalian or bird embryos will readily aggregate into clusters, migrating over the surface of the culture dish and form-

ing stable connections. Cells from different kinds of tissue were even observed to sort themselves out by cell type in forming these clusters.

The technique lent itself to the study of many previously unanswerable questions, but it fell short of being an exact and adequately controlled procedure. For one thing, it depended primarily on active movement by the cells, a highly variable capacity susceptible to a host of poorly understood conditions. The results in consequence varied unpredictably from one experiment to another.

How could one harness cell aggregation and make it into a critical tool for the study of interactions among cells? The solution turned out to be extremely simple. Most of the irrelevant chance factors that dominate the situation in a stationary cell culture can be neutralized by setting the culture in motion and thereby suspending the cells in a controlled field of force. To do so we place the culture flasks on a horizontally gyrating platform that rotates the flasks 70 times a minute. In each flask the spinning liquid forms a vortex in which the cells concentrate rapidly. They soon link into clusters, within which they construct tissues.

The formation of these clusters depends on and reflects a dynamic equilibrium among the major factors in the system: a balance between the concentrating and shearing-flow forces in the liquid; the differential capacities of the cells to cohere; and the effects of the suspension medium on the cohesiveness of the cells. In this relatively simple system all the pertinent factors—the speed of rotation, the size of the flasks, the character and volume of the medium, the kind and concentration of cells and so on—can be effectively controlled. Thus if the rotation speed and the medium are made the constants of the experiment, the results will reflect the native cohesiveness of the cells in the population tested. The more cohesive they are, the larger and fewer will be their aggregates; the less their cohesiveness, the smaller and more numerous their aggregates. In experiments employing this system we have obtained strikingly consistent results. The rate of aggregation, the number, size distribution, shape and internal structure of the aggregates are always the same when cells of a given kind are aggregated under the same set of conditions.

Such experiments yield an aggregation pattern that is characteristic of the cells in question and of the particular set

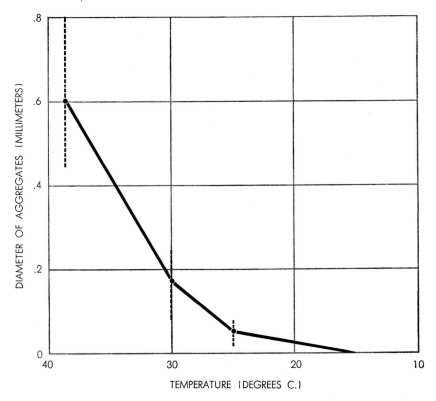

EFFECT OF LOWER TEMPERATURES on the size of aggregations of seven-day chick-embryo retina cells is plotted on this graph. The largest aggregations appear at 38 degrees centigrade; no aggregation occurs after 24 hours of rotation at 15 degrees C. The vertical broken lines show the range of size of the aggregations that build up at each temperature.

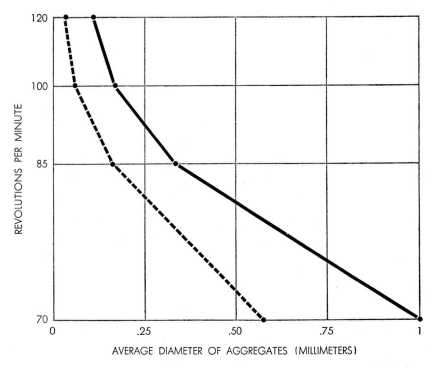

FASTER ROTATION makes the aggregations smaller, as shown by these curves. The broken curve represents chick-embryo retina cells; the solid curve, chick-embryo liver cells. The optimal aggregation size is achieved at 70 r.p.m. The vertical scale on this graph is logarithmic.

of conditions under which they are tested. These patterns can be readily described in terms of numbers, ratios and rates. The traditionally elusive subject of cell-bonding can now be reduced to laboratory prose. Moreover, since the patterns are reliably repeatable and sensitive to changes in conditions, they serve as useful base lines for the bioassay of the effects under study in a given experiment.

We soon found that aggregation patterns vary with different types and mixtures of cell. Under otherwise identical conditions, different kinds of cell "crystallize" into distinct and characteristic aggregates. Some kinds of cell consistently form a single mass; others produce numerous clusters of predictable shapes and sizes. Remarkably, those patterns that showed themselves to be characteristic of particular kinds of tissue proved to be similar for cells from different species. Whether from mouse or chick embryo, cells of the same tissue aggregate into very similar patterns. Their collective reactions seem to be guided by signals legible to both species.

For cells in general we also soon found that certain factors operate with uniform effect. The relationship of the embryonic age of the cells to their capacity for aggregation proved to be particularly striking. Under otherwise equivalent conditions, cells dissociated from tissues of older embryos are less cohesive than their counterparts from younger embryos. For each kind of cell, aggregation patterns provide a characteristic age profile. With increasing age in the donor embryo, the dissociated cells produce smaller and more numerous aggregates and eventually fail to aggregate. Cells dissociated from adult animals usually do not recohere at all.

The precise meaning of this effect of aging is not clear. There are grounds for believing that it reflects the loss by the cells of their ability to produce either the right kind or the right quantity of cell-linking substances. It may be that, as cells mature and acquire specialized functions within their stabilized associations, their metabolic machinery is gradually switched over from those processes that manufacture cell-linking materials to more pressing activities. As a result, when they have been isolated and denuded of their coatings, such cells can no longer recohere effectively. In contrast, embryonically young cells exhibit the capacity to manufacture those materials and to recohere.

If the recohesion of cells does depend

on metabolic processes, then it should be possible to inhibit it simply by lowering the temperature at which the experiment is performed, because metabolic processes are known to be dependent on temperature. This has proved to be the case. Cells that aggregate readily at the usual body temperature of 38 degrees centigrade cohere less effectively at lower temperatures; they remain separate indefinitely at 15 degrees C., even when brought together by rotation. Transferred back to 38 degrees C. after two or three days, such cooled cells aggregate well.

We do not know which of the many temperature-dependent metabolic activities that are depressed by cooling are involved in the production of cell-binding materials. But the answer, we are confident, is only a matter of time. The important point is that the problematical issue of cell-bonding can now be approached by means of concrete tests and experiments. Given the right temperature and otherwise favorable conditions, cells of suitable embryonic age construct tissues of the kind from which they have come. Aggregated liver cells make liver lobules; kidney cells reconstitute kidney tubules and corpuscles; intestinal cells produce digestive tissue; skeletal cells, cartilage and bone; retinal cells, sensory epithelium; heart cells, lumps of beating heart tissue; and so on. Although they are arbitrarily bunched by rotation, the cells rapidly organize orderly fabrics in the pattern of their original tissue. Like parts of an animated jigsaw puzzle, they re-establish a new whole in accordance with the original blueprint. At the Rockefeller Institute, Paul Weiss and Cecil A. Taylor recently grafted such aggregated cells back to embryos; the lumps became joined to the circulatory system of the embryo and developed into remarkable facsimiles of their original organs.

As in experiments with stationary cultures, mixtures of cells in rotating flasks sort themselves out by cell type. One can, for example, readily coaggregate intermingled skeletal and kidney cells. At first the cells are lumped by the spinning liquid into chaotic conglomerates, but soon they segregate by kind—skeletal cells congregating in the middle as nodules of cartilage, kidney cells lining up on the surface. Throughout these cellular maneuvers the aggregates maintain their over-all configuration. The situation obviously resembles the case of the everted hydra or of the embryo that retains its over-all configuration in spite of the extensive movement of its con-

stituent cells.

As might be expected, the final patterning of such composite aggregates reflects their cellular composition. Depending on the nature of their partners in the common aggregate, cells of the same kind may settle inside or outside.

By testing various combinations of cells one discovers a kind of hierarchical order—a "who goes where" in aggregations of various kinds. Preference as to site and competition for physiological need obviously play an important role in the patterning of aggregates. The differen-

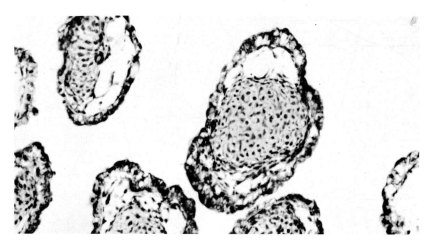

CHARACTERISTIC COMPOSITE AGGREGATIONS form when two different kinds of cell are mixed together in a rotating flask. In the resulting aggregations the cells sort out according to kind. Shown here are sections through such organized aggregations. They are composed of cartilage-forming cells surrounded by kidney cells taken from chick embryos.

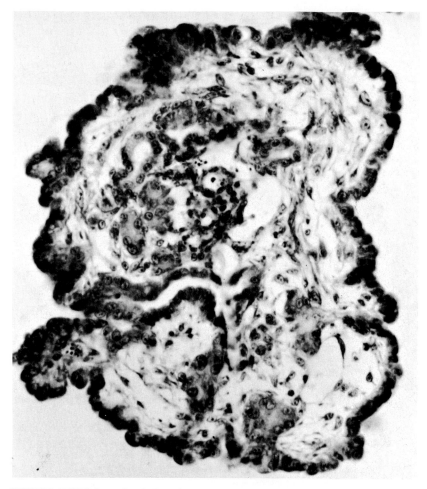

KIDNEY CELLS from the chick embryo form a complex organ-like aggregation after 24 hours of rotation in a flask. This is a highly enlarged section of such a kidney-cell aggregation.

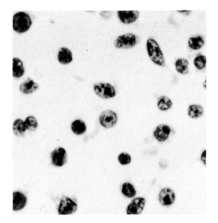
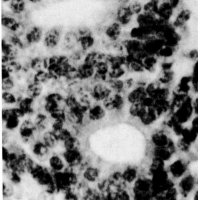
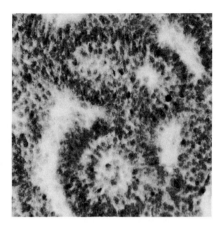

RETINA CELLS from 7-day chick embryo are already differentiated into types that will make up retina and its nerves. At left are stained dissociated cells. Center is section through aggregation formed in 24-hour rotation. 56-hour aggregation shows advanced reconstruction.

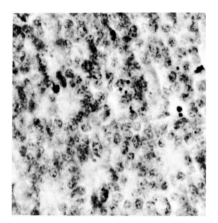
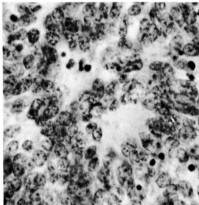
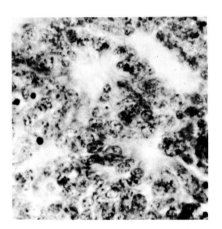

CELLS OF TWO SPECIES will form aggregations. These are sections through aggregations of chick-embryo retina cells (left), mouse-embryo retina cells (center) and mixed cells of both (right). Coming from the same kind of tissue, cells form a common fabric.

tial diffusion and availability of various constituents of the medium, of oxygen and carbon dioxide, also contribute to the outcome. But the patterning of aggregates also reflects the ability of the cells to "recognize" each other, to discriminate between self and nonself, to sort out and to associate in accordance with functional kinships.

One of the more remarkable aspects of such communication-by-contact in embryonic cells is that the signals characteristic for cells of a given tissue are not unique to a given species. One can coaggregate cells from mouse and chick embryos, either from different or from similar tissues. The cells from the dissimilar tissues aggregate separately, as might be expected. But cells of similar kind co-operate in the construction of chimeric fabrics, incorporating the cell of both species. Coaggregated kidney cells from the two species produce tubules of mouse and chicken cells. Liver, cartilage, retina and other cells likewise join in the formation of bi-specific tissues. The means by which these cells recognize each other and become effectively linked into tissues evidently transcend differences between species.

It occurred to us that one might learn more about communication among cells by trying to interfere with its specificity and effectiveness. We found that when dissociated cells are maintained in the dispersed state for some time (the time required varies from days to weeks, depending on the kind of cell and the conditions in which they are kept), they lose two significant capacities progressively and concurrently. Their cohesiveness decreases, and their precision in distinguishing between self and nonself in the organization of tissue drops markedly. These time-related changes raised the question of whether they are not also causally related; that is, whether the materials on the surfaces of cells and between them that link them together might not also play a key role in their interaction and communication.

It seemed not unlikely that these materials are bound to change in response to the novel conditions to which the cells are exposed when they are maintained in the dispersed state. The specificity of the materials could thereby become inactivated or blunted, and this would impair the ability of the cells to interact effectively. These were thin speculations, but we decided to test them by trying to reactivate modified cells by coaggregating them with freshly dissociated cells of their own kind. The fresh cells would presumably be effective producers of the cell-surface materials. When coaggregated by rotation with freshly obtained cells of the same kind, the modified cells did recover their ability to construct tissues. But when coaggregated with fresh cells of a different kind, the modified cells were largely left out of the aggregates. If they were included, they formed no clear structures.

Such findings lend themselves to different interpretations. Until we know more about the whole problem, they

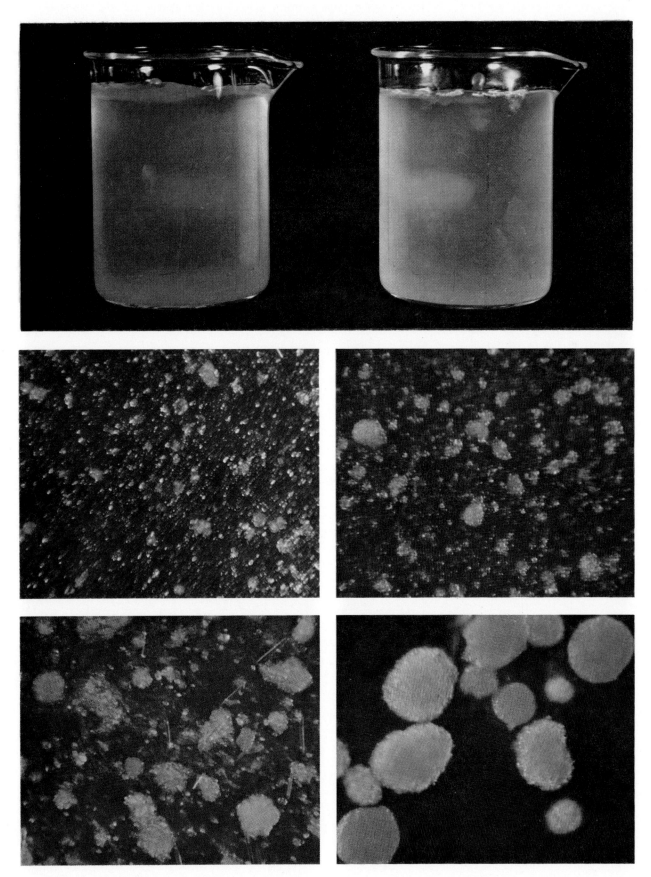

CELLULAR AGGREGATION is demonstrated in author's laboratory at the Marine Biological Laboratory in Woods Hole, Mass., using naturally orange sponge *Microciona* and yellow sponge *Cliona*. Solutions containing cells from each are in beakers at top. Cells are mixed together (*photomicrograph at middle left*). In the course of 12 hours they creep along the floor of the dish in which they have been placed and clump together by species, finally forming tiny orange sponges and yellow sponges (*bottom right*).

are at best suggestive. As such they focus attention on the possible role of cell-surface and intercellular materials in communication among cells and in their developmental association. Wherever such materials could be adequately examined they have been found to contain protein-bound carbohydrates. This fact is of considerable interest since it would seem to place them in the same chemical family with certain other cell products that have highly specific functions: the substances that determine blood groups, that compose antibodies, that are involved in the mating of microorganisms and that influence the selective susceptibility of tissue cells and bacteria to viruses. Could it be that the chemically similar materials that bind cells together also equip them with the means for mutual recognition and specific association? If so, then these cell-binding materials would be mortars of an extraordinary kind. Produced by the bricks themselves, they would serve also to co-ordinate the construction of the tissue, the organ and the organism.

The Growth of Nerve Circuits

by R. W. Sperry
November 1959

*Recent studies of the process of nerve repair have led
to a new theory of how the complex networks and
pathways of the central nervous system are formed in
the embryo*

Severe damage to the principal motor nerve of the face may leave a person afflicted with a condition known as "crocodile tears." As the injured nerve regenerates, fibers that originally activated a salivary gland can go astray and connect themselves to the lachrymal gland of one eye. Thereafter every situation calling for salivation induces weeping from that eye. Often the regenerating salivary fibers invade sweat glands and related organs in the skin, causing profuse sweating and flushing in areas of the face and temple. The random shuffling of motor-nerve connections to the muscles of the face characteristically deranges facial expression, causing a grimace-like contraction of the affected side. Sometimes, to prevent atrophy of the facial muscles when the injured facial nerve fails to regenerate, surgeons will connect the denervated facial muscles to a nearby healthy nerve: the motor nerve of the tongue or the motor nerve of the shoulder muscle. The restored facial movements still lack meaningful expression and tend to be associated with the chewing movements of the tongue or the action of the shoulder muscles.

Naturally the primary concern of the patient in such cases is whether or not normal function can be restored. If the symptoms do not clear up spontaneously, can they be corrected by training and re-education? By faithful practice in front of a mirror, for example, can a patient learn to inhibit the crocodile tears and regain control of facial expression?

Not so long ago the reply to such questions was a confident "Yes." For most of the present century investigators and physicians were agreed that the central nervous system is plastic enough so that any muscle nerve might be reconnected to any other muscle with good function-

al success. Sensory-nerve fibers were thought to be equally interchangeable within a given sensory system. It was believed that the central pathways in the nervous system were first laid down in the embryo in randomized equipotential networks. By use and learning these pathways became channelized; connections that proved adaptive in function were reinforced, while the nonadaptive ones underwent "disuse atrophy." Learning thus determined not only the function but also the structure of the nervous system. This theoretical picture was sustained by experiments on animals in which, according to the literature, the crossing of major nerve-trunks was followed by full restoration of function. In the prevailing mood of optimism physicians were able to report encouraging progress by their patients.

During the past 15 years, however, scientific and medical opinion has undergone a major shift, amounting to an almost complete about-face. No longer do physicians encourage the patient with a regenerated facial nerve to try to regain control of facial expression by training; their advice today is to inhibit all expression, to practice a "poker face" in order to make the two sides of the face match in appearance. The outlook is equally dim for restoration of coordination in cases of severe nerve injury in other parts of the body.

This changed viewpoint reflects a revision in the picture of the entire nervous system. According to the new picture, the connections necessary for normal coordination arise in embryonic development according to a biochemically determined plan that precisely connects the various nerve endings in the body to their corresponding points in the nerve centers of the brain and spinal cord. Although the higher cen-

ters in the brain are capable of extensive learning, the lower centers in the brain stem and spinal cord are quite implastic. Because their function is dictated by their structure, it cannot be significantly modified by use or learning. Nor can the disordered connections set up by the random regeneration of injured nerves be corrected by re-education.

The evidence for this view, which comes from new experiments and from exacting clinical observations, is so persuasive that it is difficult to understand how the opposite view could have prevailed so long. It appears that most of the earlier reports of the high functional plasticity of the nervous system will go down in the record as unfortunate examples of how an erroneous medical or scientific opinion, once implanted, can snowball until it biases experimental observations and crushes dissenting interpretations.

Hundreds of experiments seemed to support the now-discounted opinion. One of the experiments most frequently cited was first reported in 1912 and was repeated with concurring interpretation as recently as 1941. In several monkeys opposing pairs of eye muscles were interchanged to reverse the movement of one eye. Upon recovery from surgery, the movements of the abnormally connected eye were said to coordinate with those of the normal one. In another oft-repeated experiment the nerves that control the lifting of the foot were crossed. Instead of a reversal of foot action, the animals showed recovery of muscle coordination hardly distinguishable from that on the normal side. Even when nerves from the forelimb were cross-connected to nerves of the hindlimb, the animals appeared to make complete functional readjustment. With such results in the literature there seemed to be few limits to the restorative possibilities of peripheral nerve surgery.

The doctrine of the functional plasticity of the nervous system was sharply challenged in 1938 by Frank R. Ford and Barnes Woodhall of the Johns Hopkins School of Medicine. In an account of their clinical experience with functional disorders following the regeneration of nerves, they declared that these disorders persisted stubbornly in many of their patients for years without improvement. Their report cast serious doubt upon the accepted methods of therapy and the theory that rationalized them.

That same year I began a series of experiments in the laboratory of Paul Weiss at the University of Chicago. The initial aim of this investigation was to find out if functional plasticity was a property of the higher brain-centers only or whether it extended to the lowest levels of the spinal cord as alleged in some earlier reports. To explore the question I started to experiment on the simple reflexes involved in coordinating the foot movements of the rat. These reflexes depend upon a relatively simple circuit called a reflex arc. The fibers that activate the muscle connect to an association neuron in the spinal-reflex center; the association neuron is connected in turn to a particular type of sensory cell, the proprioceptive neuron, the terminal fibers of which are embedded in the muscle. This circuit, with the sensory nerve indicating the state of contraction or relaxation of the muscle and its orientation in space, provides the feedback necessary for proper muscle timing and coordination.

I switched the nerve connections between opposing muscles in the hindlimb of rats in such a way as to reverse the movement at the ankle joint. The nerves were cut, crossed and reunited end-to-end within tubes of dissected rat artery. Now whenever a nerve is severed, all fibers beyond the break degenerate and are absorbed. Even when they are united with the mechanical aid of an arterial tube, cut nerves do not heal together directly. New fibers sprout from the end of the central stump and grow into the muscles within the degenerate framework of the old nerve. After the surgery, I assumed, new coordinating circuits would be established and functional adjustment would follow quickly.

Much to my surprise the anticipated adjustment never occurred. The rats seemed unable to correct the reversals of motor coordination produced by the operation. When they tried to lift the

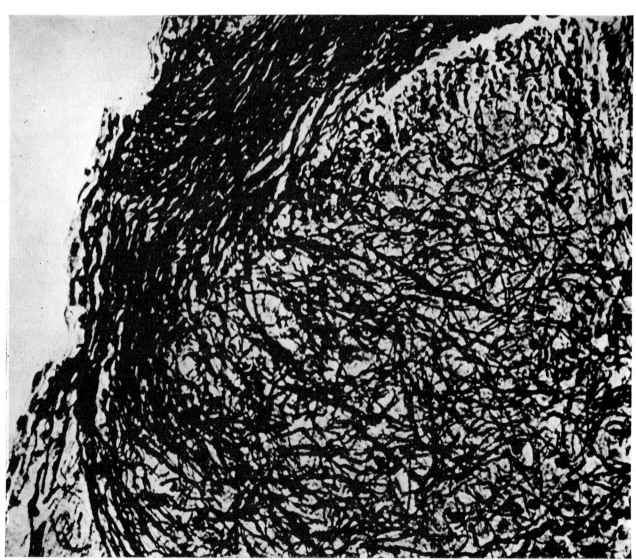

POST-MORTEM VIEW of the frog's crossed sensory root (magnified 500 diameters) shows that the cut nerve-fibers have regenerated into the spinal cord. Despite their tangled appearance, the fibers have formed the connections necessary for proper muscle timing and coordination. Because the nerves have been transposed, however, frogs display abnormal reflexes shown on following page.

affected foot, it pulled downward; when they tried to rise on the ball of the foot, their toes swung up and they fell back on their heels [*see illustration on next page*].

In a parallel experiment I presented the rats with a simpler readjustment problem by switching muscles instead of nerves. The muscles involved were transposed by cutting and crossing their tendons. Strong muscle-action was restored within two or three weeks. Yet the rats were still unable to correct the reversal of ankle movement. To check the experiment I tried crossing both muscles and nerves in a control group. Here the two reversals mutually canceled their effects, and the rat was able to raise and lower the foot in proper timing.

The rats with reversed foot-movements were put on a program of special training: they were forced to climb ladders and stretch upward on their hindlegs many times a day to get food pellets from automatic feeders. Yet the affected feet continued to work backward in machine-like fashion. We carried out a similar series of experiments on the forelimb, in which voluntary movements are under better control. But the rats still could not adapt to the rearrangement of the nerve connections.

When it became clear that re-education had little or no effect in the rat's motor system, we turned to the sensory system. In the laboratory of Karl S. Lashley at Harvard University I transposed the nerves connecting to the skin of the left and right hindfeet [*see illustration at right*]. As the crossed hindlimb nerves regenerated, the rats began to exhibit false reference of sensations. In response to a mild electric shock to the sole of the right foot, the animals withdrew their left foot. This movement shifted their weight to the right foot, thereby increasing its contact with the offending electrode.

During the course of these experiments several rats developed a persistent sore on the sole of the right foot. Until the sores responded to medication, the animals hopped about on three feet with the fourth raised protectively—but it was always the wrong, uninjured foot that they raised. When they were prompted to lick the injury, they repeatedly licked the uninjured foot. Although this accidental soreness in the re-innervated foot presented the best kind of training situation, the rats still were unable to readapt.

Having concluded that it would be

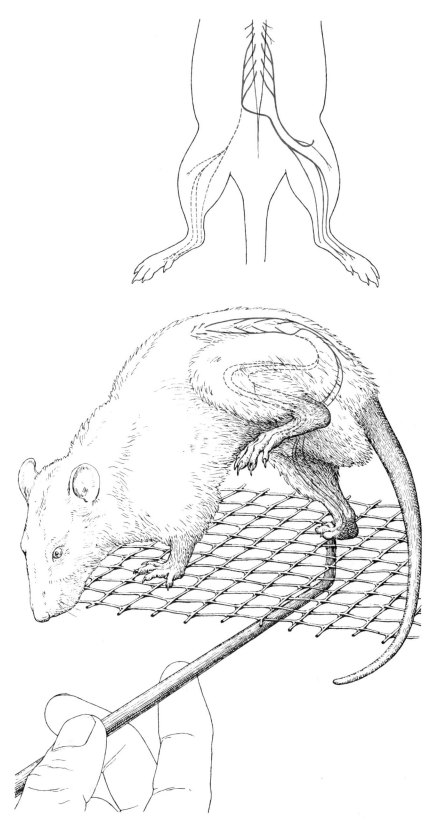

INCORRECT WITHDRAWAL REFLEX and referred sensations appear in the rat after two sensory nerves are crossed. Here the main trunk-nerve to the left foot (*broken line*) has been crossed and connected to the corresponding nerve on the right side (*colored solid line*). Afterward, when sole of right foot is stimulated electrically, rat withdraws its left foot; if the shock is strong enough to produce soreness, rat licks its **uninjured foot.**

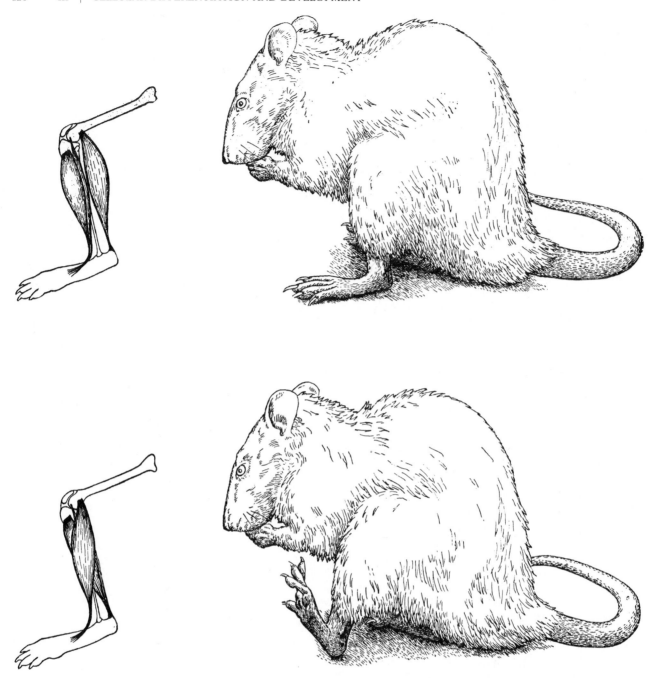

CROSSED MUSCLES produce the same effects as crossed nerves. In a normal rat (*top right*) the leg muscles are connected as shown at top left. When these connections are transposed (*bottom left*), the posture and movements of the ankle joint are reversed (*bottom right*). When the rat attempts to lift the affected foot, it pulls downward; when the animal tries to rise on the ball of the foot, the toes swing up and it falls back on its heels. Even with prolonged training, the rat can never learn to correct these movements.

difficult or impossible to demonstrate any sort of plasticity in the rat, we began a similar experiment on monkeys at the Yerkes Laboratories of Primate Biology in Orange Park, Fla. At first the monkeys seemed to make more progress than the rats: After we had transposed the nerves of the biceps and triceps muscles of the upper arm, they were quick to notice and to halt the reversed arm-movements that began to appear as the nerves regenerated. Thereafter the mon-

keys did attain a minor degree of readjustment in arm movement under the most simplified training routine. But after three years of testing and observation, they too failed to achieve any generalized positive correction in the action of the cross-innervated muscles.

The marked conflict between our results and those previously reported prompted us to reproduce the procedures of the earlier studies more closely.

Instead of crossing isolated branch-nerves to single muscles I now crossed large trunk-nerves carrying motor impulses to many different muscles. All the muscles in the region were left intact and the nerves were permitted to regenerate into their respective areas at random.

In the hindlimb of the rat this operation produced neither a reversal of movement nor good functional readjustment; instead it caused a spastic contraction of all the muscles of the lower leg. Because of the greater strength of the postural or antigravity muscles, the contraction produced a stiltlike stiffening of the ankle joint in the extended position. This result was highly illuminating. Although the extended leg-posture was clearly abnormal in the rat, it could very easily be mistaken for a return of normal coordination in an animal that walks on its toes, like a dog or a cat. The fact that dogs and cats had been used in most of the earlier investigations made it apparent that nearly all of the hundreds of earlier reports of good functional recovery were subject to reinterpretation.

We know that central nervous system plasticity can be substantiated to some extent in cases where muscles have been transplanted in human patients. But even here we must make a distinction between the degree of plasticity demonstrated after muscle transposition and that shown after nerve regeneration. When a muscle is transplanted with its nerves intact, the motor cells that activate it continue to work together as a unit. On the other hand, in human beings and higher animals the fibers of a regenerating motor-nerve become haphazardly redistributed among the muscles it previously supplied. To restore these muscles to their previous control and coordination the reflex connections in the spinal cord would have to be re-established down to the level of individual nerve cells. Even man's superb nervous system does not possess this degree of plasticity.

In most cases humans can learn to control transplanted muscles only in simple, slow, voluntary movements. The control of complex, rapid and reflex movements is limited at best, and is subject to relapse under conditions of fatigue, shock or surprise. Humans seem to have a much greater capacity for adjustment than the subhuman primates. Such re-education as does occur must therefore be due to the greater development of higher learning-centers in the human brain. Contrary to earlier supposition it does not reflect an intrinsic plasticity of nerve networks in general.

The nervous systems of reptiles, birds, and mammals other than primates show even less functional plasticity than those of primates. Farther down the evolutionary scale in the lower vertebrates, however, we find an entirely different type of neural plasticity: a structural plasticity not possessed by higher animals. Fishes, frogs and salamanders can regenerate any part of the central nervous system—even the tissue of the brain itself. Furthermore, if one of the large motor nerves of a salamander or a fish is cut, the animals re-establish normal reflex-arcs and recover coordination. This occurs even if several nerve stumps in a limb or fin are deliberately cross-connected to produce gross abnormalities in the distribution of the regenerating fibers.

In his pioneering investigations during the 1920's Paul Weiss was able to rule out the possibility that these spectacular recoveries could be based on learning or any other sort of functional plasticity. He transposed the developing forelimb buds of salamander embryos and reimplanted them with their front-to-back axes reversed. When function later appeared, the motor coordination in the transplanted limbs was perfectly normal, indicating that normal reflex-arcs had been established. Because the forelimbs were reversed in orientation, however, they pushed the animal backward when it tried to go forward, and vice versa. The perfectly coordinated reversed action persisted indefinitely without correction.

I was able to confirm the absence of any appreciable functional plasticity in amphibians in a set of experiments on the visual system of frogs and salamanders. In some I inverted the eyeballs surgically, producing upside-down vision; in others I cross-connected the eyes to the wrong sides of the brain, producing vision that was reversed from side to side. The animals never learned to correct the erroneous responses caused by this surgical rearrangement of their eyes ["The Eye and the Brain," by R. W. Sperry; SCIENTIFIC AMERICAN Offprint 1090]. Even when the eyes of frog and salamander embryos were rotated prior to the onset of vision (in later experiments by L. S. Stone at Yale University and George Szekely in Hungary), the same visual disorientation developed and persisted throughout life.

The evidence at present thus indicates that the structural plasticity observed in lower vertebrates is inherent in the growth process and is quite independent of function. It is as if the forces of embryonic development that laid down the circuits in the beginning continue to operate in regeneration. We have as yet only preliminary insight into the nature of these forces.

In studies now in progress at the California Institute of Technology Harbans Arora has made an interesting observation on the regeneration of the nerve controlling the eye muscle in fishes. His findings suggest that fibers directed by chance to their own muscles make connections more readily than foreign fibers reaching the same muscle. As a result the fibers that originally con-

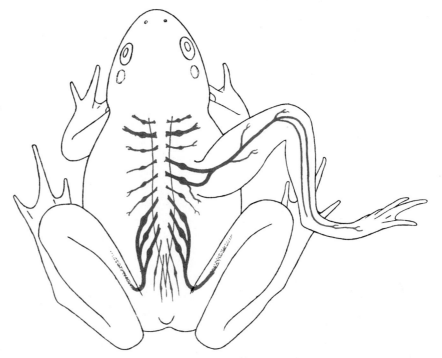

ABNORMAL REFLEX ARCS develop when a hindlimb bud is grafted onto the back of a tadpole. As the tadpole grows, sensory nerves destined for the skin of the back, flank and belly invade the nerveless extra limb and form spinal connections appropriate for limb reflexes. When grafted limb is stimulated in adult frog, muscles of the right hindlimb respond.

trolled the muscle tend to recapture control in regeneration. Such selective reaffiliation of nerve and muscle indicates that some chemical specificity must match one to the other.

Selective outgrowth of regenerating nerves to their proper end-organs seems not to be the rule, however, even in lower forms. Among mammals it has not been found at all, except on the much more gross scale that differentiates sensory from motor endings, smooth muscle from striated muscle, muscle from gland, and so on. Nor does simple selective outgrowth account for the restoration of function in salamanders. The early studies by Weiss showed that fiber outgrowth and muscle re-innervation generally proceed in these animals in a random, nonselective manner, comparable to that in mammals. Upon re-innervation, however, salamander muscles regain their former coordination and timing, even when their function is disoriented by nerve-crossing.

These observations suggest that the rearrangement of connections in the periphery of the salamander nervous system has chemical repercussions that result in a compensatory shift of reflex relations at the centers. It is postulated that the motor-nerve cells regenerating into new muscles take on a new chemical flavor, as it were. Thereupon their old central associations dissolve, and new ones form to match the new terminals in the periphery. The reflex circuit would thus be restored to its original state, with the peripheral and central terminals linked by a new pathway. Higher animals, lacking this embryonic type of

structural plasticity, show no restoration of function.

This explanation at first seemed rather far-fetched, especially from the standpoint of electrophysiology, which offers no evidence for such qualitative specificity among nerve fibers. However, the underlying idea is well supported by recent experiments on the regeneration of sensory nerves.

At the University of Chicago Nancy M. Miner, one of my former associates, is responsible for a significant series of experiments indicating the role of some sort of chemical specificity in the hookup of the nervous system. She grafted extra hindlimb buds onto the backs of tadpoles; the buds became connected to the sensory fibers that would normally innervate the skin of the belly, flank and back [see illustration on preceding page]. The grafted leg served only as a sensory field for the nearby sensory nerves because there are no nearby limb nerves to invade it. When a stimulus was applied to the grafted limb in the mature frog, the animal moved the normal hindlimb on the same side, just as it would if the normal limb had received the stimulus. The belly and trunk nerves connected to the grafted limb had evidently taken on a hindlimb "flavor" and then formed the appropriate reflex connections in the central nervous system. In another experiment Miner removed a strip of skin from the trunk of a tadpole, cut its nerves and replaced it so that the skin of the back now covered the belly, and vice versa [see illustration below]. When the grown frogs were stimulated in the grafted area of the back, they responded

by wiping at the belly with the forelimb; when they were stimulated in the grafted area on the belly, they wiped at the back with the hindlimb.

To account for these experimental findings it is necessary to conclude that the sensory fibers that made connections to the grafted tissues must have been modified by the character of these tissues. It is therefore unnecessary to postulate that each nerve fiber in embryonic development makes some predestined contact with a particular terminal point in the skin. Growing freely into the nearest area not yet innervated, the fibers establish their peripheral terminals at random. Thereafter they must proceed to form central hookups appropriate for the particular kind of skin to which they have become attached. It seems clearly to be some quality in the skin at the outer end of the circuit that determines the pattern of the reflex connections established at the center.

No attraction from a distance need be invoked in this selective patterning of the central hookup. The multiple branches of each nerve fiber undergo extensive ramification among the central nerve cells, with the tips of the branches making numerous contacts with all the cells in the vicinity. Presumably most of the contacts do not affect the growing fiber-tips. It is only when contact is made with central nerve cells which have the appropriate chemical specificity that the growing fiber adheres and forms the specialized synaptic ending capable of transmitting the nerve impulse.

In man these observations and interpretations provide the basis for the new

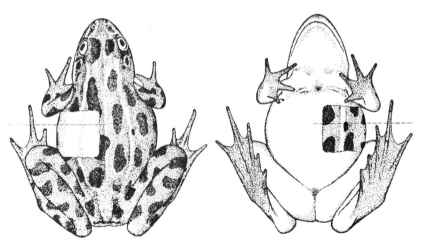
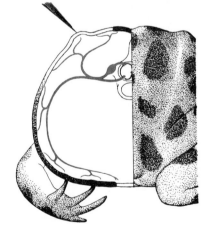

ROTATED PATCH OF SKIN demonstrates how embryonic nerves respond to the biochemical "flavor" of the tissues they innervate. Here a strip of skin was removed from a tadpole, cut free of all connections, and replaced so that the skin of the back now covered the belly and vice versa. In the grown frog the skin retained its original color and flavor despite its location, as shown in the dorsal and ventral views (*left and center, respectively*). The cutaway view at right (*made along broken lines*) shows how new nerves have invaded the graft and formed spinal reflex-arcs appropriate to the skin's flavor rather than to its location. Thus when the belly skin on the back is stimulated, the frog wipes at its belly; when the back graft on the belly is stimulated, the frog wipes at its back.

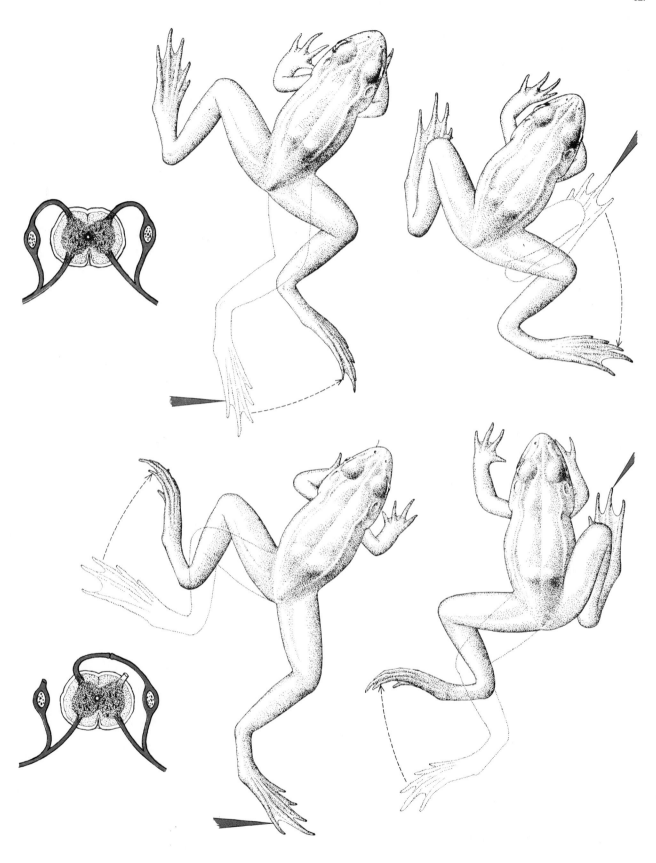

CROSSED SENSORY NERVES produce incorrect postural reflexes in the frog. The normal connections shown in the cross section of the spinal cord at top left cause the frog to withdraw an extended leg (*top center*) and to extend a flexed leg (*top right*) when a stimulus (*colored pointer*) is applied. If the sensory root entering the right side of the spinal cord is cut and surgically attached to the left side, as shown in the cross section at bottom left, the reactions of the left leg will be determined by the posture of the right.

view of the nervous system which holds that its networks are determined by biochemical processes in the course of embryonic growth. Let us consider this scheme in connection with the extreme localization of skin sensations that makes it possible to locate a pinprick, for example, anywhere on the body surface. This "local sign" quality depends upon the precise matching between the central and peripheral connection of each one of thousands upon thousands of cutaneous nerve fibers connecting the skin surfaces to the spinal cord and brain. During embryonic growth we may assume that the skin undergoes a highly refined differentiation until each spot on the skin acquires a unique chemical make-up. A mosaic is not envisaged here but rather smooth gradients of differentiation extending from front to back and from top to bottom, with local elaborations of these basic gradients in the regions of the limbs. Each skin locus becomes distinguished by a given latitude and longitude, so to speak, expressed in the tissues as a combination of biochemical properties. The cutaneous nerves, as they grow out from their central ganglia, may terminate largely at random in their respective local areas. Through intimate terminal contacts the specific local flavor is imparted to each nerve fiber. This specificity is then transmitted along the fiber to all parts of the nerve cell including its ramification within the central nervous system. In this way the local-sign properties of the skin become stamped secondarily upon the cutaneous nerves and are carried into the sensory centers of the brain and spinal cord. Precise localization is further enhanced by the overlapping of the terminal connections formed by the fibers in the skin [*see illustration at right*].

Implicit in this theory is the assumption that in the embryo the cerebral cortex and the lower relay-centers also undergo a differentiation that parallels in miniature that of the body surface. In other words, just as from the skin to the first central connection point in the spinal cord, so from relay to relay and finally to the cortex the central linkages arise on the basis of selective chemical affinities. At each of its ascending levels the nervous system forms a maplike projection of the body surface.

This mechanism presumably operates not only in the sensory system but in the nervous system in general. Since the organization of the lower nerve centers and the peripheral nerve circuits in higher animals seems to take place only in the

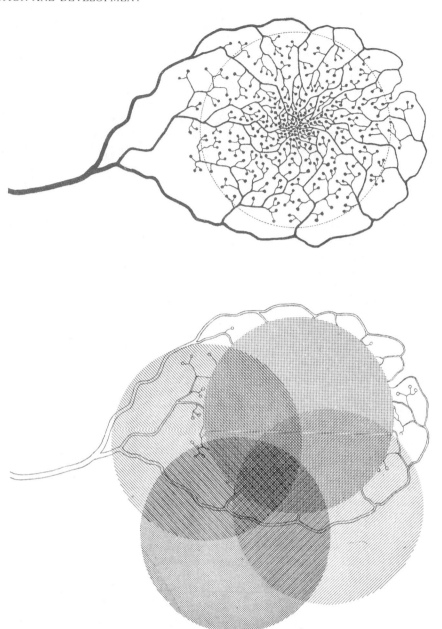

OVERLAP OF SENSORY FIBERS permits a subject to localize a pinprick accurately. The schematic diagram at top shows the terminal branches of a single cutaneous fiber; the branches are most abundant at the center of the area they innervate (*broken circle*). Bottom diagram shows how this area overlaps those of three other fibers. As shading indicates, each of the areas transmits a recognizably different signal to the central nervous system.

early plastic stages of growth, it is clear that injury to them later on cannot be repaired by any amount of re-education and training. In the structural plasticity of the system in lower animals, however, we are able to observe the processes by which our own nervous systems develop.

The new approach provides a sound biological basis for the explanation of built-in behavior mechanisms generally, from the simplest reflexes to the most complicated patterns of inherited behavior. It brings the study of behavior into the realm of experimental embryology on the same basis as other organs and organ

systems. Although learning must now yield its former monopolistic status, we must not infer that it has no role in brain development. Particularly in man, whose brain grows and matures for many years, learning is a powerful method of imposing additional organization on the higher levels of the nervous system. Until the neural basis of learning is discovered, however, we cannot say whether it produces this added organization by changing the actual layout and hookups of cerebral networks or simply by increasing the conductance of certain pre-established pathways.

IV

THE CELLULAR BASIS
OF INTEGRATION

THE CELLULAR BASIS OF INTEGRATION IV

INTRODUCTION

The outcome of the processes of differentiation described in the previous section is an assemblage of cells, each member of which has, to some degree, a special activity. For some, the activity may be a synthetic task—the manufacture of hemoglobin by a red blood cell, of keratin by an epidermal cell, or of yolk by the cells surrounding a developing egg. For other cells, the specialty may entail development of a certain *property* not shared by all cells. No cell types are more diagnostic of the basic character of complex animals than those composing nerves, muscles, and sense organs. The integration of all other systems depends upon their activities; not only are they collectively responsible for the ordered movement of the animal in its environment, but they also preside over a host of routine internal coordinations.

The first article in this section, Bernhard Katz's "How Cells Communicate" (1961), is a general introduction to the special properties of excitable membranes, some of which will receive more detailed scrutiny in the later articles by Baker and Eccles. There is good reason for beginning our account of the process of integration with an overview of "bioelectricity." Most biologists would probably agree that the central nervous systems of animals represent the highest achievement of multicellular organization. The brains of some mammals contain more than a billion cells, linked to one another in an astonishingly intricate and precise way. Each cell is capable of translating input information, which arrives in the form of chemical transmitter substances from one or many nerve cells connecting with it, into a patterned sequence of electrical impulses. These propagate without signal loss to another region, often many centimeters away, and there a cell releases the chemical that will excite another nerve cell, a muscle fiber, or perhaps a gland.

The physiological performance of this remarkable system has intrigued biologists since the days of Galvani and Volta, when knowledge of the nature of "animal electricity" was important in the understanding of purely physical systems for generating currents and voltages. Contemporary investigations have sought an understanding of nervous function at two levels; the aims and strategies of these two approaches are so different that they are really almost separate disciplines.

The first of these is fundamentally cellular; it seeks to comprehend the cellular processes that underlie electrical events in nerves, muscles, and sense organs. The second is more holistic: it focuses on the connections between nerve cells, and attempts to discover what kinds of neuronal circuitry are responsible for particular behavioral events. This latter approach will be exemplified by many of the articles in Sections VI and VII. The section you are about to begin, however, deals with the *units* that make up integrative systems.

Bernhard Katz is one of the half-dozen great figures in contemporary neuro-biology, and he has made a number of contributions to our understanding of nervous conduction and of transmission at synapses. "How Cells Communicate," like several of the other articles that begin Sections of this book, was written for the September, 1961 special issue of *Scientific American*, devoted entirely to the topic "The Living Cell." In his article Katz discusses the great successes of the 1950's in comprehending the way in which impulses are carried along the long fibers. We now know that the propagating impulse results from sudden and selective changes in the permeability of the nerve membrane, which allow first sodium ions and then potassium ions to pass. Because the concentrations of these two ions are quite different from one another in the axon and in the fluid outside it, the first change in permeability produces an inward sodium current across the membrane, temporarily reversing the normal, positive, resting potential outside the axon. The second change allows potassium to flow outward to re-establish the original resting condition. Other studies have provided a description of the events that ensue when this travel-ing sequence of changes in potential reaches the end of the nerve fiber and causes the release of a small amount of transmitter chemical. The period of the 1960's—the decade following the first publication of Katz's article—was marked by much slower progress, and a reorientation of research efforts. Current studies are aimed at understanding *how* structural changes in the fabric of the membrane allow sodium and potassium to pass with such sudden freedom, or permit the release of transmitter chemicals from nerve terminals.

The basic property of *movement* is an integrative function of cell and organ-ism that shares top billing with electrical excitability. For many decades, we have tried to understand cellular mechanics in terms of the contractile properties of muscle—the specialized cell type that exhibits cellular mechanics in most spectacular form. But in fact a host of other processes depend equally upon the performance of mechanical work by the cell. Consider, for example, the movement of the chromosomes to the two poles of the cell during a mitotic division; the formation of a budlike outgrowth from a simple epithelium at a critical point in the development of an organ; the maneuvering of secretory vesicles beneath the membrane of a cell in preparation for the release of secre-tions; the agitated comings and goings of glial cells on the bottom of a tissue-culture vessel. All these are examples of a class of cell movements more "primitive" than those involved in muscular contraction, though they may share parts of the same biochemical machinery. In "How Living Cells Change Shape" (1971), Norman K. Wessells discusses experiments in this less-sophis-ticated kind of cellular mechanics. Fine filaments frequently form a meshwork underneath the surface membranes of mechanically active parts of cells. Among the cellular events that can be attributed to the function of such fila-ments are growth of nerve cell processes, the movement of most vertebrate cells, the formation of outpocketings in embryonic tissue layers, the "stream-ing" of the cytoplasm in plant cells, and a wide range of other events.

The area discussed by Wessells is developing fast, and new findings appear almost weekly. Two aspects of the problem are of particular current interest: first, the nature of the transition between the mechanically inactive and the active state of a particular region; and second, the degree to which this "primitive" system for movement resembles (or differs from) the highly or-dered contractile systems of fully differentiated muscle cells. With regard to the first problem, it has long been known that many kinds of motile cells in tissue culture "withdraw" when the membrane at their advancing edge touches the surface of another cell. This process, known as *contact inhibition*, can occur very quickly (see "The Induction of Cancer by Viruses," by R. Dulbecco, *Scientific American* Offprint 1069). Recent work, done since the Wessells article was written, shows that thin filaments can appear within twenty seconds of such contact at the surface of a cell, and that region of the cell can then be withdrawn from contact with the other cell. Obviously a mechanism exists whereby these organelles can be rapidly assembled from components that exist in solution; equally fast means for their dissolution must be available.

As to the similarity between the various contractive systems, the evidence is still incomplete. The article by Huxley, "The Mechanism of Muscular Contraction" (1965), is an excellent summary of our current understanding of the contractile mechanism in striated muscle—an understanding to which Huxley himself has been perhaps the main contributor. Studies on the filamentous structure of such muscle cells, carried out by electron microscopy, X-ray diffraction, and other biophysical techniques, as well as by selective chemical extractions, have given us a fairly complete picture of the composition of muscle components. The thick filaments are made of the protein *myosin*, one of whose components (heavy meromyosin) is able to bind reversibly to *actin*, a second protein of which the thin filaments are composed. Excitation of the muscle causes the entry of calcium, or its release from internal stores; energy is made available at the same time, and in some fashion the thick and thin filaments slide along one another by making and breaking bonds between the actin and the cross-bridges of heavy meromyosin. Although the details of this sliding process still elude us, there is much evidence for it—especially the convincing relationship between the force generated by a muscle fiber per unit of cross-sectional area and the number of binding opportunities that exist between thick and thin filaments.

Evidence is accumulating that the microfilaments discussed by Wessells are made of actin. In most of the cells in which they are found, the microfilaments are capable of binding heavy meromyosin—good evidence that they consist in large part of a protein strongly resembling actin. Furthermore, there is some suggestion that under normal circumstances they are attached to the membrane, associated with some myosinlike protein that may be affixed to the internal surface of the membrane. This would account for the ability of the microfilament system to alter the shapes of cells so effectively. In different kinds of mature, fully differentiated muscle cells, however, the actin and myosin systems are arrayed into organized and oriented systems of longitudinal fibers with varying complexity. In "smooth" muscle cells of the gut or blood vessels, actin is distributed very similarly to the microfilaments of motile cells, while only short segments of myosin are distributed among them in a relatively haphazard way. In cardiac and skeletal muscle cells, on the other hand, actin is aligned in precise arrays relative to the ends of the cell and to the connections between the sarcomeres—the elements of contraction—which are called Z lines. Perhaps the most significant difference between the primitive contractile systems of motile cells or plant cells is not so much the contractile machinery discussed by Huxley itself, but the way that it is arranged in space within cells and subject to control by calcium ions. Thus evolution may work more with components like Z lines and calcium pumps than with the basic actomyosin machinery itself.

The article by Peter F. Baker, "The Nerve Axon" (1966), returns us to the electrical phenomena that underlie intercellular communication. In a sense, it may be viewed as an updating of the summary given by Katz in the article that began this section. Baker gives a detailed treatment of the most important experimental advances that have been made since the establishment of the basic sequence of sodium-potassium permeability-changes as the mechanism of electrical conduction. After reviewing these findings, Baker discusses the technique of removing the contents of the axon entirely, by squeezing them out, and then replacing them with solutions of known ionic composition. These results provide a convincing proof that the resting potential of the nerve cell is the passive result of ionic composition ratios between the internal and external solutions: no energy source need be added to the solution. In addition, experiments can be performed on the nature of ionic changes in the membrane during the conduction of impulses.

The last article in this section, "The Synapse" (1965), by Sir John Eccles, actually bridges the gap between integration at the cellular level and the performance of nervous *systems*—that is, of networks of nerve cells. At the same time, it emphasizes the interdependence of chemical, electrical, and mechanical work in the cell. Although one description of the nerve cell defines it as the carrier of an all-or-none electrical signal, another equally valid one calls it "a device for delivering a small amount of hormone to a restricted place." The chemical emphasis is appropriate because most—though not all—synapses operate chemically; that is, they operate by the release of a special transmitter substance from one neuron, which then affects the ionic permeability of the postsynaptic cell.

Eccles, who shared the Nobel Prize in physiology or medicine in 1963 for his researches on inhibitory and excitatory synaptic transmission in the mammalian spinal cord, emphasizes the ionic mechanisms involved in the response of the postsynaptic membrane to the chemical transmitter. Other studies, notably from the laboratory of Bernhard Katz—the author of the first article in this section—have employed the junctions between nerve cells and peripheral effector cells, especially muscle fibers. From such studies we have learned much about the presynaptic events in transmission. The basic features are described at the end of the Katz article; more recently, we have begun to learn about the details of the process by which the secretory vesicles containing the synaptic transmitter are released as a result of the arrival of an electric impulse at the nerve terminal. Just as in muscular contraction, calcium is essential: it *must* be present in the medium surrounding the terminal or the transmitter substance will not be released in response to the arriving impulse. It is an interesting, if tentative, speculation that the calcium requirement reflects a basic underlying similarity between such mechanical events as muscle contraction and the processes of transmitter release. Perhaps, as has been suggested, a meshwork of microfilaments maneuvers each vesicle up to the releasing membrane, in the manner of a cargo net.

The postsynaptic emphasis of the Eccles article, however, focuses our attention on the performance of neuronal circuits in behavior—a theme that will recur in the remainder of this book. In describing the character of transmission between nerve cells in the mammalian spinal cord, Eccles has disclosed some of the basic properties of synapses and clarified the organization of a reflex pathway mediating one of the simplest forms of mammalian behavior. The organization of the reflex pathway is only indirectly alluded to in the article: the motoneurons are excited by sensory fibers that respond to stretching of the muscle served by those same motoneurons; they are inhibited by other fibers that respond to the stretching of muscles having the opposite action at the same joint. Eccles's careful analysis of the mechanisms of synaptic excitation and inhibition has thus provided a refined understanding of the operation of a simple reflex, whereby a muscle tends to maintain its length against an applied stretch and at the same time prevents contraction in antagonistic muscles. This same bit of behavior will be looked at from the perspective of performance in the article by Merton in Section VII.

How Cells Communicate

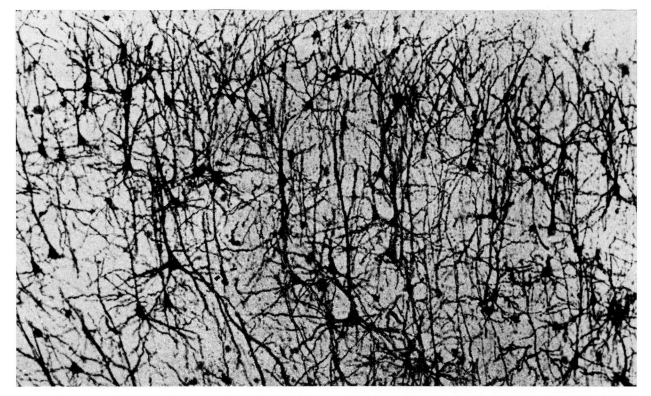

by Bernhard Katz

September 1961

The activities of cells in multicellular animals are coordinated by "chemical messengers" and nerve cells. During the past few years the character of the nerve impulse has been considerably clarified

In the animal kingdom, the "higher" the organism, the more important becomes the system of cells set aside for co-ordinating its activities. Nature has developed two distinct co-ordinating mechanisms. One depends on the release and circulation of "chemical messengers," the hormones that are manufactured by certain specialized cells and that are capable of regulating the activity of cells in other parts of the body. The second mechanism, which is in general far superior in speed and selectivity, depends on a specialized system of nerve cells, or neurons, whose function is to receive and to give instructions by means of electrical impulses directed over specific pathways. Both co-ordinating mechanisms are ancient from the viewpoint of evolution, but it is the second—the nervous system—that has lent itself to the greater evolutionary development, culminating in that wonderful and mysterious structure, the human brain.

Man's understanding of the working of his millions of brain cells is still at a primitive stage. But our knowledge is reasonably adequate to a more restricted task, which is to describe and partially explain how individual cells—the neurons—generate and transmit the electrical impulses that form the basic code element of our internal communication system.

A large fraction of the neuronal cell population can be divided into two classes: sensory and motor. The sensory neurons collect and relay to higher centers in the nervous system the impulses that arise at special receptor sites [see "How Cells Receive Stimuli," on page

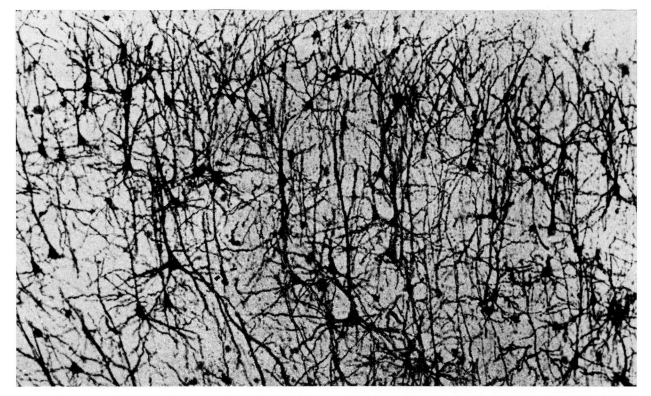

CEREBRAL CORTEX is densely packed with the bodies of nerve cells and the fibers called dendrites that branch from the cell body. This section through the sensory-motor cortex of a cat is enlarged some 150 diameters. Only about 1.5 per cent of the cells and dendrites actually present are stained and show here. The nerve axons, the fibers that carry impulses away from the cell body, are not usually shown at all by this staining method. The photomicrograph was made by the late D. A. Scholl of University College London.

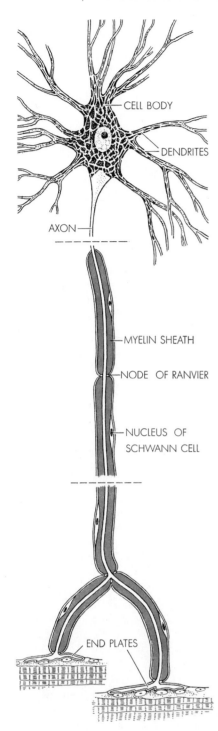

CELL BODY

DENDRITES

AXON

MYELIN SHEATH

NODE OF RANVIER

NUCLEUS OF
SCHWANN CELL

END PLATES

MOTOR NEURON is the nerve cell that carries electrical impulses to activate muscle fibers. The cell body (*top*) fans out into a number of twigs, the dendrites, which make synaptic contact with other nerve fibers (*see top illustration on opposite page*). Nerve impulses arising at the cell body travel through the axon to the motor-plate endings, which are embedded in muscle fibers. Myelin sheath is formed by Schwann cells as shown at bottom of opposite page. By insulating the axon the myelin wrapping increases the speed of signal transmission.

243], whose function is to monitor the organism's external and internal environments. The motor neurons carry impulses from the higher centers to the "working" cells, usually muscle cells, which provide the organism's response to changes in the two environments. In simple reflex reactions the transfer of signals from sensory to motor neurons is automatic and involves relatively simple synaptic mechanisms, which are fairly well understood.

When a nerve cell, either motor or sensory, begins to differentiate in the embryo, the cell body sends out a long fiber—the axon—which in some unknown way grows toward its proper peripheral station to make contact with muscle or skin. In man the adult axon may be several feet long, although it is less than .001 inch thick. It forms a kind of miniature cable for conducting messages between the periphery and the central terminus, which lies protected together with the nerve-cell body inside the spinal canal or the skull. Isolated peripheral nerve fibers probably have been subjected to more intense experimental study than any other tissue, in spite of the fact that they are only fragments of cells severed from their central nuclei as well as their terminal connections. Even so, isolated axons are capable of conducting tens of thousands of impulses before they fail to work. This fact and other observations make it clear that the nucleated body of the nerve cell is concerned with long-term maintenance of the nerve fibers—with growth and repair rather than with the immediate signaling mechanism.

For years there was controversy as to whether or not our fundamental concept of the existence of individual cell units could be applied at all to the nervous system and to its functional connections. Some investigators believed that the developing nerve cell literally grows into the cytoplasm of all cells with which it establishes a functional relationship. The matter could not be settled convincingly until the advent of high-resolution electron microscopy. It turns out that most of the surface of a nerve cell, including all its extensions, is indeed closely invested and enveloped with other cells, but that the cytoplasm of adjacent cells remains separated by distinct membranes. Moreover, there is a small extracellular gap, usually of 100 to 200 angstrom units, between adjoining cell membranes.

A fraction of these cell contacts are functional synapses: the points at which signals are transferred from one cell to

the next link in the chain. But synapses are found only at and near the cell body of the neuron or at the terminals of the axon. Most of the investing cells, particularly those clinging to the axon, are not nerve cells at all. Their function is still a puzzle. Some of these satellite cells are called Schwann cells, others glia cells; they do not appear to take any part in the immediate process of impulse transmission except perhaps indirectly to modify the pathway of electric current flow around the axon. It is significant, for example, that very few scattered satellites are to be found on the exposed cell surfaces of muscle fibers, which closely resemble nerve fibers in their ability to conduct electrical impulses from one end to the other.

One of the known functions of the axon satellites is the formation of the so-called myelin sheath, a segmented insulating jacket that improves the signaling efficiency of peripheral nerve fibers in vertebrate animals. Thanks to the electron microscope studies of Betty Ben Geren-Uzman and Francis O. Schmitt of the Massachusetts Institute of Technology, we now know that each myelin segment is produced by a nucleated Schwann cell that winds its cytoplasm tightly around the surface of the axon, forming a spiral envelope of many turns [*see bottom illustration on opposite page*]. The segments are separated by gaps—the nodes of Ranvier—which mark the points along the axon where the electrical signal is regenerated.

There are other types of nerve fiber that do not have a myelin sheath, but even these are covered by simple layers of Schwann cells. Perhaps because the axon extends so far from the nucleus of the nerve cell it requires close association with nucleated satellite cells all along its length. Muscle fibers, unlike the isolated axons, are self-contained cells with nuclei distributed along their cytoplasm, which may explain why these fibers can manage to exist without an investing layer of satellite cells. Whatever the function of the satellites, they cannot maintain the life of an axon for long once it has been severed from the main cell body; after a number of days the peripheral segment of the nerve cell disintegrates. How the nerve cell nucleus acts as a lifelong center of repair and brings its influence to bear on the distant parts of the axon—which in terms of ordinary diffusion would be years away—remains a mystery.

The experimental methods of physiology have been much more successful in dealing with the immediate processes of nerve communication than with the equally important but much more intractable long-term events. We know very little about the chemical interactions between nerve and satellite, or about the forces that guide and attract growing nerves along specific pathways and that induce the formation of synaptic contacts with other cells. Nor do we know how cells store information and provide us with memory. The rest of this article will therefore be concerned almost solely with nerve signals and the method by which they pass across the narrow synaptic gaps separating one nerve cell from another.

Much of our knowledge of the nerve cell has been obtained from the giant axon of the squid, which is nearly a millimeter in diameter. It is fairly easy to probe this useful fiber with microelectrodes and to follow the movement of radioactively labeled substances into it and out of it. The axon membrane separates two aqueous solutions that are almost equally electroconductive and that contain approximately the same number of electrically charged particles, or ions. But the chemical composition of the two solutions is quite different. In the external solution more than 90 per cent of the charged particles are sodium ions (positively charged) and chloride ions (negatively charged). Inside the cell these ions together account for less than 10 per cent of the solutes; there the principal positive ion is potassium and the negative ions are a variety of organic particles (doubtless synthesized within the cell itself) that are too large to diffuse easily through the axon membrane. Therefore the concentration of sodium is about 10 times higher *outside* the axon, and the concentration of potassium is about 30 times higher *inside* the axon. Although the permeability of the membrane to ions is low, it is not indiscriminate; potassium and chloride ions can move through the membrane much more easily than sodium and the large organic ions can. This gives rise to a voltage drop of some 60 to 90 millivolts across the membrane, with the inside of the cell being negative with respect to the outside.

To maintain these differences in ion concentration the nerve cell contains a kind of pump that forces sodium ions "uphill" and outward through the cell membrane as fast as they leak into the cell in the direction of the electrochemical gradient [*see illustration on page 142*]. The permeability of the resting cell surface to sodium is normally so low that the rate of leakage remains very small, and the work required of the

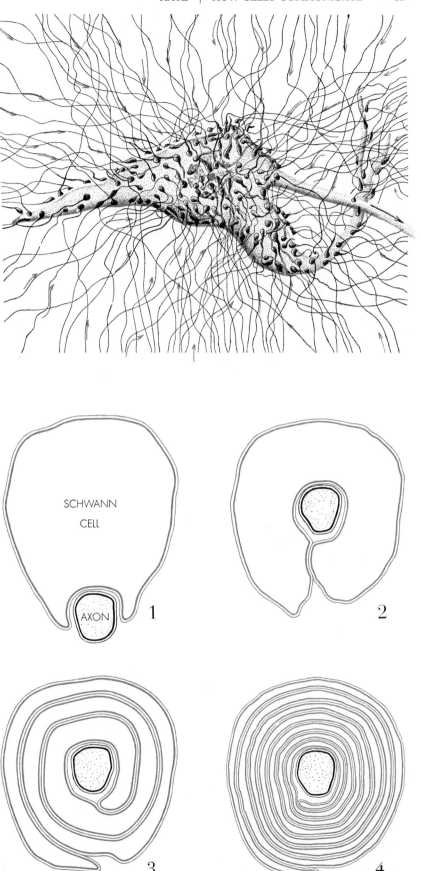

MYELIN SHEATH is created when a Schwann cell wraps itself around the nerve axon. After the enfolding is complete, the cytoplasm of the Schwann cell is expelled and the cell's folded membranes fuse into a tough, compact wrapping. Diagrams are based on studies of chick-embryo neurons by Betty Ben Geren-Uzman of Children's Medical Center in Boston.

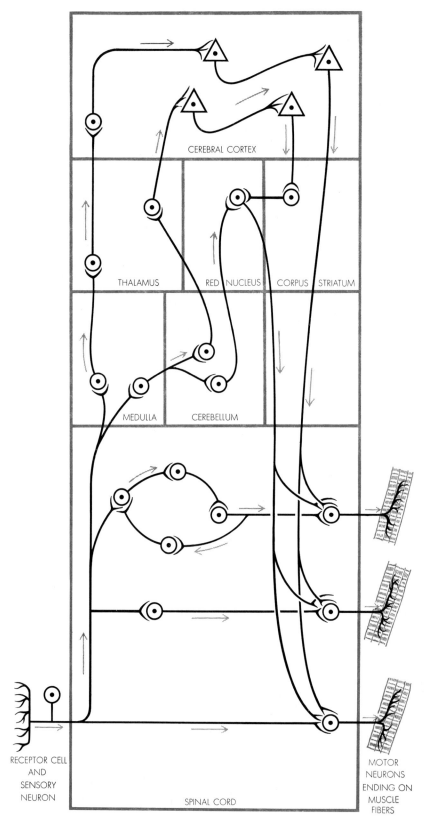

CEREBRAL CORTEX

THALAMUS

RED NUCLEUS   CORPUS STRIATUM

MEDULLA   CEREBELLUM

RECEPTOR CELL
AND
SENSORY
NEURON

MOTOR
NEURONS
ENDING ON
MUSCLE
FIBERS

SPINAL CORD

SIMPLIFIED FLOW DIAGRAM OF NERVOUS SYSTEM barely hints at the many possible pathways open to an impulse entering the spinal cord from a receptor cell and its sensory fiber. Rarely does the incoming signal directly activate a motor neuron leading to a muscle fiber. Typically it travels upward through the spinal cord and through several relay centers before arriving at the cerebral cortex. There (if not elsewhere) a "command" may be given (or withheld) that sends nerve impulses back down the spinal cord to fire a motor neuron.

pumping process amounts to only a fraction of the energy that is continuously being made available by the metabolism of the cell. We do not know in detail how this pump works, but it appears to trade sodium and potassium ions; that is, for each sodium ion ejected through the membrane it accepts one potassium ion. Once transported inside the axon the potassium ions move about as freely as the ions in any simple salt solution. When the cell is resting, they tend to leak "downhill" and outward through the membrane, but at a slow rate.

The axon membrane resembles the membrane of other cells. It is about 50 to 100 angstroms thick and incorporates a thin layer of fatty insulating material. Its specific resistance to the passage of an electric current is at least 10 million times greater than that of the salt solutions bathing it on each side. On the other hand, the axon would be quite worthless if it were employed simply as the equivalent of an electric cable. The electrical resistance of the axon's fluid core is about 100 million times greater than that of copper wire, and the axon membrane is about a million times leakier to electric current than the sheath of a good cable. If an electric pulse too weak to trigger a nerve impulse is fed into an axon, the pulse fades out and becomes badly blunted after traveling only a few millimeters.

How, then, can the axon transmit a nerve impulse for several feet without decrement and without distortion?

As one steps up the intensity of a voltage signal impressed on the membrane of a nerve cell a point is reached where the signal no longer fades and dies. Instead (if the voltage is of the right sign), a threshold is crossed and the cell becomes "excited" [*see illustrations on page 140*]. The axon of the cell no longer behaves like a passive cable but produces an extra current pulse of its own that amplifies the original input pulse. The amplified pulse, or "spike," regenerates itself from point to point without loss of amplitude and travels at constant speed down the whole length of the axon. The speed of transmission in vertebrate nerve fibers ranges from a few meters per second, for thin nonmyelinated fibers, to about 100 meters per second in the thickest myelinated fibers. The highest speeds, equivalent to some 200 miles per hour, are found in the sensory and motor fibers concerned with body balance and fast reflex movements. After transmitting an impulse the nerve is left briefly in a refractory, or inexcitable,

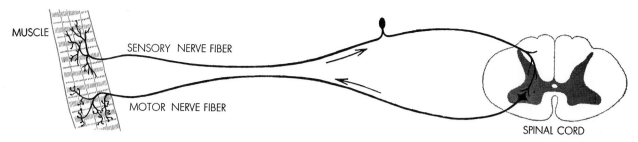

MUSCLE

SENSORY NERVE FIBER

MOTOR NERVE FIBER

SPINAL CORD

REFLEX ARC illustrates the minimum nerve circuit between stimulus and response. A sensory fiber arising in a muscle spindle enters the spinal cord, where it makes synaptic contact with a motor neuron whose axon returns to the muscle containing the spindle.

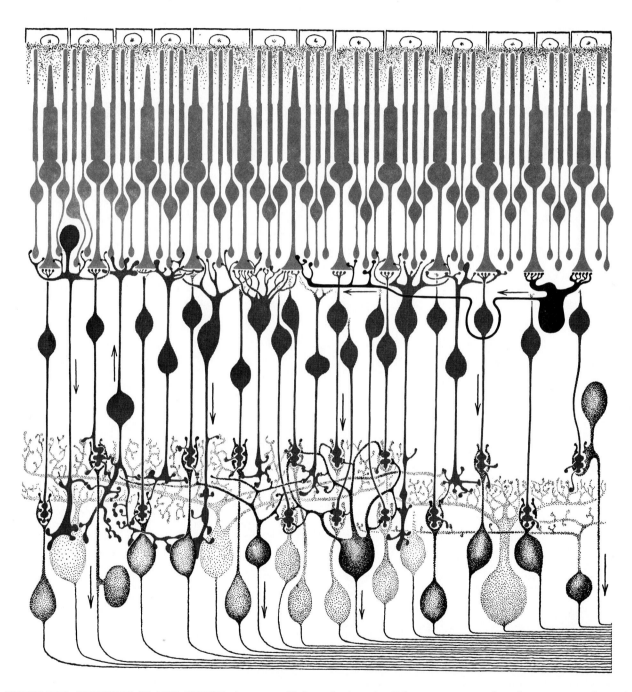

NERVE-CELL NETWORK IN THE RETINA, here magnified about 600 diameters, exemplifies the retinal complexity in man and apes. The photoreceptors are the densely packed cells shown in color; the thinner ones are rods, the thicker ones cones. To reach them the incoming light must traverse a dense but transparent layer of neurons *(dark shapes)* that have rich interconnections with the photoreceptors and with each other. The output of these neurons finally feeds into the optic nerve shown at the bottom of the diagram.

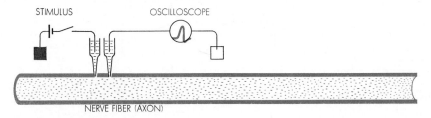

INVESTIGATION OF NERVE FIBER is carried out with two microelectrodes. One provides a stimulating pulse, the other measures changes in membrane potential (*see below*).

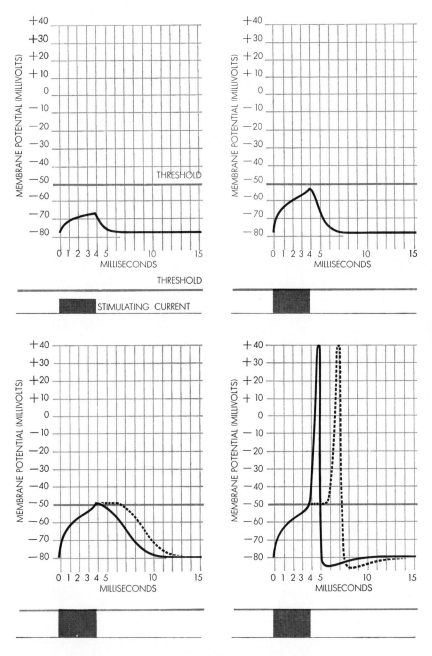

ELECTRICAL PROPERTIES OF NERVE FIBER are elucidated by measuring voltage changes across the axon membrane when stimulating pulses of varying size are applied. In the resting state the interior of the axon is about 80 millivolts negative. Subthreshold stimulating pulses (*top left and top right*) shift the potential upward momentarily. Larger pulses push the potential to its threshold, where it becomes unstable, either subsiding (*bottom left*) or flaring up into an "action potential" (*bottom right*) with a variable delay (*broken curve*).

state, but within one or two milliseconds it is ready to fire again.

The electrochemical events that underlie the nerve impulse—or action potential, as it is called—have been greatly clarified within the past 15 years. As we have seen, the voltage difference across the membrane is determined largely by the membrane's differential permeability to sodium and potassium ions. Many kinds of selective membrane, natural and artificial, show such differences. What makes the nerve membrane distinctive is that its permeability is in turn regulated by the voltage difference across the membrane, and this peculiar mutual influence is in fact the basis of the signaling process.

It was shown by A. L. Hodgkin and A. F. Huxley of the University of Cambridge that when the voltage difference across the membrane is artificially lowered, the immediate effect is to increase its sodium permeability. We do not know why the ionic insulation of the membrane is altered in this specific way, but the consequences are far-reaching. As sodium ions, with their positive charges, leak through the membrane they cancel out locally a portion of the excess negative charge inside the axon, thereby further reducing the voltage drop across the membrane. This is a regenerative process that leads to automatic self-reinforcement; the flow of some sodium ions through the membrane makes it easier for others to follow. When the voltage drop across the membrane has been reduced to the threshold level, sodium ions enter in such numbers that they change the internal potential of the membrane from negative to positive; the process "ignites" and flares up to create the nerve impulse, or action potential. The impulse, which shows up as a spike on the oscilloscope, changes the permeability of the axon membrane immediately ahead of it and sets up the conditions for sodium to flow into the axon, repeating the whole regenerative process in a progressive wave until the spike has traveled the length of the axon [*see illustration on opposite page*].

Immediately after the peak of the wave other events are taking place. The "sodium gates," which had opened during the rise of the spike, are closed again, and the "potassium gates" are opened briefly. This causes a rapid outflow of the positive potassium ions, which restores the original negative charge of the interior of the axon. For a few milliseconds after the membrane voltage has been driven toward its initial level it is difficult to displace the voltage and

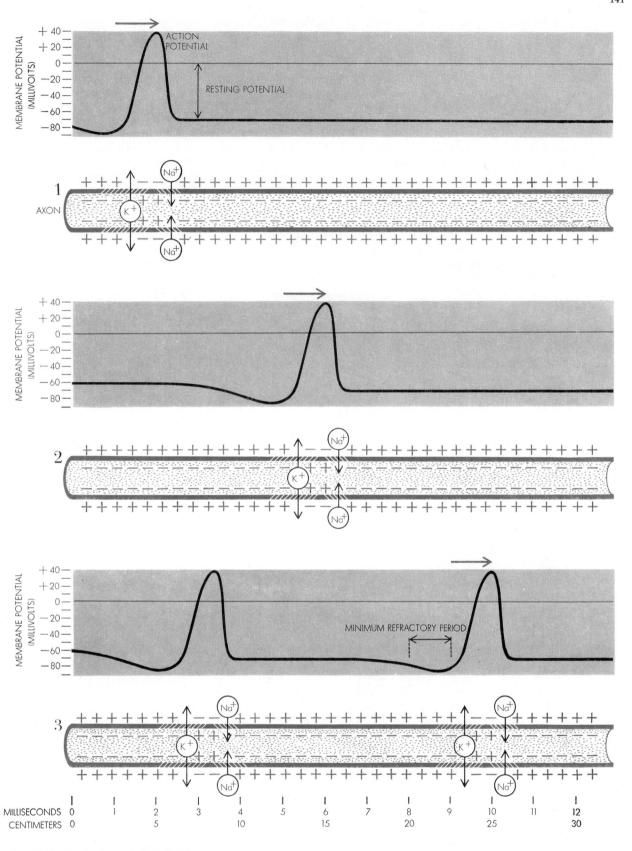

PROPAGATION OF NERVE IMPULSE coincides with changes in the permeability of the axon membrane. Normally the axon interior is rich in potassium ions and poor in sodium ions; the fluid outside has a reverse composition. When a nerve impulse arises, having been triggered in some fashion, a "gate" opens and lets sodium ions pour into the axon in advance of the impulse, making the axon interior locally positive. In the wake of the impulse the sodium gate closes and a potassium gate opens, allowing potassium ions to flow out, restoring the normal negative potential. As the nerve impulse moves along the axon (1 and 2) it leaves the axon in a refractory state briefly, after which a second impulse can follow (3). The impulse propagation speed is that of a squid axon.

set up another impulse. But the ionic permeabilities quickly return to their initial condition and the cell is ready to fire another impulse.

The inflow of sodium ions and subsequent outflow of potassium ions is so brief and involves so few particles that the over-all internal composition of the axon is scarcely affected. Even without replenishment the store of potassium ions inside the axon is sufficient to provide tens of thousands of impulses. In the living organism the cellular enzyme system that runs the sodium pump has no difficulty keeping nerves in continuous firing condition.

This intricate process—signal conduction through a leaky cable coupled with repeated automatic boosting along the transmission path—provides the long-distance communication needs of our nervous system. It imposes a certain stereotyped form of "coding" on our signaling channels: brief pulses of almost constant amplitude following each other at variable intervals, limited only by the refractory period of the nerve cell. To make up for the limitations of this simple coding system, large numbers of axon channels, each a separate nerve cell, are provided and arranged in parallel. For example, in the optic nerve trunk emerging from the eye there are more than a million channels running close together, all capable of transmitting separate signals to the higher centers of the brain.

Let us now turn to the question of what happens at a synapse, the point at which the impulse reaches the end of one cell and encounters another nerve cell. The self-amplifying cable process that serves within the borders of any one cell is not designed to jump automatically across the border to adjacent cells. Indeed, if there were such "cross talk" between adjacent channels, for instance among the fibers closely packed together in our nerve bundles, the system would become quite useless. It is true that at functional synaptic contacts the separation between the cell membranes is only 100 to a few hundred angstroms. But from what we know of the dimensions of the contact area, and of the insulating properties of cell membranes, it is unlikely that an effective cable connection could exist between the terminal of one nerve cell and the interior of its neighbor. This can easily be demonstrated by trying to pass a subthreshold pulse—that is, one that does not trigger a spike—across the synapse that separates a motor nerve from a muscle fiber. A recording probe located just inside the muscle detects no signal when a weak pulse is applied to the motor nerve close to the synapse. Clearly the cable linkage is broken at the synapse and some other process must take its place.

The nature of this process was discovered some 25 years ago by Sir Henry Dale and his collaborators at the National Institute for Medical Research in London. In some ways it resembles the hormonal mechanism mentioned at the beginning of this article. The motor nerve terminals act rather like glands secreting a chemical messenger. Upon arrival of an impulse, the terminals release a special substance, acetylcholine, that quickly and efficiently diffuses across the short synaptic gap. Acetylcholine molecules combine with receptor molecules in the contact area of the muscle fiber and somehow open its ionic gates, allowing sodium to flow in and trigger an impulse. The same result can be obtained by artificially applying ace-

OUTSIDE   MEMBRANE   INSIDE

20 MILLIVOLTS

POTASSIUM IONS

DIFFUSION

SODIUM PUMP

METABOLIC DRIVE

SODIUM IONS

130 MILLIVOLTS

DIFFUSION

"SODIUM PUMP," details unknown, is required to expel sodium ions from the interior of the nerve axon so that the interior sodium-ion concentration is held to about 10 per cent of the exterior fluid. At the same time the pump drives potassium ions "uphill" from a low external concentration to a 30-times-higher internal concentration. The pumping rate must keep up with the "downhill" leakage of the two kinds of ion. Since both are positively charged, sodium ions have the higher leakage rate (expressed in terms of millivolts of driving force) because they are attracted to the negatively charged interior of the axon, whereas potassium ions tend to be retained. But there is still a net outward leakage of potassium.

tylcholine to the contact region of the muscle fiber. It is probable that similar processes of chemical mediation take place at the majority of cell contacts in our central nervous system. But it is most unlikely that acetylcholine is the universal mediator at all these points, and an intensive search is being made by many workers for other naturally occurring transmitter substances.

Synaptic transmission presents two quite distinct sets of problems. First, exactly how does a nerve impulse manage to cause the secretion of the chemical mediator? Second, what are the physicochemical factors that decide whether a mediator will stimulate the next cell to fire in some cases or inhibit it from firing in others? So far we have said nothing about inhibition, even though it occurs throughout the nervous system and is one of the most curious modes of nervous activity. Inhibition takes place when a nerve impulse acts as a brake on the next cell, preventing it from becoming activated by excitatory messages that may be arriving along other channels at the same time. The impulse that travels along an inhibitory axon cannot be distinguished electrically from an impulse traveling in an excitatory axon. But the physicochemical effect that it induces at a synapse must be different in kind. Presumably inhibition results from a process that in some way stabilizes the membrane potential (degree of electrification) of the receiving cell and prevents it from being driven to its unstable threshold, or "ignition" point.

There are several processes by which such a stabilization could be achieved. One of them has already been mentioned; it occurs in the refractory period immediately after a spike has been generated. In this period the membrane potential is driven to a high stable level (some 80 to 90 millivolts negative inside the membrane) because, to put it somewhat crudely, the potassium gates are wide open and the sodium gates are firmly shut. If the transmitter substance can produce one or both of these states of ionic permeability, it will undoubtedly act as an inhibitor. There are good reasons for believing that this is the way impulses from the vagus nerve slow down and inhibit the heartbeat; incidentally, the transmitter substance released from the vagus nerve is again acetylcholine, as was discovered by Otto Loewi 40 years ago. Similar effects occur at various inhibitory synapses in the spinal cord, but there the chemical nature of the transmitter has so far eluded identification.

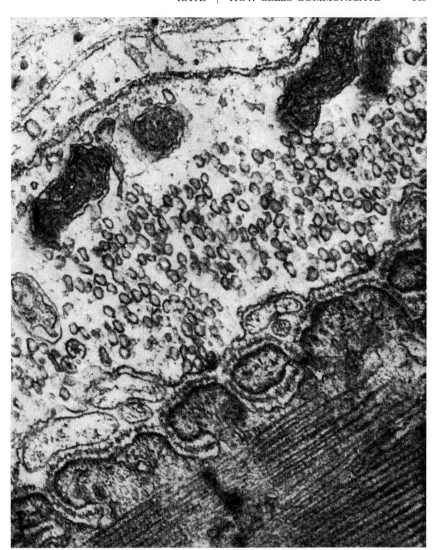

NERVE-MUSCLE SYNAPSE is the site at which a nerve impulse activates the contraction of a muscle fiber. In this electron micrograph (made by R. Birks, H. E. Huxley and the author) the region of the synapse is enlarged 53,000 diameters. Motor nerve terminal runs diagonally from lower left to upper right, being bounded at upper left by a Schwann cell. Muscle fiber is the dark striated area at lower right, with a folded membrane. Nerve terminal is populated with "synaptic vesicles" that may contain acetylcholine, which is released into the synaptic cleft by a nerve impulse and evokes electrical activity in the muscle.

Inhibition would also result if two "antagonistic" axons converged on the same spot of a third nerve cell and released chemically competing molecules. Although a natural example of this kind has not yet been demonstrated, the chemical and pharmacological use of competitive inhibitors is well established. (For example, the paralyzing effect of the drug curare arises from its competitive attachment to the region of the muscle fiber that is normally free to react with acetylcholine.) Alternatively, a substance released by an inhibitory nerve ending could act on the excitatory nerve terminal in such a way as to reduce its secretory power, thereby causing less of the excitatory transmitter substance to be released.

This brings us back to the question:

How does a nerve impulse lead to the secretion of transmitter substances? Recent experiments on the nerve-muscle junction have shown that the effect of the nerve impulse is not to initiate a process of secretion but rather, by altering the membrane potential, to change the rate of a secretory process that goes on all the time. Even in the absence of any form of stimulation, packets of acetylcholine are released from discrete spots of the nerve terminals at random intervals, each packet containing a large number—probably thousands—of molecules.

Each time one of these quanta of transmitter molecules is liberated spontaneously, it is possible to detect a sudden minute local response in the muscle fiber on the other side of the synapse.

Within a millisecond there is a drop of .5 millivolt in the potential of the muscle membrane, which takes about 20 milliseconds to recover. By systematically altering the potential of the membrane of the nerve ending it has been possible to work out the characteristic relation between the membrane potential of the axon terminal and the rate of secretion of transmitter packets. It appears that the rate of release increases by a factor of about 100 times for each 30-millivolt lowering of membrane potential. In the resting condition there is a random discharge of about one packet per second at each nerve-muscle junction. But during the brief 120-millivolt change associated with the nerve impulse the frequency rises momentarily by a factor of nearly a million, providing a synchronous release of a few hundred packets within a fraction of a millisecond.

It is significant that the transmitter is released not in independent molecular doses but always in multimolecular parcels of standard size. The explanation of this feature is probably to be found in the microstructural make-up of the nerve terminals. They contain a characteristic accumulation of so-called vesicles, each about 500 angstroms in diameter, which may contain the transmitter substance parceled and ready for release [*see illustration on page 143*]. Conceivably when the vesicles collide with the axon membrane, as they often must, the collision may sometimes cause the vesicular content to spill into the synaptic cleft. Such ideas have yet to be proved by direct evidence, but they provide a reasonable explanation of all that is known about the quantal spontaneous release of acetylcholine and its accelerated release under various natural and experimental conditions. At any rate, the ideas provide an interesting meeting point between the functional and structural approaches to a common problem.

Because of the sparseness of existing knowledge, we have left out of this discussion many fascinating problems of the long-term interactions and adaptive modifications that must certainly take place in nerve pathways. For handling such problems investigators will probably have to develop very different methods from those followed in the past. It may be that our preoccupation with the techniques that have been so successful in illuminating the brief reactions of excitable cells has prevented us from making inroads on the problems of learning, of memory, of conditioning and of the structural and operating relations between nerve cells and their neighbors.

13

How Living Cells
Change Shape

by Norman K. Wessells
October 1971

*All embryonic cells and many mature ones bend,
bulge, stretch out and even travel from place to place.
In doing so the cells use two distinct kinds of filament
that seem to act as skeleton and muscle*

Single-celled animals such as the amoeba are not the only cells that are capable of self-propelled movement. Even in highly organized multicelled animals, including man, many cells can creep about and engage in movements that change their shape. Clearly the individual cells must have built-in machinery that enables them to do this. The nature of the machinery is not obvious, as these cells do not possess cilia or other appendages that might account for their locomotion or shape-altering movements. With the help of the electron microscope and certain drugs, however, the mechanisms responsible for the performance and control of these movements are now being investigated in detail. The subject is of much practical interest, since cell movement plays a crucial role in the normal development of animal embryos, and abnormal movements of cells may be a critical factor in certain disease conditions.

When a single cell of certain tissues is isolated and placed in a culture medium, the cell can be seen to wander about over the bottom of the culture dish. Studying such movement, Michael Abercrombie and his colleagues at University College London found that the moving cell thrusts the forward edge of its membrane ahead; as the thrust occurs, the edge appears to flutter up and down. The undulating, advancing part of the membrane attaches itself to the substratum and apparently contracts to draw the cell forward. The cell can readily change the direction of its travel by activating a different part of its perimeter; the side that had been moving forward becomes quiescent and the newly activated side begins to flutter and extend itself, drawing the cell off in the new direction. A migrating cell often changes direction in this way when it encounters other cells or asymmetries in

its path.

A related kind of movement is involved in the alteration of the shape of the cells that make up certain tissues. Usually a cell in a tissue remains in a fixed position in relation to its neighbors; nevertheless, it can change in shape by elongation or by a widening or narrowing of some part of the cell body, apparently through a process of contraction. When such a narrowing or elongation of cells takes place, the tissue itself assumes a new configuration; a flat sheet, for example, may be converted into a ball-like, hollow structure. This is the process that forms organs such as the lungs and the pancreas during the development of an embryo.

What kind of system could account for a cell's ability to engage in the movements we have described? Thinking about the situation in strictly biological terms, one can start from the basic features of the machinery for movement of an animal as a whole. Broadly speaking, this system has two principal components: the skeleton, which gives shape, rigidity and support to the body or its appendages, and the muscles, which provide the power that moves the skeleton and thus the organism itself. Can we find structures in the individual cell that are analogous to these two components? Investigation has now shown that the cell does indeed possess such structures. It turns out that this cellular machinery apparently accounts for the two very different types of movement discussed above—locomotion and change of shape—and also for a third kind of move-

ment within the cell. An example is the action of the mitotic spindle in pulling the chromosomes apart when a cell divides.

The two basic components of the machinery for movement discovered in the cell are microtubules, which correspond to a skeleton, and microfilaments, which correspond to muscles. Let us consider the microtubules first.

The microtubules are very fine tubes averaging 250 angstroms in diameter. They are found in cilia, in the tail of sperm cells, in the mitotic spindle of a dividing cell and in the cytoplasm of many types of cell (where they have been studied by Keith R. Porter, Lewis G. Tilney and J. Richard McIntosh of Harvard University). A clue to a possible method of examining their function was provided by the response of the tubules in the mitotic spindle to the drug colchicine. This substance, an alkaloid extracted from plants of the lily family, has been used for many years as an inhibitor of cell division in plants and animals; division stops because the spindle microtubules are disrupted and therefore prevented from pulling the chromosomes apart. In order to analyze the roles of microtubules in cells we applied colchicine to cells growing in cultures in our laboratory at Stanford University. Following the lead of Henry Wisniewski of the Albert Einstein College of Medicine in New York, who had worked with nerves in whole organisms, we started with nerve cells.

The axon, or principal fiber, of a nerve

CONTRACTING CORTEX of the fertilized egg of a frog is magnified 2,000 times in the photomicrograph on the following page. The contraction was induced by injecting calcium ions through the surface membrane of the embryo. The presence of a distinctive band of dense material, consisting largely of microfilaments, in the cytoplasm at the site of the injection suggests that injection of calcium caused network of filaments in egg to contract.

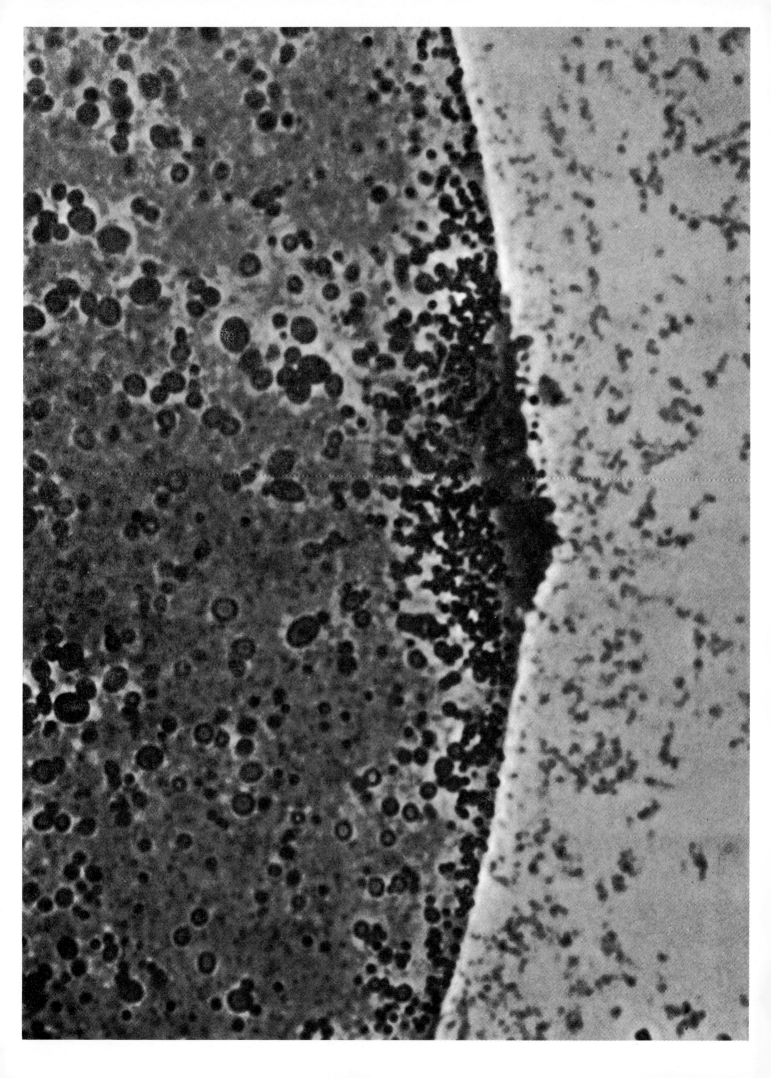

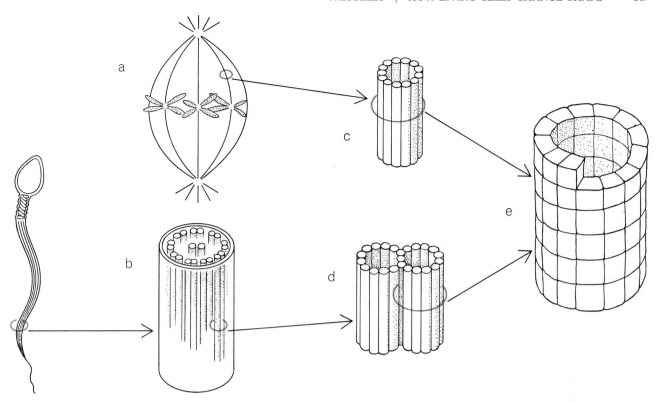

MICROTUBULES are found in such dissimilar cellular structures as the mitotic spindle (*a*) that controls the separation of the chromosomes during cell division and the flagellum (*b*) that propels the sperm cell. The spindle tubules (*c*) form a circular array totaling 13. The flagellum tubules form a double circle (*d*) and number 23. The tubules (*e*) consist of a series of linked protein subunits.

cell is a long, rather rigid cylinder containing many microtubules and tipped at its end with a mobile "growth cone." In the culture of a single nerve cell on a plate the growth tip of the axon advances over the surface of the plate. We found that when colchicine was added to the culture medium, the cell went on growing in this way for about 30 minutes; then the walls of the axon began to look crinkled and the axon shrank back toward the cell body. Eventually the axon pulled its tip free of the plate and collapsed completely into the cell body.

The microtubules had disappeared; instead the cytoplasm of the cell body contained masses of a filament, averaging only about 100 angstroms in diameter, that apparently was a different form of the protein that had composed the microtubules. The results of the experiment indicated that the microtubules when intact had acted as a "skeleton" supporting the axon, much as the long bones support a human arm or leg. The axon had apparently collapsed when these microtubules were broken down through the action of colchicine. The microtubules themselves were evidently not the direct agents for the movement of the growth cone, but in the absence of the axon's stabilizing "skeleton" the tip could not grow and advance.

Colchicine experiments with other types of cell added significant further information. These were migratory cells from the embryonic heart and the

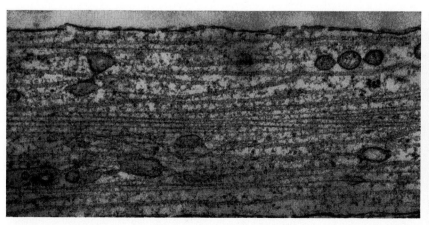

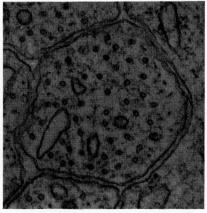

NERVE-CELL AXON is seen in longitudinal section (*left*) and transverse section (*right*). The long parallel structures near the top and bottom of the longitudinal section are microtubules. Near the center of the axon are thinner structures; they are called neurofilaments, and their function is unknown. In the transverse section the microtubules have the appearance of the cut ends of straws.

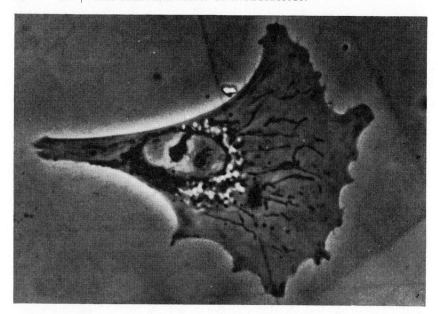

LIVING CELL travels from left to right across the bottom of a culture dish. The central oval is the nucleus of the cell; the wormlike dark structures are mitochondria. The "tail" of the cell (*left*) is supported by a "skeleton" of microtubules. The membrane that is advancing (*right*) will contract after attachment to the surface of the dish, thus pulling the cell forward. The movement may be powered by microfilaments, the "muscles" of the cell.

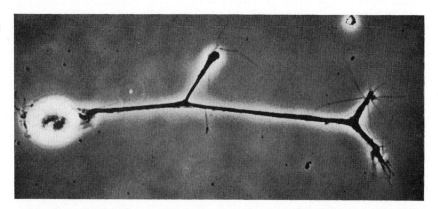

GROWING NERVE CELL in a culture dish consists of a cell body (*left*), containing the nucleus, and a long axon that has one upward branch and two more branches at the tip. The thin spikelets at the end of the branches arise from the "growth cone" regions of the cell. When the microtubules that form the "skeleton" of the axon are exposed to the alkaloid colchicine, microtubules are broken down and the axon soon withdraws into cell body.

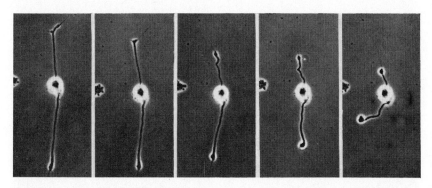

EFFECT OF COLCHICINE on the axons of a nerve cell is dramatic. Growth continues for perhaps half an hour; the first micrograph shows the cell 17 minutes after the alkaloid was added to the culture dish. The axons begin to collapse after 25 minutes (*second micrograph*), shrink further after 32 and 38 minutes (*third and fourth*) and are mostly withdrawn into the cell body after 45 minutes (*fifth*). Thin filaments are found in cell but no tubules.

nervous system. Such cells often have a trailing "tail," drawn out into a long, stiff process, that contains microtubules. When the cell is treated with colchicine, the disruption of the microtubules causes the "tail" to be retracted into the cell body, and thereafter any part of the membrane perimeter may act as an undulating leading edge, so that the cell moves to and fro in various directions—forward, backward or sideways with equal impartiality. My colleague Brian Spooner observed that the cell behaves as if it had lost its "steering wheel" while the engine (the undulating membrane system) continues to operate. Experiments conducted by the Russian biologist J. M. Vasiliev support this conclusion, particularly with respect to the treated cells' inability to change the direction of travel in response to "roadbed" asymmetries. It seems likely that in these cells the microtubules normally serve as a skeleton that stabilizes the sides of the cell but not the front. Therefore the front of the membrane normally is free to act as an undulating leading edge, causing the cell to move in a directed manner rather than at random.

What is the nature of the "engine" that powers cell movement? Here again the use of a drug, applied to cell cultures, helped to identify the mechanism. It was known that many cells contain bundles or networks of microfilaments. The filaments are only 40 to 60 angstroms in diameter—considerably smaller than the microtubules that serve as cellular skeleton. During the division of a cell there is a ring of such filaments just below the contracting "furrow" at the area of cleavage. At the Woods Hole Marine Biological Laboratory, Thomas Schroeder examined this system by means of a drug called cytochalasin, a compound secreted by certain fungi. He found that the drug caused the microfilaments to disappear, and the furrow ceased to contract. When the drug was removed from the culture, the filaments reappeared and the cells resumed their division! Apparently the microfilaments were the contractile agents, or "muscles," that drove the cleavage process.

My associate Kenneth Yamada has used cytochalasin to study the movement of nerve cells and migratory cells. Both of these cell types are rich in microfilaments at the scene of action. At the growing tip of a nerve axon there is a network of microfilaments within the growth cone and in the long, thin processes, called microspikes, that extend ahead of the cone like a set of antennas and wave back and forth as if they were probing the area in front.

Similarly, in a migrating cell the undulating membrane that pulls the cell forward is filled with a network of microfilaments.

In the case of the nerve cell, when Yamada added cytochalasin to the culture medium, the axon stopped elongating within a few minutes. The microspikes wilted and retracted into the growth cone, and the tip as a whole rounded up and ceased to move over the substratum. Examination with the electron microscope showed that the filamentous network had become a mass of short, densely packed filaments. Treatment of migrating cells with cytochalasin altered the network of microfilaments in the undulating membrane in a similar way, stopping the cell's movement and causing the surface membrane to sink back toward the cell nucleus.

The most extraordinary feature of these experiments is the reversibility of the drug's effect. Even after hours of immobilization under cytochalasin's influence the cells recover their mobility when the drug is removed from the medium. The filamentous network is restored, the nerve cell's growth tip and the migratory cell's undulating membrane regain their normal structure and movement is resumed.

The accumulating evidence, although indirect and circumstantial, leaves little doubt that the microfilaments are indeed the "muscles" that actuate the cell movements, particularly since there are no other visible structures in the cells that could plausibly play that role. What is the chemical mechanism that gives the microfilaments the power to contract? This is still an unanswered question. Studies of primitive organisms suggest a possible answer. Certain biochemical investigations of slime molds and protozoans indicate that the microfilaments may be composed of a substance like actin, the contractile protein that is an important component of muscle in the higher animals.

When we turn to investigation of what is involved in the change of shape in cells and cell populations, it becomes apparent that here too the active agents are the microfilaments. Their role has been demonstrated in events such as the shaping of a bird's oviduct and of a mammal's salivary gland during development of the embryo.

Consider the form of the oviduct in a mature bird that has reached the egg-producing stage. The wall of the organ bulges outward in many places in finger-like structures known as tubular glands. The glands form originally as small knoblike bulges in the oviduct wall. Microscopic inspection shows that on the inner side of such a bulge some of the cells are narrowed at one end. The narrowness accounts for the curvature and rounded shape of the bulges. If the cells are examined at the time of narrowing, bands of microfilaments are seen running across the cell. In all likelihood it is the contraction of these filaments that causes the narrowing. The filaments are the only observable structures in the cell that are oriented in a manner that could produce such a result.

My associate Joan Wrenn investigated development of the oviduct, speeded up by artificial means, in very young chicks. An injection of the female hormone estrogen can stimulate the oviduct to develop prematurely. Mrs. Wrenn found that after estrogen was administered to a five-day-old chick, within 24 hours bands of microfilaments formed across the inner ends of the cells in the embryonic oviduct, and tubular glands soon began to bulge from the tissue. It turned out that the drug cytochalasin could halt this process just as it did the movement of a migrating cell or a cell's mitotic division. The drug disrupted the bands of microfilaments in the cells and caused the glands to sink back into the oviduct structure so that it reverted to a knobless cylinder.

The formation of a mammal's salivary glands is physically opposite to that of a bird's oviduct; instead of there being bulges, a number of folds and clefts develop in the corpus of the salivary gland, with the result that it becomes a branching, treelike structure. This process is due to narrowing of the outer ends of cells at the point of the cleft; as in the bird oviduct bands of microfilaments are found at the narrowed cell ends. Cytochalasin will reverse this development too: it disrupts the microfilament bands, smooths out the clefts and folds in the salivary gland and causes the entire organ to flatten into a thin, cookie-like sheet. When the drug is removed from the culture medium, the gland recovers in a spectacular fashion. Within 18 hours it re-forms deep clefts, with thick bands of microfilaments reappearing at the bases of the clefts. Interestingly, the recovery takes place even if the cells are treated with drugs that inhibit the synthesis of new protein. The re-formation of microfilament networks also occurs in

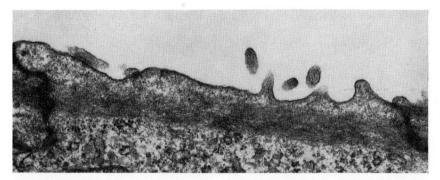

CELL "MUSCLE," which can produce changes in the shape of cells, is made up of microfilaments arrayed in bundles. Seen in the micrograph is such a bundle, lying parallel to the surface of an epithelial cell in the oviduct of a bird. The drug cytochalasin, the secretion of a fungus, has the property of disrupting microfilament "muscles" (see *illustration below*).

FILAMENT BUNDLES have disappeared in the surface region of an oviduct cell similar to the one at top. Cytochalasin was added to medium in the culture dish. The filaments will reorganize themselves into bundles once cytochalasin is removed even though the cell can no longer make protein, suggesting that the filaments were dispersed, not destroyed.

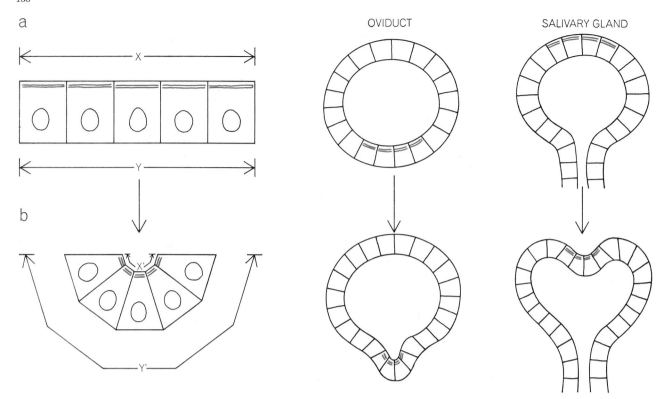

OVIDUCT

SALIVARY GLAND

CHANGES IN CELL SHAPE produce movement in a sheetlike array of adjacent cells. In an ideal sheet of cells (*a*) distances at top and bottom of the array (*x, y*) are equal as long as the line of microfilaments (*color*) does not contract. Following contraction (*b*), however, the distance at top is reduced (*x'*) and the distance at bottom (*y'*) is the same or greater. Similar contractions make cells push out into the tissue surrounding a maturing oviduct or intrude into the central cavity of the salivary gland (*right*).

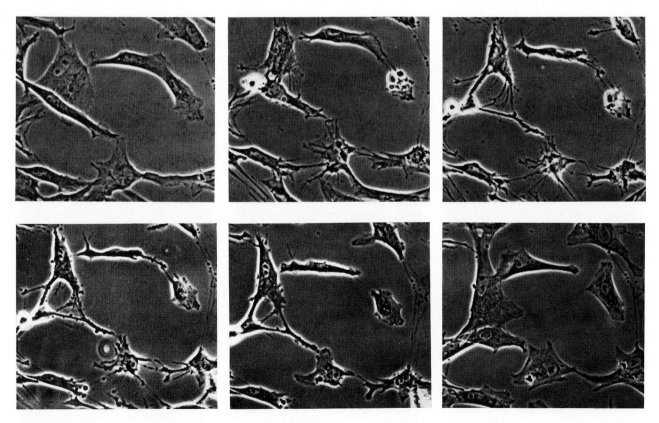

REVERSIBILITY of cytochalasin disruption is demonstrated in these micrographs of cells from embryo heart tissue. The top three micrographs show cells in untreated medium (*left*) and cells after six minutes' exposure (*center*) and 11 minutes' exposure (*right*) to cytochalasin. The bottom three micrographs record the recovery of the cells. The first (*left*) shows the cells two minutes after the cytochalasin was removed from the culture dish, the next (*center*) shows them after five minutes and the last (*right*) an hour later.

other situations in which protein is not being produced. It seems likely, therefore, that cytochalasin does not destroy the microfilaments themselves but only disrupts their cooperative aggregation and activity.

We arrive at the apparently inescapable conclusion that the normal shaping of organs, as well as the migration and other movements of cells, depends on the integrity and action of microfilament systems. In the case of the developing oviduct we know what elicits the formation of the microfilament bundle; the inciting agent is the hormone estrogen. We are still in the dark, however, as to the mechanism that triggers microfilament systems to contract. Obviously it is important to understand this process, because it determines when organs are formed and what their shape will be as well as when cells may migrate or nerve axons may elongate.

What might the control mechanism be? We know that the contraction of muscles is activated by nerve impulses. A single migrating cell is not connected to nerves, however, nor does a cell population that is being shaped into an organ usually have any contact with nerves. A hint as to the identity of a possible activating agent in cells was furnished by experiments performed by the British investigator David Gingell.

He found that an injection of calcium ions into an amphibian's egg caused the surface of the egg at the injection site to contract. Short lengths of filamentous material showed up near the site.

Now, it is well known that calcium ions are involved in the contraction of muscle. Might they also perform a similar function at the cell level? My colleague J. F. Ash proceeded to experiment with the calcium treatment, using early embryos of the African clawed toad (*Xenopus laevis*). He inserted the tip of a very fine hollow electrode into cells just through the surface membrane. When a current was applied, calcium ions flowed from within the electrode into the cells. The egg cortex rapidly contracted, and within minutes the electrode was surrounded by a dense black halo composed of pigment granules. Examination of the individual cells with the light microscope and the electron microscope revealed a distinctive band of dense material, largely composed of microfilaments, running through the cytoplasm across the site where the ions had entered the cell. A distinct network of microfilaments could not be discerned, but this may be attributable to the fact that amphibian cells are notoriously difficult to preserve intact for examination with the electron microscope. The observed presence of microfilaments, however, was consistent with the

hypothesis that a network of filaments had been formed in the area where calcium had instigated contractions.

An intriguing question immediately presented itself: Would treatment with cytochalasin counteract the effect of calcium? The answer was clear. Within 10 minutes after the drug was applied the cells of this amphibian embryo lost all capacity to respond to calcium ions. The injection of calcium produced no contraction of the cell cortex, nor did it cause a dark halo to form at the site of injection. Here, then, was a link to the behavior and possible contraction mechanism of other cells that exhibit movement: the migratory cells and those that change shape. Since they are sensitive to cytochalasin, it seems likely that in those cells, as in the cells of the *Xenopus* embryo, calcium is the agent that controls contraction.

Further experiments with the *Xenopus* embryo brought forth another interesting finding. The cells of this embryo, like those of most other early embryos, have the capacity to heal a wound in the cell surface. When a small rip is made in the surface membrane with a needle, the edges of the wound first spread apart and then within minutes are drawn together by contraction to close the opening. This contraction and healing takes place only if calcium is present in the bathing medium. On the other hand, treatment with cytochalasin prevents the healing response. The wound gradually enlarges, even in the presence of calcium, and eventually the cell bursts.

A uniform and consistent picture is beginning to emerge. It appears that the various forms of cell movement—the healing of wounds in normal embryonic cells, the migration of individual cells, the division of cells at mitosis, the shaping of cell populations into organs—are all produced by a common mechanism: systems of contractile microfilaments. It also appears that calcium may play a key role in instigating contraction of this cellular system, just as it does in producing the contraction of skeletal muscles.

If this hypothesis is correct, a crucial question remains to be answered. How is the calcium control called into play? Is this activating substance stored away somewhere in the cell when the cell has no need for movement or operation of the microfilaments? If so, how is it released on demand to spark the important cell movements on which normal development and life depend? And what happens if, through some aberration, the calcium activator is released needlessly, initiating cell movements that can serve no good purpose?

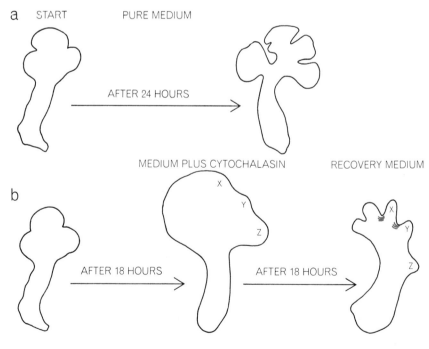

NORMAL AND ABNORMAL BEHAVIOR of an epithelium from a salivary gland was traced in the presence and absence of cytochalasin. The control tissue (*a*) grew in a medium lacking the drug; 24 hours later its clefts had grown deeper and its ramifications were more numerous. After 18 hours in medium that included the drug the test tissue (*b*) showed greatly reduced clefts. With drug removed clefts soon reappeared, as did filaments (*color*).

14

The Mechanism
of Muscular Contraction

by H. E. Huxley
December 1965

*When a muscle contracts, one kind of filament
within it slides past another kind. Electron microscopy
and other techniques have begun to disclose how the
filaments exert a force on each other*

An outstanding characteristic of all animals is their ability to move voluntarily by contracting their muscles. When I summarized our understanding of muscle contraction seven years ago [see "The Contraction of Muscle," by H. E. Huxley; SCIENTIFIC AMERICAN Offprint 19]. it had already been determined that during contraction two kinds of filament in voluntary muscle—thick filaments and thin ones—slide past each other so as to produce changes in the length of the muscle. At that time one could offer only a hypothetical description of contraction at a more detailed level; it was assumed that a relative force is somehow exerted between the thick and thin filaments at sites where they are connected by tiny cross-bridges. Now, thanks to advances in electron microscopy and allied techniques, we have been able to substantiate that hypothesis and to learn considerably more about the nature of the interaction of the thick filaments (composed mainly of the protein myosin) and the thin ones (composed of another protein, actin). It appears that at each site where the proteins of the two kinds of filament are in contact one of them (probably myosin) acts as an enzyme to split a phosphate group from adenosine triphosphate (ATP) and thus provide the energy for contraction. The basic problem is to understand how the conversion of chemical into mechanical energy takes place.

Let us briefly review what is known about the structure and function of muscle. Under the microscope voluntary muscles—for example those that can move the leg of a frog—appear regularly striated at right angles to their length. The muscles responsible for the slow and regular movements of organs that work involuntarily, such as the gut, appear smooth. For reasons of technical convenience most investigations of muscle have dealt with striated muscle, and so our discussion will refer specifically to muscle of that type. A good deal of what has been learned about striated muscle, however, may apply to smooth muscle as well.

Striated muscle can shorten at speeds equal to several times its length per second; it can generate a tension of some 40 pounds per square inch of its cross section; it can contract or relax in a very small fraction of a second. A muscle consists of individual fibers with a diameter of between 10 and 100 microns (a micron is a thousandth of a millimeter); the fibers run the length of the muscle, or a good part of it. Each fiber is surrounded by an electrically po-

STRIATED MUSCLE from the leg of a frog is shown in longitudinal section in an electron micrograph (*top*) and the overlap of filaments that gives rise to its band pattern is illustrated schematically (*bottom*). Parts of two myofibrils (long parallel strands organized into muscle fiber) are enlarged some 23,000 diameters in the micrograph. The myofibrils are separated by a gap running horizontally across the micrograph. The major features of the sarcomere (a functional unit enclosed by two membranes, the *Z* lines) are labeled. The *I* band is light because it consists only of thin filaments. The *A* band is dense (and thus dark) where it consists of overlapping thick and thin filaments; it is lighter in the *H* zone, where it consists solely of thick filaments. The *M* line is caused by a bulge in the center of each thick filament, and the pseudo *H* zone by a bare region immediately surrounding the bulge. The electron micrograph and others illustrating this article were made by the author.

larized membrane, the inside of which is generally a tenth of a volt negative with respect to the outside. Contraction is signaled by an impulse that travels down a nerve to a motor "end plate" in contact with the fiber. The arrival of the impulse depolarizes the membrane and causes the release throughout the fiber of an activating substance, probably calcium. It is this activation that enables one of the muscle proteins to act as an enzyme and split a phosphate group from ATP. The muscle stays contracted until nerve impulses cease (or until it becomes exhausted), at which point the activating substance withdraws, probably by being bound to the sarcoplasmic reticulum, a network of tiny channels within the fiber.

An individual fiber is made up of a number of parallel elements called myofibrils, each about a micron in diameter. Each myofibril itself consists of parallel actin and myosin filaments that, when they are viewed from the end, are seen to lie a few hundred angstrom units apart in a remarkably regular array. (An angstrom unit is a ten-thousandth of a micron.) The myosin filaments are some 160 angstroms in diameter but often appear somewhat thinner when fixed for electron microscopy. They are about a micron and a half in length. The actin filaments are only about a micron in length and 50 to 70 angstroms in diameter. The overlap between the arrays of thick and thin filaments gives rise to the pattern of striations visible in the microscope. The pattern is characterized by a succession of dense bands (called A bands) and light bands (I bands). In the A bands the myosin filaments lie in register in hexagonal array and are responsible for the bands' high density. The actin filaments are attached in register on each side of a narrow, dense structure that traverses the I band: the Z line. In a relaxed muscle the distance between Z lines—one sarcomere—is such that about half of the length of a thin filament and two-thirds of the length of an adjacent thick filament overlap. In the region of overlap in a relaxed fiber the array contains twice as many thin filaments as thick ones. The thin filaments terminate at the edge of the H zone, a region of low density in the center of the A band. In the center of the H zone lies the "pseudo H zone," a region of even lower density that maintains its width no matter how the length of the muscle changes. This light zone surrounds a thin, dark strip known as the M line, which is now thought to be caused by a slight bulge in the center of each thick filament.

When a longitudinal section of muscle is viewed in the electron microscope, it can be seen that the cross-bridges between a given pair of thick and thin filaments come at fairly regular intervals. The cross-bridges are the only mechanical linkage between the filaments, and they are responsible for the structural and mechanical continuity along the whole length of a muscle. It is the cross-bridges that must generate or sustain the tension developed by a muscle. As the sarcomere changes its length, either actively during contraction or passively (stretching or shortening while at rest), the filaments themselves do not perceptibly change in length but slide past one another; the thin filaments move farther into the A bands during shortening and farther out of them during stretching.

Since normal contractions involve changes in the length of the sarcomere of 20 percent or more, the thin filaments in each half of the A band must move distances of at least a quarter of a micron while maintaining tension. It seems physically impossible that the cross-bridges could remain attached to the same point on the actin filament throughout this process. We supposed, therefore, that they are attached to one site on the filament for part of the contraction, then detach and reattach themselves at a new site farther along. Moreover, we assumed that at each site

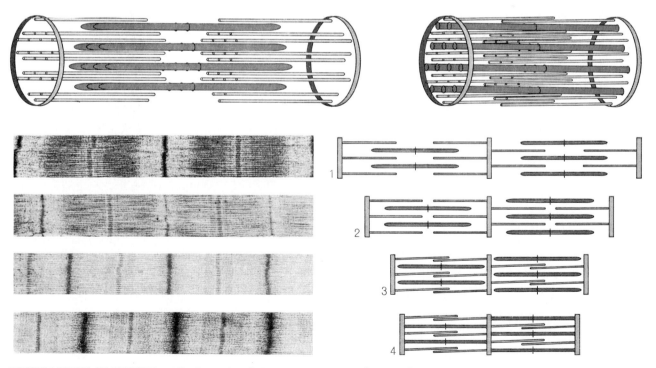

CONTRACTION OF MUSCLE entails change in relative position of the thick and thin filaments that comprise the myofibril (*top left and right*). The effect of contraction on the band pattern of muscle is indicated by four electron micrographs and accompanying schematic illustrations of muscle in longitudinal section, fixed at consecutive stages of contraction. First the *H* zone closes (*1*), then a new dense zone develops in the center of the *A* band (*2, 3 and 4*) as thin filaments from each end of the sarcomere overlap.

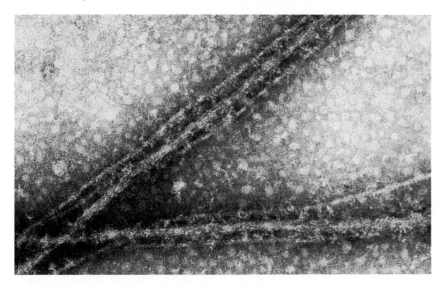

CROSS-BRIDGES between thick and thin filaments are enlarged 180,000 diameters in electron micrograph made by "negative staining." Technique involves surrounding very small objects with a dense salt (*white substance in background*) so that they stand out by contrast.

where cross-bridge and filament interact, one molecule of ATP is split to generate a sliding force between the two kinds of filament (and hence between arrays of filaments).

This general description of the structural changes associated with contraction is the sliding-filament hypothesis put forward a decade ago by Jean Hanson of the Medical Research Council unit at King's College in London and me, and independently by A. F. Huxley and R. Niedergerke of the University of Cambridge. Our hypothesis was partly based on observations of muscle prepared by what is called the thin-sectioning technique. That method involved steeping a chemically fixed (for preservation) and stained (for contrast) piece of muscle in liquid plastic and then cutting the solidified plastic into slices as thin as 100 angstroms. It turned out, however, that the thin-sectioning technique was not adequate to the task of illuminating many details of the hypothesis. In order to ascertain how a force might be developed between thick and thin filaments we needed information about the detailed structure of actin and myosin, and such information was not forthcoming until the arrival of the technique of electron microscopy known as negative staining.

This new method, in which the specimen under examination is embedded in a thin film of some very dense material such as uranyl acetate, has in recent years revealed much about the structure of small spherical viruses and particles of similar size. As adapted in our

laboratory at the Medical Research Council unit in Cambridge, the technique involves applying a drop in which particles of muscle are suspended to an electron microscope grid covered by a thin film of carbon. Many particles adhere to this film; the excess is washed away with a few drops of solvent. Before the preparation dries a shallow drop of the negative-staining material—a heavy metal salt in dilute solution—is applied. It is allowed to dry around the particles. The regions of the particles that are not penetrated by the stain show up clearly by negative contrast because they consist of protein and are much less dense than the salt that surrounds them.

The negative-staining method brings to light far more detail than the conventional positive-staining technique (which artificially increases the density of objects with respect to their background). Its disadvantage is that it can only be applied to very thin specimens; thick ones and the associated thick deposit of negative stain would impair the resolution of the electron microscope image. Thus the method is not directly applicable to whole pieces of tissue such as muscle. The muscle must first be broken down into fragments of suitable thickness (such as individual filaments), which is not easily accomplished.

The usual method of breaking down muscle tissue for purposes of investigation is to homogenize it in the Waring blendor; under this treatment it disperses readily into its constituent myofibrils but no further. In fact, the myo-

fibrils strongly resist further breakdown, probably because the cross-bridges between thick and thin filaments bind the whole structure together in a very robust fashion. Making the assumption that the cross-bridges are the sites where actin and myosin combine, we wondered if we might weaken the structure by suspending the fibrils in certain salt solutions that tend to dissociate the two proteins. We were delighted to find that if muscles, either freshly isolated or preserved in a deep freezer in a solution of water and glycerol, were placed in the appropriate salt solution and then homogenized in the blendor, they indeed broke down into their constituent filaments. Thus they could provide excellent material for examination by the negative-staining technique.

The first specimens we prepared by this method consisted of filaments from the psoas muscle in the back of a rabbit. In the electron micrographs the layer of negative stain was thickest in the region immediately surrounding the filaments; accordingly the filaments have a dense outline [*see illustrations at bottom of opposite page*]. We could at once recognize the thick filaments by their resemblance to the thick filaments in earlier preparations of striated muscle. The diameter of the filaments was the expected 160 angstroms; their length was apparently about 1.5 microns. (Longer structures were never observed but shorter ones, presumably fragments, were.) Small projections, extending sideways from the filaments along most of their length, seemed to correspond to the cross-bridges. Thinner filaments 50 to 70 angstroms in diameter could also be seen, and in places we noticed a large group of such filaments extending for about a micron on each side of a Z line, to which they were still attached [*see illustration at bottom left on opposite page*]. These observations of thick and thin filaments of characteristic size, lying side by side and sometimes still connected by cross-bridges, confirmed the conclusions about the structure of the myofibrils reached earlier by X-ray diffraction techniques and by conventional light microscopy and electron microscopy. We subsequently considered the appearance of the individual filaments more closely.

A regular feature of the thick filaments is a short region, midway along their length, from which the projections we believe to be cross-bridges are absent. This differentiated, projection-free area, some .15 to .2 micron long, can be seen not only in negatively stained

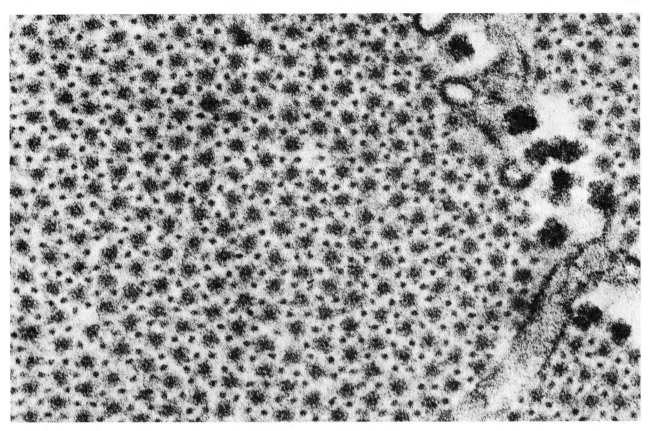

TRANSVERSE SECTION through a frog's leg muscle in its un-
contracted state shows how thick and thin filaments are arrayed in
a regular hexagonal pattern. Breaks in the pattern at the right side
of the micrograph are channels of the sarcoplasmic reticulum. From
the end thick and thin filaments look like large and small dots.
This electron micrograph enlarges them some 200,000 diameters.

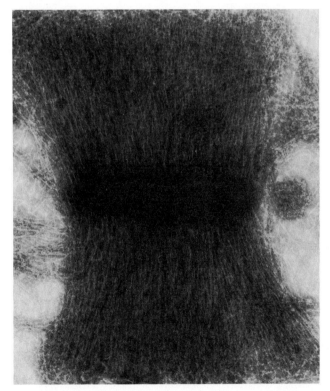

FILAMENTS IN REGISTER at Z line (*membrane in center*) are
the thin filaments of actin, which alone comprise the I segment.
This sample was obtained by homogenizing muscle from the back
of a rabbit in the Waring blendor; it was prepared for the micrograph by negative staining. Magnification is some 47,000 diameters.

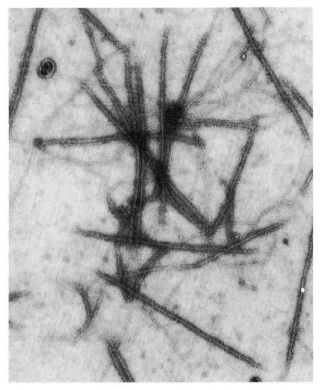

SEPARATED FILAMENTS are from rabbit muscle that has been
homogenized in a Waring blendor. The dark, thick strands are filaments of myosin. The very faint thin strands are filaments of actin.
Thin filaments are still attached to remnant of Z line (*dark patch
at top center*). Filaments are enlarged some 35,000 diameters.

material but also in sectioned specimens. It is now apparent that the absence of cross-bridges from this region is responsible for the mysterious pseudo *H* zone. This zone maintains its uniform size at various muscle lengths because it is a structural feature of the filaments themselves and is not created by their pattern of overlap.

At first sight the projection-free middle region of the thick filaments did not seem particularly significant. It was conceivable that the region was composed of some other protein constituent of muscle. The situation was transformed, however, when we found that filaments of virtually the same appearance could be synthesized from purified solutions

of myosin, the protein that is the main component of the thick filament.

The myosin molecule is known to be an elongated structure with a length of about 1,500 angstroms and a diameter of 20 to 40 angstroms. It can be split (by the enzyme trypsin) into two well-defined fragments; the fragments were named light meromyosin and heavy meromyosin by Andrew G. Szent-Györgyi of the Institute for Muscle Research in Woods Hole, Mass. The heavy-meromyosin fragment has the ability to split a phosphate group from ATP and the ability to combine with actin. The light-meromyosin fragment possesses neither of these attributes but retains

the solubility properties that enable it to form the same kind of structure that intact myosin does. The molecule of heavy meromyosin appears to be more globular than the molecule of light meromyosin. The dimensions of the fragments suggest that before cleavage of the myosin molecule they are arranged in simple end-to-end fashion.

Isolated myosin molecules have been examined under the electron microscope, first by Robert V. Rice of the Mellon Institute in Pittsburgh and subsequently by other workers, by means of the technique known as shadow-casting. This entails treating particles on a film in a vacuum by spraying them at an angle with a vaporized heavy metal.

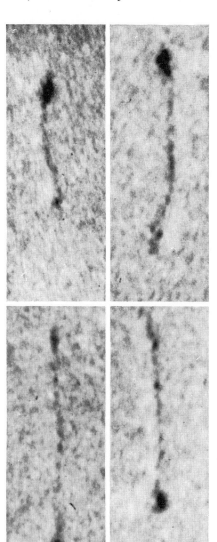

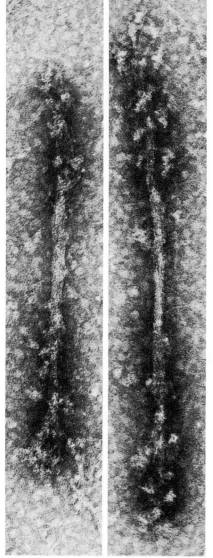

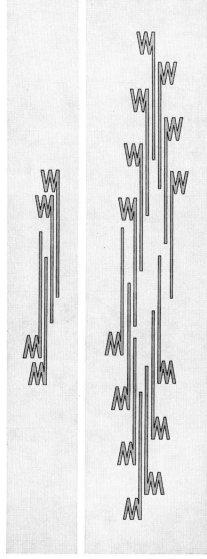

MYOSIN MOLECULES appear in electron micrographs prepared by shadow-casting method. The wide head has enzymatic properties and combines with actin. The straight tail can aggregate with other myosin molecules. Magnification is 300,000 diameters.

AGGREGATIONS of several molecules from a precipitate of pure myosin were negatively stained and magnified 175,000 diameters to reveal their characteristic appearance: a thick strand with projections near the ends and a bare region in the middle.

MOLECULAR STRUCTURE of myosin makes it aggregate in the manner shown here. Head of molecule is schematically represented by zigzag line, tail by straight line. Tails join in center; heads extend as projections at ends, oppositely pointed at each end.

In forming a layer over the sample the metal builds up the particles on the near side and leaves a shadow on the far side, where it is blocked from landing on the underlying film. When myosin molecules are prepared by this method for viewing in the electron microscope, they appear as linear structures with a globular region at one end [*see illustration at left on opposite page*]. Heavy-meromyosin fragments are seen to consist of a large globular head with a short tail. Light-meromyosin fragments appear as simple linear strands. It therefore seems that the intact myosin molecule is asymmetric—a molecule with a head and a tail. The sites (perhaps a single site) responsible for its enzymatic activity and its affinity for actin are located in its globular head, and the sites responsible for its affinity for other myosin molecules are in its tail. The head, which is 40 angstroms in diameter, accounts for about a sixth of the length of the molecule; the tail, 20 angstroms in diameter, accounts for the rest.

It is known that under certain conditions purified myosin in potassium chloride solution will precipitate. When we examined such a precipitate by the negative-staining technique, we were delighted to find that it consisted entirely of filaments. They varied somewhat in length and diameter but generally bore a most remarkable resemblance to the thick filaments prepared directly from muscle. Systematic examination of these synthetic filaments, first of short filaments only two or three times the length of a single myosin molecule and then of longer ones, turned up an even more remarkable feature. The shortest filaments were straight rods some 1,500 to 2,500 angstroms long, with clusters of globular projections at both ends. It occurred to us that we were looking at a small number of myosin molecules arranged in two opposite directions, with their globular heads forming the projections and their linear tails overlapping [*see middle illustration on opposite page*]. Longer filaments had longer clusters of projections, but the projection-free region in the middle of each filament was the same length as the corresponding region in the shorter filaments. The longest synthetic filaments we observed closely imitated the appearance of thick filaments extracted from muscle. It seems clear that myosin filaments grow by the addition of molecules parallel to the molecules that have already aggregated. The molecules are oriented in one of two opposite directions, depending on which

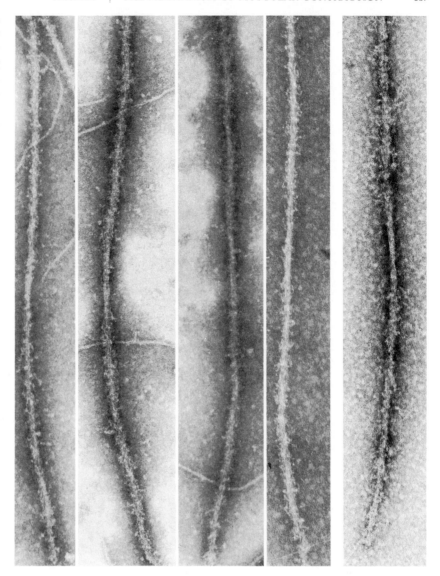

MYOSIN FILAMENTS were obtained directly from muscle homogenized in a blendor (*four electron micrographs at left*) for comparison with a synthetic filament from precipitate of pure myosin (*micrograph at right*). The thick filaments from muscle and the synthetic filament have the same form, characterized by the bridge-free zone in the center and the projections clustered at each end. The filaments are enlarged some 105,000 diameters.

end of the filament a given molecule is joining. It is this method of construction that gives rise to the projection-free region in the synthetic filaments and of course to the same feature in the natural filaments.

This study of myosin molecules and the way they aggregate impressed on us two features that explain the role of these molecules in muscle. First, the head of the molecule has the enzymatic and actin-binding properties we have long assumed the cross-bridges must have. Second, because the molecules aggregate with their heads pointed in one direction along half of the filament and in the opposite direction along the other half, they have an inherent direction-

ality. The first observation leads us to conclude that the heads of myosin molecules serve as the cross-bridges connecting the thick and thin filaments in muscle. The second is important because it explains a crucial feature of the sliding-filament hypothesis at the molecular level.

In a sliding-filament system in which a relative force is developed between actin and myosin molecules located in the two types of filament, it is essential that the appropriate directionality of sliding be built into the filaments in some way. In a striated muscle the thin filaments move toward each other in the center of the A bands, so that it is required that all the elements of force generated by the cross-bridges in one

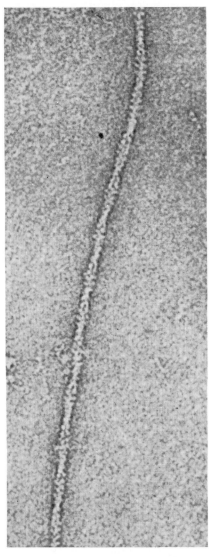

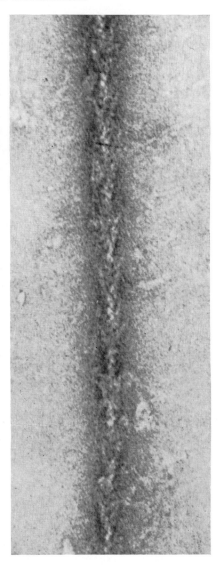

ACTIN FILAMENT has a characteristic structure, visible in micrograph in which filament is enlarged some 420,000 diameters. The filament has the appearance of two coils of globular units wound in a double helix.

"ARROWHEADS" point in one direction along each filament of actin labeled with heavy meromyosin (extract of the globular halves of myosin molecules), implying that actin has an inherent polarity of its own.

muscle show an identical structure; it can often be seen even when they are still attached to a Z line. Filaments made from actin prepared by standard biochemical techniques again show the same pattern. Thus we can confirm that the thin filaments of striated muscle do contain actin, as we had supposed. We can also deduce that the globular subunits are molecules of actin that aggregate to build up the filament. The structure itself might resemble two strings of beads twisted around each other; its alternating high points and low points suggest a general arrangement for the successive active sites on the filament to which the cross-bridges may attach themselves (assuming that each globular unit has one site). We cannot directly view enough of the internal structure or shape of the subunits to make any deductions about their directionality. To reveal such polarity we have used a natural marker, namely heavy meromyosin, the fragment of myosin that combines with actin.

When actin filaments are treated with a solution of heavy meromyosin and examined in the electron microscope by negative staining, they assume a complex appearance that we do not yet understand in full detail. Nevertheless, one salient feature stands out immediately: the filaments of the resulting compound have a well-defined structural polarity that manifests itself in an obvious arrowhead pattern [see illustration at right on this page]. The arrows always point in the same direction over the length of a given filament, even when only dilute solutions of heavy meromyosin have been applied and the arrow pattern is interrupted by long stretches of normal uncombined actin. If the polarity were imposed by some local condition such as the direction along the actin filament at which a series of heavy meromyosin molecules were attached during the formation of the compound filament, one would expect the pattern of arrowheads to lack such consistency. Therefore it would seem that it is the underlying structure of the actin that imposes the pattern. Precisely which feature of the myosin-actin combination gives rise to the arrowhead effect is unclear; it may well be that the pattern reveals the actual orientation of some part of the heavy-meromyosin fragments. A general feature can be deduced, however: all the actin molecules in a filament will combine with heavy meromyosin in precisely the same way [see top illustration on page 160]. We can conclude that all

half of the A band be oriented in the same direction, and that the direction of the force be reversed in the other half. The direction of the force developed as a result of the interaction of actin and myosin would depend either on the orientation of the myosin molecules, the orientation of the actin molecules or both. Our electron microscope observations suggest strongly that all or part of this directionality is achieved by the fact that the myosin molecules are arranged so that they point in the same direction in half of each thick filament (and hence in each A band) and in the opposite direction in the other half [see illustration at right on page 156]. Moreover, we have shown that filaments with this essential reversal of polarity at their midpoint will assemble themselves auto-

matically in vitro from purified preparations of myosin; this finding has obvious relevance to problems of how muscle develops its structure.

Let us now turn to the thin filaments. It was first noticed by Jean Hanson and J. Lowy of the Medical Research Council unit in London that the thin filaments from the smooth muscle of clams had a characteristic beaded appearance. They were able to show that the filament had the form of a double helix consisting of two chains of roughly globular subunits, the chains twisted around each other so that viewed from a given direction the crossover points were about 360 angstroms apart [see illustration at left on this page].

The thin filaments from striated

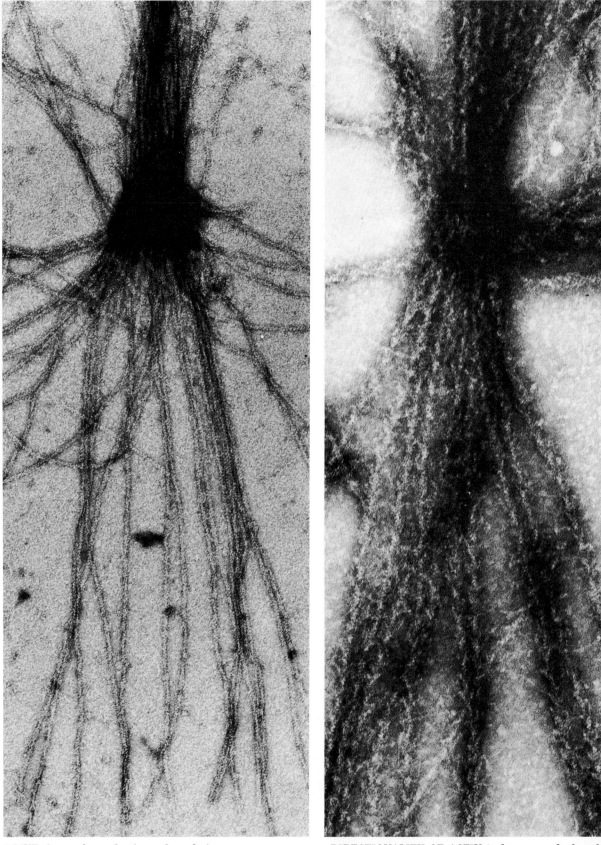

Z LINE, the membrane that forms the end of a sarcomere, appears as dark region from which strands radiate. The strands are thin actin filaments that comprise the *I* band. At times, as in this instance, they remain attached even after muscle has been homogenized. This micrograph and one at right were made by negative staining. They both have magnification of some 165,000 diameters.

DIRECTIONALITY OF ACTIN is demonstrated when thin filaments attached at the *Z* line are labeled with heavy meromyosin. Arrowheads form, pointing away from the *Z* line on each side. In this micrograph they point up at top of *Z* line, down at bottom. Opposite orientation of the two *I* segments of a sarcomere enables filaments from left and right *I* segments to converge on center.

the actin molecules in a given thin filament are oriented in the same sense and that they can all interact in identical fashion with a given myosin cross-bridge.

We have used the same technique to investigate the way in which the actin filaments are attached at the Z lines. As I have mentioned, preparations of thin filaments from homogenized muscle frequently contain groups of filaments still connected to both sides of a Z line. We find, in examining such assemblies after treatment with heavy meromyosin, that the arrows on all filaments always point away from the Z lines. The filaments forming the I substance on one side of a Z line are all similarly oriented; on the opposite side of the Z line the orientation is reversed.

This is exactly the arrangement we require in order for the same relative orientation of the actin and myosin molecules to obtain in the two halves of the A bands but for the absolute orientation to be reversed. The direction of the forces developed will consequently be reversed and the actin filaments can move in opposite directions, that is, toward each other in the middle of the sarcomere.

We conclude that both the thick and the thin filaments in a striated muscle are assembled and oriented in such a way that if a relative force were developed between a given actin and myosin molecule in either filament, all the elements of force in the whole system would be added together in the appropriate manner to give rise to the organized behavior we have observed. Several years ago we tentatively proposed the analogy of a ratchet to describe the interaction of sliding filaments. Now our understanding of the way in which cross-bridges of myosin seem to hook onto consecutive active sites on the actin filament makes the analogy seem even more appropriate.

Recently we have examined with our improved electron microscope techniques sections of muscle fixed at various stages of contraction. The filament lengths, measured by Sally G. Page of University College London, appear to remain constant and equal to the corresponding lengths in resting muscle (discounting small changes in length from tension during fixation and other preparative steps). The most interesting feature of the contraction sequences we have studied is that at the shorter sarcomere lengths a dense zone appears in the center of the A band; the zone progressively increases in width as the muscle shortens. This zone first appears after the H zone has closed up completely. We had shown previously that the closing of the H zone during shortening is caused by the ends of the thin filaments sliding toward each other in the center of the A band. Now when we measured the distance from the Z line to the opposite end of the new dense zone, we found that it was still equal to the length of the thin filaments. We therefore suspected that the new zone might correspond to a region where the thin filaments from each end of the sarcomere overlap [see illustration on page 153].

This view was confirmed when we examined cross sections of muscle cut through the region of supposed double

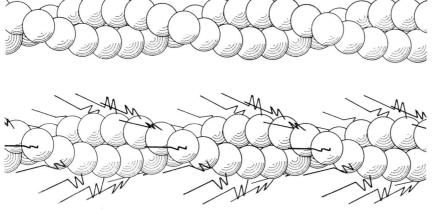

STRUCTURE OF ACTIN is represented by two chains of beads twisted into a double helix (top). The way in which actin might combine with heavy-meromyosin fragments to give rise to arrowheads apparent in micrograph at right on page 158 is suggested at bottom.

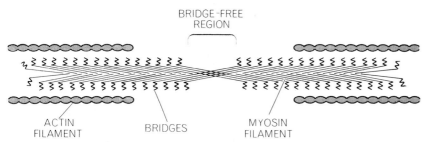

CONTACT OF ACTIN AND MYOSIN in muscle might be made in the manner schematically illustrated here. The thin actin filaments at top and bottom are so shaped that certain sites are closest to thick myosin filament in the middle. The heads of individual myosin molecules (zigzag lines) extend as cross-bridges to the actin filament at these close sites.

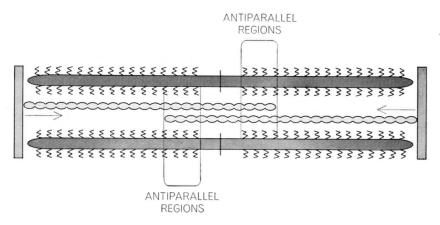

DOUBLE OVERLAP of thin filaments from each side of the sarcomere would result if the sliding-filament hypothesis is essentially correct. It is now assumed that muscle generates maximum tension when thin filaments reach center of A band, and that tension falls when thin filaments cross the center and interact with improperly oriented cross-bridges.

overlap. Instead of the normal pattern of thick and thin filaments in regular hexagonal array with twice as many thin filaments as thick ones, the pattern was less regular and there were four times as many thin filaments as thick ones [*see illustration at right*]. Apparently we were seeing the thin filaments from both ends of the sarcomere at the same time, and these must have slid past one another during the shortening. This finding confirmed that the simple sliding process describes the behavior of the thin filaments under all conditions. (Objectors had proposed, for instance, that the thin filaments might coil up within the A bands.) The finding also suggested to our group and to A. F. Huxley, who is now at University College London, a possible explanation for the observable decrease in tension generated by striated muscles at sarcomere lengths shorter than resting length. Tension would fall off in the double-overlap region because there is a progressively increasing penetration of the thin filaments from one Z line into the "wrong" end of the A bands. We know that actin molecules in part of a thin filament penetrating the center of the A band would have an abnormal orientation with respect to the adjacent cross-bridges [*see bottom illustration on opposite page*]. Such a region would not be expected to contribute to the development of tension by the muscle, and by interfering mechanically and chemically with the interaction of the correctly oriented actin and myosin molecules it might reduce the tension.

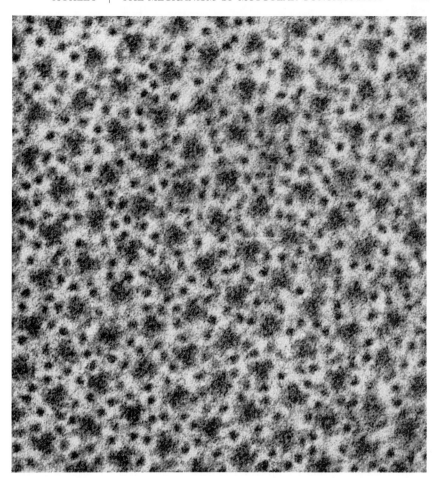

CONTRACTED MUSCLE viewed end on in this electron micrograph has four times as many thin filaments (*small dots*) as thick (*large dots*). The regular array of thick filaments is well preserved; the array of thin filaments is not. Since the ratio of thin to thick filaments in relaxed muscle (*evident in micrograph at top of page 155*) is two to one, it appears that actin filaments from each end of the sarcomere overlap during contraction. This transverse section, made by cutting through the center of an A band of a muscle from the leg of a frog (the method used in making section at top of page 155), is enlarged 250,000 diameters.

Can we extend our investigations to consider changes in the arrangement or configuration of the actin and myosin molecules in muscles that are actually contracting? This goal has indeed been attained, thanks to the sophistication of a technique long used in the study of muscle: X-ray diffraction. Untreated muscle reflects X rays in a regular pattern. We can compare the way in which contracted and relaxed muscles reflect X rays and thus determine if activation and the development of tension are associated with appreciable changes in the length of the repeating units of pattern formed by the arrangement of actin and myosin molecules within the filaments (and hence with possible changes in filament length). Two groups of workers—W. Brown, K. C. Holmes and I in Cambridge, and G. F. Elliott, Lowy and B. M. Millman at the Medical Research Council Biophysics Research Unit at King's College in London—have independently conducted such studies, and both groups report that no such changes in length occur during contraction.

Another exciting finding, reported by our group, is that the relative intensity of some of the X-ray reflections associated with myosin filaments changes greatly during contraction. (Subsequently the London group reported observations consistent with our findings.) These effects have still to be analyzed in detail, but they indicate a substantial movement of the cross-bridges during contractile activity. Very recently members of our group and a group of investigators under J. W. S. Pringle at the University of Oxford have demonstrated a movement of the cross-bridges associated with the contraction of insect flight muscle. These latest findings open up new possibilities. Now that we know that measurable changes in the X-ray reflections do in fact occur during contraction, we have a method of distinguishing steps in the process by which energy for contraction is obtained.

A contracting muscle offers a uniquely favorable system for studying the outstanding problems of protein structure and function. In muscle we have now clearly identified the interacting protein molecules, the high concentration in which they are present and their regularity of arrangement. The major unsolved question about contractility is a general question of biochemistry: how do proteins act as catalysts for biochemical reactions, and what happens to them in the process? It is interesting in this regard to recall that ATP itself was first identified as the source of energy in the contraction of muscle and subsequently as the universal carrier of chemical energy in the living cell. We expect that the study of the precise basis of contractility will also lead to broadly applicable results.

5

The Nerve Axon

by Peter F. Baker
March 1966

*The fiber that conducts the nerve impulse is a tubelike
structure. Its operation can be studied by squeezing
the contents out of the giant axon of the squid and
replacing them with various solutions*

Axons are the communication lines of the nervous system, and along them message-bearing electrical impulses travel from one part of the body to another. They are sometimes compared to electric cables, but they do not carry an electric current the way a wire does. Whereas in a copper wire electricity travels at a speed approaching the velocity of light, in an axon an impulse moves at only about 100 meters per second at best. The interior of an axon is about 100 million times more resistant to the flow of electricity than a copper wire is. Moreover, the membrane sheathing the axon is about a million times leakier to electric current than the sheath of a good cable. If the propagation of electricity in an axon depended on conduction alone, a current fed into it would die out within a few millimeters. The fact is that the axon propagates a current not by simple conduction but by means of a built-in amplifying and relay system.

How does the system work? This is a central question in the investigation of the nervous system, and it has long intrigued physiologists. This article is primarily an account of one of the new techniques developed for investigating the process of nerve-impulse conduction. The technique can be described briefly as experimenting with perfused axons.

The electrical activity of an axon is based on the interaction of three elements: (1) the fluid contents of the axon, called the axoplasm, (2) the membrane that encloses these contents and (3) the outside fluid that bathes the axon. The key to the axon's propagation of an impulse lies in the membrane. Essentially our technique consists in emptying the axon of its contents and perfusing the emptied tube with various experimental solutions, the ob-

ject being to determine just what factors are required to make an electric current travel along the axon.

These studies, like many others that have been done on the transmission of nerve impulses, are made possible by that remarkably convenient gift of nature: the giant axon of the squid. This axon measures up to a millimeter in diameter. The axons of human nerve cells, in contrast, are only about a hundredth of a millimeter in diameter. For the squid the giant axon represents an adaptation to a vital need in its particular way of life. For the physiologist this axon provides an ideal experimental preparation. All the available evidence suggests that experimental results obtained with squid axons are applicable to all other nerve fibers.

Why some animals and not others have evolved giant axons is not fully understood, but such axons appear to be involved in escape responses. The giant axon of the squid is part of the mechanism by which the animal flees its enemies in the water. In order to dart away rapidly the squid uses a jet-propulsion system, squirting water out of a tube at one end of its body. This calls for the synchronous contraction of muscles located throughout its body mantle, and therefore all those muscles must receive the message from the brain simultaneously. The device that takes care of this timing is a variation in the size of the axons radiating from the brain to the various muscles. The farther the muscle is from the brain, the thicker is the axon leading to it, and experiments have shown that the thicker the axon, the faster it conducts impulses. Hence the diameter of the axon is adjusted to the length of the route to be covered, and this ensures that the

signal will reach all the muscles at the same time.

The giant axon is easily dissected out of the squid. It is by probing the isolated axon with microelectrodes and by other means that Kenneth S. Cole in the U.S., A. L. Hodgkin, A. F. Huxley and Bernhard Katz in Britain and other investigators have since 1939 developed an outline of the main events that take place when an electrical impulse passes along a nerve fiber. Much of this story has already been related in *Scientific American*. I shall only briefly review the principal features of the picture.

Between the axoplasm inside the axon and the fluid bathing the axon on the outside there are distinct chemical and electrical differences. Chemically the axoplasm is distinguished by the presence of various organic molecules and a comparatively high content of potassium. The outside fluid, on the other hand, is quite similar to seawater: it principally contains sodium ions and chloride ions. The concentration of potassium is about 30 times higher inside the axon than outside; the concentration of sodium is about 10 times higher outside than inside. Because of these differences potassium ions tend to leak out of the axon (by diffusion) and sodium ions to leak in, to the extent that the membrane will allow the ions to pass. (We can disregard the axoplasm's organic molecules and the outside fluid's chloride in this connection, because the membrane is highly impermeable to them.)

The electric-charge situation complicates the picture. Inside the axon we have a high concentration of positively charged potassium ions; outside, a high concentration of positively charged so-

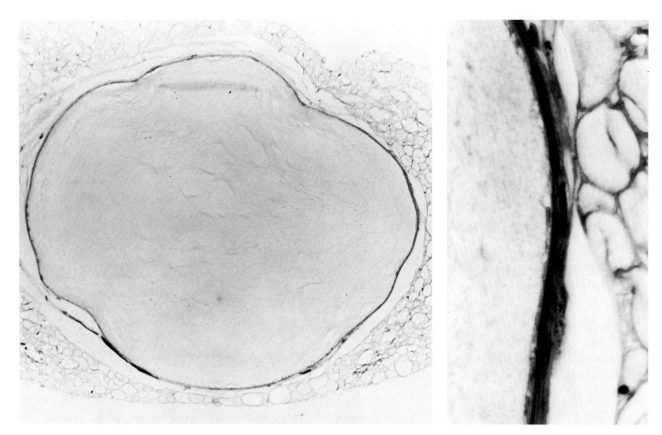

INTACT SQUID AXON is seen in transverse section in these photomicrographs made by J. S. Alexandrowicz of the Plymouth Marine Laboratory. The section is enlarged 140 diameters (*left*). A segment of the axon's perimeter is enlarged 1,150 diameters (*right*). The gray material inside the axon is the axoplasm. The dark boundary of the axon is composed primarily of a layer of Schwann cells.

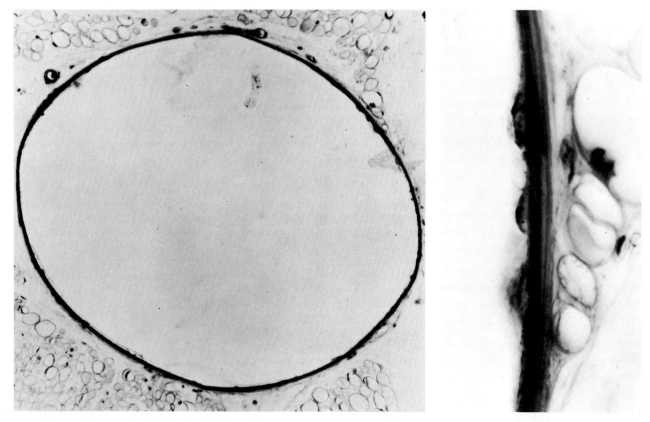

PERFUSED AXON (*left*) and part of its perimeter (*right*) are enlarged as in the top micrographs. The axoplasm has been replaced with a potassium sulfate solution. The small amount of grayish residual axoplasm is thickest near a nucleus of a Schwann cell (*right*).

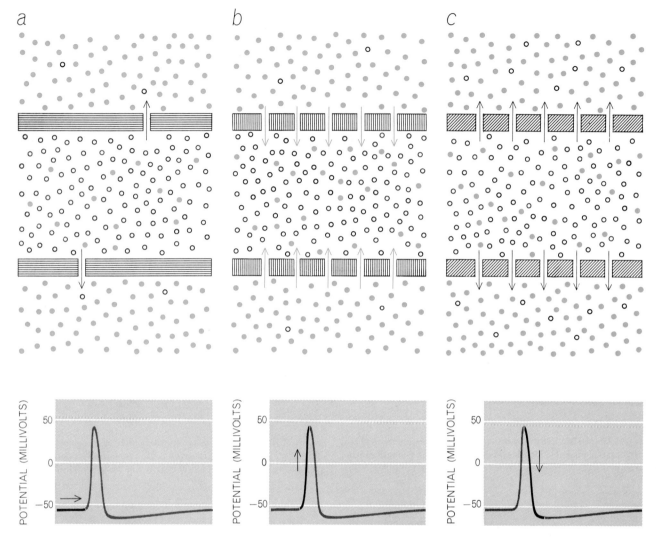

INTERNAL POTENTIAL of an axon is established by the concentration gradients of potassium and sodium ions and by the axon membrane's changing permeability to these two positive ions. Because it is more concentrated inside the axon than outside, potassium (*open circles*) tends to diffuse out; sodium (*colored dots*), more concentrated outside than inside, diffuses in. A normal resting axon membrane is more permeable to potassium than to sodium. The resulting net outward diffusion of potassium establishes the inside-negative resting potential (*a*). An action potential occurs when the nerve is depolarized (by a nerve impulse or artificial stimulation). The action potential has two phases. First the membrane becomes very permeable to sodium, which enters the axon and makes the inside positive (*b*). Then the sodium "gates" close; the membrane thereupon becomes very permeable to potassium, which moves out, making the interior negative again (*c*). A reduction in potassium permeability reestablishes the resting condition.

dium ions. When the membrane is in its unstimulated resting condition, it is highly permeable to potassium but only slightly permeable to sodium. Because of potassium's relative freedom to pass through the membrane and its steep concentration gradient, it tends to leak out of the axon at a high rate. The result is that the inside of the axon becomes electrically negative with respect to the outside. There is a limit to this process; an equilibrium is reached when the tendency for potassium to diffuse out is balanced by the electric field that has been set up. At this point the interior of the axon is about 60 millivolts (thousandths of a volt) negative in relation to the external fluid. That difference—a negative potential of 60 milli-

volts—is called the resting potential of the nerve cell. It is created, in effect, by a potassium battery.

It is relevant to inquire what would happen to the potential if the membrane were highly permeable not to potassium but to one of the other ions. The concentration gradient of the positively charged sodium ions is from outside to inside, acting to make the inside of the axon positive. The gradient of the negatively charged organic molecules is from inside to outside and will also act to make the inside positive. The negatively charged chloride ions are more concentrated outside the axon than inside and will act to make the inside negative. Altering the permeability of the membrane to these ions can estab-

lish a wide range of potentials, the inside being negative (because of potassium or chloride) or positive (because of sodium or organic ions) or anything in between (because of a mixture). The possibilities this variability offers for the control of membrane potential have been thoroughly exploited by the cells of the central nervous system.

Obviously any change in the membrane's permeability to one of the ions can change the potential. This is precisely what happens when an electrical impulse passes along the axon. The current reduces the potential at the point of its arrival, and the resting potential there drops toward zero; the membrane is said to be depolarized. In response to the drop in potential the membrane's permeability to sodium suddenly increases; this further reduces the potential, which in turn makes the membrane still more permeable to sodium. As if a door were suddenly opened wide, sodium ions from the surrounding fluid rush into the axon. The result is that within a small fraction of a second the interior of the axon switches from a *negative* potential of about 60 millivolts to a *positive* potential of about 50 millivolts. The new condition is the first phase of what is called an action potential.

The local region within the axon is now positive, whereas the next adjacent section, which still has a normal resting potential, is negative. Consequently a current flows from the positive to the negative region, completing the circuit by returning to the positive region through the conducting solution outside the axon. The current arriving in the region of normal resting potential opens the membrane door to sodium and thus triggers the generation of an action potential like that in the region it has just left. In this manner the impulse is regenerated from point to point along the axon and flows from one end of it to the other. In each region, shortly after an action potential has been generated, the membrane's permeability to sodium is switched off and its permeability to potassium increases, and as a result that section of the axon returns to the resting potential. The entire local action potential lasts only about a millisecond.

For many years physiologists have been looking into this process experimentally. What would happen if the concentration of ions on one side of the membrane or the other was changed artificially? Isolated axons remain functional for many hours when they are immersed in a simple salt solution containing the major ions present in seawater. It is an easy matter to vary the concentration of these ions in order to study their influence on the process of impulse conduction. Experimenters found that when they added potassium to the medium, thereby reducing the potassium gradient between the inside of the axon and the outside, the resting potential dropped. When they removed sodium from the medium (replacing it with an osmotically equivalent amount of sugar or the positive organic molecule choline), the axon became incapable of propagating an electrical impulse. These results supported the general view of the critical roles normally played by the sodium and potassium

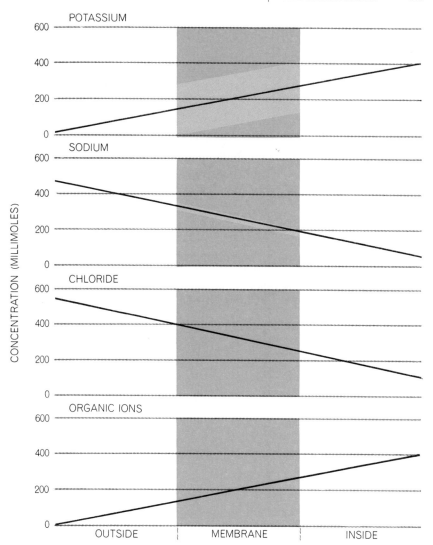

CONCENTRATION GRADIENTS for the major nerve-axon ions are shown. The concentration of each ion on the outside (*left*) and inside (*right*) of the membrane (*color*) is given; the slope shows the direction of the concentration gradient, down which the ions have a tendency to diffuse. This diffusion is blocked by the membrane, which resists mixing of internal and external ions except as it is permeable to one or another of them. The permeability of the membrane during the normal resting phase is suggested here: it is 10 times as permeable to sodium as to organic ions or chloride, 10 times as permeable to potassium as to sodium. Sodium and potassium ions are positive, chloride and organic ions negative.

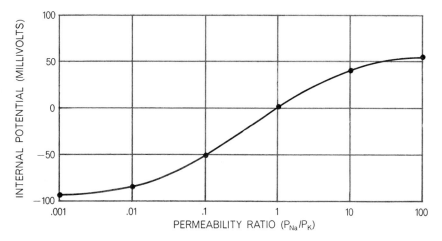

MEMBRANE PERMEABILITY to potassium and sodium ions can be varied. The curve shows, for the concentrations given in the chart at the top of the page, the effect on the internal potential of changes in the membrane's relative permeability to sodium and potassium.

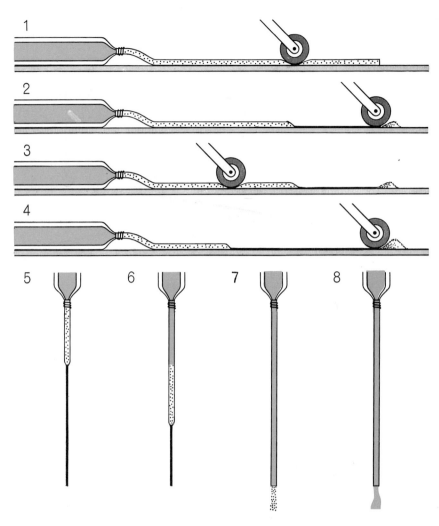

AXOPLASM is removed from a piece of squid axon and replaced with an experimental solution. A piece of axon with a cannula fixed to its smaller end is placed on a rubber pad and axoplasm is squeezed out of it by successive passes with a rubber roller (1–4). Then the nearly empty sheath is suspended vertically in seawater (5) and the perfusion fluid is forced through it, removing the remaining plug of axoplasm and refilling the axon (6–8).

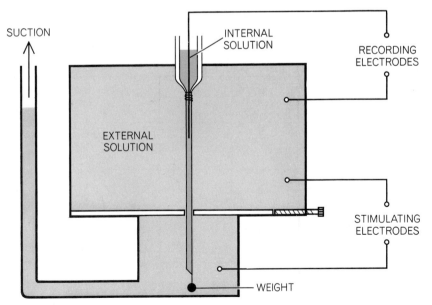

PERFUSED AXON is mounted so as to allow solutions to be changed with an internal electrode in position. To fill it with a new solution, suction is applied to the external fluid.

gradients in the generation of the action potential and the resting potential.

Experiments were also performed in changing the composition of the axoplasm within the axon. To inject extra sodium or potassium into the intact fiber a special microsyringe was used by Harry Grundfest, C. Y. Kao and M. Altamirano at the Columbia University College of Physicians and Surgeons, and also by Hodgkin and Richard D. Keynes at the University of Cambridge. It is not possible to change the inside-outside gradients of ions by any great amount in this way, but when extra sodium was injected into the axon, raising its concentration closer to the level of the sodium outside, the action potential produced by an electrical impulse was smaller than normal, presumably because the usual inrush of sodium ions was checked.

Could one find a way to remove the core of the axon and substitute a strictly experimental solution inside the fiber? Recently two different techniques for doing this were developed almost simultaneously, one by Ichiji Tasaki and his co-workers at the Marine Biological Laboratory in Woods Hole, Mass., and the other by T. I. Shaw and the author at the Plymouth Marine Laboratory in Britain. In Tasaki's method a fine capillary tube is used to ream out some of the axoplasm, and then a fluid is run through the tube to wash out as much of the remaining material as possible. It is an extremely tricky operation, particularly on the comparatively small "giant" axons of the squid found in North American waters, and a considerable amount of axoplasm is left in the fiber.

Our technique is crude by comparison, but we have found that it effectively removes most of the axoplasm. It is based on an old stratagem: in 1937 Richard S. Bear, Francis O. Schmitt and J. Z. Young, working at the Marine Biological Laboratory in Woods Hole, discovered that if an end of an axon was cut, they could squeeze the axoplasm out of it much as one squeezes toothpaste out of a tube. We were led to wonder about the casing left by this operation. Did the squeezing spoil the membrane's properties or could it still conduct an electrical impulse? We undertook an experiment to answer the question. We laid a section of axon about eight centimeters long (a little more than three inches) on a rubber pad and squeezed axoplasm out of the wider end by a series of fairly firm strokes with a rubber-covered roller. When the axoplasm had been extruded from half of the axon's length, leaving that part a flattened sheath, we rolled the other half to push some of its axoplasm into the emptied section. To our surprise we found that the roughly handled membrane could still conduct and boost electrical impulses. This experiment showed that the excitable properties of the membrane are not destroyed by extrusion and encouraged us to try to replace the axoplasm with artificial solutions.

Our procedure was to insert a small glass cannula into the narrower end of the axon and squeeze out as much of the axoplasm as possible; then, attaching the cannula to a motor-driven syringe, we suspended the nerve in seawater and forced "artificial axoplasm" through the flattened sheath. When the nerve was refilled, its excitability was tested, and in about 75 percent of the experiments the preparation was found to function like a normal nerve. With this method about 95 percent of the axoplasm is extruded and a long length of axon can be perfused. Once perfused, the axon can be tied off at both ends and handled like a normal axon. It is more convenient, however, to mount it in such a way that a microelectrode can be inserted into the axon and, with the electrode in place, the perfusion fluid can be changed repeatedly [see illustrations on opposite page]. It is also possible to change the external solution.

We now undertook a series of perfusion experiments in collaboration with Hodgkin. The first question we asked was: What substances must the axoplasm contain in order to generate a resting potential and an action potential? The requirements turned out to be simple indeed. The only essentials are that the solution be rich in potassium ions and poor in sodium ions, and that it have about the same osmotic pressure and concentration of hydrogen ions as normal axoplasm. As for the negatively charged ions (which normally are ionized organic molecules), their nature is not critical; we have used a wide variety of negative ions successfully. A particularly convenient solution is buffered potassium sulfate. When axons are perfused with this solution, they produce resting and action potentials that are almost identical with those generated by intact fibers [see top illustration at right].

It appears, then, that the bulk of the natural axoplasm is not necessary for the propagation of impulses. Are crucial remnants of the original material left in our axons—substances that could not be dispensed with and that make conduction possible? It is not easy to answer categorically; all that can be said is that the axons are still fully functional and can conduct up to half a million impulses, even after they have been washed by a flow of artificial solution amounting to 100 times the volume of the original fluid. If essential molecules were diffusing from the remaining axoplasm to the membrane, they should have been completely washed out by such a massive flow. There remains the possibility that something essential for impulse conduction is supplied by a layer of cells that surrounds the giant axon. The function of these cells, called Schwann cells, is obscure. Various kinds of evidence suggest, however, that they are not directly involved in electrical conduction. For instance, when the Venezuelan workers R. Villegas, Maximo Gimenez and L. Villegas inserted two microelectrodes into a squid nerve—one into a Schwann cell and the other into the axon—they were unable to detect any electrical change in the Schwann cell when an action potential passed along the axon.

Perhaps the most remarkable finding is the fact that apparently no source of energy other than the difference in ion concentrations on the two sides of the membrane is needed to amplify an electical impulse and propagate it along the axon. Our artificial axoplasm contained no sugar, adenosine triphosphate (ATP) or other chemical source of energy, and it is unlikely that any ATP was produced by the traces of original axoplasm left in the sheath. Yet the axon could generate both resting and action potentials whenever the inside and outside solutions had the right concentrations of potassium ions and sodium ions.

To produce the resting potential we need only make sure that the fluid perfusing the experimental axon is primed with a sufficiently high concentration of potassium. If we substitute sodium for potassium in the potassium sulfate perfusion fluid, the resting potential drops, and the amount of this drop depends on the extent of the substitution; when the concentration of potassium inside the axon is reduced to the same level as that outside, the potential drops to zero. If we make the potassium concentration outside much higher than that inside, the inside of the axon becomes positive,

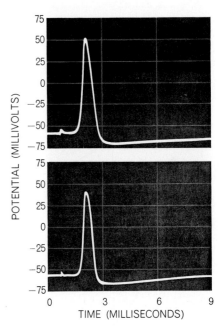

ACTION POTENTIALS from an axon perfused with potassium sulfate (*top*) and from an intact axon (*bottom*) are quite similar.

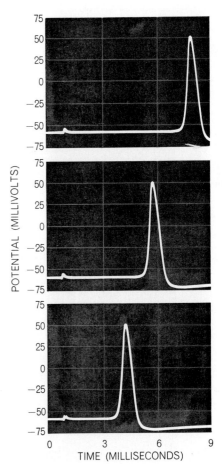

DEGREE OF INFLATION with perfusion fluid does not change the action potential but does increase the conduction velocity. These potentials were recorded from an empty axon (*top*), a partly inflated one (*middle*) and a fully inflated one (*bottom*).

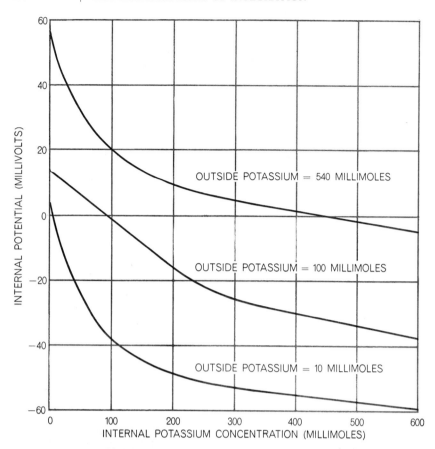

RESTING POTENTIALS of perfused axons depend on the potassium concentration gradient. The internal potassium content was changed by substituting sodium chloride for potassium chloride inside the axons. The lower the potassium gradient, the less negative the internal potential became. With the gradient reversed the internal potential became positive.

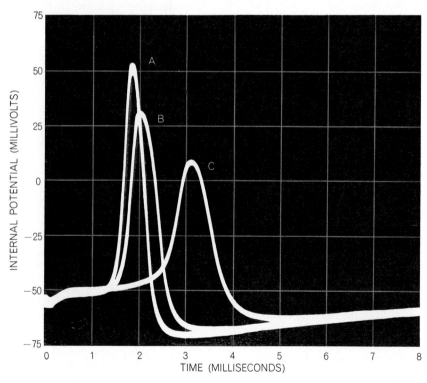

"OVERSHOOT," the amount by which the action potential rises above zero, decreases as the sodium gradient is reduced. These potentials are from a normally perfused axon (A) and axons in which a quarter (B) and half (C) of the potassium was replaced with sodium.

that is, the sign of the resting potential is plus instead of minus. These observations are what would be expected if the resting nerve membrane is freely permeable to potassium ions but not to other ions. The magnitude and the sign of the potential generated are dependent on the steepness and the direction of the potassium concentration gradient.

These experiments also show that the inward diffusion of the negatively charged chloride ions contributes very little to the resting potential. When there is no potassium concentration gradient, there is still a steep gradient for chloride, and yet the measured resting potential is close to zero. This type of experiment suggests that the resting membrane is highly impermeable to negative ions but freely permeable to potassium. A similar argument applied to sodium indicates that the resting membrane is also impermeable to this ion.

During the action potential, on the other hand, the permeability to sodium ions is markedly increased, but the amount by which the action potential overshoots zero and becomes positive is dependent on the sodium concentration gradient. The progressive replacement of potassium inside the axon by sodium reduces the overshoot and finally abolishes the axon's ability to conduct impulses; the ability is restored as the sodium is washed out and replaced with potassium. A rather interesting corollary is the change in action potential that occurs when most of the potassium sulfate in the perfusion fluid is replaced by an osmotically equivalent amount of sugar. Under these conditions the overshoot of the action potential is increased, only to fall again when the axon is refilled with potassium sulfate. This suggests that when potassium is present within the axon, it acts to some extent as a barrier to the inflow of sodium. That is to say, the potassium ions serve in some degree as if they were sodium ions, and when they are absent, the sodium gradient from outside to inside is steepened.

In performing these experiments we noticed that, although the substitution of sugar for potassium sulfate enhanced the action potential, it slowed the axon's conduction of impulses. Presumably this could be attributed to the low electrical conductivity of sugar, which would shorten the range of each stage of propagation (that is, each local circuit) and thereby slow the overall rate of travel. Indeed, we found by experiment that the higher the electrical conductivity of the material we used for perfusing the axon, the faster the speed of impulse propagation.

Recent experiments on perfused axons have been directed toward a detailed analysis of the properties of the action-potential mechanism. These investigations have depended to a large extent on the device known as the voltage clamp. This technique was devised almost 20 years ago by Cole and his co-workers and by Hodgkin, Huxley and Katz, and in their hands it provided almost all the detailed evidence for the sequence of changes in membrane permeability that occurs during the action potential.

The technique is simple in principle but often very difficult in practice. The idea is to produce a sudden displacement of the membrane potential from its resting value and to hold the potential at this new fixed level by means of a feedback amplifier. The current that flows through a definite area of membrane under the influence of the impressed voltage is measured with a separate amplifier [*see illustration at right*].

When the membrane of an intact axon is depolarized and the potential is held at some value close to zero, the current that flows is at first directed inward, but it rapidly reverses its direction and continues to flow outward as long as the membrane is kept depolarized. There is every indication that the initial inward current is carried by a rapid inflow of sodium ions resulting from a transient increase in the membrane's permeability to sodium, and that the delayed outward current results from a prolonged increase in its permeability to potassium. For one thing, experiments with the same technique show that if sodium is absent from the outside medium, so that the downward gradient of sodium concentration is from inside to outside instead of the other way around, there is a small initial outward flow of current instead of an inward one. Sodium ions now move from inside to outside through the door of increased permeability opened by depolarization of the membrane. The same result can be obtained, even when the axon is bathed on the outside by seawater, by reversing the resting potential, that is, by making the inside of the axon positive instead of negative and holding the potential at the positive value. The potential difference then drives sodium ions out of the axon against the chemi-

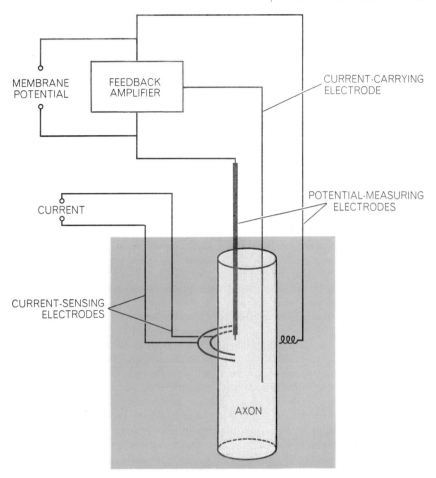

"VOLTAGE CLAMP" is set up as shown in this schematic diagram. A change in membrane potential is produced by the current-carrying electrode. Sensed by two other electrodes, the potential is thereafter maintained by a small current regulated by feedback. The current that flows through a known area of membrane at this fixed potential is then measured.

cal gradient, and a current flows outward. At some intermediate value the applied potential just balances the tendency of sodium ions to enter the axon and there is no detectable early current. This potential is called the sodium equilibrium potential.

The technique of perfusing axons allows the voltage clamp to be applied to situations in which the interior of the axon, as well as the external environment, can be varied. What will happen if sodium is completely absent both outside and inside the axon? In such a case we might expect that, if the axon contains only potassium, depolarization of the membrane should simply bring about a lasting outward current, carried by potassium ions diffusing out of the axon. When Knox Chandler and Hans Meves of our laboratory applied the voltage clamp, they found that depolarization always resulted promptly in a small outward current that was identical in its prop-

erties with the one normally carried by sodium ions. Were potassium ions in the axon acting like sodium ions in this case, passing out through the same membrane channel that opened up for sodium ions during depolarization?

Chandler and Meves tested this surmise by adding sodium to the formerly sodium-free outside medium. If the potassium ions within the axon did not behave like sodium, the downward gradient of sodium concentration from outside to inside would be very steep; hence a large inside-positive potential should be required to prevent a flow of current across the membrane. Actually the experiments showed that the equilibrium potential (the potential required to abolish the early inward current) had the value that would be expected if the potassium inside the axon acted like a small amount of sodium. These results indicate that the channel by which sodium ions enter the axon during an action potential is not completely specific for sodium ions. Chandler and

Meves were able to calculate from their measurements that the "sodium channel" enables sodium ions to pass about 12 times more easily than potassium ions.

This conclusion has recently been strengthened by a different approach. The fish known as the puffer manufactures a potent nerve poison called tetrodotoxin; analysis by means of the voltage clamp of how this poison acts has shown that it very specifically blocks the increase in permeability to sodium that occurs immediately following depolarization. It has no effect on the delayed increase in permeability to potassium. When tetrodotoxin was applied to perfused axons, Chandler and Meves found that it also blocks the early current carried by potassium ions, thus confirming the view that the potassium is passing through the "sodium channel."

The axon membrane loses its capacity for increasing its permeability to sodium when its resting potential is kept at progressively less negative levels. Hence as the resting potential is reduced toward zero (for instance by the replacement of potassium ions inside the axon with sodium or choline) the mechanism for admitting sodium is progressively inactivated, and therefore the axon's capacity for producing an action potential becomes progressively smaller. If, however, the potassium in the perfusion fluid is replaced by sugar instead of another ion, this loss is not so sharp; the membrane maintains its activity, or ability to increase its permeability to sodium, at much lower levels of resting potential. Chandler, Hodgkin and Meves have proposed a possible explanation for this puzzling result. They suggest that the reduction or elimination of ions in the perfusion fluid uncovers negatively charged groups of atoms on the inner face of the membrane. This process would increase the electric field within the membrane without altering the total potential difference between the internal and external solutions. Accordingly a charged molecule within the membrane will experience an imposed electric field identical with that which it experiences in an intact axon at the resting potential.

Experiments with perfused axons have yielded results that would have been impossible to obtain with intact axons and thus represent a considerable advance, but in general they have not produced any revolutionary changes in ideas about the mechanism of propagation of the nerve impulse. They have, however, served to define more sharply the questions to be answered about the basic chemistry of the process. How do ions pass through the membrane? What makes the resting membrane so much more permeable to potassium than to sodium? What specific change in the membrane (brought about by a drop in potential) causes it suddenly to open its doors to sodium?

There are many hypotheses on these questions but so far no convincing items of evidence. The difficulties facing those who are trying to solve such problems in chemical terms are immense. Many of the unique properties of living membranes are dependent on the potential that normally exists across them. The application of the routine biochemical technique of homogenizing cells in order to isolate the cell membrane would break up the membrane and destroy the potential across it; this might so alter the architecture of the molecular groups involved in nerve activity that they would be unrecognizable. Moreover, the relevant groups are probably quite thinly scattered through the membrane material and hence much diluted by less interesting molecules. Perhaps artificial membranes, synthesized from substances extracted from natural ones, will yield some clues, but a complete explanation of the behavior of the nerve membrane in molecular terms is probably still a long way off.

Further experiments with perfused axons may tell us something about the process of recovery from nerve activity. During the passage of an impulse in an intact nerve there is a small net gain of sodium and a small net loss of potassium. If this were to continue unchecked, the nerve would lose potassium and gain sodium, and the concentration gradients on which nerve activity depends would be destroyed. The tendency is counteracted by the mechanism known as the "sodium pump," which uses metabolic energy in the form of ATP to extrude sodium ions from the axon in exchange for potassium ions. Although an intact nerve can function for some time without an operating sodium pump, the pump is essential for the long-term maintenance of nerve activity. In perfused axons this is not the case, since fresh fluid can be constantly passed through the axon. There is evidence, however, that the pump mechanism survives perfusion and that it can be activated by adding ATP and small amounts of magnesium and sodium to the perfusion fluid.

The Synapse

by Sir John Eccles
January 1965

*How does one nerve cell transmit the nerve impulse to
another cell? Electron microscopy and other methods
show that it does so by means of special extensions
that deliver a squirt of transmitter substance*

The human brain is the most highly organized form of matter known, and in complexity the brains of the other higher animals are not greatly inferior. For certain purposes it is expedient to regard the brain as being analogous to a machine. Even if it is so regarded, however, it is a machine of a totally different kind from those made by man. In trying to understand the workings of his own brain man meets his highest challenge. Nothing is given; there are no operating diagrams, no maker's instructions.

The first step in trying to understand the brain is to examine its structure in order to discover the components from which it is built and how they are related to one another. After that one can attempt to understand the mode of operation of the simplest components. These two modes of investigation—the morphological and the physiological—have now become complementary. In studying the nervous system with today's sensitive electrical devices, however, it is all too easy to find physiological events that cannot be correlated with any known anatomical structure. Con-

versely, the electron microscope reveals many structural details whose physiological significance is obscure or unknown.

At the close of the past century the Spanish anatomist Santiago Ramón y Cajal showed how all parts of the nervous system are built up of individual nerve cells of many different shapes and sizes. Like other cells, each nerve cell has a nucleus and a surrounding cytoplasm. Its outer surface consists of numerous fine branches—the dendrites—that receive nerve impulses from other nerve cells, and one relatively long branch—the axon—that transmits nerve impulses. Near its end the axon divides into branches that terminate at the dendrites or bodies of other nerve cells. The axon can be as short as a fraction of a millimeter or as long as a meter, depending on its place and function. It has many of the properties of an electric cable and is uniquely specialized to conduct the brief electrical waves called nerve impulses [see the article "How Cells Communicate," by Bernhard Katz, beginning on page 135]. In very thin axons these impulses travel at less

than one meter per second; in others, for example in the large axons of the nerve cells that activate muscles, they travel as fast as 100 meters per second.

The electrical impulse that travels along the axon ceases abruptly when it comes to the point where the axon's terminal fibers make contact with another nerve cell. These junction points were given the name "synapses" by Sir Charles Sherrington, who laid the foundations of what is sometimes called synaptology. If the nerve impulse is to continue beyond the synapse, it must be regenerated afresh on the other side. As recently as 15 years ago some physiologists held that transmission at the synapse was predominantly, if not exclusively, an electrical phenomenon. Now, however, there is abundant evidence that transmission is effectuated by the release of specific chemical substances that trigger a regeneration of the impulse. In fact, the first strong evidence showing that a transmitter substance acts across the synapse was provided more than 40 years ago by Sir Henry Dale and Otto Loewi.

It has been estimated that the hu-

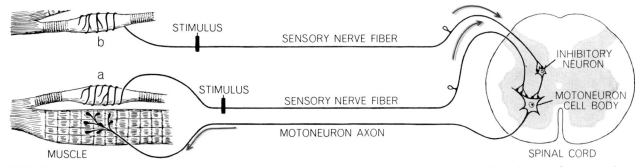

REFLEX ARCS provide simple pathways for studying the transmission of nerve impulses from one nerve cell to another. This transmission is effectuated at the junction points called synapses. In the illustration the sensory fiber from one muscle stretch receptor (*a*) makes direct synaptic contact with a motoneuron in the spinal cord. Nerve impulses generated by the moto-

neuron activate the muscle to which the stretch receptor is attached. Stretch receptor *b* responds to the tension in a neighboring antagonistic muscle and sends impulses to a nerve cell that can inhibit the firing of the motoneuron. By electrically stimulating the appropriate stretch-receptor fibers one can study the effect of excitatory and inhibitory impulses on motoneurons.

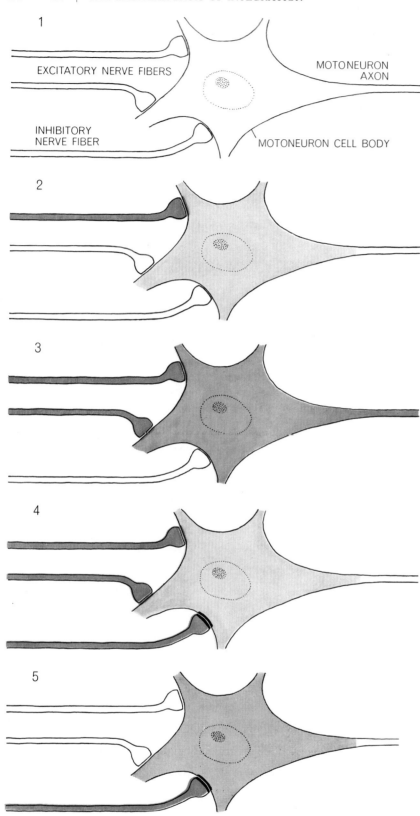

EXCITATION AND INHIBITION of a nerve cell are accomplished by the nerve fibers that form synapses on its surface. Diagram *1* shows a motoneuron in the resting state. In *2* impulses received from one excitatory fiber are inadequate to cause the motoneuron to fire. In *3* impulses from a second excitatory fiber raise the motoneuron to firing threshold. In *4* impulses carried by an inhibitory fiber restore the subthreshold condition. In *5* the inhibitory fiber alone is carrying impulses. There is no difference in the electrical impulses carried by excitatory and inhibitory nerve fibers. They achieve opposite effects because they release different chemical transmitter substances at their synaptic endings.

man central nervous system, which of course includes the spinal cord as well as the brain itself, consists of about 10 billion (10^{10}) nerve cells. With rare exceptions each nerve cell receives information directly in the form of impulses from many other nerve cells—often hundreds—and transmits information to a like number. Depending on its threshold of response, a given nerve cell may fire an impulse when stimulated by only a few incoming fibers or it may not fire until stimulated by many incoming fibers. It has long been known that this threshold can be raised or lowered by various factors. Moreover, it was conjectured some 60 years ago that some of the incoming fibers must inhibit the firing of the receiving cell rather than excite it [*see illustration at left*]. The conjecture was subsequently confirmed, and the mechanism of the inhibitory effect has now been clarified. This mechanism and its equally fundamental counterpart—nerve-cell excitation—are the subject of this article.

Probing the Nerve Cell

At the level of anatomy there are some clues to indicate how the fine axon terminals impinging on a nerve cell can make the cell regenerate a nerve impulse of its own. The top illustration on the next page shows how a nerve cell and its dendrites are covered by fine branches of nerve fibers that terminate in knoblike structures. These structures are the synapses.

The electron microscope has revealed structural details of synapses that fit in nicely with the view that a chemical transmitter is involved in nerve transmission [*see lower two illustrations on next page*]. Enclosed in the synaptic knob are many vesicles, or tiny sacs, which appear to contain the transmitter substances that induce synaptic transmission. Between the synaptic knob and the synaptic membrane of the adjoining nerve cell is a remarkably uniform space of about 20 millimicrons that is termed the synaptic cleft. Many of the synaptic vesicles are concentrated adjacent to this cleft; it seems plausible that the transmitter substance is discharged from the nearest vesicles into the cleft, where it can act on the adjacent cell membrane. This hypothesis is supported by the discovery that the transmitter is released in packets of a few thousand molecules.

The study of synaptic transmission was revolutionized in 1951 by the introduction of delicate techniques for recording electrically from the interior of single nerve cells. This is done by inserting into the nerve cell an extremely fine glass pipette with a diameter of .5 micron—about a fifty-thousandth of an inch. The pipette is filled with an electrically conducting salt solution such as concentrated potassium chloride. If the pipette is carefully inserted and held rigidly in place, the cell membrane appears to seal quickly around the glass, thus preventing the flow of a short-circuiting current through the puncture in the cell membrane. Impaled in this fashion, nerve cells can function normally for hours. Although there is no way of observing the cells during the insertion of the pipette, the insertion can be guided by using as clues the electric signals that the pipette picks up when close to active nerve cells.

When my colleagues and I in New Zealand and later at the John Curtin School of Medical Research in Canberra first employed this technique, we chose to study the large nerve cells called motoneurons, which lie in the spinal cord and whose function is to activate muscles. This was a fortunate choice: intracellular investigations with motoneurons have proved to be easier and more rewarding than those with any other kind of mammalian nerve cell.

We soon found that when the nerve cell responds to the chemical synaptic transmitter, the response depends in part on characteristic features of ionic composition that are also concerned with the transmission of impulses in the cell and along its axon. When the nerve cell is at rest, its physiological makeup resembles that of most other cells in that the water solution inside the cell is quite different in composition from the solution in which the cell is bathed. The nerve cell is able to exploit this difference between external and internal composition and use it in quite different ways for generating an electrical impulse and for synaptic transmission.

The composition of the external solution is well established because the solution is essentially the same as blood from which cells and proteins have been removed. The composition of the internal solution is known only approximately. Indirect evidence indicates that the concentrations of sodium and chloride ions outside the cell are respectively some 10 and 14 times higher than the concentrations inside the cell. In contrast, the concentration of potassium ions inside the cell is about 30 times higher than the concentration outside.

How can one account for this re-

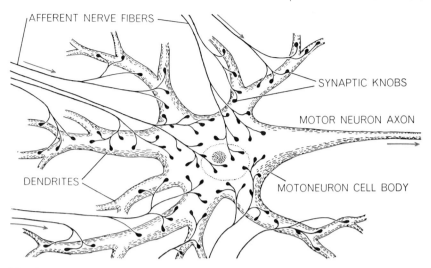

MOTONEURON CELL BODY and branches called dendrites are covered with synaptic knobs, which represent the terminals of axons, or impulse-carrying fibers, from other nerve cells. The axon of each motoneuron, in turn, terminates at a muscle fiber.

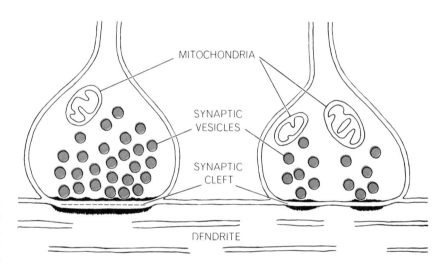

SYNAPTIC KNOBS are designed to deliver short bursts of a chemical transmitter substance into the synaptic cleft, where it can act on the surface of the nerve-cell membrane below. Before release, molecules of the chemical transmitter are stored in numerous vesicles, or sacs. Mitochondria are specialized structures that help to supply the cell with energy.

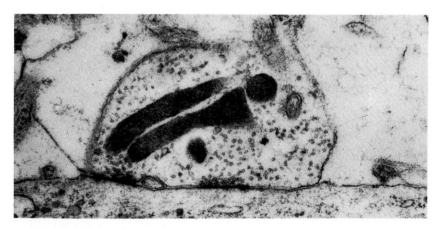

ASSUMED INHIBITORY SYNAPSE on a nerve cell is magnified 28,000 diameters in this electron micrograph by the late L. H. Hamlyn of University College London. Synaptic vesicles, believed to contain the transmitter substance, are bunched in two regions along the synaptic cleft. The darkening of the cleft in these regions is so far unexplained.

markable state of affairs? Part of the explanation is that the inside of the cell is negatively charged with respect to the outside of the cell by about 70 millivolts. Since like charges repel each other, this internal negative charge tends to drive chloride ions (Cl⁻) outward through the cell membrane and, at the same time, to impede their inward movement. In fact, a potential difference of 70 millivolts is just sufficient to maintain the observed disparity in the concentration of chloride ions inside the cell and outside it; chloride ions diffuse inward and outward at equal rates. A drop of 70 millivolts across the membrane therefore defines the "equilibrium potential" for chloride ions.

To obtain a concentration of potassium ions (K⁺) that is 30 times higher inside the cell than outside would require that the interior of the cell membrane be about 90 millivolts negative with respect to the exterior. Since the actual interior is only 70 millivolts nega-

tive, it falls short of the equilibrium potential for potassium ions by 20 millivolts. Evidently the thirtyfold concentration can be achieved and maintained only if there is some auxiliary mechanism for "pumping" potassium ions into the cell at a rate equal to their spontaneous net outward diffusion.

The pumping mechanism has the still more difficult task of pumping sodium ions (Na⁺) out of the cell against a potential gradient of 130 millivolts. This figure is obtained by adding the 70 millivolts of internal negative charge to the equilibrium potential for sodium ions, which is 60 millivolts of internal *positive* charge [*see illustrations on page 176*]. If it were not for this postulated pump, the concentration of sodium ions inside and outside the cell would be almost the reverse of what is observed.

In their classic studies of nerve-impulse transmission in the giant axon of the squid, A. L. Hodgkin, A. F. Huxley

and Bernhard Katz of Britain demonstrated that the propagation of the impulse coincides with abrupt changes in the permeability of the axon membrane. When a nerve impulse has been triggered in some way, what can be described as a gate opens and lets sodium ions pour into the axon during the advance of the impulse, making the interior of the axon locally positive. The process is self-reinforcing in that the flow of some sodium ions through the membrane opens the gate further and makes it easier for others to follow. The sharp reversal of the internal polarity of the membrane constitutes the nerve impulse, which moves like a wave until it has traveled the length of the axon. In the wake of the impulse the sodium gate closes and a potassium gate opens, thereby restoring the normal polarity of the membrane within a millisecond or less.

With this understanding of the nerve impulse in hand, one is ready to follow

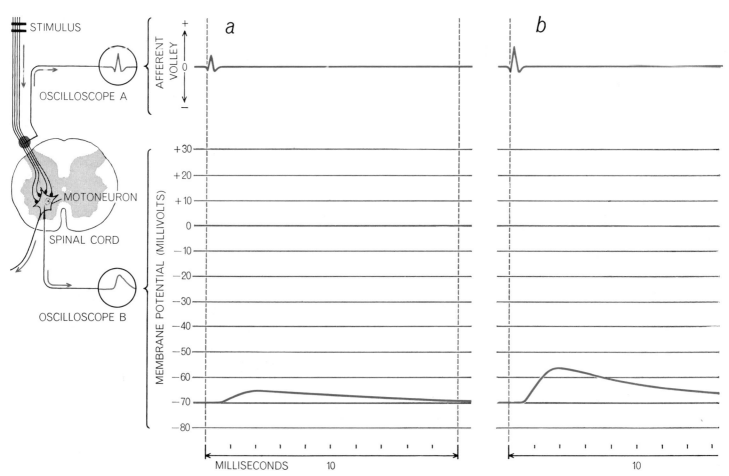

EXCITATION OF A MOTONEURON is studied by stimulating the sensory fibers that send impulses to it. The size of the "afferent volleys" reaching the motoneuron is displayed on oscilloscope *A*. A microelectrode implanted in the motoneuron measures the changes in the cell's internal electric potential. These changes, called excitatory postsynaptic potentials (EPSP's), appear on os-

cilloscope *B*. The size of the afferent volley is proportional to the number of fibers stimulated to fire. It is assumed here that one to four fibers can be activated. When only one fiber is activated (*a*), the potential inside the motoneuron shifts only slightly. When two fibers are activated (*b*), the shift is somewhat greater. When three fibers are activated (*c*), the potential reaches the threshold

the electrical events at the excitatory synapse. One might guess that if the nerve impulse results from an abrupt inflow of sodium ions and a rapid change in the electrical polarity of the axon's interior, something similar must happen at the body and dendrites of the nerve cell in order to generate the impulse in the first place. Indeed, the function of the excitatory synaptic terminals on the cell body and its dendrites is to depolarize the interior of the cell membrane essentially by permitting an inflow of sodium ions. When the depolarization reaches a threshold value, a nerve impulse is triggered.

As a simple instance of this phenomenon we have recorded the depolarization that occurs in a single motoneuron activated directly by the large nerve fibers that enter the spinal cord from special stretch-receptors known as annulospiral endings. These receptors in turn are located in the same muscle that is activated by the motoneuron under

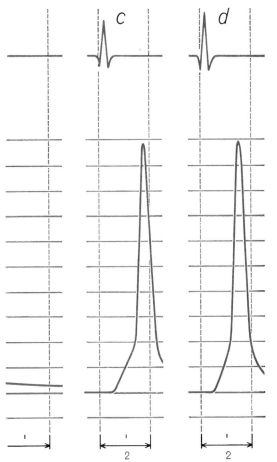

at which depolarization proceeds swiftly and a spike appears on oscilloscope *B*. The spike signifies that the motoneuron has generated a nerve impulse of its own. When four or more fibers are activated (*d*), the motoneuron reaches the threshold more quickly.

study. Thus the whole system forms a typical reflex arc, such as the arc responsible for the patellar reflex, or "knee jerk" [*see illustration on page 171*].

To conduct the experiment we anesthetize an animal (most often a cat) and free by dissection a muscle nerve that contains these large nerve fibers. By applying a mild electric shock to the exposed nerve one can produce a single impulse in each of the fibers; since the impulses travel to the spinal cord almost synchronously they are referred to collectively as a volley. The number of impulses contained in the volley can be reduced by reducing the stimulation applied to the nerve. The volley strength is measured at a point just outside the spinal cord and is displayed on an oscilloscope. About half a millisecond after detection of a volley there is a wavelike change in the voltage inside the motoneuron that has received the volley. The change is detected by a microelectrode inserted in the motoneuron and is displayed on another oscilloscope.

What we find is that the negative voltage inside the cell becomes progressively less negative as more of the fibers impinging on the cell are stimulated to fire. This observed depolarization is in fact a simple summation of the depolarizations produced by each individual synapse. When the depolarization of the interior of the motoneuron reaches a critical point, a "spike" suddenly appears on the second oscilloscope, showing that a nerve impulse has been generated. During the spike the voltage inside the cell changes from about 70 millivolts negative to as much as 30 millivolts positive. The spike regularly appears when the depolarization, or reduction of membrane potential, reaches a critical level, which is usually between 10 and 18 millivolts. The only effect of a further strengthening of the synaptic stimulus is to shorten the time needed for the motoneuron to reach the firing threshold [*see illustration at left*]. The depolarizing potentials produced in the cell membrane by excitatory synapses are called excitatory postsynaptic potentials, or EPSP's.

Through one barrel of a double-barreled microelectrode one can apply a background current to change the resting potential of the interior of the cell membrane, either increasing it or decreasing it. When the potential is made more negative, the EPSP rises more steeply to an earlier peak. When the potential is made less negative, the EPSP rises more slowly to a lower peak. Finally, when the charge inside the cell

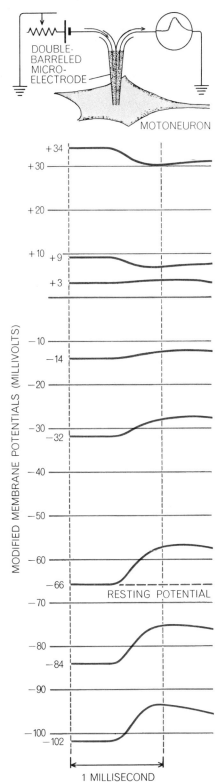

MANIPULATION of the resting potential of a motoneuron clarifies the nature of the EPSP. A steady background current applied through the left barrel of a microelectrode (*top*) shifts the membrane potential away from its normal resting level (minus 66 millivolts in this particular cell). The other barrel records the EPSP. The equilibrium potential, the potential at which the EPSP reverses direction, is about zero millivolts.

is reversed so as to be positive with respect to the exterior, the excitatory synapses give rise to an EPSP that is actually the reverse of the normal one [*see illustration at right on page 175*].

These observations support the hypothesis that excitatory synapses produce what amounts virtually to a short circuit in the synaptic membrane potential. When this occurs, the membrane no longer acts as a barrier to the passage of ions but lets them flow through in response to the differing electric potential on the two sides of the membrane. In other words, the ions are momentarily allowed to travel freely down their electrochemical gradients, which means that sodium ions flow into the cell and, to a lesser degree, potassium ions flow out. It is this net flow of positive ions that creates the excitatory postsynaptic potential. The flow of negative ions, such as the chloride ion, is apparently not involved. By artificially

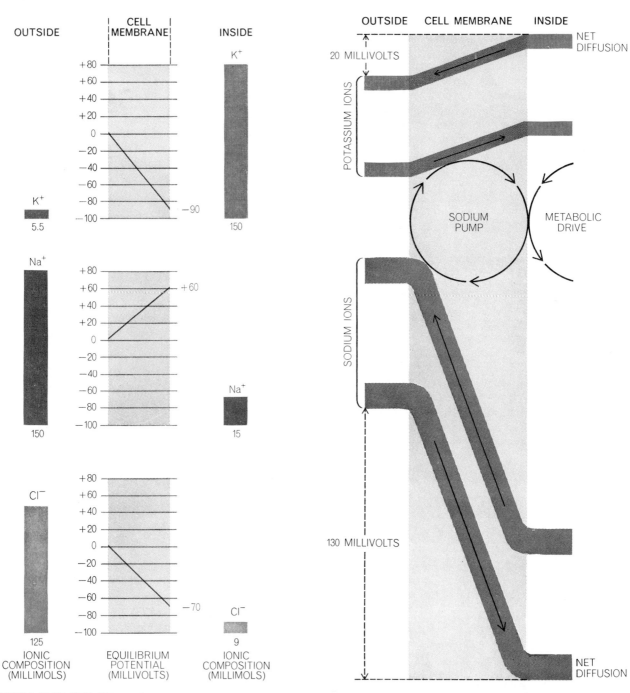

IONIC COMPOSITION outside and inside the nerve cell is markedly different. The "equilibrium potential" is the voltage drop that would have to exist across the membrane of the nerve cell to produce the observed difference in concentration for each type of ion. The actual voltage drop is about 70 millivolts, with the inside being negative. Given this drop, chloride ions diffuse inward and outward at equal rates, but the concentration of sodium and potassium must be maintained by some auxiliary mechanism (*right*).

METABOLIC PUMP must be postulated to account for the observed concentrations of potassium and sodium ions on opposite sides of the nerve-cell membrane. The negative potential inside is 20 millivolts short of the equilibrium potential for potassium ions. Thus there is a net outward diffusion of potassium ions that must be balanced by the pump. For sodium ions the potential across the membrane is 130 millivolts in the wrong direction, so very energetic pumping is needed. Chloride ions are in equilibrium.

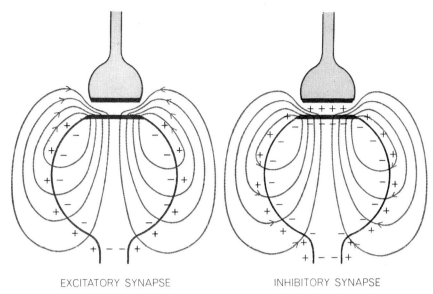

EXCITATORY SYNAPSE   INHIBITORY SYNAPSE

CURRENT FLOWS induced by excitatory and inhibitory synapses are respectively shown at left and right. When the nerve cell is at rest, the interior of the cell membrane is uniformly negative with respect to the exterior. The excitatory synapse releases a chemical substance that depolarizes the cell membrane below the synaptic cleft, thus letting current flow into the cell at that point. At an inhibitory synapse the current flow is reversed.

altering the potential inside the cell one can establish that there is no flow of ions, and therefore no EPSP, when the voltage drop across the membrane is zero.

How is the synaptic membrane converted from a strong ionic barrier into an ion-permeable state? It is currently accepted that the agency of conversion is the chemical transmitter substance contained in the vesicles inside the synaptic knob. When a nerve impulse reaches the synaptic knob, some of the vesicles are caused to eject the transmitter substance into the synaptic cleft [*see illustration below*]. The molecules of the substance would take only a few microseconds to diffuse across the cleft and become attached to specific receptor sites on the surface membrane of the adjacent nerve cell.

Presumably the receptor sites are as-

sociated with fine channels in the membrane that are opened in some way by the attachment of the transmitter-substance molecules to the receptor sites. With the channels thus opened, sodium and potassium ions flow through the membrane thousands of times more readily than they normally do, thereby producing the intense ionic flux that depolarizes the cell membrane and produces the EPSP. In many synapses the current flows strongly for only about a millisecond before the transmitter substance is eliminated from the synaptic cleft, either by diffusion into the surrounding regions or as a result of being destroyed by enzymes. The latter process is known to occur when the transmitter substance is acetylcholine, which is destroyed by the enzyme acetylcholinesterase.

The substantiation of this general picture of synaptic transmission requires the solution of many fundamental problems. Since we do not know the specific transmitter substance for the vast majority of synapses in the nervous system we do not know if there are many different substances or only a few. The only one identified with reasonable certainty in the mammalian central nervous system is acetylcholine. We know practically nothing about the mechanism by which a presynaptic nerve impulse causes the transmitter substance to be injected into the synaptic cleft. Nor do we know how the synaptic vesicles not immediately adjacent to the synaptic cleft are moved up to the firing line to replace the emptied vesicles. It is conjectured that the vesicles contain the enzyme systems needed to recharge themselves. The entire process must be swift and efficient: the total amount of transmitter substance in synaptic terminals is enough for only a few minutes of synaptic activity at normal operating rates. There are also knotty problems to be solved on the other side of the synaptic cleft. What, for example, is the nature of the receptor sites? How are the ionic channels in the membrane opened up?

The Inhibitory Synapse

Let us turn now to the second type of synapse that has been identified in the nervous system. These are the synapses that can inhibit the firing of a nerve cell even though it may be receiving a volley of excitatory impulses. When inhibitory synapses are examined in the electron microscope, they look very much like excitatory synapses.

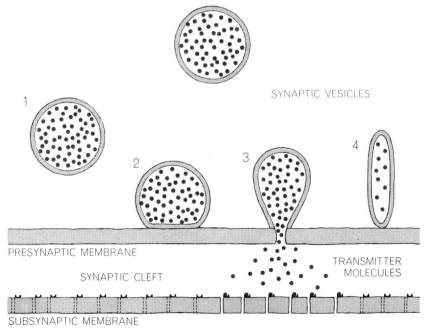

SYNAPTIC VESICLES

PRESYNAPTIC MEMBRANE

SYNAPTIC CLEFT

TRANSMITTER MOLECULES

SUBSYNAPTIC MEMBRANE

SYNAPTIC VESICLES containing a chemical transmitter are distributed throughout the synaptic knob. They are arranged here in a probable sequence, showing how they move up to the synaptic cleft, discharge their contents and return to the interior for recharging.

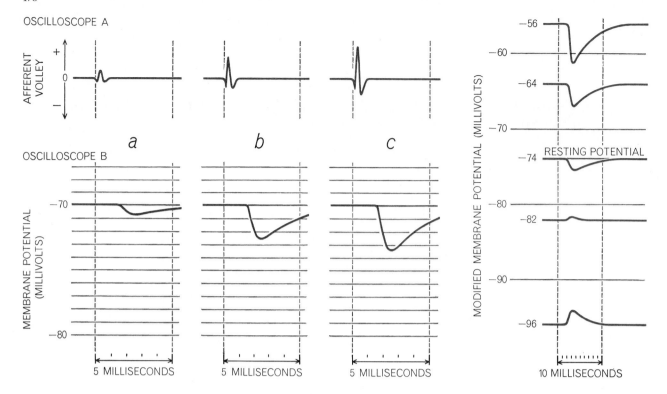

OSCILLOSCOPE A

AFFERENT VOLLEY

a *b* *c*

OSCILLOSCOPE B

MEMBRANE POTENTIAL (MILLIVOLTS)

−70

−80

5 MILLISECONDS 5 MILLISECONDS 5 MILLISECONDS

MODIFIED MEMBRANE POTENTIAL (MILLIVOLTS)

−56
−60
−64
−70
−74 RESTING POTENTIAL
−80
−82
−90
−96

10 MILLISECONDS

INHIBITION OF A MOTONEURON is investigated by methods like those used for studying the EPSP. The inhibitory counterpart of the EPSP is the IPSP: the inhibitory postsynaptic potential. Oscilloscope *A* records an afferent volley that travels to a number of inhibitory nerve cells whose axons form synapses on a nearby motoneuron (*see illustration on page 171*). A microelec- trode in the mononeuron is connected to oscilloscope *B*. The se- quence *a*, *b* and *c* shows how successively larger afferent volleys produce successively deeper IPSP's. Curves at right show how the IPSP is modified when a background current is used to change the motoneuron's resting potential. The equilibrium potential where the IPSP reverses direction is about minus 80 millivolts.

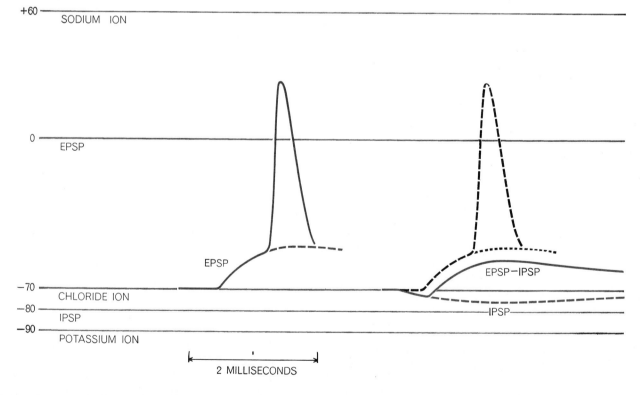

+60 ————— SODIUM ION

0 ———— EPSP

EPSP

EPSP−IPSP

−70 —— CHLORIDE ION
−80 —— IPSP
−90 —— POTASSIUM ION

IPSP

2 MILLISECONDS

INHIBITION OF A SPIKE DISCHARGE is an electrical sub- traction process. When a normal EPSP reaches a threshold (*left*), it will ordinarily produce a spike. An IPSP widens the gap be- tween the cell's internal potential and the firing threshold. Thus if a cell is simultaneously subjected to both excitatory and inhibitory stimulation, the IPSP is subtracted from the EPSP (*right*) and no spike occurs. The five horizontal lines show equilibrium potentials for the three principal ions as well as for the EPSP and IPSP.

(There are probably some subtle differences, but they need not concern us here.) Microelectrode recordings of the activity of single motoneurons and other nerve cells have now shown that the inhibitory postsynaptic potential (IPSP) is virtually a mirror image of the EPSP [*see top illustration on preceding page*]. Moreover, individual inhibitory synapses, like excitatory synapses, have a cumulative effect. The chief difference is simply that the IPSP makes the cell's internal voltage more negative than it is normally, which is in a direction opposite to that needed for generating a spike discharge.

By driving the internal voltage of a nerve cell in the negative direction inhibitory synapses oppose the action of excitatory synapses, which of course drive it in the positive direction. Hence if the potential inside a resting cell is 70 millivolts negative, a strong volley of inhibitory impulses can drive the potential to 75 or 80 millivolts negative. One can easily see that if the potential is made more negative in this way the excitatory synapses find it more difficult to raise the internal voltage to the threshold point for the generation of a spike. Thus the nerve cell responds to the algebraic sum of the internal voltage changes produced by excitatory and inhibitory synapses [*see bottom illustration on preceding page*].

If, as in the experiment described earlier, the internal membrane potential is altered by the flow of an electric current through one barrel of a double-barreled microelectrode, one can observe the effect of such changes on the inhibitory postsynaptic potential. When the internal potential is made less negative, the inhibitory postsynaptic potential is deepened. Conversely, when the potential is made more negative, the IPSP diminishes; it finally reverses when the internal potential is driven below minus 80 millivolts.

One can therefore conclude that inhibitory synapses share with excitatory synapses the ability to change the ionic permeability of the synaptic membrane. The difference is that inhibitory synapses enable ions to flow freely down an electrochemical gradient that has an equilibrium point at minus 80 millivolts rather than at zero, as is the case for excitatory synapses. This effect could be achieved by the outward flow of positively charged ions such as potassium or the inward flow of negatively charged ions such as chloride, or by a combination of negative and positive ionic flows such that the interior reaches equilibrium at minus 80 millivolts.

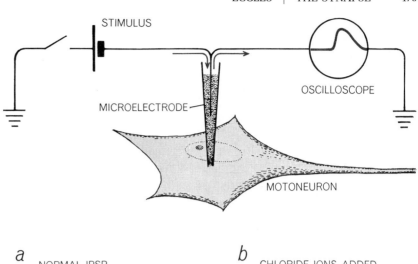

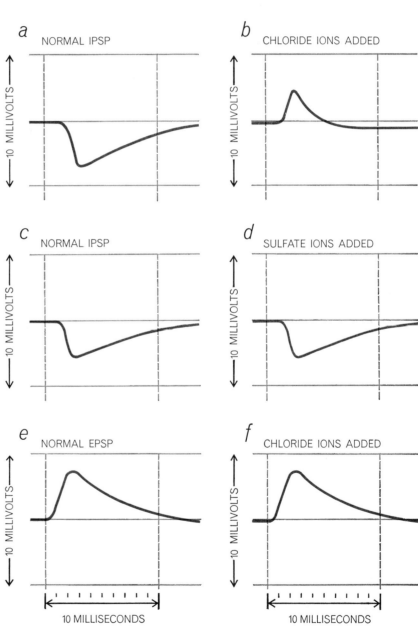

MODIFICATION OF ION CONCENTRATION within the nerve cell gives information about the permeability of the cell membrane. The internal ionic composition is altered by injecting selected ions through a microelectrode a minute or so before applying an afferent volley and recording the EPSP or IPSP. In the first experiment a normal IPSP (*a*) is changed to a pseudo-EPSP (*b*) by an injection of chloride ions. When sulfate ions are similarly injected, the IPSP is practically unchanged (*b, c*). The third experiment shows that an injection of chloride ions has no significant effect on the EPSP (*e, f*).

In an effort to discover the permeability changes associated with the inhibitory potential my colleagues and I have altered the concentration of ions normally found in motoneurons and have introduced a variety of other ions that are not normally present. This can be done by impaling nerve cells with micropipettes that are filled with a salt solution containing the ion to be injected. The actual injection is achieved by passing a brief current through the micropipette.

If the concentration of chloride ions within the cell is in this way increased as much as three times, the inhibitory postsynaptic potential reverses and acts as a depolarizing current; that is, it resembles an excitatory potential. On the other hand, if the cell is heavily injected with sulfate ions, which are also negatively charged, there is no such reversal [see illustration on preceding page]. This simple test shows that under the influence of the inhibitory transmitter substance, which is still unidentified, the subsynaptic membrane becomes permeable momentarily to chloride ions but not to sulfate ions. During the generation of the IPSP the outflow of chloride ions is so rapid that it more than outweighs the flow of other ions that generate the normal inhibitory potential.

My colleagues have now tested the effect of injecting motoneurons with more than 30 kinds of negatively charged ion. With one exception the hydrated ions (ions bound to water) to which the cell membrane is permeable under the influence of the inhibitory transmitter substance are smaller than the hydrated ions to which the membrane is impermeable. The exception is the formate ion (HCO_2^-), which may have an ellipsoidal shape and so be able to pass through membrane pores that block smaller spherical ions.

Apart from the formate ion all the ions to which the membrane is permeable have a diameter not greater than 1.14 times the diameter of the potassium ion; that is, they are less than 2.9 angstrom units in diameter. Comparable investigations in other laboratories have found the same permeability effects, including the exceptional behavior of the formate ion, in fishes, toads and snails. It may well be that the ionic mechanism responsible for synaptic inhibition is the same throughout the animal kingdom.

The significance of these and other studies is that they strongly indicate that the inhibitory transmitter substance opens the membrane to the flow of potassium ions but not to sodium ions. It

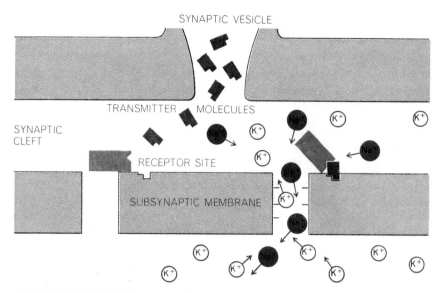

EXCITATORY SYNAPSE may employ transmitter molecules that open large channels in the nerve-cell membrane. This would permit sodium ions, which are plentiful outside the cell, to pour through the membrane freely. The outward flow of potassium ions, driven by a smaller potential gradient, would be at a much slower rate. Chloride ions (not shown) may be prevented from flowing by negative charges on the channel walls.

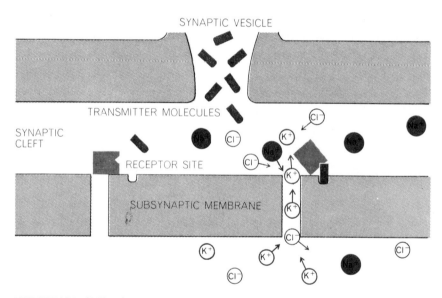

INHIBITORY SYNAPSE may employ another type of transmitter molecule that opens channels too small to pass sodium ions. The net outflow of potassium ions and inflow of chloride ions would account for the hyperpolarization that is observed as an IPSP.

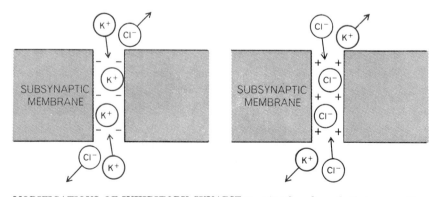

MODIFICATIONS OF INHIBITORY SYNAPSE may involve channels that carry either negative or positive charges on their walls. Negative charges (left) would permit only potassium ions to pass. Positive charges (right) would permit only chloride ions to pass.

is known that the sodium ion is somewhat larger than any of the negatively charged ions, including the formate ion, that are able to pass through the membrane during synaptic inhibition. It is not possible, however, to test the effectiveness of potassium ions by injecting excess amounts into the cell because the excess is immediately diluted by an osmotic flow of water into the cell.

As I have indicated, the concentration of potassium ions inside the nerve cell is about 30 times greater than the concentration outside, and to maintain this large difference in concentration without the help of a metabolic pump the inside of the membrane would have to be charged 90 millivolts negative with respect to the exterior. This implies that if the membrane were suddenly made porous to potassium ions, the resulting outflow of ions would make the inside potential of the membrane even more negative than it is in the resting state, and that is just what happens during synaptic inhibition. The membrane must not simultaneously become porous to sodium ions, because they exist in much higher concentration outside the cell than inside and their rapid inflow would more than compensate for the potassium outflow. In fact, the fundamental difference between synaptic excitation and synaptic inhibition is that the membrane freely passes sodium ions in response to the former and largely excludes the passage of sodium ions in response to the latter.

Channels in the Membrane

This fine discrimination between ions that are not very different in size must be explained by any hypothesis of synaptic action. It is most unlikely that the channels through the membrane are created afresh and accurately maintained for a thousandth of a second every time a burst of transmitter substance is released into the synaptic cleft. It is more likely that channels of at least two different sizes are built directly into the membrane structure. In some way the excitatory transmitter substance would selectively unplug the larger channels and permit the free inflow of sodium ions. Potassium ions would simultaneously flow out and thus would tend to counteract the large potential change that would be produced by the massive sodium inflow. The inhibitory transmitter substance would selectively unplug the smaller channels that are large enough to pass potassium and chloride ions but not sodium ions [*see upper two illustrations on previous page*].

To explain certain types of inhibition other features must be added to this hypothesis of synaptic transmission. In the simple hypothesis chloride and potassium ions can flow freely through pores of all inhibitory synapses. It has been shown, however, that the inhibition of the contraction of heart muscle by the vagus nerve is due almost exclusively to potassium-ion flow. On the other hand, in the muscles of crustaceans and in nerve cells in the snail's brain synaptic inhibition is due largely to the flow of chloride ions. This selective permeability could be explained if there were fixed charges along the walls of the channels. If such charges were negative, they would repel negatively charged ions and prevent their passage; if they were positive, they would similarly prevent the passage of positively charged ions. One can now suggest that the channels opened by the excitatory transmitter are negatively charged and so do not permit the passage of the negatively charged chloride ion, even though it is small enough to move through the channel freely.

One might wonder if a given nerve cell can have excitatory synaptic action at some of its axon terminals and inhibitory action at others. The answer is no. Two different kinds of nerve cell are needed, one for each type of transmission and synaptic transmitter substance. This can readily be demonstrated by the effect of strychnine and tetanus toxin in the spinal cord; they specifically prevent inhibitory synaptic action and leave excitatory action unaltered. As a result the synaptic excitation of nerve cells is uncontrolled and convulsions result. The special types of cell responsible for inhibitory synaptic action are now being recognized in many parts of the central nervous system.

This account of communication between nerve cells is necessarily oversimplified, yet it shows that some significant advances are being made at the level of individual components of the nervous system. By selecting the most favorable situations we have been able to throw light on some details of nerve-cell behavior. We can be encouraged by these limited successes. But the task of understanding in a comprehensive way how the human brain operates staggers its own imagination.

V

CHEMICAL
COMMUNICATION

V CHEMICAL COMMUNICATION

INTRODUCTION

The preceding sections have already emphasized that the basic medium of communication within and between cells is a chemical one; biological systems speak in molecular language. Our understanding of chemical communication as a general process, however, is distressingly spotty. For example, as the section on development emphasized, we have not yet been able to translate a single specific message by which differentiation in one cell population is induced by the presence of a neighboring population. Yet we know that such inductive effects are widespread, and that they play an important role in the formation of organs during embryonic development.

This is just one example of how incomplete our knowledge is about the general nature of chemical communication. Our grasp is much better in a few selected areas; oddly enough, these involve phenomena at very disparate levels of organization. At the molecular level, recent work has provided a more complete picture of the mechanisms underlying genetic regulation. That progress is summarized in another collection of *Scientific American* articles, *The Chemical Basis of Life*, with introductions by Philip C. Hanawalt and Robert H. Haynes (1973). Different portions of the genetic material of a cell can be called into activity, or turned off (repressed), depending upon the cell's need for a particular gene product. Experiments on microbial cells have demonstrated that a product of a specific gene, called the *repressor*, is able to bind reversibly to a site (the *operator*) that controls the transcription of a region of DNA (the *operon*). In bacteria, an operon typically contains the templates for several enzymes that function along a single metabolic pathway—for the synthesis of a particular amino acid, for example, or for the degradation of a certain kind of sugar. The repressor substance—the molecule made by the repressor gene—is located at a distance from the operator. It prevents transcription of the operon by binding to the operator site; but in some cases it can be removed through combination with a particular small molecule. In such cases, the small molecule is said to act as an *inducer*: that is, its presence causes the appropriate region of the genetic material to become active in making its product. The sugar lactose functions as just such an inducer; when added to a culture of bacteria, lactose activates the machinery for producing the enzymes that degrade it. But there are cases in which the *presence* of a small molecule will enable the repressor molecule to bind to the operator. This results in an inhibition of the production of the end product. The final product of a metabolic pathway, if it occurs in excessive amounts, acts to shut down the manufacture of those enzymes necessary for its production. The process of repression thus permits the bacterial cell to produce only the amount of gene product (enzyme) necessary for supplying a minimum amount of a particular metabolite.

A major effort in molecular biology is the extension of this analysis from bacterial cells to the more complex eucaryotic cells characteristic of multi-cellular organisms. The task is a very difficult one, because genetic regulation is accomplished by very small numbers of molecules. Although it is easy to raise large numbers of bacterial cells on a simple, chemically defined medium, the task is much harder with animal cells in tissue culture. Being more complex, animal cells take much longer to reproduce; moreover, they have more fastidious nutritional requirements and are difficult to analyze chemically because there is so much more there. However, it is in just such cells that the mechanism of genetic regulation ought to be most crucial. Recall that a differentiated mammalian cell has a thousand times as much genetic material as a bacterium; yet it has no more complex a basic set of metabolic tasks. The "extra" DNA in the mammalian cell is there because each single cell in a multicellular organism must carry the entire genetic encyclopedia for the organism. To repeat an example used in the introduction to Section III, the retinal cell does not make hemoglobin, nor the red blood cell rhodopsin. But the nucleus of the developing retinal cell must, as we know from experiments on nuclear potency, have the genetic information for hemoglobin—and indeed for each of the multitude of specialized proteins produced by every cell in the body. How is the developing cell regulated so that it produces only its *own* proteins? This is the major challenge posed for molecular biologists by multicellular organization.

In "Chromosome Puffs" (1964), Wolfgang Beermann and Ulrich Clever recount what has become one of the most promising approaches to this problem. The "puffs" observed on giant chromosomes in certain insect cells appear to represent selective synthetic activity at specific genetic loci. Beermann and Clever cite evidence that these puffs are produced only in a certain kind of cytoplasm, or at a specific stage of development, or under the influence of a hormone. The genome of each cell is presumed to contain the information required for all the syntheses performed by the organism during its life; but the experiments on local gene activity indicate that this encyclopedia is read with great selectivity during establishment of the differentiated state.

A second area in which knowledge is now growing rapidly is study of the control of metabolic and other cellular activities by hormones, both during development and in adulthood. In the article by Beermann and Clever, hormones active during insect development are implicated in the direct control of genetic expression. The subsequent article, "Hormones and Genes" (1965), by Eric H. Davidson, extends this concept to a variety of vertebrate hormones, and attempts to evaluate evidence suggesting that hormones control gene activity directly. Davidson's article was written in 1965. Although it remains a remarkably foresighted and well-balanced analysis, it is not surprising that subsequent research in such an active field has wrought some changes. These changes have been in two areas especially: first, in attempts to learn the cellular locus for action of certain hormones, and the nature of hormonal effects on gene activity; and second, in the study of "intermediate" chemical messengers.

Recent experiments on the action of steroid hormones tends to confirm Davidson's view of the directness of their action upon the material of the nucleus. Labeled hormone is not only bound to target cells; the label actually appears in the cell nucleus, and furthermore the chromosomal material can bind steroids selectively. Moreover, other workers have been able to measure directly the influence of a steroid hormone on the synthesis of specific messenger RNA, in cells in which the RNA encodes a protein whose production is consequently controlled by the hormone. Not only does the hormone actually get to the genetic material; the amount of RNA produced by cultured cells is proportional to the amount of hormone added to the culture medium.

This result is hard to interpret unless we suppose that the hormone directly influences the transcription of a specified region of DNA.

The second area that has opened up dramatically since the Davidson article was first published is the operation of a *second messenger* as a cellular intermediary between the hormone and its nuclear target. This substance and its actions are described in the article that follows Davidson's in this section—"Cyclic AMP" (1972), by Ira Pastan. Pastan describes the function of the second messenger in the metabolic response of the cell to epinephrine (adrenalin), a scheme that was first worked out by Earl Sutherland at Vanderbilt University. He goes on to describe experiments on microbial cells that suggest that cyclic AMP influences gene activation. As Pastan points out, the amount of cyclic AMP in the cell is influenced by a variety of different hormones. Most of these are small molecules, or proteins; it may be that the steroids are especially adapted for direct control, but proof of this proposition awaits further study.

A quite different question concerns the movement of possible messenger molecules between cells, especially during development. We have always tended to think of cells as relatively isolated from one another at all times: the membrane of a typical cell is actually quite impermeable to most substances. Even the common small ions like potassium and chloride, to which we think of the cell as quite permeable, pass the membrane with a million times more difficulty than they would an equally thick layer of water.

Given these facts, it was a real surprise to discover that the cells in many epithelia, and in many embryonic tissues, are actually connected to one another with "bridges" that are quite permeable: chemical compounds with molecules of molecular weights in the vicinity of several thousand pass readily across these special places, and small ions move so freely that the electrical resistance *separating* the cells is as low as it is *within* a single cell. Werner R. Loewenstein discusses the transmission route and its significance for cellular coordination in his article "Intercellular Communication" (1970). Work in several laboratories has now shown that these low-resistance intercellular junctions are common in embryonic (and several adult) tissues. In some instances, their function is clearly related to the process of electrical transmission itself: junctions of this kind are found in the nervous system as well as in epithelia, where the propagation of electrical signals seems to be important. But in other systems we can see no reason why lowered electrical resistance should be useful to the organism, and can only speculate that the exchange of small molecules plays some coordinating role.

The final article in this section deals with a different kind of control entirely: communication *between individuals*, usually of the same species, by specific chemicals that are effective in low concentrations. The specific chemical compounds are given the name *pheromones*, which is also the title of the article by Edward O. Wilson. Unlike hormones, which transmit signals from one tissue to another in the same organism, pheromones provide an avenue for the transfer of chemical information between individuals. This communication may be part of courtship and mating behavior, as it is in those female moths that secrete a volatile attractant detected by chemoreceptors on the male's antennae; or it may enable one individual to control the development of another, as termite "kings" and "queens" do in termite nests when they secrete substances that inhibit the sexual maturation of other members of the colony. Pheromones may even provide orientation and guidance, as do the ant "trail substances" described in detail by Wilson.

Since "Pheromones" was first published in 1963 a number of additional examples of communication by pheromones have been described. In particular, our knowledge of the chemistry of the insect pheromones used in sexual attraction has improved rapidly; this is due in part to the promising solutions that these compounds offer to problems of pest control in agriculture and public health. Thus, pheromones that influence mating behavior have been isolated and characterized for such economically important insects as the gypsy moth, the pink bollworm (a devastating pest in the cotton-growing areas of the southwestern United States, and others. To date, man has used the compounds primarily as insect attractants, to estimate the density of their populations. Another active subject of pheromone research has, rather surprisingly, been mammals. We do not usually think of higher vertebrates as placing much reliance on chemical signals; yet many species of mammals mark territorial boundaries with the products of special scent glands, or they emit chemical sexual signals. Recently it has been shown that a male pheromone in rodents affects reproductive success in the female, and work on this aspect of mammalian reproductive biology has been greatly accelerated as a result.

17 Chromosome Puffs

by Wolfgang Beermann and Ulrich Clever
April 1964

These enlarged regions on the giant chromosomes found in some insect cells have been shown to be active genes. They probably produce the nucleic acid that translates the genetic information

The genetic material performs two functions that are basic to life: it replicates itself and it ultimately directs all the manifold chemical activities of every living cell. The first function is expressed at the time of cell division in the manufacture of more of the genetic material: deoxyribonucleic acid (DNA). The second is accomplished during the "interphase" between cell divisions; DNA directs the synthesis of ribonucleic acid (RNA), which in turn directs the synthesis of proteins, which as enzymes in turn catalyze the other reactions of the cell. In this way RNA translates the genetic information of DNA into the language of physiology and growth, into the everyday processes of synthesis and metabolism.

As readers of SCIENTIFIC AMERICAN are aware, the work of elucidating the genetic code is now being carried out by investigators in laboratories throughout the world, largely by the breeding and statistical study of certain bacteria and the viruses that infect them. In recent years our laboratory at the Max Planck Institute for Bi-ology in Tübingen and several other laboratories have adopted somewhat different techniques for investigating the relation between DNA and RNA in the genetic material of higher organisms—those belonging to the insect order Diptera, such as the fruit fly *Drosophila* and the midge *Chironomus*. In these insects, as in all higher organisms, the DNA resides in the structures called chromosomes. In certain exceptionally large cells of *Drosophila* and *Chironomus* we have found that we can actually see the ultimate units of heredity—the genes—at work. These active genes take the form of "puffs" scattered here and there along the giant chromosomes of the giant cells. We have found that the puffs produce RNA and that the RNA made in one puff differs from the RNA made in another. Observations of the puffs have also enabled us to trace the time patterns of gene activity in several tissues of developing insect larvae. Furthermore, by administering hormones and other substances we can start, stop and prevent some of these activities.

The giant chromosomes were first ob-served late in the last century, but it was not until 1933 that Emil Heitz and Hans Bauer of the University of Hamburg recognized them as chromosomes. By 1933 breeding studies of the fruit fly had resulted in detailed "maps" on which genes were placed in relation to each other along the chromosomes. The genes, however, were still conceptions rather than physical entities, and the chromosomes had been recognized only during cell division, when they are coiled like a spring and present a condensed, rodlike appearance. During interphase, when they are directing cellular activity, the chromosomes in typical cells are virtually invisible because, although they are long, they are so thin that they can be seen only at the extremely high magnifications provided by the electron microscope, a comparatively recent invention.

Heitz and Bauer realized that giant chromosomes, which are clearly visible in the light microscope, are the equivalent of the interphase chromosomes of typical cells. In the words of T. S. Painter of the University of Texas, the giant salivary-gland chromosomes of fruit fly larvae were "the material of which every geneticist had been dreaming. The way led to the lair of the gene." Intensive work by Painter and others in the U.S., including H. J. Muller, Calvin B. Bridges and Milislav Demerec, soon identified specific characteristics of flies with particular loci, or bands, on the giant chromosomes. Since then the bands have been considered the material equivalent of the conceptual Mendelian genes.

The giant chromosomes are found primarily in well-differentiated organs that are engaged in vigorous metabolic activity, such as salivary glands, intestines and the Malpighian tubules (excretory

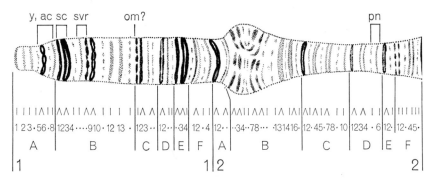

TIP OF A GIANT CHROMOSOME from the salivary gland of the fruit fly *Drosophila melanogaster* is shown in this diagram. The reference system below it was devised by Calvin B. Bridges of the California Institute of Technology. The letters and brackets above it mark certain sites known to be associated with specific bodily characteristics. For example, the "*y*" at left denotes the band or gene responsible for yellow body color.

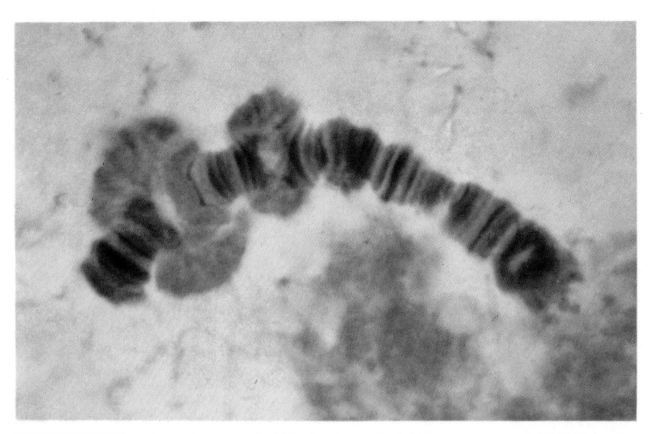

CHROMOSOME PUFFS are the protuberances on the left-hand portion of the giant chromosome in this photomicrograph. Very large puffs, of which two are seen, are called Balbiani rings. Protein has been stained green, deoxyribonucleic acid (DNA) brown.

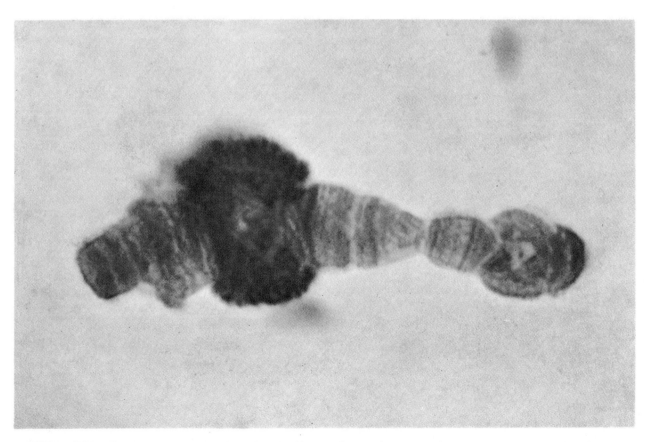

PRODUCT OF PUFFS, ribonucleic acid (RNA), is reddish-violet when dyed with toluidine blue. Here the DNA is blue. The photomicrographs on this page show two different specimens of the giant chromosome IV from the salivary gland of the midge *Chironomus tentans*. Both were made at the Max Planck Institute for Biology in Tübingen. The magnification in each is some 2,500 diameters.

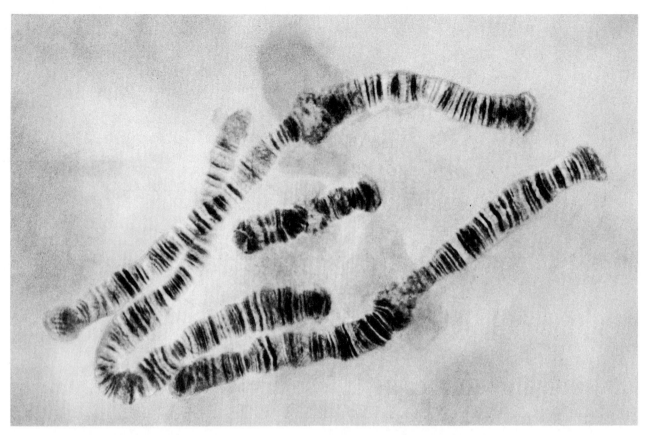

SET OF FOUR GIANT CHROMOSOMES from a cell in the salivary gland of *Ch. tentans* is here magnified some 700 diameters. The enlarged regions on two of the long chromosomes are nucleoli.

Chromosome IV is the shortest of the four; it has a Balbiani ring. The banding pattern on each of the four chromosomes is visible in corresponding giant chromosomes from entirely different tissues.

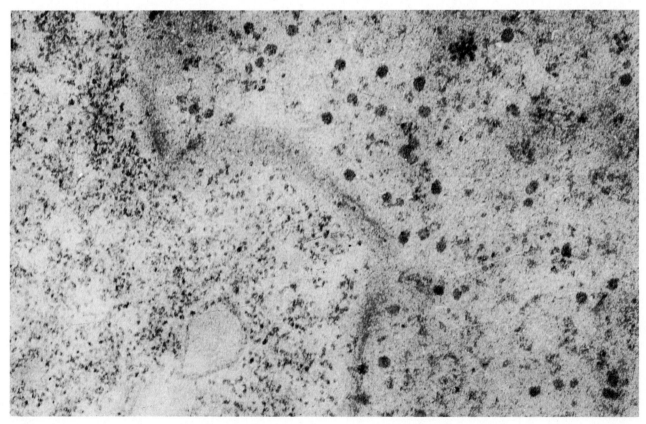

MIGRATING GRANULES, consisting of ribonucleic acid and protein (*right*), are penetrating pores in the nuclear membrane (*bottom center*) in this electron micrograph. Cytoplasm of cell is

to left of membrane. The small particles in it are ribosomes, the sites of protein synthesis. The RNA in the large particles may be on its way to the ribosomes to act as a template for proteins.

organs). These tissues grow by an increase in cell size rather than in cell number. Apparently the giant cells require more genetic material than typical cells do; as they expand, the chromosomes replicate again and again and also increase in length. Along individual chromosome fibers there are numerous dense spots where presumably the structure is drawn into tight folds. These locations are called chromomeres.

As the chromosome filaments in giant cells increase in number, those of a particular chromosome remain tightly bound together; each chromomere is fastened to the homologous, or matching, chromomere of the neighboring filaments. Such locations become the bands, which are also known as chromomeres. The chromosome that results from this growth process is said to be "polytene": it has a multistrand structure resembling a rope. At full size the giant chromosomes are almost 100 times thicker and more than 10 times longer than the chromosomes of typical cells at cell division.

The bands, which vary in thickness, contain a high concentration of DNA and histone, a protein associated with DNA. The spaces between the bands, known as interbands, contain a very low concentration of these substances. It was discovered in 1933 that each giant chromosome in a set within a cell has its own characteristic sequence, or pattern, of banding and that, even more striking, every detail of the pattern recurs with the utmost precision in the homologous giant chromosome of every individual of the species.

In the past most cell geneticists were so occupied with localizing the genes in the salivary-gland chromosomes that they did not investigate the giant chromosomes in other tissues. Yet the presence of such chromosomes in cells with quite different functions poses an obvious challenge to the biologist interested in development and differentiation. It had long been held that every cell of an individual possesses exactly the same set of chromosomes and the same pattern of genes. Giant chromosomes in a variety of tissues provided an opportunity for testing this idea, that is, for determining if the special metabolic condition or function of a cell influences in any way the state of its chromosomes and genes. For example, in spite of the constancy of the banding pattern found in salivary-gland chromosomes, the same chromosomes in other organs of the same species might present a different banding pattern. If this were true, the

localization of genes in specific bands would lose all general meaning.

Assertions that different tissues have different banding patterns were actually made 15 years ago by Curt Kosswig and Atif Şengün of the University of Istanbul. One of us (Beermann, then working in the laboratory of Hans Bauer at the Max Planck Institute for Marine Biology in Wilhelmshaven) checked these claims by a detailed comparative study of the banding of giant chromosomes from four different tissues of the midge *Chironomus tentans*. Independently Clodowaldo Pavan and Martha E. Breuer of the University of São Paulo carried out similar investigations on the fly *Rhynchosciara angelae*. We could not find any detectable variation in the arrangement and sequence of bands along the chromosomes in different tissues. The uniformity of chromosome banding lends strong support to the basic concept that the linear arrangement of the genes as mapped in breeding experiments corresponds to the pattern of the bands on giant chromosomes.

At the same time, however, we found that chromosomal differentiation of a very interesting kind does exist. The fine structure of individual bands can differ with respect to puffs that are in one location on a chromosome in one tissue and in another location on the same chromosome at another time or in another tissue. These localized modifications in chromosome structure of various Diptera had been noted many years earlier, but their possible significance was overlooked.

The coherence of the chromosome filaments is loosened at the puffed regions. The loosening always starts at a single band. In small puffs a particular band simply loses its sharp contour and presents a diffuse, out-of-focus appearance in the microscope. At other loci or at other times a band may look as though it had "exploded" into a large ring of loops around the chromosome [*see top illustrations on next two pages*]. Such doughnut-like structures are called Balbiani rings, after E. G. Balbiani of the Collège de France, who first described them in 1881. Puffing is thought to be due to the unfolding or uncoiling of individual chromomeres in a band. On observing that specific tissues and stages of development are characterized by

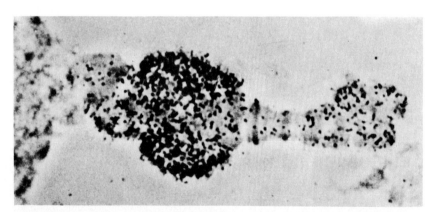

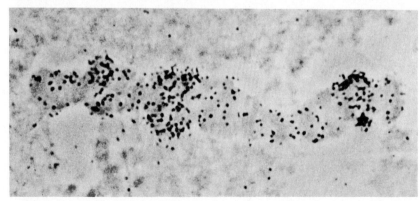

INHIBITION OF PUFFING and of RNA synthesis is accomplished by treatment with the antibiotic actinomycin D. At top an autoradiogram of a chromosome IV of *Ch. tentans* shows the incorporation of much radioactive uridine (*black spots*), which takes place during the production of RNA, as explained in the text. Another chromosome IV (*bottom*) that had been puffing shows puff regression and little radioactivity after half an hour of treatment with minute amounts of actinomycin D, which inhibits RNA synthesis by DNA.

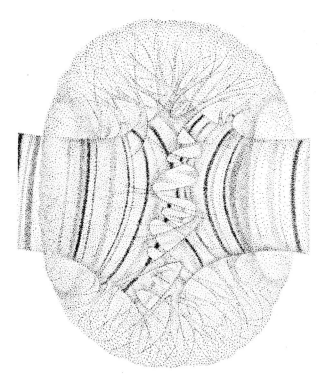

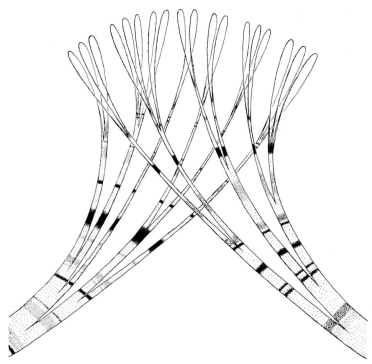

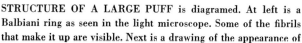

STRUCTURE OF A LARGE PUFF is diagramed. At left is a Balbiani ring as seen in the light microscope. Some of the fibrils that make it up are visible. Next is a drawing of the appearance of a few of the fibrils at very high magnification in the light microscope. The much greater magnification provided by the electron microscope (*third from left*) shows two puff fibrils with granules

definite puff patterns, one of us (Beermann) postulated in 1952 that a particular sequence of puffs represents a corresponding pattern of gene activity. At about the same time, Pavan and Breuer arrived at a comparable conclusion based on their experiments with *Rhynchosciara.*

If differential gene activation does in fact occur, one would predict that genes in a specific type of cell will regularly puff whereas the same gene in another type of cell will not. A gene of exactly this kind has been discovered in *Chironomus.* A group of four cells near the duct of the salivary gland of the species *Chironomus pallidivittatus* produces a granular secretion. The same cells in the closely related species *Ch. tentans* give off a clear, nongranular fluid. In hybrids of the two species this characteristic follows simple Mendelian laws of heredity. We have been able to localize the difference in a group of fewer than 10 bands in one of *Chironomus'* four chromosomes; the chromosome is designated IV. The granule-producing cells of *Ch. pallidivittatus* have a puff associated with this group of bands, a puff that is entirely absent at the corresponding loci of chromosome IV in *Ch. tentans.* In hybrids the puff appears only on the chromosome coming from the *Ch. pallidivittatus* parent; the hybrid produces a far smaller number of granules than

that parent. Moreover, the size of the puff is positively correlated with the number of granules. This reveals quite clearly the association between a puff and a specific cellular product.

Such analysis can demonstrate only that a specific relation exists between certain puffed genes and certain cell functions. We therefore sought to find a biochemical method for showing that puffing patterns along chromosomes are in fact patterns of gene activity. According to the current hypothesis the sequence of the four bases that characterize DNA—guanine, adenine, thymine and cytosine—represents a code for the sequence of the 20 kinds of amino acid unit that make up a protein. Most, if not all, protein synthesis takes place not in the nucleus of the cell but in the surrounding cytoplasm. The DNA always remains in the nucleus. As a result the instructions supplied by DNA must be carried to the cytoplasm, where the translation is made. The carrier and translator of the DNA information is thought to be the special form of RNA called messenger RNA. Each DNA molecule serves as a template for a specific messenger RNA molecule, which then acts as a template in the synthesis of a particular protein. Hence what we have termed gene activity becomes equivalent to the rate of production of messenger RNA at each gene.

It has been known for some time that chromosome puffs contain significant amounts of RNA. As we have noted, the normal, unpuffed bands chiefly contain DNA and histone. In general the amount of these compounds remains unchanged in the transition from a band to a puff, whereas the amount of RNA increases considerably. The presence of RNA is beautifully demonstrated by metachromatic dyes such as toluidine blue, which simultaneously stains RNA red-violet and DNA a shade of blue. A great increase in the amount of RNA, however, is not sufficient to demonstrate that RNA synthesis is the main function of puffs. For one thing, some dyes show that a protein other than histone accumulates in the puffs along with RNA. Perhaps it too is made there.

In order to find out if RNA is the major puff product, Claus Pelling of our laboratory employed the technique of autoradiography. His "tracer" was uridine, a substance the cell tends to use to make RNA rather than DNA, that had been labeled with the radioactive isotope hydrogen 3 (tritium). He injected the uridine into *Chironomus* larvae, which he later killed. When giant cells from the larvae were placed in contact with a photographic emulsion, the radioactive loci in their chromosomes darkened the emulsion [*see illustration*

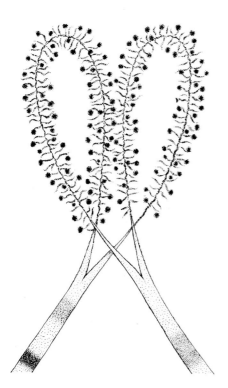

that are believed to be messenger RNA produced by the genes. In particularly small puffs the loops cannot actually be observed.

SCHEMATIC REPRESENTATION of how a large puff is formed shows fibrils untwisted and "popped out" of the cable-like structure. A giant chromosome in reality contains thousands of fibrils. Those untwisted here are tightly coiled when in the form of bands.

on page 191]. In every case in which Pelling killed the larvae soon after injection, sometimes as quickly as two minutes afterward, only the puffs, the Balbiani rings and the nucleoli were labeled. (Nucleoli are large deposits of RNA and protein that are formed in all types of cells by chromosomal regions known as nucleolar organizers.

Presumably they are involved in the formation of ribosomes, which are the sites of protein synthesis in the cytoplasm.) The rest of the chromosomal material and the cytoplasm showed very little radioactive label until long after the injection.

When the preparations were treated with an enzyme that decomposes RNA

before placing them in contact with the emulsion, the label was absent. Pelling demonstrated further that the rate of RNA synthesis is closely correlated with the relative size of the puffs. The administration of the antibiotic actinomycin D, a specific inhibitor of any RNA synthesis that depends directly on DNA, stopped the formation of RNA.

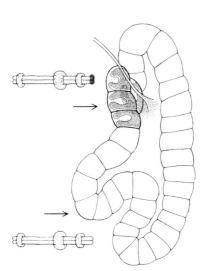

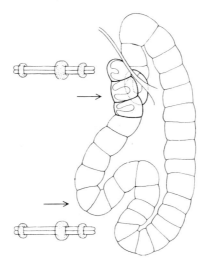

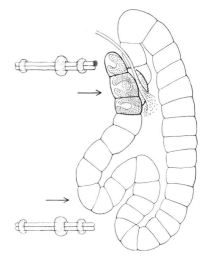

DIFFERENTIAL GENE ACTIVATION occurs on homologous, or matching, chromosomes of *Chironomus.* Four salivary-gland cells in the species *Ch. pallidivittatus* (*left*) produce granules (*colored stippling*). The species *Ch. tentans* (*center*) makes no granules. Chromosome IV from the four granule-producing cells (*at left of cells*) has a puff at one end (*color*), whereas the same chromosome from other cells (*lower left*) of the same gland and from all salivary-gland cells of *Ch. tentans* have no puff there. (In each case the chromosome inherited from both parents is shown.) Hybrids of the two species (*right*) have a puff only on the chromosome from the *Ch. pallidivittatus* parent in the four granule-producing cells. As indicated in the drawing, they make far fewer granules.

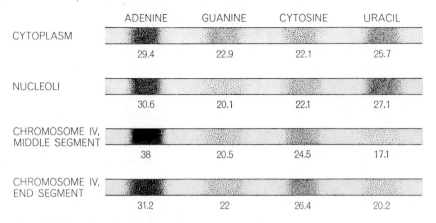

	ADENINE	GUANINE	CYTOSINE	URACIL
CYTOPLASM	29.4	22.9	22.1	25.7
NUCLEOLI	30.6	20.1	22.1	27.1
CHROMOSOME IV, MIDDLE SEGMENT	38	20.5	24.5	17.1
CHROMOSOME IV, END SEGMENT	31.2	22	26.4	20.2

RNA'S FROM VARIOUS REGIONS of salivary-gland cells of *Chironomus* differ from one another in percentages of the four RNA bases. The RNA from each place was decomposed by an enzyme and the bases were then separated by electrophoresis on rayon threads. Samples from chromosome IV differ widely from nucleolar and cytoplasmic RNA's.

This proved that the synthesis was taking place at the site of the DNA in the chromosome. All these results agree with the assumption that the pattern of puffing along chromosomes is a quantitative reflection of the pattern of synthetic activities from gene to gene.

The protein in the puffs, in contrast to RNA, takes up little or no radioactive material if we inject the larvae with radioactively labeled leucine or another labeled amino acid. The labeled protein always appears first in the cytoplasm and does not reach the chromosomes for at least an hour. One of us (Clever) obtained the same result when he injected labeled leucine together with the hormone ecdysone, which elicits puffing at several sites in a short time. We have concluded, therefore, that the puff protein is made elsewhere than in the chromosome. Probably some of this protein is the enzyme RNA polymerase, which presides at the synthesis of RNA.

In order to learn if the RNA made in the puffs is messenger RNA, we collaborated with Jan-Erik Edström of the University of Göteborg in Sweden. He has developed an elegant microelectrophoretic technique that makes it possible to determine the base composition of very small amounts of RNA. He applies RNA that has been decomposed by an enzyme to moist rayon threads, which are then laid between two electric poles. Thereafter the different bases move different distances along the threads in a given time. Their quantities can be determined by photometry and their relative proportions established. We made separate analyses of the proportions of bases in RNA's from various parts of the salivary-gland cells of *Chironomus,* including the cytoplasm, the nucleoli,

the entire chromosome I and the three large Balbiani rings of chromosome IV. This involved, among other things, cutting several hundred IV chromosomes into three pieces. The base compositions of all these RNA's differ from one another. The RNA's of the cytoplasm and the nucleoli appear to be nearly identical, but both differ from the RNA of the entire chromosome I and particularly from the RNA of puffs. In addition, there are slight but significant differences among the RNA's of the three Balbiani rings. One conclusion is that puff RNA certainly represents a special type of RNA. Is it therefore messenger RNA? An unusual feature of its base composition suggests that it is.

The RNA of salivary-gland chromosome puffs consistently contains more adenine than uracil—twice as much in the case of one Balbiani ring. (RNA contains uracil in place of the thymine in DNA.) Deviations from a one-to-one ratio are also found with respect to guanine and cytosine. In typical DNA the ratios of adenine to thymine and of guanine to cytosine invariably equal one because the bases are paired in the double-strand helix of the DNA molecule. RNA, being single-stranded, is not subject to this rule. In the case of messenger RNA, however, if one assumes that both strands of DNA make complementary copies of RNA, the ratios of adenine to uracil and guanine to cytosine should also be one. Most investigators confirm this expectation. Our data, on the other hand, strongly suggest that puff RNA is a copy of only one DNA strand. This appears to us to be a more reasonable way to make messenger RNA, since in protein synthesis only one of the two putative RNA copies of double-strand DNA could serve as a tem-

plate. Messenger RNA fractions similar in composition to ours have now been discovered in other organisms.

Evidence for the physical movement of our messenger RNA has been found recently in electron micrographs of sections through the Balbiani rings that reveal the presence of ribonucleoprotein (RNA and protein) particles. In other electron micrographs such particles are seen floating freely in the nuclear sap and through pores in the nuclear membrane [*see bottom illustration on page 190*]. They break up in the cytoplasm. We believe these particles carry the messenger RNA to the ribosomes, where it would serve as the template for the synthesis of proteins.

In the hope of delineating at least some of the forces that control the behavior of genes, one of us (Clever) set out to learn about the conditions under which puffs are produced or changed. Since insect metamorphosis has been studied rather fully, a good starting point seemed to be the changes of the puff pattern in the course of metamorphosis.

Insect metamorphosis is the transformation from the larva to the adult. In the higher insects, to which the Diptera belong, it begins with the molting of the larva into the pupa and ends with the molting of the pupa into the imago, or adult. The moltings are caused by the hormone ecdysone, which is produced by the prothorax gland located in the thorax. So far this is the only insect hormone that has been purified. Because ecdysone affects single cells directly, injection of it induces changes related to molting in all cells of the insect body.

First we examined the time relation between the changes in puffing of individual loci and the metamorphic processes in the larvae. In the great majority of the puffed loci in the salivary glands, phases in which a puff is produced alternate with phases in which a puff is absent [*see illustration on opposite page*]. Some of the phases of puff formation have no recognizable connection with the molting process. Other puffs, however, appear regularly only after the molting of the larva has begun; some at the start of molting, others later. Apparently these chromosomal sites participate in metabolic processes that take place in the cell only during the molting stage. Finally, a third group of puffs, which are found in larvae of all ages, always become particularly large during metamorphosis. This indicates that some components of the metabolic process not specific to molting are intensified at that time.

Further experiments and observations have given some indication of how ecdysone regulates the activity of single sites during molting. In the first place, the hormone not only initiates the process; it must also be present continuously in the hemolymph, or blood, of the insect if molting is to continue. The secretion of ecdysone may stop for a time in *Chironomus* larvae that had begun to molt. In such larvae all the puffs characteristic of molting are absent, which shows that the hormone controls the pattern of gene activity specific to molting. Hans-Joachim Becker of the University of Marburg confirmed this by knotting a thread around *Drosophila* larvae at the start of metamorphosis so that the prothorax gland and part of the salivary gland were in front of the knot and another part of the salivary gland was behind it, cut off from the prothorax secretions. After a time he killed the larvae and found that the puff pattern of metamorphosis was absent in the salivary-gland cells behind the knot but present in cells in front of it.

In detail ecdysone affects the puffs in a variety of ways. If we inject the hormone into *Chironomus* larvae, most of the puffs do not react until long afterward. For some the interval is a few hours, for others one or more days, and this is independent of the quantity of ecdysone. Two puffs, on the other hand, appear quite soon after the injection of ecdysone into larvae that have not begun to molt. One puff arises in 15 to 30 minutes at locus 18-C of chromosome I, the other in 30 to 60 minutes at locus 2-B of chromosome IV. These are the earliest observable gene activations produced so far by the administration of ecdysone. At both loci the higher the dosage of hormone, the longer the puffs last. The injection of more hormone slows the regression of the puffs at both loci, and if ecdysone is injected after the puffs have regressed, they swell up again. From this we conclude that the cause of puff regression is the elimination of the hormone.

The two loci exhibit different reaction thresholds. At locus 18-C on chromosome I a minimum ecdysone concentration of about 10^{-7} microgram (one ten-trillionth of a gram) per milligram of larval weight is required to induce puffing. The locus 2-B on chromosome IV reacts only to about 10^{-6} microgram per milligram of larval weight. In these concentrations there can be no more than 100 ecdysone molecules at each of the chromosome strands in a puff, assuming that each giant chromosome

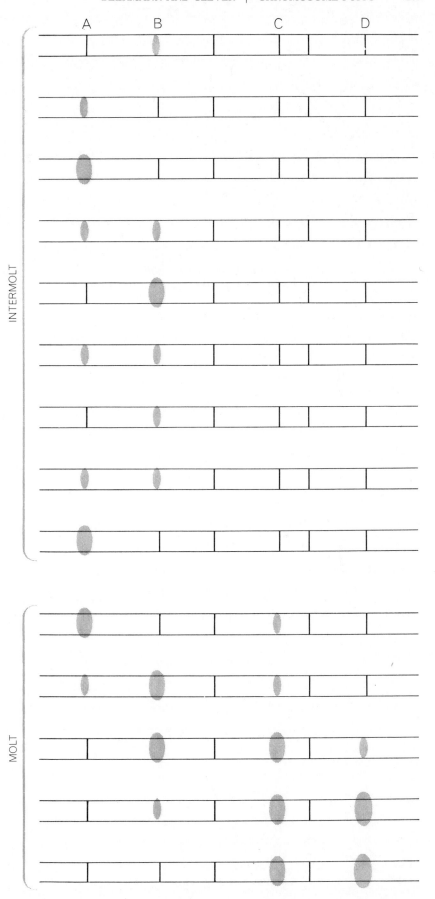

SEQUENCE OF PUFFING at four sites (*A, B, C, D*) of one chromosome in the salivary gland of *Ch. tentans* is diagramed. Some bands that do not puff are also shown. Starting from the top, the changes occur before and during the molt that begins pupation.

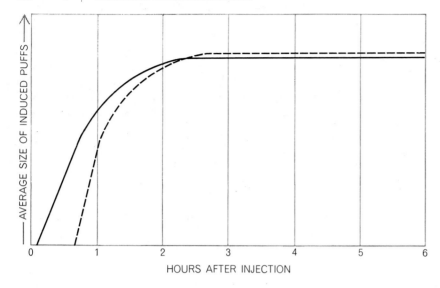

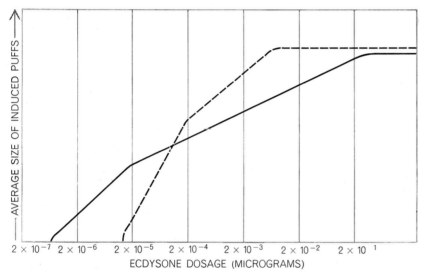

INDUCED PUFFING follows injection of the hormone ecdysone at locus 18-C of chromosome I in *Ch. tentans* (*solid curves*) and locus 2-B of chromosome IV (*broken curves*). Upper diagram shows time schedule of puffs, lower diagram the relation of puff size to quantity of hormone. Dosage is in micrograms per milligram of total weight of larva.

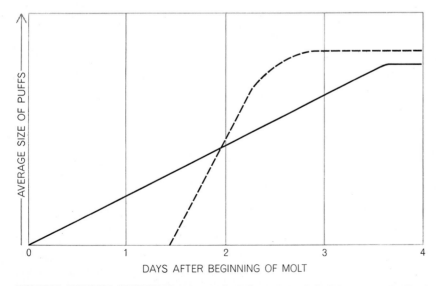

NORMAL PUFFING SEQUENCE at loci 18-C of chromosome I (*solid curve*) and 2-B of chromosome IV (*broken curve*) follows a schedule different from that of induced puffing.

consists of 10,000 to 20,000 single strands and that the hormone is distributed evenly throughout the larva. The puff at locus 2-B attains maximum size at a lower concentration than that at locus 18-C.

By applying these findings to a record of the growth and change of the two sites during normal molting, we find that the hormone level apparently increases gradually during metamorphosis. Thus the different activity patterns of these two loci can be explained as responses to the same factor: the changing hormone concentration.

In the case of locus 2-B, however, ecdysone is not the only active agent. The puff at this locus begins to regress during the second half of the prepupal phase (the last larval stage before pupation) even though the puff at locus 18-C persists to the end of the phase. In larvae that are ready to pupate, the puff at locus 2-B has usually regressed altogether. Yet when we inject hemolymph from these prepupae into young larvae, puffing is induced both at locus 18-C and locus 2-B. The former puff is quite large, which indicates that the hemolymph from the older larvae still contains ecdysone in high concentration. Evidently in these larvae, although not in their hemolymph, there is an antagonistic factor that actively represses puffing at locus 2-B in spite of the presence of ecdysone. This demonstrates that in higher organisms the activity of genes falls under the control of more than one factor.

Whereas gene activity at the loci 18-C and 2-B is subject to stringent regulation by very specific factors, other tests show that later puffs elsewhere are not the result of changes in the concentration of ecdysone. Rather they behave in an all-or-nothing manner that appears to depend on the duration and size of puffing at the two sites of earliest reaction.

We know nothing as yet about the mechanism by which the hormone regulates the genes. We are not even certain of the exact point of action of the hormone, although we would like to believe that it is the gene itself. The induction of puffing can be prevented with inhibitors of nucleic acid metabolism such as actinomycin or mitomycin, but inhibitors of protein synthesis, such as chloramphenicol and puromycin, have no apparent effect on the puffing. Thus ecdysone does not seem to act through the stimulation of protein synthesis in the cytoplasm, or to depend for its action on this synthesis. Only further investigation will solve such problems.

Hormones and Genes

18

by Eric H. Davidson

June 1965

*One of the traditional questions of biology is: How do
hormones exert their powerful effects on cells?
Evidence is accumulating that many of these effects
are due to the activation of genes*

In the living cell the activities of life proceed under the direction of the genes. In a many-celled organism the cells are marshaled in tissues, and in order for each tissue to perform its role its cells must function in a cooperative manner. For more than a century biologists have studied the ways in which tissue functions are controlled, providing the organism with the flexibility it needs to adapt to a changing environment. Gradually it has become clear that among the primary controllers are the hormones. Thus whereas the genes control the activities of individual cells, these same cells constitute the tissues that respond to the influence of hormones.

New experimental evidence is now making it possible to complete this syllogism: it is being found that hormones can affect the activity of genes. Hormones of the most diverse sources, molecular structure and physiological influence appear able to rapidly alter the pattern of genetic activity in the cells responsive to them. The establishment of a link between hormones and gene action completes a conceptual bridge stretching from the molecular level to ecology and animal behavior.

In order to understand the nature of the link between hormones and genes it will be useful to review briefly what is known of how genes function in differentiated, or specialized, cells. One of the most striking examples of cell specialization in animals is the red blood cell, the protein content of which can be more than 90 percent hemoglobin. It has been shown that in man the ability to manufacture a given type of hemoglobin is inherited; this provides a clear case of a differentiated-cell function under genetic control. Hemoglobin also furnishes an example of another

principle that is fundamental to the study of differentiation: the specialized character of a cell depends on the type and quantity of proteins in it, and therefore the process of differentiation is basically the process of developing a specific pattern of protein synthesis. Some cells, such as red blood cells and the cells of the pancreas that produce digestive enzymes, specialize in synthesizing one kind of protein; other cells specialize in synthesizing an entire set of protein enzymes to manufacture nonprotein end products, for example glycogen, or animal starch (which is made by liver cells), and steroid hormones (which are made by cells of the adrenal cortex).

If one understood the means by which the type and quantity of protein made by cells was controlled, one would have taken a long step toward understanding the nature of the differentiated cell. Part of this objective has been attained: we now know something of how genes act and how proteins are synthesized. A protein owes its properties to the sequence of amino acid subunits in its chainlike molecule. The genes of most organisms consist of deoxyribonucleic acid (DNA), the chainlike molecules of which are made up of nucleotide subunits. The sequence of nucleotides in a single gene determines the sequence of amino acids in a single protein.

The protein is not assembled directly on the gene; instead the cell copies the sequence of nucleotides in the gene by synthesizing a molecule of ribonucleic acid (RNA). This "messenger" RNA moves away from the gene to the small bodies called ribosomes. On the ribosomes, which contain their own unique kind of RNA, the amino acids are assembled into protein. In the assembly

process each molecule of amino acid is identified and moved into position through its attachment to a specific molecule of a third kind of RNA: "transfer" RNA. It can therefore be said that the characteristics of the cell are determined at the level of "gene transcription"—the synthesis of messenger and ribosomal RNA.

Each differentiated cell in a many-celled organism contains a complete set of the organism's genes. It is obvious, however, that in such a cell only a small fraction of the genes are actually functioning; the gene for hemoglobin is not active in a skin cell and the assortment of genes active in a liver cell is not the same as the assortment active in an adrenal cell. The active genes release their information in the form of messenger RNA and the inactive genes do not. Exactly how the inactive genes are repressed is not clearly understood, but the repression seems to involve a chemical combination between DNA and the proteins called histones; it has been shown that histones inhibit the synthesis of messenger RNA in the isolated nuclei of calf-thymus cells, and similar results have been obtained with the nuclei of other kinds of cell. In any case it is clear that the characteristics of the cell are the result of variable gene activity. The prime question becomes: How are the genes selectively turned on or selectively repressed during the life of the cell?

Gene action is often closely linked to cell function in terms of time. It has been demonstrated that genes can exercise immediate control over the activities of differentiated cells—particularly very active or growing cells—and over cells that are going through some change of state. In many specialized cells at least part of the messenger RNA

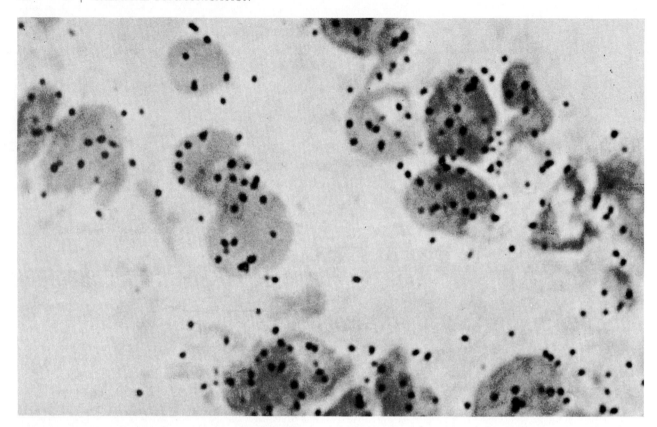

HORMONE IS LOCALIZED IN NUCLEI of cells in this radio-autograph made by George A. Porter, Rita Bogoroch and Isidore S. Edelman of the University of California School of Medicine (San Francisco). The hormone aldosterone was radioactively labeled and administered to a preparation of toad bladder tissue. When the tissue was radioautographed, the hormone revealed its presence by black dots. The dots appear predominantly in the nuclei (*dark gray areas*) of the cells rather than in the cytoplasm (*light gray areas*).

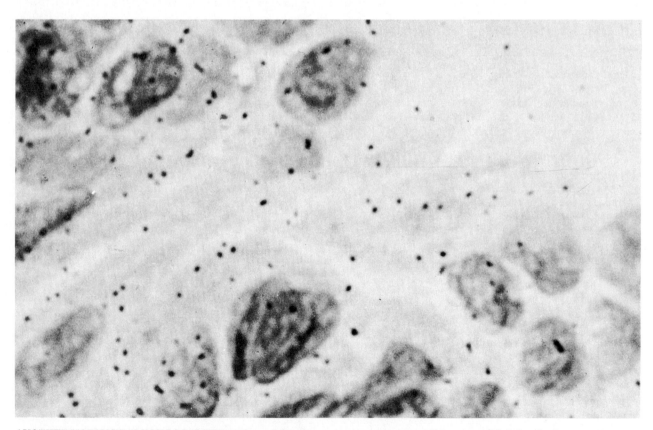

ANOTHER HORMONE IS NOT LOCALIZED in the nuclei in this radioautograph made by the same investigators. Here the hormone was progesterone, and it too was labeled and administered to toad bladder tissue. The dots are distributed more or less at random.

HORMONE	SOURCE		CHEMICAL NATURE	FUNCTION
ECDYSONE	INSECT PROTHORACIC GLAND		STEROID	Causes molting, initiation of adult development and puparium formation.
GLUCOCORTICOIDS (CORTISONE)	ADRENAL CORTEX		STEROID	Causes glycogen synthesis in liver. Causes redistribution of fat throughout organism. Alters nitrogen balance. Causes complete revision of white blood cell type frequencies. Is required for muscle function. Alters central nervous system excitation threshold. Affects connective tissue differentiation. Promotes healing. Induces appearance of new enzymes in liver. Affects almost all tissues.
INSULIN	PANCREAS (ISLETS OF LANGERHANS)		POLYPEPTIDE	Affects entry rate of carbohydrates, amino acids, cations and fatty acids into cells. Promotes protein synthesis. Affects glycogen synthetic activity. Stimulates fat synthesis. Stimulates acid mucopolysaccharide synthesis. Affects almost all tissues.
ESTROGEN	OVARY		STEROID	Promotes appearance of secondary sexual characteristics. Increases synthesis of contractile and other proteins in uterus. Increases synthesis of yolk proteins in fowl liver. Increases synthesis of polysaccharides. Affects rates of glycolysis, respiration and substrate uptake into cells. Probably affects almost all tissues.
ALDOSTERONE	ADRENAL CORTEX		STEROID	Controls sodium and potassium excretion and cation flux across many internal body membranes.
PITUITARY ACTH	ANTERIOR PITUITARY		POLYPEPTIDE	Stimulates glucocorticoid synthesis by adrenal cortex. Stimulates adrenal protein synthesis and glucose uptake. Inhibits protein synthesis in adipose tissue. Stimulates fat breakdown.
PITUITARY GH	ANTERIOR PITUITARY		PROTEIN	Stimulates all anabolic processes. Affects nitrogen balance, water balance, growth rate and all aspects of protein metabolism. Stimulates amino acid uptake and acid mucopolysaccharide synthesis. Affects fat metabolism. Probably affects all tissues.
THYROXIN	THYROID		THYRONINE DERIVATIVE	Affects metabolic rate, growth, water and ion excretion. Promotes protein synthesis. Is required for normal muscle function. Affects carbohydrate levels, transport and synthesis. Probably affects all tissues.

HORMONES DISCUSSED IN THIS ARTICLE are listed according to their source, their chemical nature and their effects, which are usually quite diverse. Pituitary GH is the pituitary growth hormone. The steroid hormones share a basic molecular skeleton consisting of adjoining four-ring structures. The polypeptide hormones and the protein hormones consist of chains of amino acid subunits.

produced by the active genes decays in a matter of hours, and therefore the genes must be continuously active for protein synthesis to continue normally. Other differentiated cells display the opposite characteristic, in that gene activity occurs at a time relatively remote from the time at which the messenger RNA acts. The very existence of this time element in gene control of cell function indicates how extensive that control is. Furthermore, certain genes can be alternately active and inactive over a short period; for example, if a leaf is bleached by being kept in the dark and is then exposed to light, it immediately begins to manufacture messenger RNA for the synthesis of chlorophyll.

The sum of such observations is that the patterns of gene activity in the living cell are in a state of continuous flux. For a cell in a many-celled organism, however, it is essential that the genetic apparatus be responsive to external conditions. The cell must be able to meet changing situations with altered metabolism, and if all the cells in a tissue are to alter their metabolism in a coordinated way, some kind of organized external control is needed. Evidence obtained from experiments with a number of biological systems suggests that such control is obtained by externally modulating the highly variable activity of the cellular genetic apparatus. The studies that will be reviewed here are cases of this general proposition; in these cases the external agents that alter the pattern of gene activity are hormones.

Many efforts have been made to explain the basis of hormone action. It has been suggested that hormones are coenzymes (that is, cofactors in enzymatic reactions), that they activate key enzymes, that they modify the outer membrane of cells and that they directly affect the physical state of structures within the cell. For each hypothesis there is evidence from studies of one or several hormones. As an example, experiments with the pituitary hormone vasopressin, which causes blood vessels to constrict and decreases the excretion of urine by the kidney, strongly support the conclusion that the hormone attaches itself to the outer membrane of the cells on which it acts.

To these hypotheses has been added the new one that hormones act by regulating the genetic apparatus, and many investigators have undertaken to study the effects of hormones on gene activity. It turns out that the gene-regulation

hypothesis is more successful than the others in explaining some of the most puzzling features of hormone activity, such as the time lag between the administration of some hormones and the initial appearance of their effects, and also the astonishing variety of these effects [*see illustration on preceding page*]. There can be no doubt that some hormone action is independent of gene activity, but it has now been shown that a wide variety of hormones can affect such activity. This conclusion is strongly supported by the fact that each of these same hormones is powerless to exert some or all of its characteristic effects when the genes of the cells on which it acts are prevented from functioning.

The genes can be blocked by the remarkably specific action of the antibiotic actinomycin D. The antibiotic penetrates the cell and forms a complex with the cell's DNA; once this has happened the DNA cannot participate in the synthesis of messenger RNA. The specificity of actinomycin is indicated by the fact that it does not affect other activities of the cell: protein synthesis, respiration and so on. These activities continue until the cellular machinery stops because it is starved for messenger RNA. In high concentrations actinomycin totally suppresses the synthesis of messenger RNA; in lower concentrations it depresses this synthesis and appears to prevent it from developing at new sites.

So far the greatest number of studies of the effects of hormones on genes have been concerned with the steroid hormones, particularly the estrogens produced by the ovaries. This work has been carried forward by many investigators in many laboratories. It has been found that when the ovaries are removed from an experimental animal and then estrogen is administered to the animal at a later date, the synthesis of protein by cells in the uterus of the animal increases by as much as 300 percent. The increase is detected by measuring the incorporation of radioactively labeled amino acids into uterine protein, or by testing the capacity for protein synthesis of homogenized uterine tissue removed from the animal at various times after the administration of estrogen. Added proof that these observations have to do with the synthesis of protein is provided by the fact that the stimulating effects of estrogen are blocked by the antibiotic puromycin, which specifically inhibits protein synthesis.

In these experiments the principal rise in protein synthesis is first observed between two and four hours after estrogen treatment. Less than 30 minutes after the treatment, however, there is a dramatic increase in the rate of RNA synthesis. When actinomycin is used to block the rise in RNA synthesis, the administration of estrogen has no effect on protein synthesis! What this means is that since the diverse metabolic changes brought about in uterine cells by estrogen are all mediated by protein enzymes, none of the changes can occur unless the estrogen has induced gene action. Among the changes are the increased synthesis of amino acids from glucose, the increased evolution of carbon dioxide and the increased synthesis of the fatty lipids and phospholipids. It is not surprising to find that none of these metabolic changes in uterine cells can be detected when estrogen is administered to an animal that has first been treated with actinomycin.

The effect of estrogen on the synthesis of RNA is not limited to messenger RNA. There is also an increase in the manufacture of the other two kinds of RNA: transfer RNA and ribosomal RNA. The administration of estrogen first stimulates the production of messenger RNA and transfer RNA. The genes responsible for the synthesis of ribosomal RNA become active somewhat later, and the number of ribosomes per cell increases. One of the earliest changes brought about by estrogen, however, is an increase in the activity of the enzyme RNA-DNA polymerase. This enzyme appears to be responsible for all RNA synthesis in such cells.

Two main conclusions can be drawn from these various observations. First, there can be no reasonable doubt that treatment with estrogenic hormones results in activation at the gene level, and that many of the well-known effects of estrogen on uterine cells result from this gene activation. Second, it is clear that a considerable number of genes must be activated in order to account for the many different responses of the cells to estrogen. Consider only the fact that estrogen stimulates the production of three different kinds of RNA. At least two different genes are known to be associated with the synthesis of ribosomal RNA, and each cell needs to manufacture perhaps as many as 60 species of transfer RNA. As for messenger RNA, the variety of the changes induced by estrogen implies that under such influences it too must be produced

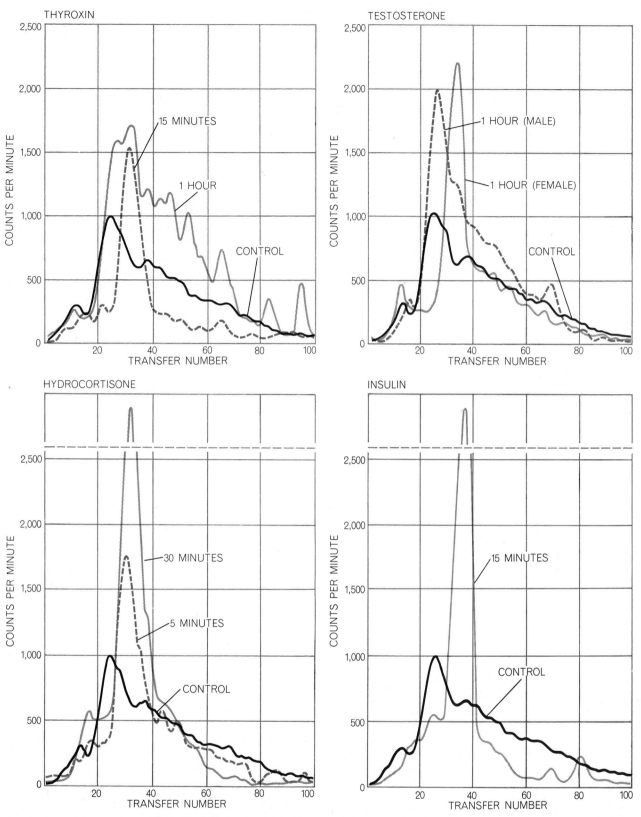

GENETIC ACTIVITY OF SEVERAL HORMONES is indicated by measurements made by Chev Kidson and K. S. Kirby of the Chester Beatty Research Institute in London. Their basic technique was first to administer to rats radioactively labeled orotic acid, which is a precursor of RNA. The tissues of the rat then incorporated the radioactive label into new RNA. Next liver tissue was removed from the rat and the species of RNA called "messenger" RNA was extracted from its cells. When the messenger RNA was analyzed by the method of countercurrent distribution, it gave rise to a characteristic curve (*black "Control" curve in each graph*); "Transfer number" refers to a stage of transfer in the countercurrent-distribution process and "Counts per minute" to the radioactivity of the solution at that point. Then, in separate measurements, rats were first given one of a number of hormones (*top left of each graph*) and shortly thereafter radioactively labeled orotic acid. The curves (*color*) of the messenger RNA obtained from such rats were entirely different, depending on the time that had elapsed before the administration of the orotic acid or on the sex of the animal (*top right*).

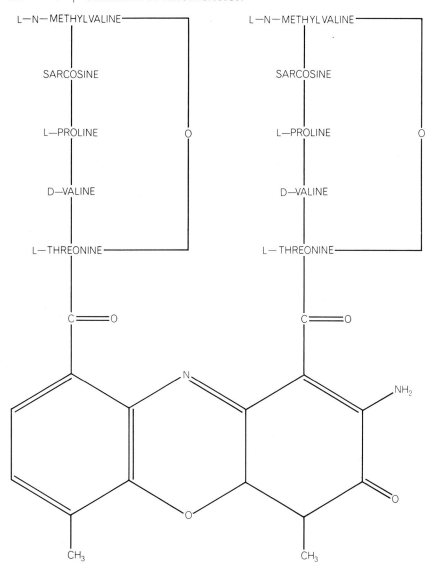

ANTIBIOTIC ACTINOMYCIN D has a complex chemical structure. The antibiotic blocks the participation of the genetic material in the synthesis of ribonucleic acid (RNA); thus it can be used in studies to determine whether or not a given hormone stimulates gene activity.

recognized that hormonal specificity resides less in the hormone than in the "target" cell. We are now, however, able to ask new questions: How are the sets of genes that are activated by a given hormone selected? Are these genes somehow preset for hormonal activation? How does the hormone interact not only with the gene itself but also with the cell's entire system of genetic regulation?

The male hormone testosterone has also been shown to operate by gene activation. Like the estrogens, the male sex hormones can give rise to dramatic increases of RNA synthesis in various cells. In experiments on male and female rats it has been found that the effect of testosterone on the liver cells of a female is somewhat different from that on the liver cells of a castrated male. In both cases the hormone causes an increase in the *amount* of messenger RNA produced, but in the female it also brings about the synthesis of a new *variety* of messenger RNA. This effect, like the ability of estrogen to stimulate a rooster's liver cells to produce egg-yolk proteins, provides a new approach for examining the whole question of sexual differentiation.

Apart from the sex hormones, the principal steroids in mammals are those secreted by the adrenal cortex. One group of adrenocortical hormones is typified by cortisone; this hormone and its relatives are known for their quite different effects in different tissues. Only a fraction of these effects have been studied from the standpoint of gene activation, and there is much evidence to indicate that some of them are not mediated by the genes. Some responses to cortisone, however, do appear to be the consequence of gene activation.

If the adrenal glands are removed from an experimental animal and cortisone is administered later, the hormone induces in the liver cells of the animal the production of a number of new proteins. Among these proteins are enzymes required for the synthesis of glucose (but not the breakdown of glucose) and enzymes involved in the metabolism of amino acids. Moreover, cortisone steps up the total production of protein by the liver cells. The effect of cortisone on the synthesis of messenger RNA is apparent as soon as five minutes after the hormone has been administered; within 30 minutes the amount of RNA produced has increased two to three times and probably includes not

in a number of molecular species. We are therefore confronted with a major mystery of gene regulation: How can a single hormone activate an entire set of functionally related but otherwise quite separate genes, and activate them in a specific sequence and to a specific degree?

The question can be sharpened somewhat by considering the effect of estrogen not on uterine cells but on the cells of the liver. When an egg is being formed in a hen, the estrogen produced by the hen's ovaries stimulates its liver to produce the yolk proteins lipovitellin and phosvitin. Obviously a rooster does not need to synthesize these proteins, but if it is treated with estrogen, its liver will make them in large amounts! A more unequivocal example of the

selective activation of repressed genes by a hormone could scarcely be imagined. What is more, experiments by E. N. Carlsen and his co-workers at the University of California School of Medicine (Los Angeles) have demonstrated that this gene-activating effect of estrogen is remarkably specific. Phosvitin is an unusual protein in that nearly half of its subunits are of one kind: they are residues of the amino acid serine. Carlsen and his colleagues found that estrogen most strongly stimulates liver cells to produce the particular species of transfer RNA that is associated with the incorporation of this amino acid into protein.

The effect of estrogen on liver cells is thus quite different from its effect on uterine cells. Indeed, it has long been

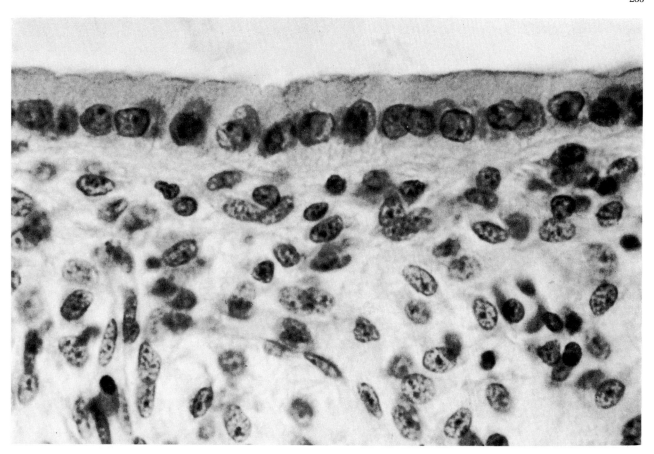

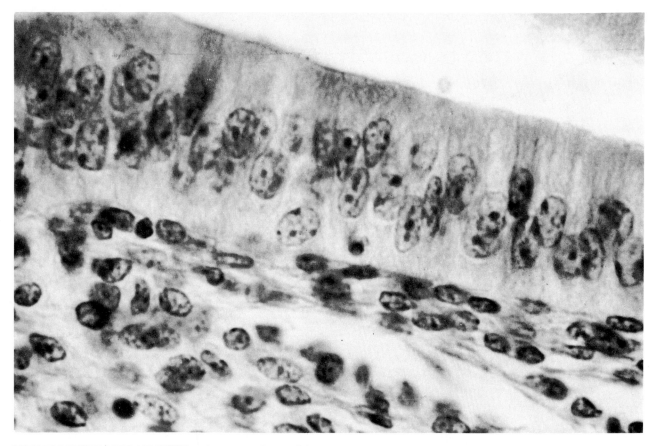

EFFECT OF ESTROGEN ON CELLS in the uterus of rats is demonstrated in these photomicrographs made by Sheldon J. Segal and G. P. Talwar of the Rockefeller Institute. The photomicrograph at top shows uterine cells from a rat that had not been treated with estrogen; the layer of cells at the surface of the tissue is relatively thin. The photomicrograph at bottom shows uterine cells from a rat that had been treated with the hormone; the layer of cells is much thicker. The effect involves enhanced synthesis of protein.

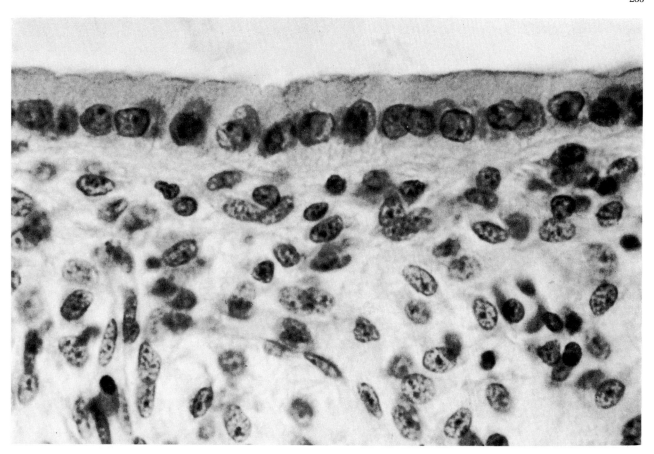 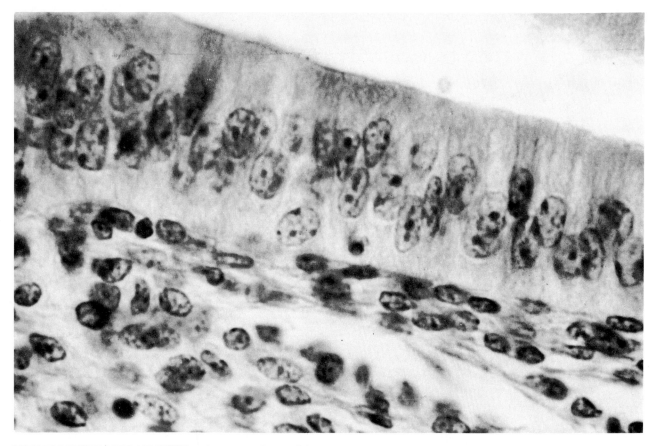

ROOSTER TREATED WITH ESTROGEN (*bottom*) is compared with a normal rooster (*top*). The signs of femaleness induced by estrogen include changes in comb and plumage.

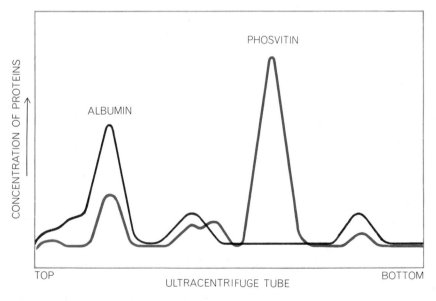

ULTRACENTRIFUGE PATTERNS show that phosvitin, a yolk protein found only in hens, is present in serum extracted from a bird that had been injected with estrogen (*colored curve*) but not in serum from a bird used as a control (*black curve*). Each curve gives the concentration of proteins as they are separated out of a mixture by an ultracentrifuge.

only messenger RNA but also ribosomal RNA. These events are followed by the increase in enzyme activity. Olga Greengard and George Acs of the Institute for Muscle Disease in New York have shown that if the animal is treated with actinomycin before cortisone is administered, the new enzymes fail to appear in its liver cells.

Another clear case of the activation of genes by an adrenocortical hormone has been demonstrated by Isidore S. Edelman, Rita Bogoroch and George A. Porter of the University of California School of Medicine (San Francisco). They employed the hormone aldosterone, which regulates the passage through the cell membrane of sodium and potassium ions. Tracer studies with radioactively labeled aldosterone showed that when the bladder cells of a toad were exposed to the hormone, the molecules of hormone penetrated all the way into the nuclei of the cells [*see illustrations on page 198*]. About an hour and a half after the aldosterone has reached its peak concentration within the cells the movement of sodium ions across the cell membrane increases. It appears that this facilitation of sodium transport is brought about by proteins the cell is induced to make, because it will not occur if the cells have been treated beforehand with puromycin, the drug that blocks the synthesis of protein. Moreover, treatment of the cells with actinomycin will block the aldosterone-induced increase in sodium transport through the membrane. Thus the experiments indicate that aldosterone activates genes in the nucleus and gives rise to proteins—that is, enzymes—that speed up the passage of sodium ions across the membrane.

Ecdysone, a steroid hormone of insects, is also believed to be a gene activator. The evidence for this conclusion has been provided by Wolfgang Beermann and his colleagues at the Max Planck Institute for Biology in Tübingen [*see the article "Chromosome Puffs," by Wolfgang Beermann and Ulrich Clever, beginning on page 188*]. If the larva of an insect lacks ecdysone, the development of the larva is indefinitely arrested at a stage preceding its metamorphosis into a pupa. Only when, in the course of normal development, the concentration of ecdysone in the tissues of the larva begins to rise does further differentiation take place; the larva then advances to metamorphosis. Ecdysone has been of especial interest to cell biologists because it has been observed to cause startling changes in the chromosomes within the nuclei of the cells affected by it. Studies of this kind are possible in insects because the cells of certain insect tissues have giant chromosomes that can easily be examined in the microscope. These "polytene" chromosomes develop in many kinds of differentiated cell by means of a process in which the chromosomes repeatedly replicate but do not separate.

In some polytene chromosomes genetic loci, or specific regions, have a distended, diffuse appearance [see illustration below]. Biologists regard these regions, which have been named "puffs," as sites of intense gene activity. Evidence for this conclusion is provided by radioautograph studies, which show that the puffs are localized sites of intense RNA synthesis. In such studies a molecular precursor of RNA is radio-actively labeled and after it has been incorporated into RNA reveals its presence as a black dot in the emulsion of the radioautograph. According to the view of differentiation presented in this article, different genes should be active in different types of cell, and this appears to be the case in insect cells with polytene chromosomes. In many different kinds of cell—salivary-gland cells, rectal-gland cells and excretory-tubule cells—the giant chromosomes have a different constellation of puffs; this suggests that different sets of genes are active, a given gene being active in one cell and quiescent in another.

On the polytene chromosomes of insect salivary-gland cells new puffs develop as metamorphosis begins. This is where ecdysone comes into the picture: the hormone seems to be capable of inducing the appearance of specific new puffs. When a minute amount of ecdysone is injected into an insect larva, a specific puff appears on one of its salivary-gland chromosomes; when a slightly larger amount of ecdysone is injected, a second puff materializes at a different chromosomal location. In the normal course of events the concentration of ecdysone increases as the larva nears metamorphosis; therefore there exists a mechanism whereby the more sensitive genetic locus can be aroused first. This example of hormone action at the gene level, which is directly visible to the investigator, seems to have provided some of the strongest evidence for the regulation of gene action by hormones. The effect of ecdysone, which is clearly needed for differentiation, appears to be to arouse quiescent genes to visible states of activity. In this way the specific patterns of gene activity required for differentiation are provided.

What about nonsteroid hormones? Here the overall picture is not as clearcut. The effects of some hormones are quite evidently due to gene activation, and yet other effects of the same hormones are not blocked by the administration of actinomycin; a small sample of these effects is listed in the illustration on the opposite page. As for the hormonal effects that are quite definitely not genetic, they fall into one of the following categories.

(1) Some hormones act on specific enzymes; for example, the thyroid hormone thyroxin promotes the dissociation of the enzyme glutamic dehydrogenase. (2) Other hormones, for instance insulin and vasopressin, act on systems that transport things through cell mem-

"PUFF" ON A GIANT CHROMOSOME from the salivary gland of the midge *Chironomus tentans* appears after administration of the insect hormone ecdysone. In the radioautograph at left the round area at top center is a puff. The black dots result from the fact that the midge was given radioactively labeled uridine, which is a precursor of RNA. The concentration of dots in the puff indicates that it is actively synthesizing RNA. In the radioautograph at right is a chromosome from a fly that had been treated with actinomycin before receiving ecdysone. No puff has occurred and RNA synthesis appears to be muted. The radioautographs were made by Claus Pelling of the Max Planck Institute for Biology in Tübingen.

HORMONE	EVIDENCE FOR HORMONAL ACTION BY GENE ACTIVATION.	EVIDENCE THAT HORMONAL ACTION IS CLEARLY INDEPENDENT OF IMMEDIATE GENE ACTIVATION.
PITUITARY GROWTH HORMONE	General stimulation of protein synthesis. Stimulation of rates of synthesis of ribosomal RNA, transfer RNA and messenger RNA within 90 minutes in liver. Effect blocked with actinomycin.	
PITUITARY ACTH	Stimulates adrenal protein synthesis. Messenger RNA and total RNA synthesis stimulated.	Steroid synthesis in isolated adrenal sections is independent of RNA synthesis and is insensitive to actinomycin D.
THYROXIN	Promotes new messenger RNA synthesis within 10 to 15 minutes of administration, promotes stimulation of all classes of RNA by 60 minutes. Promotes increase in RNA–DNA polymerase at 10 hours, later promotes general increase in protein synthesis.	Causes isolated, purified glutamic dehydrogenase to dissociate to the inactive form. Affects isolated mitochondria in vitro.
INSULIN	Promotes 100 percent increase in rate of RNA synthesis. Causes striking change in messenger RNA profile within 15 minutes of administration to rat diaphragm; effect blocked with actinomycin. Actinomycin-sensitive induction of glucokinase activity.	Actinomycin-insensitive increase in ATP synthesis and in glucose transport into cells; mechanism appears to involve insulin binding to cell membrane, occurs at 0 degrees C.
VASOPRESSIN		Actinomycin-insensitive promotion of water transport in isolated bladder preparation under same conditions in which aldosterone action is blocked by actinomycin.

SUMMARY OF EXPERIMENTAL EVIDENCE is given in table. Facts indicating that hormones activate the genes (*middle column*) are compared with facts suggesting that hormonal action does not entail the immediate activation of the genes (*column at right*).

branes; indeed, it is believed that both of these hormones attach themselves directly to the membranes whose function they affect. (3) Still other hormones rapidly activate a particular enzyme; phosphorylase, a key enzyme in determining the overall rate at which glycogen is broken down, is converted from an inactive form by several hormones, including epinephrine, glucagon and ACTH.

This does not alter the fact that many nonsteroid hormones operate at the gene level. Some of the best evidence for this statement is provided by studies of several hormones made by Chev Kidson and K. S. Kirby of the Chester Beatty Research Institute of the Royal Cancer Hospital in London. They separately injected rats with thyroxin, testosterone, cortisone and insulin and then mea-

sured the synthesis of messenger RNA by the rats' liver cells [*see illustration on page 203*]. The most striking aspect of their measurements is the extremely short time lag between the administration of the hormone and the change in the pattern of gene activity. The activation of genes in the nuclei of the affected cells occurs so quickly that one is tempted to assume that it is an initial effect of the hormone.

Here, however, we come face to face with a basic problem that must be solved in any attempt to explain the exact molecular mechanism of hormone action. The problem is simply that of identifying the initial site of reaction in a cell exposed to a hormone. Does a hormone move directly to the chromosome and exert its effect, so to speak, "in person"? As we have seen, aldoste-

rone does appear to enter the nucleus, but there is little real evidence that other hormones do so.

For many years biologists have been looking for the "receptor" substance of various hormones. The discovery that hormones ultimately act on genes makes this search all the more interesting. The evidence presented here only goes as far as to prove that an early stage in the operation of many hormones is the selective stimulation of genetic activity in the target cell. The molecules of the hormones range in size and structure from the tiny molecule of thyroxin to the unique multi-ring molecule of a steroid and the giant molecule of a protein; how these various molecules similarly affect the genetic apparatus of their target cells remains an intriguing mystery.

Cyclic AMP

by Ira Pastan
August 1972

This comparatively small molecule is a "second messenger" between a hormone and its effects within the cell. It operates in cells as diverse as bacteria and cancerous animal cells

The chemical reactions that proceed in the living cell are catalyzed by the large molecules called enzymes. If all the enzymes found within cells were working at top speed, the result would be chaos, and many mechanisms have evolved that control the speed at which these enzymes function. A small molecule that plays a key role in regulating the speed of chemical processes in organisms as distantly related as bacteria and man is cyclic-3'5'-adenosine monophosphate, more widely known as cyclic AMP. ("Cyclic" refers to the fact that the atoms in the single phosphate group of the molecule are arranged in a ring.) Among the many functions served by cyclic AMP in man and other animals is acting as a chemical messenger that regulates the enzymatic reactions within cells that store sugars and fats. Cyclic AMP has also been shown to control the activity of genes. Moreover, a precondition for one of the kinds of uncontrolled cell growth we call cancer appears to be an inadequate supply of cyclic AMP.

The first steps leading to the discovery of cyclic AMP were taken by Earl W. Sutherland, Jr., at Washington University some 25 years ago. For this work Sutherland was awarded the Nobel prize in physiology and medicine for 1971. Sutherland was trying to trace the sequence of events in a well-known physiological reaction whereby the hormone epinephrine (as adrenalin is generally known in the U.S.) causes an increase in the amount of glucose, or blood sugar, in the circulatory system. It is this reaction, usually a response to pain, anger or fear, that provides an animal with the energy either to fight or to flee. It is not a simple one-step process. Glycogen, a polymeric storage form of glucose, is held in reserve in the cells of the liver. What transforms the glycogen into glucose, which can then leave the liver and

enter the bloodstream, is a series of steps involving intermediate substances. Sutherland measured the levels of these intermediates and concluded that only the initial step in the series (the transformation of glycogen into the intermediate sugar glucose-1-phosphate) was mediated by epinephrine. The transformation itself is actually catalyzed by an enzyme known as phosphorylase. Observing the activity of phosphorylase in cell-free extracts of liver tissue, Sutherland was able to enhance the enzyme's performance by first exposing to epinephrine the cells he used to make his extract.

Sutherland began to examine the properties of phosphorylase in more detail. He found that the enzyme could exist in two forms: one that degraded glycogen rapidly and one that had no effect on it. The conversion of the enzyme from the active to the inactive form was catalyzed by a second enzyme, whose only action was to remove inorganic phosphate from the phosphorylase molecule. The conversion is worth noting; it is an important example of how the activity of an enzyme can be controlled by a relatively small change in its structure.

In collaboration with the first of a number of talented co-workers, Walter D. Wosilait, Sutherland next found still another liver enzyme, phosphorylase *b* kinase, which could restore the inactive form of phosphorylase to the active state. As one might expect, the reversal was accomplished by replacement of the missing phosphate; the donor of the phosphate was a close chemical relative of cyclic AMP, adenosine triphosphate (ATP). At about the same time Edwin G. Krebs and Edmond H. Fischer of the University of Washington detected a similar activating kinase in muscle tissue. ("Kinase" is the name reserved for trans-

formations where ATP is the phosphate-donor.)

Having established the existence of two forms of phosphorylase, Sutherland and his colleagues concluded that the speed at which glycogen was broken down in the liver was a function of the amount of the enzyme present in its active form. Sutherland and Theodore W. Rall now made preparations of ruptured liver cells. When they added epinephrine to these broken cells, they found that in spite of the damage the hormone still increased the activity of the enzyme. The experiment, although simple, was extremely significant. Never before had a hormone been observed to

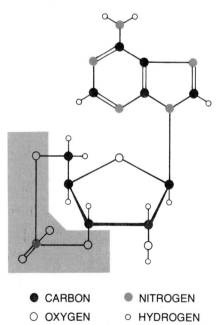

● CARBON   ● NITROGEN
○ OXYGEN   ○ HYDROGEN
● PHOSPHORUS

CYCLIC AMP is so named because the phosphate group in its molecule (*colored area*) forms a ring with the carbon atoms to which it is attached. Earl W. Sutherland, Jr., and his colleagues isolated substance in 1958.

ADENOSINE
TRIPHOSPHATE

CYCLIC AMP

ADENYLATE
CYCLASE

FORMATION OF CYCLIC AMP takes place when an enzyme in the cell membrane, adenyl-ate cyclase, responds to the arrival of a hormone at the membrane. The enzyme transforms molecules of adenosine triphosphate, or ATP (*left*), within the cell into cyclic AMP (*right*).

function in a preparation that contained no intact cells. Once such a reaction can be shown to occur in a cell-free preparation the investigator can go on to test various cell components one at a time to determine just which ones are affected by the hormone.

This Sutherland and his co-workers proceeded to do. They knew that the phosphorylase was present in liver-cell cytoplasm: that part of the cell outside the nucleus and inside the cell membrane. When they added epinephrine to preparations composed of cytoplasm alone, however, there was no increase in enzyme activity. The absence of response suggested that the hormone exerted its effect on some other component of the liver cell. In due course they found that this component was the cell membrane.

Exactly what was the hormone doing to the cell membrane? In an effort to find out Sutherland employed a stratagem commonly used in biochemistry. He incubated a preparation of cell membrane (which itself contains no phosphorylase) with epinephrine. He then brought the mixture to the boiling point, expecting the heat to destroy the activity of the enzyme or enzymes in the membrane that were required for epinephrine action. When he added the now denatured mixture to a cell-free preparation that contained phosphorylase but no cell membrane, the activity of the phosphorylase was increased. The epinephrine had interacted with the cell membrane during the initial incubation period, evidently causing some enzyme in the membrane to produce a heat-stable factor that enhanced the activity of phosphorylase. Unfortunately the factor—whatever it was—was present in very small amounts and was therefore difficult to identify.

Sutherland eventually collected a large enough sample of the factor to de-termine that it belonged to the group of small molecules known as nucleotides. It did not, however, appear to be any of the known nucleotides. He wrote to Leon A. Heppel of the National Institutes of Health, who had developed many of the methods used in preparing and identifying a number of nucleotides, asking him for a quantity of an enzyme that breaks nucleotides into their component parts and thus facilitates their identification. This request set the stage for a remarkable coincidence.

It is Heppel's habit to let letters that do not require an immediate answer accumulate on his desk. He covers each few days' correspondence with a fresh sheet of wrapping paper, and every few months he clears his desk. Heppel immediately sent Sutherland the enzyme but left the letter of request on his desk. By chance the stratum just below contained a chatty letter from a friend and former colleague, David Lipkin of Washington University, describing an experiment where ATP was treated with a solution of barium hydroxide. The result was the formation of an unusual nucleotide.

Heppel remembers coming into his office one Saturday to clear his desk. He found Sutherland's and Lipkin's letters in adjacent strata and consequently re-read them together. It seemed likely to him that both men had isolated the same substance, and he proceeded to put them in touch. Lipkin's chemical synthesis readily produced the nucleotide in large quantities. This made it easy to establish that the synthetic substance was structurally identical with natural cyclic AMP. It also provided an abundant supply of synthetic cyclic AMP for experimental purposes.

Taking advantage of the demonstrated ability of cyclic AMP to increase phosphorylase activity in cell-free preparations, Sutherland and his co-workers were able to measure the amount of cyclic AMP present in a wide variety of cells. They found that it was 1,000 times less abundant than ATP, being present in cell water in a ratio of about one part per million in contrast to ATP's one part per 1,000. Although scanty in amount, cyclic AMP was present in virtually every organism they examined, from bacteria and brine-shrimp eggs to man. Among mammals it was present in almost every type of body cell. Sutherland's group went on to examine a number of tissues that are characterized by their secretion of various substances following stimulation by hormones. He discovered that the level of cyclic AMP in such cells rose soon after exposure to the hormone. Moreover, when the tissues were exposed to nothing but cyclic AMP, or to derivatives of cyclic AMP that enter cells rapidly, they produced secretions just as readily.

The cyclic AMP molecule is formed from ATP by the action of a special enzyme: adenylate cyclase. The enzyme is located in the membrane of the cell wall. Normally its activity is low and the transformation of ATP into cyclic AMP takes place at a slow rate. Let us consider what happens, however, when a hormone enters the bloodstream. The hormone acts as a "first messenger." It travels to the target cell and then binds to specific receptor sites on the outside of the cell wall. Thyroid cells have receptors that "recognize" thyrotrophin, adrenal cells have receptors that recognize adrenocorticotrophin, and so forth. The binding of the hormone to the receptor site increases the activity of the adenylate cyclase in the cell membrane; just how this occurs has not yet been established. In any event, cyclic AMP is produced as a result, utilizing the abundant supply of ATP on the inner side of the cell membrane. The cyclic AMP is then free to diffuse throughout the cell, where it acts as a "second messenger," instructing the cell to respond in a characteristic way. For example, a thyroid cell responds to this second message by secreting more thyroxine, whereas an adrenal cell responds by producing and secreting steroid hormones. In the cells of the liver the instruction results in the conversion of glycogen into glucose.

Because cyclic AMP is such a powerful regulator of cell functions the cell must be able to control its level or concentration. In most cells control is accomplished by regulating the rate of synthesis of cyclic AMP and by the actions of one or more enzymes, known as phosphodiesterases, that degrade cyclic

AMP into an inert form of adenosine monophosphate. The deactivation results from a splitting of the ester bond that joins the phosphate to the 3′ carbon of the ribose ring. The quantity of the degradative enzymes in the cell is not kept constant; apparently more can be made whenever the level of cyclic AMP in the cell is elevated for more than a few minutes. The level of cyclic AMP is also controlled by diffusion through the cell wall; this is the mechanism operating in addition to enzyme degrada-

tion in bacteria and in the cells of some animal tissues.

Krebs, whose earlier work with Fischer had established the presence of active and inactive forms of phosphorylase in muscle tissue, was able in 1968 to specify the role played by cyclic AMP in activating the enzyme. Working with Donal Walsh, he found that the cyclic AMP binds to yet another enzyme, protein kinase, which is inactive until cyclic AMP is present. The activated kinase then performs the same function for a

related enzyme, phosphorylase kinase. It is this second enzyme that at last activates the phosphorylase. The result of the final activation is the breakdown of glycogen, the storage form of glucose, in a series of steps similar to those that proceed in liver cells.

Whenever glycogen is being degraded in order to satisfy the organism's need for glucose, it would be a waste of energy to continue the synthesis of additional glycogen. This waste is avoided. A specific enzyme mediates the syn-

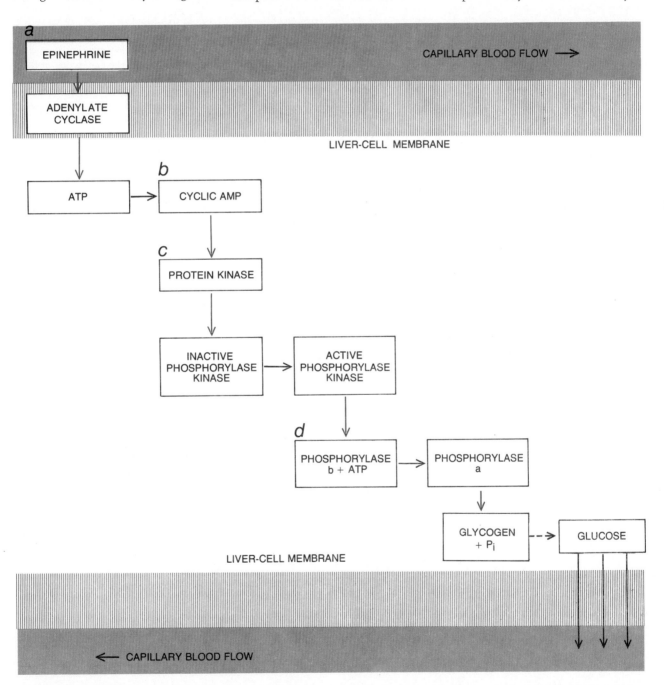

RELEASE OF BLOOD SUGAR by a glycogen storage cell in the liver is mediated by cyclic AMP. In the schematic diagram arrows in color symbolize actions and black arrows the results. "First messenger," epinephrine, arrives at the cell membrane and activates the enzyme, adenylate cyclase (*a*), causing it to convert some of the ATP present in the cytoplasm into cyclic AMP (*b*), the second messenger. The cyclic AMP then activates a protein kinase (*c*), which activates a second kinase. The second kinase (*d*) triggers a four-step sequence (*not shown*) that converts the glycogen into the assimilable sugar glucose, which then passes into the bloodstream.

thesis of glycogen. It is glycogen synthetase, and like phosphorylase and many other enzymes it has an active and an inactive form. At the time when some molecules of cyclic AMP are initiating the chain of events that leads to the breakdown of glycogen, other cyclic AMP molecules are at work converting glycogen synthetase from the active to the inactive form.

Just as cells that store glycogen have the task of supplying the fasting organism with glucose, so the task of fat cells is to satisfy the organism's need for fatty acids. Fatty acids are present in the fat cell in the storage form triglyceride. The triglyceride can ocupy as much as 90 percent of the cell's volume. Here again it is cyclic AMP that initiates the breakdown of the stored fat. In response to any of several hormonal stimuli the level of cyclic AMP in the fat cell begins to rise. As with muscle tissue, this activates a protein kinase. The kinase in turn activates a second enzyme, triglyceride lipase. On being converted to the active form the second enzyme begins to degrade the stored triglyceride into the re-

quired fatty acids. It should be noted here that protein kinases are present in the cells of many tissues other than fat and muscle; numerous other actions of cyclic AMP presumably also involve these ATP-powered enzymes.

In addition to such effects within cells cyclic AMP has been observed to stimulate the expression of genetic information. How this stimulation is accomplished in animal cells remains obscure. Almost all the detailed observations of the regulation of gene activity involve a single microorganism: the common intestinal bacterium *Escherichia coli*. There are many good reasons for molecular biologists who are engaged in genetic studies choosing to work with this simple organism. One of them is that a typical animal cell contains enough DNA to account for 10 million individual genes; *E. coli* has only enough DNA for about 10,000 genes.

Sutherland and another colleague, Richard Makman, established the presence of cyclic AMP in *E. coli* cells in 1965. By then the substance was already

known to control a variety of cell processes, and Robert Perlman and I at the National Institutes of Health guessed that it must also play an important role in the bacterium. But what role?

There were two clues. First, cultures of *E. coli* that are nourished exclusively with glucose show low levels of cyclic AMP. Second, such cultures can synthesize only very small amounts of a number of enzymes, including those needed to metabolize sugars other than glucose; this inhibition is called the "glucose effect." Putting these clues together, we reasoned that cyclic AMP was a chemical switch, so to speak; if it was present in *E. coli* in adequate quantities, it would activate the expression of those genes that are necessary for the synthesis of the missing enzymes. In order to test this speculation we added cyclic AMP to cultures of inhibited *E. coli*. The cells were then able to metabolize such sugars as maltose, lactose and arabinose in addition to a variety of other nutrients.

Now, factors that operate at the level of the gene do so by stimulating the synthesis of the messenger RNA that in

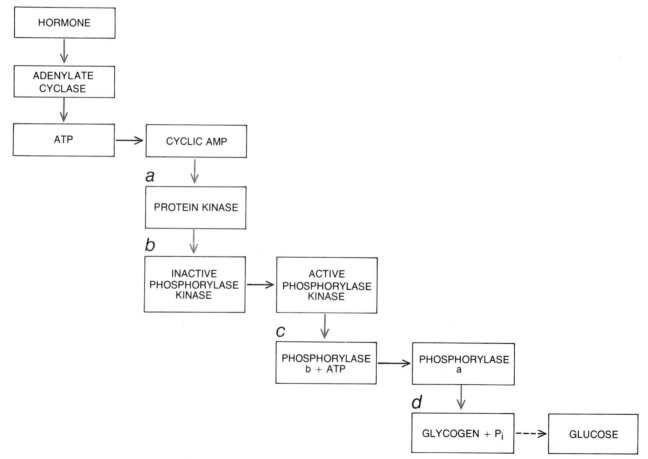

INITIAL SEQUENCE that triggers the transformation of glycogen into glucose in muscle tissue also involves cyclic AMP. First (*a*) the cyclic AMP that has been formed from ATP in the cell cytoplasm activates a protein kinase. The kinase in turn activates a second kinase (*b*) that is capable of transforming phosphorylase *b* into phosphorylase *a*. When this takes place (*c*), the transformed phosphorylase then starts the sequence (*d*) that converts glycogen into glucose. Edwin Krebs and Edmond Fischer of the University of Washington discovered the phosphorylase alteration in muscle at the same time that Sutherland found the same process in the liver.

effect reproduces the information contained in the DNA of the gene. The messenger RNA then provides the ribosomes of the cell with the information they need to construct the appropriate enzyme. In our laboratory by this time we had isolated a mutant form of *E. coli* that lacked the cell-membrane enzyme required for the synthesis of cyclic AMP. Curiously, the mutant cells were viable; apparently the presence of cyclic AMP is not absolutely crucial to survival. For our next experiment we selected cultures of the mutant strain that contained a known amount of the two messenger RNA's needed for the metabolism of two sugars: lactose and galactose. When we added cyclic AMP to the mutant cultures, we found that the quantity of the two messenger RNA's was increased but that the expression of most of the other genes in the cells was not affected. In order to learn exactly how the increase took place we now needed to study the reaction in a cell-free preparation.

At Columbia University, Geoffrey L. Zubay and his co-workers were working with a complex cell-free preparation made from *E. coli*. When they added to the preparation DNA that was greatly enriched in the genes for lactose metabolism, small amounts of one of the enzymes of lactose metabolism, beta-galactosidase, were formed. The DNA was enriched because it was derived from a bacterial virus that had acquired the lactose genes from the *E. coli* host it had once lived in and now contained it permanently. The virus is a hybrid of two other viruses, designated λ and 80, and the DNA derived from it is λ80*lac* (for lactose) DNA.

When Zubay and his co-worker Donald Chambers added cyclic AMP to the cell-free preparation, synthesis of the enzyme for lactose metabolism was greatly stimulated. Soon thereafter my colleagues B. de Crombrugghe and H. Varmus showed that the synthesis of *lac* messenger RNA was also increased. John Parks in our laboratory, following Zubay's procedure, developed a cell-free *E. coli* preparation that responded to the addition of λ*gal* (for galactose) DNA by producing one of the enzymes of galactose metabolism. Like Zubay's preparation, Parks's synthesized much more of the enzyme when cyclic AMP was added.

One might now have expected that the same result could be achieved without even using the complex cell-free preparation. It would be necessary only to mix together in the test tube appropriate quantities of DNA rich in lactose genes and of the special enzyme that

TISSUE	HORMONE	PRINCIPAL RESPONSE
FROG SKIN	MELANOCYTE-STIMULATING HORMONE	DARKENING
BONE	PARATHYROID HORMONE	CALCIUM RESORPTION
MUSCLE	EPINEPHRINE	GLYCOGENOLYSIS
FAT	EPINEPHRINE	LIPOLYSIS
	ADRENOCORTICO-TROPHIC HORMONE	LIPOLYSIS
	GLUCAGON	LIPOLYSIS
BRAIN	NOREPINEPHRINE	DISCHARGE OF PURKINJE CELLS
THYROID	THYROID-STIMULATING HORMONE	THYROXIN SECRETION
HEART	EPINEPHRINE	INCREASED CONTRACTILITY
LIVER	EPINEPHRINE	GLYCOGENOLYSIS
KIDNEY	PARATHYROID HORMONE	PHOSPHATE EXCRETION
	VASOPRESSIN	WATER REABSORPTION
ADRENAL	ADRENOCORTICO-TROPHIC HORMONE	HYDROCORTISONE SECRETION
OVARY	LUTEINIZING HORMONE	PROGESTERONE SECRETION

FOURTEEN EXAMPLES of hormonal activities that affect many different target tissues (*left*) have one factor in common: each causes an increase in the level of cyclic AMP in the tissue. It seems probable that all the responses are set in train by the sudden increase in level.

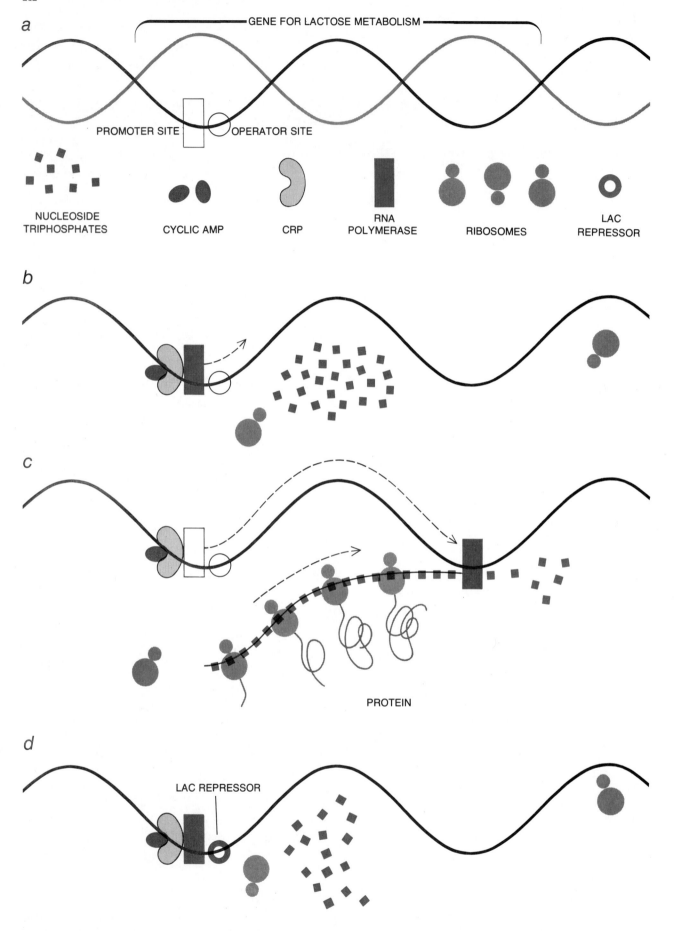

a

GENE FOR LACTOSE METABOLISM

PROMOTER SITE   OPERATOR SITE

NUCLEOSIDE TRIPHOSPHATES   CYCLIC AMP   CRP   RNA POLYMERASE   RIBOSOMES   LAC REPRESSOR

b

c

PROTEIN

d

LAC REPRESSOR

copies DNA into RNA, add some ATP and other nucleoside triphosphates as building blocks and cyclic AMP as mediator and the result would be the production of *lac* messenger RNA. The expectation proved to be false. Some RNA was formed in the test tube but none of it corresponded to the *lac* gene. Clearly some factor was missing. Fortunately there was a clue to its nature.

In our search for mutant strains of *E. coli* that could not make various enzymes known to be controlled by cyclic AMP we had found some mutants that produced cyclic AMP in abundance but were still unable to metabolize either lactose or several other sugars. Similar *E. coli* mutants had been isolated at the Harvard Medical School by Jonathan Beckwith and his colleagues. What was lacking in the mutants was a protein that has the ability to bind cyclic AMP. The protein, which is known either as catabolite gene activation protein (abbreviated CAP) or as cyclic AMP receptor protein (CRP), was difficult to purify but a pure form was finally prepared by my colleague Wayne B. Anderson.

De Crombrugghe added some of the pure protein to the test-tube mixture that had failed to yield *lac* messenger RNA. This time the effort was successful. Moreover, when CRP was added to a similar test-tube mixture that included λ*gal* DNA, *gal* messenger RNA was formed. These experiments, incidentally, were the first to achieve the transcription of a bacterial gene in a system containing only purified components. We could now proceed to examine in detail how the gene activity was controlled.

The initial step in the process involved the combination of cyclic AMP with CRP. The complex of the two substances was then bound to the DNA; once this took place the enzyme that copies DNA into RNA was enabled to bind to a specific site, called the "pro-

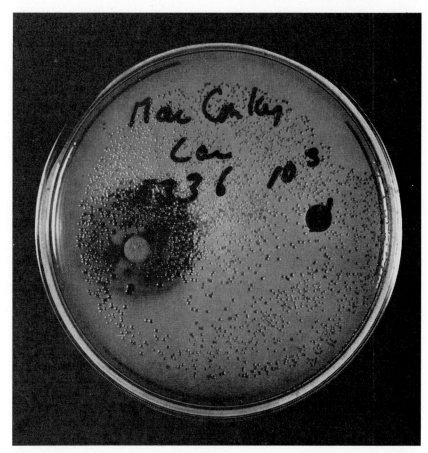

MUTANT STRAIN of the bacterium *Escherichia coli* lacks the enzyme that turns ATP into cyclic AMP. As the different reactions show, addition of cyclic AMP to one disk in the dish (*left*) enabled the colonies of bacteria nearby to metabolize certain sugars. The bacteria near the second disk (*right*), to which only 5' AMP was added, were not able to do this.

moter," at the beginning of the *lac* gene in the bacterial DNA. After that the nucleoside triphosphates initiated the *lac* transcription process. We soon found that the transcription of the genes for galactose metabolism followed the same steps.

Of the 10,000 or so *E. coli* genes, perhaps some hundreds are regulated in this way. How the others are controlled remains a mystery. The regulation of even those genes that are controlled by

cyclic AMP involves other substances as well. The *lac* transcription process provides an example. A specific protein, *lac* repressor, is bound to the DNA at a site (called the operator) at the beginning of the *lac* gene. The operator site is near the site on the DNA where the complex of cyclic AMP and CRP binds. The cell cannot now produce *lac* messenger RNA even in the presence of CRP and cyclic AMP because the repressor prevents the copying enzyme from beginning the transcription [*see illustration on opposite page*]. Only when the repressor is removed by the action of a substance closely related to lactose can the transcription take place.

This pattern of events, where cyclic AMP is able to stimulate the transcription of many genes while at the same time individual repressors can prevent certain transcriptions, gives the *E. coli* cell the flexibility it needs for the efficient utilization of the foodstuffs in its environment. *E. coli* does not store large amounts of carbohydrate and must live on what is in its immediate vicinity. If the cell is exposed, say, both to glucose (which it can already metabolize) and to

TRANSCRIPTION OF A GENE, the sequence of events shown schematically on the opposite page, occurred in a cell-free medium that was supplied (*a*) with RNA polymerase (*solid bar*), nucleoside triphosphates (*colored squares*), the protein CRP (*crescent shape*), which reversibly binds cyclic AMP (*colored ovals*), and quantities of DNA enriched with the gene for lactose metabolism (*bracketed area of helix*). Transcription begins when a combined unit of CRP and cyclic AMP activates a promoter site at the beginning of the *lac* operon (*b*); RNA polymerase now binds to the promoter site, ready to link the nucleoside triphosphates together in the correct sequence. As transcription proceeds (*c*), the RNA polymerase arranges the nucleoside triphosphates in the correct sequence. Ribosomes meanwhile begin the process of assembling the proteins that comprise the enzymes for lactose metabolism. If still another protein, *lac* repressor (*colored doughnut*), attaches itself to the operator site at the beginning of the *lac* operon (*d*), then the RNA polymerase can bind but cannot transcribe in spite of the presence of the CRP-cyclic AMP unit. This pattern, whereby cyclic AMP stimulates the transcription of many different operons, whereas the repressors are specific for a particular gene, makes for flexibility in the utilization of foodstuffs.

lactose (which it cannot), there is nothing to be gained by expending energy to produce the enzymes for lactose metabolism. As long as its supply of cyclic AMP remains at a low level the *E. coli* cell conserves its capacity for protein synthesis.

That so many different cell functions should be under the control of a single substance is remarkable. Cyclic AMP also participates in the process of visual excitation and regulates the aggregation of certain social amoebae so that they can form complex reproductive structures [see "Hormones in Social Amoebae and Mammals," by John Tyler Bonner; SCIENTIFIC AMERICAN Offprint 1145]. A good example of the substance's versa-

tility is provided by the contrast between the mechanisms of glycogen and lipid degradation on the one hand and the stimulation of gene transcription on the other. The degradation reactions involve enzymes and depend on protein phosphorylation. In *E. coli* gene transcription no phosphorylation is involved. It will be interesting to learn whether the *E. coli* mechanism is employed to control gene activity elsewhere in nature and whether cyclic AMP acts in still other ways. Most of the substance's actions in animal cells remain to be explored.

Abnormalities in the metabolism of cyclic AMP may explain the nature of certain diseases. For example, in chol-

era the bacteria responsible for the disease produce a toxin that stimulates the intestinal cells to secrete huge amounts of salt and water; it is the resulting dehydration, if it is unchecked, that is fatal [see "Cholera," by Norbert Hirschhorn and William B. Greenough III; SCIENTIFIC AMERICAN, August, 1971]. It appears that the toxin first stimulates the intestinal cells to accumulate excess cyclic AMP, whereupon the cyclic AMP instructs the cells to secrete the salty fluid.

In our own laboratory we are interested in understanding the difference between normal cells and cancer cells. Since cancer cells typically grow in an uncontrolled manner and lose their ability to carry out specialized functions, it seemed possible to us that some of their abnormal properties might be attributable to their inability to accumulate normal amounts of cyclic AMP. My colleague George Johnson and I have begun to investigate this possibility.

The cells that are commonly used for such studies are fibroblasts: cells that contribute to the formation of connective tissue. They are usually taken from the embryos of chickens, mice or other animals and allowed to grow in a nutrient fluid. After a short period of culture in a medium approximating blood serum in composition, the embryonic cells take on the appearance of the normal fibroblast cells in connective tissue and grow in bottles or dishes in a controlled manner. When the cultured cells are exposed to cancer-producing viruses or to chemical carcinogens, however, they begin to grow in an abnormal manner. They take on the appearance of tumor cells, and when they are injected into a suitable host, they usually produce tumors. The process of changing a normal cell to a cancer cell in culture is termed transformation. Among the properties of transformed cells are a change in appearance, accelerated growth, looser adherence to the container surface, alterations in the rate of production of specialized large molecules such as mucopolysaccharides and clumping on exposure to certain agglutinative plant proteins.

Now, cells that have been transformed and are then grown in the presence of cyclic AMP tend to return to normal. They grow more slowly, adhere more tenaciously to the container, synthesize certain large molecules at a faster rate and clump less when they are exposed to agglutinative plant protein. Their appearance also frequently returns toward normal. As far as we now know, the morphologic reversal occurs only in embryo cells and those derived from con-

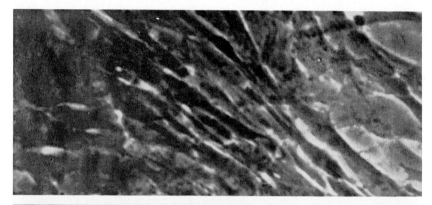

CHICK EMBRYO CELLS infected with a temperature-sensitive mutant strain of Rous sarcoma virus maintain a normal appearance (*top micrograph*) when they are cultured at a temperature of 40.5 degrees Celsius. When the temperature is reduced to 36 degrees, however, they quickly develop an abnormal appearance (*middle*). If cyclic AMP is added to the cells, their appearance remains normal (*bottom*) even after the temperature is reduced.

nective tissue; tumors from a few other kinds of cell do not show the same morphologic response.

In any event the reversal suggested to us that transformed cells might contain abnormally low levels of cyclic AMP. One of our co-workers, Jack Otten, investigated this possibility. He found that cells from chick embryos that had been transformed by exposure to Rous sarcoma virus contained much less cyclic AMP than normal chick-embryo cells. We could not be certain, however, which came first: the abnormal appearance of the transformed cells or the low level of cyclic AMP.

To settle the question we needed a way to transform cells very rapidly. We had at our disposal a mutant variety of the Rous sarcoma virus that made rapid cell transformation possible. For example, chick cells infected with the mutant virus (which had been isolated by our colleague John Bader of the National Cancer Institute) remain normal in appearance as long as the culture is kept at a temperature of 40.5 degrees Celsius. When the temperature is reduced to 36 degrees, however, the cells are rapidly transformed.

We grew cultures infected with this temperature-sensitive virus and kept them at the "normal" temperature. We incubated some of the cultures with a potent derivative of cyclic AMP and some without it. When we lowered the temperature of the cells incubated with cyclic AMP to the transformation level, they continued to look normal for some time [*see illustration on opposite page*]. The cultures without cyclic AMP, once the temperature was lowered, developed the characteristic transformed appearance within a few hours.

Otten measured the level of natural cyclic AMP in the readily transformed cells after their exposure to the lower temperature. He found that as soon as 20 minutes later the level had fallen greatly; this was well before the cells began to develop a transformed appearance.

This finding obviously leaves many questions unanswered. Is the phenomenon confined to tumors of connective tissue? How many of the abnormal properties of the transformed cells result from the low level of cyclic AMP? Which enzymes are responsible for lowering the level? Are there other tumor-forming viruses and chemical carcinogens that similarly lower the cell's supply of cyclic AMP? Not least, can the findings to date be exploited in a therapeutically useful manner? The search for answers continues in each of these areas.

20

Intercellular
Communication

by Werner R. Loewenstein
May 1970

The cells of living tissue act in concert in various ways. A newly observed pathway may allow signal molecules to travel from one cell interior to the next through special junctions in the cell membrane

A multicellular organism is a community of cells that act in a coordinated way and are ordered according to a pattern. To say this is to state an obvious truth, but when we ask how the order comes about, we are immediately plunged into one of the mysteries of life. Order requires communication. Certain forms of communication are fairly well known; for example, a hormone carries a message to a cell that is specifically sensitive to it, and an antibody recognizes a cell surface antigen by its molecular shape. Recently a form of communication has come to light wherein the interiors of cells are directly connected through special molecular arrangements in the cells' surface membranes. This kind of communication seems ideally suited for controlling the patterning of cells in a developing organism.

My story starts with a chance observation. In 1962 Yoshinobu Kanno and I were studying the permeability of the membrane around the cell nucleus. For this work we used the large cells of the salivary gland of the fruit fly larva. We injected an electric current into the nucleus of one cell with a micropipette and discovered to our great surprise that the resulting electrical potential in the adjacent cell was nearly as high as the potential in the injected cell. Evidently a large fraction of the current had passed from one cell to another through regions in the cell membrane where the cells were in contact. At about the same time Stephen W. Kuffler and David D. Potter at the Harvard Medical School made a similar observation on nerve satellite cells of the leech. Half a year later Kanno and I injected relatively large molecules into a salivary-gland cell and found that they also passed across regions where cells were in contact.

These results were startling. The prevailing belief was that cells are separate units. This tenet was at the center of cell theory, which has had a strong hold on biologists, and for good reason. Cell theory originated in the first half of the 19th century, when cells were everywhere found to be self-replicating anatomical units. If any doubts remained about the anatomical independence of cells, they were dispelled during the past two decades, when the electron microscope detected a continuous membrane bounding all cells, even where the membrane had not shown up in the light microscope. It was assumed that this membrane is a continuous barrier; similar electron-microscope membrane images were seen in skeletal muscle cells, nerve cells and red blood cells, and the boundary of such cells had been shown conclusively, by osmotic and electrical

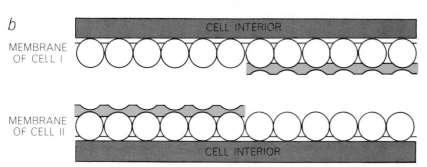

AREAS OF CONTACT between the membranes of two liver cells are shown enlarged 140,-000 times in the electron micrograph on the opposite page. Stanley Bullivant of the University of Auckland prepared the specimen by the "freeze-fracturing" process; the specimen was broken at extremely low temperature and shadowed with platinum to reveal three-dimensional detail. The diagrams (*above*) show the specimen schematically in transverse section, before fracturing along the plane of contact (*a*) and afterward (*b*). The colored area represents intercellular space. Only the lower fragment is seen in the micrograph; all the upper membrane has been broken away except for a part lying in the fracture plane (*left*). The break appears in the micrograph as a bright vertical "escarpment." To the left of it is an inside aspect of Cell I membrane as seen from that cell's interior; the closely spaced "pockmarks" indicate a zone in intimate contact with the membrane of Cell II. To the right is an inside aspect of Cell II membrane as seen from intercellular space. The small oval cluster of particles (*far right*) is a zone formerly in intimate contact with the membrane of Cell I above it. Contact zones may provide the junctions for intercellular communication.

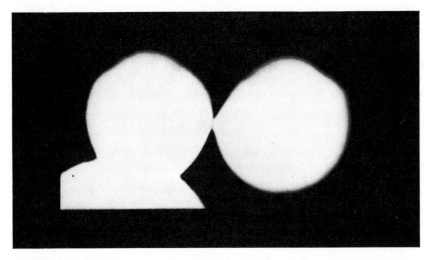

MOMENT OF CELL JUNCTION is captured in this micrograph, which shows two giant cells, isolated from a newt embryo, being manipulated into contact. The cells are .3 millimeter in diameter; measurements show that they communicate fully in a few seconds.

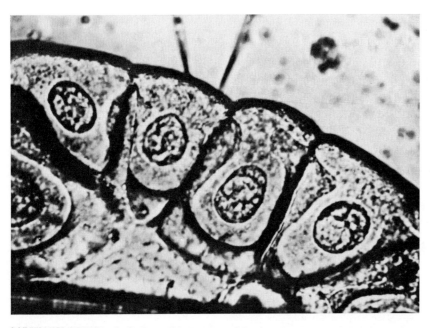

SALIVARY GLAND of a living midge is prepared for a study of intercellular communication by inserting micropipettes into individual cells; two of the pipettes are seen entering the cell second from left. An experiment using the technique is shown on the opposite page.

techniques, to be a continuous barrier. The membrane of these cells thus became the prototype of the cell surface membrane. The only misfits were heart cells and certain nerve fibers, which Silvio Weidman (working in 1952 at the University of Cambridge) and Edwin J. Furshpan and Potter (working in 1957 at University College London) had shown to be electrically coupled. Since these cells specialize in communication by electrical impulses, however, such phenomena were properly considered to be specializations adapted to this mode of communication.

What was thought to be the cell-membrane prototype turned out to be an exception instead. Work over the past few years in our laboratory at the Columbia University College of Physicians and Surgeons has shown that cells in a wide variety of animal tissues that are not engaged in electrical signaling communicate through permeable membrane junctions. We found this to be true of all epithelial cells we examined: cells of salivary gland, liver, kidney, thyroid, skin, urinary bladder, stomach, gut and sensory epithelia.

Electrical measurements in these cell systems showed that the small inorganic ions that carried our probing currents can move rather freely from one cell interior to another but not to the exterior.

Where the cells are in contact the membrane regions (junctional membranes) are 1,000 to 10,000 times more permeable to these ions than the membrane regions facing the exterior (nonjunctional membranes). Moreover, the system's interior is effectively isolated from the exterior at the level of the cell junction. The electrical measurements thus told us that the membrane junctions must be organized to form passageways between cells. We still know very little about the structure of the passageways, but we can guess at their general location on the cell membranes. With the high resolution of the electron microscope several investigators have obtained pictures showing organized connecting regions in the areas of membrane contact. Known as "tight junctions," "gap junctions" and "septate junctions," these regions (particularly the last two) are likely sites for communication because they show some structured continuity between the cells and because they are commonly present in communicating cell systems. In order to resolve the structural organization, however, we shall need more refined techniques that do not alter the membrane structure as preparation for electron microscopy currently does.

Since epithelial cells normally are not engaged in communication with electrical signals, the question naturally came up of whether or not particles larger than potassium or chloride ions, the carriers of our probing electric currents, pass through the junction too. To explore this question we turned to molecules tagged with fluorescence or color so that they could be traced inside the cells. We injected these molecules with micropipettes under hydraulic or electric pressure. Our first tracer particle was fluorescein, a negative ion with a molecular weight of 330 that fluoresces intensely when it is excited by ultraviolet radiation or blue light. When fluorescein was injected into a cell of a salivary gland while the cell system was observed in the microscope under the exciting illumination, we found that the molecules passed from one cell interior to another without significant leakage out of the cells or their junctions. Moreover, the passage of fluorescein was blocked by treatments that block the passage of the small ions, which leads us to believe the relatively large fluorescein particle takes the same route through the cell junction that the smaller ions take.

More recently the passage of fluorescein was also shown in experiments on fibroblast cells by Furshpan, Potter and Edwin S. Lennox at Harvard University and the Salk Institute for Biological

Studies, and on an electrically transmitting nerve-cell junction by Michael Van der Laan Bennett at the College of Physicians and Surgeons. In further work in our own laboratory we have found that molecules with a molecular weight of the order of 1,000 and in some instances of the order of 10,000 (but not higher) go through the cell junctions in salivary-gland cells. This leaves ample room for the cell-to-cell passage of a wide variety of cellular substances.

How does a communicating cell junction form? Here the first clue came from experiments Shizuo Ito and I did with individual cells that we manipulated into contact while monitoring the electrical resistance of their surface membranes. Cells paired up in this way produced a communicating junction in a surprisingly short time: within minutes in the case of sponge cells, and within seconds in the case of cells from early newt embryos. When the cells were separated again by manipulation or chemical treatment, the formerly permeable junctional membranes sealed rapidly and became as impermeable as the membrane of a normal free-floating cell. A new communicating junction could then be made by pushing the cells into contact again. Since the contact occurred at random at different places on the membranes, a large part of the total cell surface membrane must have the potential for forming a communicative junction.

A second clue came from experiments done in collaboration with Muhamed Nakas and Sidney J. Socolar showing that junctional-membrane permeability somehow depended on the calcium ion. Ordinarily the concentration of calcium ions in a cell's cytoplasm is several orders of magnitude lower than the concentration outside the cell. We injected calcium ions into the salivary-gland cells of midges and found that the permeability of the junctional membranes fell rapidly. Several other procedures that raised the level of calcium ions in the cytoplasm produced the same result. Under conditions where the level of calcium inside must have approached the level outside, the junctional membranes were no longer distinguishable in their permeability from the nonjunctional membranes; communication was effectively interrupted.

Another clue was provided by experiments showing that junctional-membrane permeability depended on the insulation around the junction. In these experiments I assembled pairs of sponge cells and played with the concentrations of three factors known to promote the process by which cells stick to one another: a glycoprotein-containing material at the cell surface (discovered by Tom D. Humphreys and Aron A. Moscona at the University of Chicago) and calcium and magnesium in the medium. When the concentrations of these factors in the medium were above certain values, good insulation developed around the junction, and the membranes in the junction became permeable. When the medium was deficient in these factors, no insulation formed and the junctional-membrane regions stayed as impermeable as they were when the cells were fully separated. Furthermore, when I spoiled the insulation at an established junction by removing some, but not all, of the calcium and magnesium from the medium, the formerly permeable junctional membranes sealed themselves off. This change from high to low permeability was also shown in several more organized and cohesive cells of higher animals, including mammals, by treating the tissues with certain enzymes that spoil the insulation.

On the basis of these results we have put forward the simple hypothesis that communication comes about by a conversion in permeability in the membranes owing to change in the concentration of calcium ions. We assume first that the experimental decrease in membrane permeability with an increase in calcium is one side of a two-way process, the other being an increase of permeability with a decrease in calcium. Stated differently, the assumption is that certain binding sites on the surface membrane, occupied by calcium in the impermeable state, are unoccupied in the permeable state. This is a familiar supposition to investigators in the field of biological membranes; perhaps the only unusual

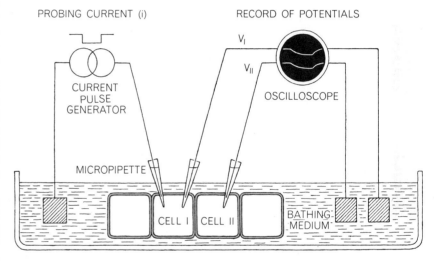

PROBING CURRENT (i)   RECORD OF POTENTIALS

CURRENT PULSE GENERATOR

MICROPIPETTE

V_I V_{II}

OSCILLOSCOPE

CELL I   CELL II

BATHING MEDIUM

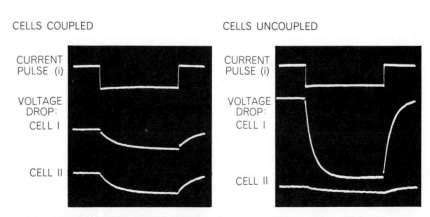

CELLS COUPLED   CELLS UNCOUPLED

CURRENT PULSE (i)   CURRENT PULSE (i)

VOLTAGE DROP: CELL I   VOLTAGE DROP: CELL I

CELL II   CELL II

INTERCELLULAR COMMUNICATION is demonstrated by passing small inorganic ions from one cell to the next (top). Two micropipettes containing a salt solution that conducts electricity are placed in Cell I (left) and a similar pipette is placed in Cell II. An electrical pulse is applied to the first cell and the resulting changes in voltage across the surface membranes of both cells are recorded by the deflection of two oscilloscope beams (bottom). When the cells are communicating fully and the ions move quite freely from cell to cell, the pulse produces nearly identical voltage changes in both (left). When the cells are uncoupled (right), the pulse that produces a large drop in Cell I is barely detectable in Cell II.

feature here is the wide span of the membrane permeability. From our three experimental clues the hypothesis follows: On formation of the insulating seals around the portions of the cell membranes in contact these portions are incorporated into the intracellular compartment, where they face a calcium-ion concentration several orders of magnitude lower than they did when they

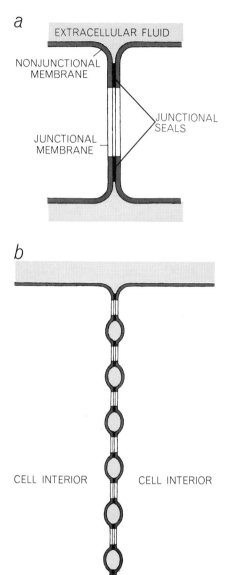

a

EXTRACELLULAR FLUID

NONJUNCTIONAL MEMBRANE

JUNCTIONAL SEALS

JUNCTIONAL MEMBRANE

b

CELL INTERIOR   CELL INTERIOR

PASSAGEWAY UNITS in a portion of a cell junction are represented schematically. The elements of each unit, as identified by means of intracellular probing, are shown at top (*a*), and an array of units between two cell membranes in close contact is shown at bottom (*b*). The three passageway elements are nonjunctional membrane (which is relatively impermeable), junctional membrane (which is permeable) and junctional "seals" where the nonjunctional membranes of two cells meet. Seals may simply be formed by the close contact between two membranes or they may be some kind of separate structure.

faced the outside medium. The chemical gradients now favor the detachment of calcium from these portions of the membrane, and their conversion from low permeability to high permeability ensues [*see top illustration on page 222*].

This has been our working hypothesis for two years. The hypothesis is attractive because it requires no special properties of the membrane regions that are to become junctional and no special switching mechanism for turning communication on. The essential element for triggering the permeability transformation and restricting it to the junctional-membrane region is the insulating seal, and the driving forces behind the transformation are the same forces that keep the calcium level low inside the cells. The fate of the membrane calcium inside the cell system is the same as that of all cytoplasmic calcium ions. The calcium goes to cellular depots—the mitochondria and other organelles—or to the outside through the nonjunctional membranes. As for the fate of the calcium ion on the outside of the membranes, this needs no special consideration. If the electron microscope gives anything like the true picture, there is no room to trap a significant amount of calcium ions between the intimately joined portions of membrane.

Several important points in the calcium hypothesis still lack proof. We have no direct evidence that the calcium-induced decrease of permeability in the junctional membrane is reversible or that the calcium ion reacts directly with the membrane. Even so, the hypothesis has led to new experiments telling us more about the mechanisms that promote communication between cells.

Central to the hypothesis is the idea that the low level of calcium ions within the cells is responsible for the state of high permeability in the junctional membranes. The maintenance of the low calcium level is known, in several types of cell, to require continuous transport of the ion to the calcium-rich outside by energy ultimately coming from the metabolism of the cell. One would therefore expect inhibition of metabolism to reduce the permeability of the junctional membranes. This consideration prompted Alberto Politoff, Socolar and me to investigate the effects of metabolic inhibitors of junctional communication in midge salivary-gland cells. In one set of experiments we inhibited the cell's metabolism by cooling the tissue or by adding to the bathing medium small amounts of a poison (such as cyanide, dinitrophenol or oligomycin) that blocks the synthesis of adenosine triphosphate

(ATP) and other sources of energy in the cell. This treatment produced a decrease in junctional-membrane permeability, effectively uncoupling communication between the cells. The onset of uncoupling took some time: from 15 minutes to two hours. The uncoupling by some of the treatments was reversible; repetition of the treatment then produced uncoupling with practically no delay. Presumably on repetition of the treatment the cells no longer had a reserve of ATP such as they had before the initial experiment. In another set of experiments we combined the metabolism-inhibiting treatment (dinitrophenol) with the injection of ATP into the cells. This prevented or reversed uncoupling in a fair proportion of the trials [*see bottom illustration on page 222*].

A further upshot of the calcium hypothesis was the finding that substitution of lithium for sodium in the extracellular medium causes uncoupling. Such substitution was known to produce an increase in cytoplasmic calcium in various tissues. Peter Baker, Mordecai Blaustein, Alan L. Hodgkin and Robert Steinhardt at the University of Cambridge had just shown that in squid nerve fibers this increase was due to a slowing of outflow and a speeding up of inflow of calcium ions. Birgit Rose in our laboratory now undertook to study the effects of lithium substitution on cell communication. She found that within three hours of replacing all the sodium with lithium in the medium bathing the salivary glands of the midge, the cells became uncoupled. That this was not simply due to the lack of sodium ions was shown by experiments in which choline instead of lithium was the sodium substitute; then communication was maintained just as well as when sodium was present.

We come now to the question of what kind of information may be conveyed through the cell junctions. The most interesting finding in this connection was that fairly large molecules can pass by this route from cell to cell. The range in the size of particles permeating the junction is wide enough to include most molecules involved in metabolism (metabolites) and many other molecules that regulate cellular activities. Thus the exciting possibility presented itself that the size range also included substances that regulate gene activity, or, to put it more bluntly, that the junction is instrumental in conveying substances that control the growth and differentiation of cells. This possibility was appealing because the preceding 10 years of microbial genetics had shown that metabolites

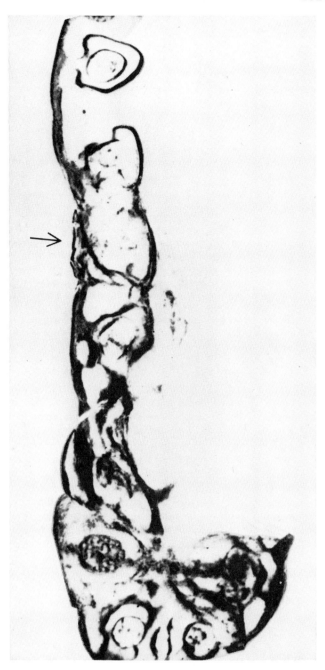

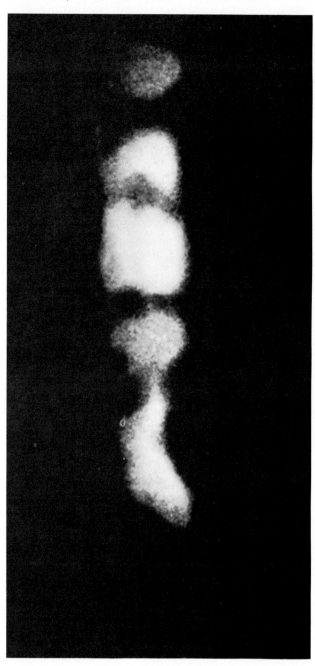

CELL-TO-CELL FLOW of relatively large molecules is demonstrated in the salivary gland of a midge. A fluorescent tracer is injected into one cell (*left, arrow*). When the tracer is excited by blue
light 10 minutes later it has spread into the cells on both sides.

and other molecules of comparable size
can regulate genes in bacteria, and because the past 40 years of experimental
embryology had firmly left the lesson
that differentiation in the embryo involves some form of close-range interaction between cells with diffusible
molecules. In junctional communication
we had an obvious candidate for close-
range interaction: a closed system in
which molecules can diffuse from cytoplasm to cytoplasm with little loss. There
are, of course, other possible close-range
forms of intercellular communication;
for instance, molecules can be passed
between cells through the extracellular

liquid, and this form of communication
is undoubtedly important in embryonic
development. Junctional communication
has an unusual potential, however, in
developmental processes where information about the number and position of
cells in a cell community must be conveyed. The cell community has a finite
volume and a sharp peripheral diffusion
boundary (the nonjunctional membranes
and the junctional seals), and it thus has
the potential for obtaining such information on the basis of simple clues in
the concentration of diffusing signal molecules.

When we found that a particle as

large as fluorescein could pass through
a cell junction, we were immediately
drawn to the exploration of the possibility that the communication system dis-
seminates signal molecules for the regulation of cellular growth and differentiation. We would have liked to play with
the communication signals, but in no instance of close-range embryonic development had a signal molecule been identified. We therefore had to resort to the
indirect tactic of examining the communication situation in aberrant cells
where the control of growth had obviously gone wrong.

We began with experiments intended

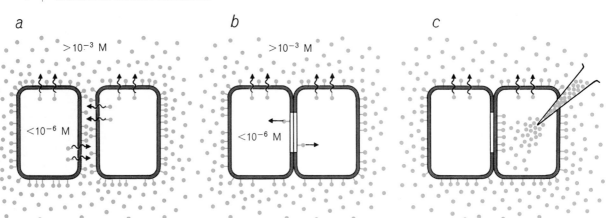

a   *b*   *c*

$>10^{-3}$ M

$<10^{-6}$ M

$>10^{-3}$ M

$<10^{-6}$ M

CONVERSION HYPOTHESIS assumes that the permeability of cell membrane is determined by the concentration of calcium ions ("*a*," *colored dots*) in the medium on both sides of the membrane. On a membrane surface exposed to normal extracellular medium, with a calcium concentration greater than a thousandth of a mole per liter, calcium binding sites are largely occupied. Inside the cell calcium ions are comparatively rare; they are constantly being expelled (*wavy arrows*) by energy derived from cell metabolism and the calcium binding sites on the membrane interior are sparsely occupied. When two cells come together (*b*), junctional seals enclose regions of the membrane surface that were formerly exposed to the extracellular medium. The regions lose their former quota of ions (*straight arrows*) and become permeable (*white area*). The reverse process may be demonstrated experimentally (*c*). Junctional membranes between communicating cells become impermeable when one cell interior is artificially loaded with calcium ions.

to determine if cells in a developing organism actually do communicate by means of junctions. Our first attempts in this direction were discouraging. Robert Ashman, Kanno and I spent the summer of 1964 at the Marine Biological Laboratory in Woods Hole studying the communication between the earliest cells of a starfish embryo: the first daughters of the egg cell. In these large and transparent cells we could monitor communication electrically throughout the division of the egg cell, from the time the cell begins to cleave until the first cell pair in the organism is formed. We found good communication early in cleavage, when there was presumably still protoplasmic continuity between the cells. When cleavage was completed, however, the cell pair ceased to communicate. This was the opposite of what we had expected. An attempt the following summer by Ito and myself on the first cell pair of another marine organism, the sand dollar, gave the same disappointing result.

A year later my colleague Ito and also Furshpan, Potter and Lennox at Harvard had more exciting results with older embryos. Ito, using newts, had evidence that most, and perhaps all, cells of the embryo are communicating by the time the embryo has become the globular cluster of cells called the morula. The Harvard group, working with the squid embryo and later, when joined by Judson Sheridan, with the chick embryo, found extensive junctional communication at even later stages, when the cells had visibly differentiated. The results of the Harvard group were particularly interesting; they showed that there was communication between various kinds of embryonic tissue whose fully differentiated cellular descendants are clearly unconnected in the adult animal, and that certain cells lose their connections in the course of development. Recently Bennett and J. P. Trinkhaus also found communication in the early embryos of the fish *Fundulus,* as did Christina Slack and J. F. Palmer at the Middlesex Hospital in London in the toad *Xenopus.* The demonstrations in

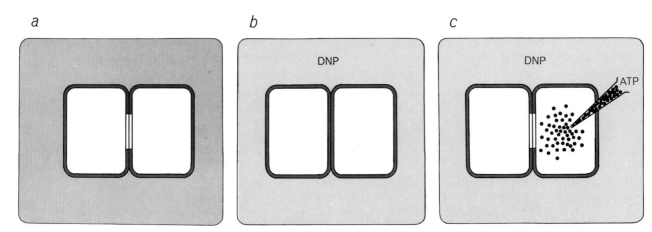

a   *b*   *c*

DNP

DNP

ATP

ROLE OF METABOLIC ENERGY in maintaining the permeability of junctional membrane is demonstrated in the case of two communicating cells (*a*). Dinitrophenol, a poison that blocks synthesis of the energetic substance adenosine triphosphate (ATP), is added to the extracellular medium. Fifteen minutes later junctional membrane permeability has fallen markedly (*b*); the cells are no longer communicating. Uncoupling presumably occurs because ATP is no longer available to power the expulsion of calcium ions from the cell interior. As the number of these ions within the cytoplasm rises the membrane becomes impermeable (*see illustration at top of page*). Next (*c*) one cell is artificially loaded with ATP. Calcium expulsion and communication resume in spite of poisoned medium.

these five animals, belonging to widely different groups, established that junctional communication is present during the period of active growth and differentiation.

Our work with aberrant cells was prompted by the consideration that if the junction was indeed a gateway for growth-controlling substances, one could expect that unregulated cell growth such as cancer might in some cases be due to poor junctional communication. We hoped that our electrical techniques for probing the junctional passage of small inorganic ions would be useful for detecting any such alterations in communication.

A suitable cell material for this work was the mammalian liver. My colleague Richard D. Penn had already found that cells of normal liver have good junctional communication, and various kinds of mammalian liver tumors, produced experimentally, were available from several laboratories. Kanno and I examined four types of such tumors, representative of a broad spectrum of growth rates and differentiation; in none could we detect communication. In later experiments the same proved to be true of tumors of thyroid in the rat and of human stomach epithelium. My colleagues Carmia Borek and Shoji Higashino performed experiments along similar lines with liver cells in culture. In these experiments cells from normal and cancerous liver were grown in glass dishes where electrical measurements of communication could be made under conditions closely matched for all the cell types. Here again only the junction between normal cells turned out to be communicating; there was no communi-

cation detectable between one cancerous cell and another, or between a cancerous cell and a normal one. Apparently the membrane of these cancerous cells, unlike their normal counterparts, is rather impermeable even where the cells are in contact. The contact regions also appear different in the electron microscope; several investigators have found that cancerous liver and certain other cancerous tissues lack some of the structures that join normal cells.

The lack of junctional communication appears to be a manifestation of cancerous growth but not of growth in general. Penn and I showed this in a study of regenerating liver. The liver of an adult rat—like the liver of the legendary Prometheus—is capable of regenerating itself rapidly. If a part of the liver, say two-thirds, is cut off, the stump grows at the rate of a billion cells per day (a rate more rapid than that of the fastest tumor we have examined) during the first four days. The growth then slows down. By the seventh day, when the liver has regenerated to nearly its original size, growth is down to less than 8 percent of the maximum rate. This pattern of regenerative growth mirrors the picture of normal growth in an organ of the embryo. It is the picture of a cell population that knows when and where to stop growing. We found that such a regenerating cell population in fact possesses junctional communication.

Not all cancerous cell types lack junctional communication. Over the past four years the Harvard group and our own have encountered a variety of cancerous cells that show no defects in the passage of small inorganic ions or (in one case) the passage of fluorescein. This was not surprising. Defects in junctional com-

COMMUNICATION between embryonic cells is monitored electrically by measuring the resistance to ion movement across the plane of cleavage. Electrodes are inserted (*top*) into a fertilized starfish egg. When cell division is complete (*bottom*), communication between the two cells has virtually halted. It begins again at a later stage.

munication are of course not the only possible cause of uncontrolled growth. Cancerous growth may arise from defects in the production or in the reception of growth-controlling substances, quite apart from defects in their intercellular passage. The finding that some forms of cancer are associated with failure in intercellular communication, however, is an encouraging development. It opens up a new approach in the story of cell growth, and research in this direction is now going forward.

21   Pheromones

by Edward O. Wilson
May 1963

A pheromone is a substance secreted by an animal that influences the behavior of other animals of the same species. Recent studies indicate that such chemical communication is surprisingly common

It is conceivable that somewhere on other worlds civilizations exist that communicate entirely by the exchange of chemical substances that are smelled or tasted. Unlikely as this may seem, the theoretical possibility cannot be ruled out. It is not difficult to design, on paper at least, a chemical communication system that can transmit a large amount of information with rather good efficiency. The notion of such a communication system is of course strange because our outlook is shaped so strongly by our own peculiar auditory and visual conventions. This limitation of outlook is found even among students of animal behavior; they have favored species whose communication methods are similar to our own and therefore more accessible to analysis. It is becoming increasingly clear, however, that chemical systems provide the dominant means of communication in many animal species, perhaps even in most. In the past several years animal behaviorists and organic chemists, working together, have made a start at deciphering some of these systems and have discovered a number of surprising new biological phenomena.

In earlier literature on the subject, chemicals used in communication were usually referred to as "ectohormones." Since 1959 the less awkward and etymologically more accurate term "pheromones" has been widely adopted. It is used to describe substances exchanged among members of the same animal species. Unlike true hormones, which are secreted internally to regulate the organism's own physiology, or internal environment, pheromones are secreted externally and help to regulate the organism's external environment by influencing other animals. The mode of influence can take either of two general forms. If the pheromone produces a more or less immediate and reversible change

in the behavior of the recipient, it is said to have a "releaser" effect. In this case the chemical substance seems to act directly on the recipient's central nervous system. If the principal function of the pheromone is to trigger a chain of physiological events in the recipient, it has what we have recently labeled a "primer" effect. The physiological changes, in turn, equip the organism with a new behavioral repertory, the components of which are thenceforth evoked by appropriate stimuli. In termites, for example, the reproductive and soldier castes prevent other termites from developing into

their own castes by secreting substances that are ingested and act through the *corpus allatum*, an endocrine gland controlling differentiation [see "The Termite and the Cell," by Martin Lüscher; SCIENTIFIC AMERICAN, May, 1953].

These indirect primer pheromones do not always act by physiological inhibition. They can have the opposite effect. Adult males of the migratory locust *Schistocerca gregaria* secrete a volatile substance from their skin surface that accelerates the growth of young locusts. When the nymphs detect this substance with their antennae, their hind legs,

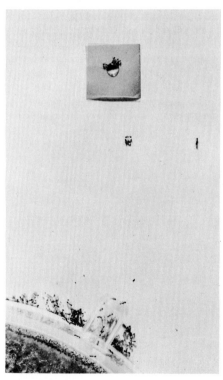

INVISIBLE ODOR TRAILS guide fire ant workers to a source of food: a drop of sugar solution. The trails consist of a pheromone laid down by workers returning to their nest after finding a source of food. Sometimes the chemical message is reinforced by the touching of antennae if a returning worker meets a wandering fellow along the way. This is hap-

some of their mouth parts and the antennae themselves vibrate. The secretion, in conjunction with tactile and visual signals, plays an important role in the formation of migratory locust swarms.

A striking feature of some primer pheromones is that they cause important physiological change without an immediate accompanying behavioral response, at least none that can be said to be peculiar to the pheromone. Beginning in 1955 with the work of S. van der Lee and L. M. Boot in the Netherlands, mammalian endocrinologists have discovered several unexpected effects on the female mouse that are produced by odors of other members of the same species. These changes are not marked by any immediate distinctive behavioral patterns. In the "Lee-Boot effect" females placed in groups of four show an increase in the percentage of pseudopregnancies. A completely normal reproductive pattern can be restored by removing the olfactory bulbs of the mice or by housing the mice separately. When more and more female mice are forced to live together, their oestrous cycles become highly irregular and in most of the mice the cycle stops completely for long periods. Recently W. K. Whitten of the Australian National University has discovered that the odor of a male mouse can initiate and synchronize the oestrous cycles of female mice. The male odor also reduces the frequency of reproductive abnormalities arising when female mice are forced to live under crowded conditions.

A still more surprising primer effect has been found by Helen Bruce of the National Institute for Medical Research in London. She observed that the odor of a strange male mouse will block the pregnancy of a newly impregnated female mouse. The odor of the original stud male, of course, leaves pregnancy undisturbed. The mouse reproductive pheromones have not yet been identified chemically, and their mode of action is only partly understood. There is evidence that the odor of the strange male suppresses the secretion of the hormone prolactin, with the result that the *corpus luteum* (a ductless ovarian gland) fails to develop and normal oestrus is restored. The pheromones are probably part of the complex set of control mechanisms that regulate the population density of animals [see "Population Density and Social Pathology," by John B. Calhoun; SCIENTIFIC AMERICAN, Offprint 506].

Pheromones that produce a simple releaser effect—a single specific response mediated directly by the central nervous system—are widespread in the animal kingdom and serve a great many functions. Sex attractants constitute a large and important category. The chemical structures of six attractants are shown on page 231. Although two of the six—the mammalian scents muskone and civetone—have been known for some 40 years and are generally assumed to serve a sexual function, their exact role has never been rigorously established by experiments with living animals. In fact, mammals seem to employ musklike compounds, alone or in combination with other substances, to serve several functions: to mark home ranges, to assist in territorial defense and to identify the sexes.

The nature and role of the four insect sex attractants are much better understood. The identification of each represents a technical feat of considerable magnitude. To obtain 12 milligrams of esters of bombykol, the sex attractant of the female silkworm moth, Adolf F. J. Butenandt and his associates at the Max Planck Institute of Biochemistry in Munich had to extract material from 250,000 moths. Martin Jacobson, Morton Beroza and William Jones of the U.S. Department of Agriculture processed 500,000 female gypsy moths to get 20 milligrams of the gypsy-moth attractant gyplure. Each moth yielded only about .01 microgram (millionth of a gram) of

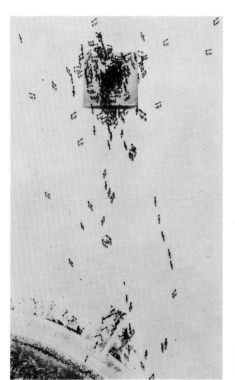
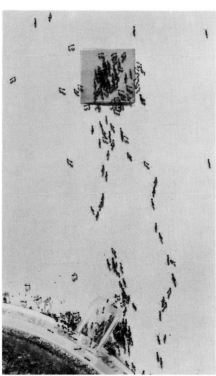

pening in the photograph at the far left. A few foraging workers have just found the sugar drop and a returning trail-layer is communicating the news to another ant. In the next two pictures the trail has been completed and workers stream from the nest in increasing numbers. In the fourth picture unrewarded workers return to the nest without laying trails and outward-bound traffic wanes. In the last picture most of the trails have evaporated completely and only a few stragglers remain at the site, eating the last bits of food.

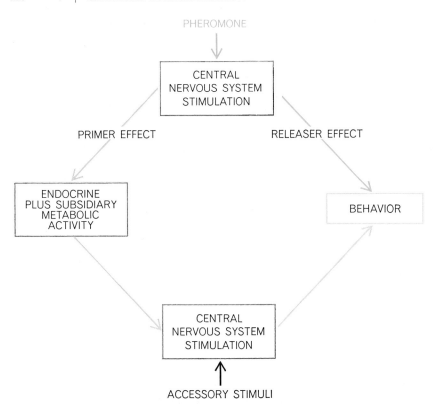

PHEROMONE

CENTRAL
NERVOUS SYSTEM
STIMULATION

PRIMER EFFECT   RELEASER EFFECT

ENDOCRINE
PLUS SUBSIDIARY
METABOLIC
ACTIVITY

BEHAVIOR

CENTRAL
NERVOUS SYSTEM
STIMULATION

ACCESSORY STIMULI

PHEROMONES INFLUENCE BEHAVIOR directly or indirectly, as shown in this schematic diagram. If a pheromone stimulates the recipient's central nervous system into producing an immediate change in behavior, it is said to have a "releaser" effect. If it alters a
set of long-term physiological conditions so that the recipient's behavior can subsequently
be influenced by specific accessory stimuli, the pheromone is said to have a "primer" effect.

French naturalist Jean Henri Fabre, speculating on sex attraction in insects, could not bring himself to believe that the female moth could communicate over such great distances by odor alone, since "one might as well expect to tint a lake with a drop of carmine." We now know that Fabre's conclusion was wrong but that his analogy was exact: to the male moth's powerful chemoreceptors the lake is indeed tinted.

One must now ask how the male moth, smelling the faintly tinted air, knows which way to fly to find the source of the tinting. He cannot simply fly in the direction of increasing scent; it can be shown mathematically that the attractant is distributed almost uniformly after it has drifted more than a few meters from the female. Recent experiments by Ilse Schwinck of the University of Munich have revealed what is probably the alternative procedure used. When male moths are activated by the pheromone, they simply fly upwind and thus inevitably move toward the female. If by accident they pass out of the active zone, they either abandon the search or fly about at random until they pick up the scent again. Eventually, as they approach the female, there is a slight increase in the concentration of the chemical attractant and this can serve as a guide for the remaining distance.

gyplure, or less than a millionth of its body weight. Bombykol and gyplure were obtained by killing the insects and subjecting crude extracts of material to chromatography, the separation technique in which compounds move at different rates through a column packed with a suitable adsorbent substance. Another technique has been more recently developed by Robert T. Yamamoto of the U.S. Department of Agriculture, in collaboration with Jacobson and Beroza, to harvest the equally elusive sex attractant of the American cockroach. Virgin females were housed in metal cans and air was continuously drawn through the cans and passed through chilled containers to condense any vaporized materials. In this manner the equivalent of 10,000 females were "milked" over a nine-month period to yield 12.2 milligrams of what was considered to be the pure attractant.

The power of the insect attractants is almost unbelievable. If some 10,000 molecules of the most active form of bombykol are allowed to diffuse from a source one centimeter from the antennae of a male silkworm moth, a characteristic sexual response is obtained in most cases. If volatility and diffusion rate

are taken into account, it can be estimated that the threshold concentration is no more than a few hundred molecules per cubic centimeter, and the actual number required to stimulate the male is probably even smaller. From this one can calculate that .01 microgram of gyplure, the minimum average content of a single female moth, would be theoretically adequate, if distributed with maximum efficiency, to excite more than a billion male moths.

In nature the female uses her powerful pheromone to advertise her presence over a large area with a minimum expenditure of energy. With the aid of published data from field experiments and newly contrived mathematical models of the diffusion process, William H. Bossert, one of my associates in the Biological Laboratories at Harvard University, and I have deduced the shape and size of the ellipsoidal space within which male moths can be attracted under natural conditions [*see bottom illustration on opposite page*]. When a moderate wind is blowing, the active space has a long axis of thousands of meters and a transverse axis parallel to the ground of more than 200 meters at the widest point. The 19th-century

If one is looking for the most highly developed chemical communication systems in nature, it is reasonable to study the behavior of the social insects, particularly the social wasps, bees, termites and ants, all of which communicate mostly in the dark interiors of their nests and are known to have advanced chemoreceptive powers. In recent years experimental techniques have been developed to separate and identify the pheromones of these insects, and rapid progress has been made in deciphering the hitherto intractable codes, particularly those of the ants. The most successful procedure has been to dissect out single glandular reservoirs and see what effect their contents have on the behavior of the worker caste, which is the most numerous and presumably the most in need of continuing guidance. Other pheromones, not present in distinct reservoirs, are identified in chromatographic fractions of crude extracts.

Ants of all castes are constructed with an exceptionally well-developed exocrine glandular system. Many of the most prominent of these glands, whose function has long been a mystery to entomologists, have now been identified as the source of pheromones [*see illustra-*

tion on page 229]. The analysis of the gland-pheromone complex has led to the beginnings of a new and deeper understanding of how ant societies are organized.

Consider the chemical trail. According to the traditional view, trail secretions served as only a limited guide for worker ants and had to be augmented by other kinds of signals exchanged inside the nest. Now it is known that the trail substance is extraordinarily versatile. In the fire ant (*Solenopsis saevissima*), for instance, it functions both to activate and to guide foraging workers in search of food and new nest sites. It also contributes as one of the alarm signals emitted by workers in distress. The trail of the fire ant consists of a substance secreted in minute amounts by Dufour's gland; the substance leaves the ant's body by way of the extruded sting, which is touched intermittently to the ground much like a moving pen dispensing ink. The trail pheromone, which has not yet been chemically identified, acts primarily to attract the fire ant workers. Upon encountering the attractant the workers move automatically up the gradient to the source of emission. When the substance is drawn out in a line, the workers run along the direction of the line away from the nest. This simple response brings them to the food source or new nest site from which the trail is laid. In our laboratory we have extracted the pheromone from the Dufour's glands of freshly killed workers and have used it to create artificial trails. Groups of workers will follow these trails away from the nest and along arbitrary routes (including circles leading back to the nest) for considerable periods of time. When the pheromone is presented to whole colonies in massive doses, a large portion of the colony, including the queen, can be drawn out in a close simulation of the emigration process.

The trail substance is rather volatile, and a natural trail laid by one worker diffuses to below the threshold concentration within two minutes. Consequently outward-bound workers are able to follow it only for the distance they can travel in this time, which is about 40 centimeters. Although this strictly limits the distance over which the ants can communicate, it provides at least two important compensatory advantages. The more obvious advantage is that old, useless trails do not linger to confuse the hunting workers. In addition, the intensity of the trail laid by many workers provides a sensitive index of the amount of food at a given site and the rate of its depletion. As workers move to and from

ANTENNAE OF GYPSY MOTHS differ radically in structure according to their function. In the male (*left*) they are broad and finely divided to detect minute quantities of sex attractant released by the female (*right*). The antennae of the female are much less developed.

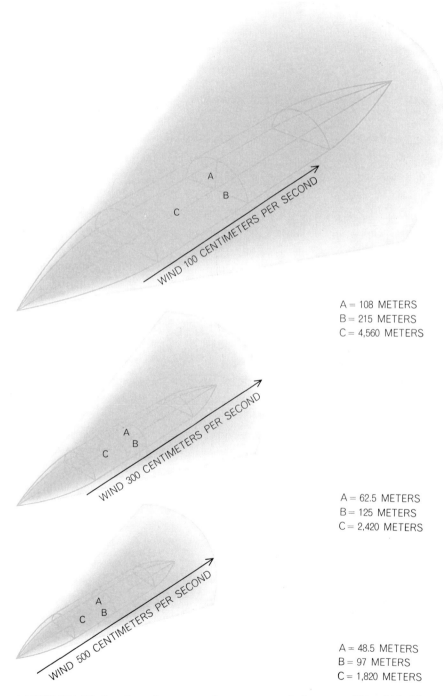

A = 108 METERS
B = 215 METERS
C = 4,560 METERS

A = 62.5 METERS
B = 125 METERS
C = 2,420 METERS

A = 48.5 METERS
B = 97 METERS
C = 1,820 METERS

ACTIVE SPACE of gyplure, the gypsy moth sex attractant, is the space within which this pheromone is sufficiently dense to attract males to a single, continuously emitting female. The actual dimensions, deduced from linear measurements and general gas-diffusion models, are given at right. Height (*A*) and width (*B*) are exaggerated in the drawing. As wind shifts from moderate to strong, increased turbulence contracts the active space.

FIRE ANT WORKER lays an odor trail by exuding a pheromone along its extended sting. The sting is touched to the ground periodically, breaking the trail into a series of streaks.

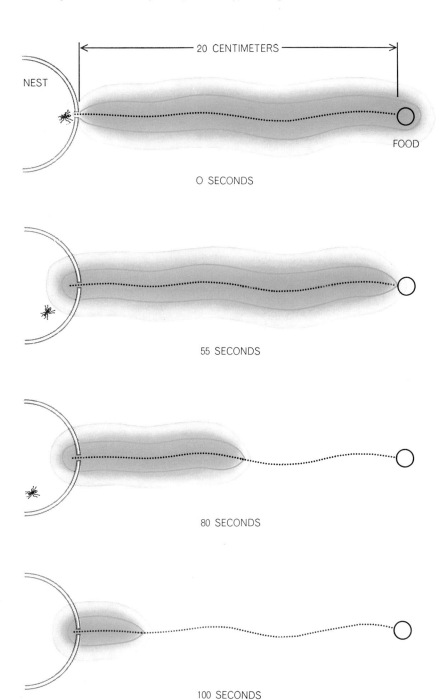

ACTIVE SPACE OF ANT TRAIL, within which the pheromone is dense enough to be perceived by other workers, is narrow and nearly constant in shape with the maximum gradient situated near its outer surface. The rapidity with which the trail evaporates is indicated.

the food finds (consisting mostly of dead insects and sugar sources) they continuously add their own secretions to the trail produced by the original discoverers of the food. Only if an ant is rewarded by food does it lay a trail on its trip back to the nest; therefore the more food encountered at the end of the trail, the more workers that can be rewarded and the heavier the trail. The heavier the trail, the more workers that are drawn from the nest and arrive at the end of the trail. As the food is consumed, the number of workers laying trail substance drops, and the old trail fades by evaporation and diffusion, gradually constricting the outward flow of workers.

The fire ant odor trail shows other evidences of being efficiently designed. The active space within which the pheromone is dense enough to be perceived by workers remains narrow and nearly constant in shape over most of the length of the trail. It has been further deduced from diffusion models that the maximum gradient must be situated near the outer surface of the active space. Thus workers are informed of the space boundary in a highly efficient way. Together these features ensure that the following workers keep in close formation with a minimum chance of losing the trail.

The fire ant trail is one of the few animal communication systems whose information content can be measured with fair precision. Unlike many communicating animals, the ants have a distinct goal in space—the food find or nest site—the direction and distance of which must both be communicated. It is possible by a simple technique to measure how close trail-followers come to the trail end, and, by making use of a standard equation from information theory, one can translate the accuracy of their response into the "bits" of information received. A similar procedure can be applied (as first suggested by the British biologist J. B. S. Haldane) to the "waggle dance" of the honeybee, a radically different form of communication system from the ant trail [see "Dialects in the Language of the Bees," by Karl von Frisch; SCIENTIFIC AMERICAN Offprint 130]. Surprisingly, it turns out that the two systems, although of wholly different evolutionary origin, transmit about the same amount of information with reference to distance (two bits) and direction (four bits in the honeybee, and four or possibly five in the ant). Four bits of information will direct an ant or a bee into one of 16 equally probable sectors of a circle and two bits will identify one of four equally probable distances. It is conceivable that these information values represent the maximum that can be achieved with the insect brain and sensory apparatus.

Not all kinds of ants lay chemical trails. Among those that do, however, the pheromones are highly species-specific in their action. In experiments in which artificial trails extracted from one species were directed to living colonies of other species, the results have almost always been negative, even among related species. It is as if each species had its own private language. As a result there is little or no confusion when the trails of two or more species cross.

Another important class of ant pheromone is composed of alarm substances. A simple backyard experiment will show that if a worker ant is disturbed by a clean instrument, it will, for a short time, excite other workers with whom it comes in contact. Until recently most students of ant behavior thought that the alarm was spread by touch, that one worker simply jostled another in its excitement or drummed on its neighbor with its antennae in some peculiar way. Now it is known that disturbed workers discharge chemicals, stored in special glandular reservoirs, that can produce all the characteristic alarm responses solely by themselves. The chemical structure of four alarm substances is shown on page 233. Nothing could illustrate more clearly the wide differences between the human perceptual world and that of chemically communicating animals. To the human nose the alarm substances are mild or even pleasant, but to the ant they represent an urgent tocsin that can propel a colony into violent and instant action.

As in the case of the trail substances, the employment of the alarm substances appears to be ideally designed for the purpose it serves. When the contents of the mandibular glands of a worker of the harvesting ant (*Pogonomyrmex badius*) are discharged into still air, the volatile material forms a rapidly expanding sphere, which attains a radius of about six centimeters in 13 seconds. Then it contracts until the signal fades out completely some 35 seconds after the moment of discharge. The outer shell of the active space contains a low concentration of pheromone, which is actually attractive to harvester workers. This serves to draw them toward the point of disturbance. The central region of the active space, however, contains a concentration high enough to evoke the characteristic frenzy of alarm. The "alarm sphere" expands to a radius of about three centimeters in eight seconds and, as might be expected, fades out more quickly than the "attraction sphere."

The advantage to the ants of an alarm signal that is both local and short-lived becomes obvious when a *Pogonomyrmex* colony is observed under natural conditions. The ant nest is subject to almost innumerable minor disturbances. If the

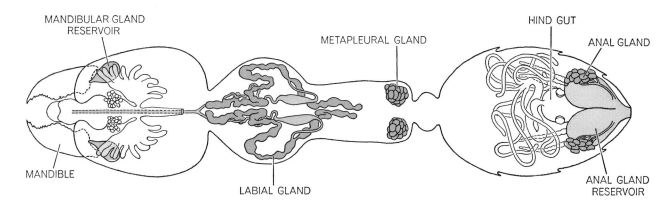

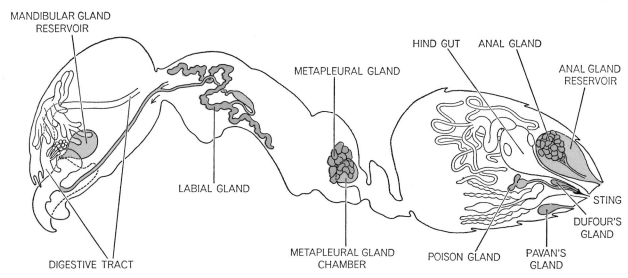

EXOCRINE GLANDULAR SYSTEM of a worker ant (*shown here in top and side cutaway views*) is specially adapted for the production of chemical communication substances. Some pheromones are stored in reservoirs and released in bursts only when needed; others are secreted continuously. Depending on the species, trail substances are produced by Dufour's gland, Pavan's gland or the poison glands; alarm substances are produced by the anal and mandibular glands. The glandular sources of other pheromones are unknown.

FORAGING INFORMATION conveyed by two different insect communication systems can be represented on two similar "compass" diagrams. The honeybee "waggle dance" (*top*) transmits about four bits of information with respect to direction, enabling a honeybee worker to pinpoint a target within one of 16 equally probable angular sectors. The number of "bits" in this case remains independent of distance, given in meters. The pheromone system used by trail-laying fire ants (*bottom*) is superior in that the amount of directional information increases with distance, given in centimeters. At distances *c* and *d*, the probable sector in which the target lies is smaller for ants than for bees. (For ants, directional information actually increases gradually and not by jumps.) Both insects transmit two bits of distance information, specifying one of four equally probable distance ranges.

alarm spheres generated by individual ant workers were much wider and more durable, the colony would be kept in ceaseless and futile turmoil. As it is, local disturbances such as intrusions by foreign insects are dealt with quickly and efficiently by small groups of workers, and the excitement soon dies away.

The trail and alarm substances are only part of the ants' chemical vocabulary. There is evidence for the existence of other secretions that induce gathering and settling of workers, acts of grooming, food exchange, and other operations fundamental to the care of the queen and immature ants. Even dead ants produce a pheromone of sorts. An ant that has just died will be groomed by other workers as if it were still alive. Its complete immobility and crumpled posture by themselves cause no new response. But in a day or two chemical decomposition products accumulate and stimulate the workers to bear the corpse to the refuse pile outside the nest. Only a few decomposition products trigger this funereal response; they include certain long-chain fatty acids and their esters. When other objects, including living workers, are experimentally daubed with these substances, they are dutifully carried to the refuse pile. After being dumped on the refuse the "living dead" scramble to their feet and promptly return to the nest, only to be carried out again. The hapless creatures are thrown back on the refuse pile time and again until most of the scent of death has been worn off their bodies by the ritual.

Our observation of ant colonies over long periods has led us to believe that as few as 10 pheromones, transmitted singly or in simple combinations, might suffice for the total organization of ant society. The task of separating and characterizing these substances, as well as judging the roles of other kinds of stimuli such as sound, is a job largely for the future.

Even in animal species where other kinds of communication devices are prominently developed, deeper investigation usually reveals the existence of pheromonal communication as well. I have mentioned the auxiliary roles of primer pheromones in the lives of mice and migratory locusts. A more striking example is the communication system of the honeybee. The insect is celebrated for its employment of the "round" and "waggle" dances (augmented, perhaps, by auditory signals) to designate the location of food and new nest sites. It is not so widely known that chemical signals play equally important roles in other aspects of honeybee life. The mother queen regulates the reproductive cycle of the colony by secreting from her mandibular glands a substance recently identified as 9-ketodecanoic acid. When this pheromone is ingested by the worker bees, it inhibits development of their ovaries and also their ability to manufacture the royal cells in which new queens are reared. The same pheromone serves as a sex attractant in the queen's nuptial flights.

Under certain conditions, including the discovery of new food sources, worker bees release geraniol, a pleasant-smelling alcohol, from the abdominal Nassanoff glands. As the geraniol diffuses through the air it attracts other workers and so supplements information contained in the waggle dance. When a worker stings an intruder, it discharges, in addition to the venom, tiny amounts of a secretion from clusters of unicellular glands located next to the basal plates of the sting. This secretion is responsible for the tendency, well known to beekeepers, of angry swarms of workers to sting at the same spot. One component, which acts as a simple attractant, has been identified as isoamyl acetate, a compound that has a banana-like odor. It is possible that the stinging response is evoked by at least one unidentified alarm substance secreted along with the attractant.

Knowledge of pheromones has advanced to the point where one can make some tentative generalizations about their chemistry. In the first place, there appear to be good reasons why sex attractants should be compounds that contain between 10 and 17 carbon atoms and that have molecular weights between about 180 and 300—the range actually observed in attractants so far identified. (For comparison, the weight of a single carbon atom is 12.) Only compounds of roughly this size or greater can meet the two known requirements of a sex attractant: narrow specificity, so that only members of one species will respond to it, and high potency. Compounds that contain fewer than five or so carbon atoms and that have a molecular weight of less than about 100 cannot be assembled in enough different ways to provide a distinctive molecule for all the insects that want to advertise their presence.

It also seems to be a rule, at least with insects, that attraction potency increases with molecular weight. In one series of esters tested on flies, for instance, a doubling of molecular weight resulted in as much as a thousandfold increase in efficiency. On the other hand, the molecule cannot be too large and complex or it will be prohibitively difficult for the insect to synthesize. An equally important limitation on size is

SIX SEX PHEROMONES include the identified sex attractants of four insect species as well as two mammalian musks generally believed to be sex attractants. The high molecular weight of most sex pheromones accounts for their narrow specificity and high potency.

the fact that volatility—and, as a result, diffusibility—declines with increasing molecular weight.

One can also predict from first principles that the molecular weight of alarm substances will tend to be less than those of the sex attractants. Among the ants there is little specificity; each species responds strongly to the alarm substances of other species. Furthermore, an alarm substance, which is used primarily within the confines of the nest, does not need the stimulative potency of a sex attractant, which must carry its message for long distances. For these reasons small molecules will suffice for alarm purposes. Of seven alarm substances known in the social insects, six have 10 or fewer carbon atoms and one (dendrolasin) has 15. It will be interesting to see if future discoveries bear out these early generalizations.

Do human pheromones exist? Primer pheromones might be difficult to detect, since they can affect the endocrine system without producing overt specific behavioral responses. About all that can be said at present is that striking sexual differences have been observed in the ability of humans to smell certain

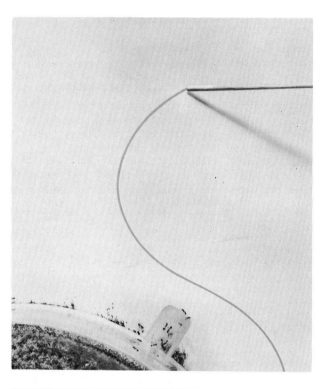

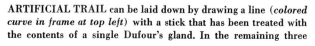

ARTIFICIAL TRAIL can be laid down by drawing a line (*colored curve in frame at top left*) with a stick that has been treated with the contents of a single Dufour's gland. In the remaining three frames, workers are attracted from the nest, follow the artificial route in close formation and mill about in confusion at its arbitrary terminus. Such a trail is not renewed by the unrewarded workers.

DENDROLASIN (*LASIUS FULIGINOSUS*)

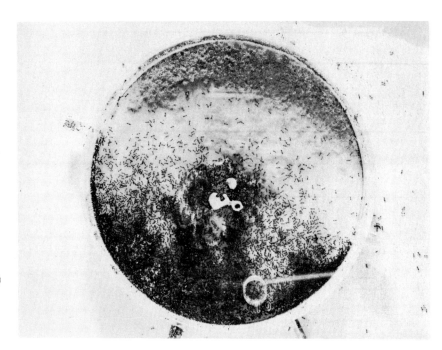

CITRAL (*ATTA SEXDENS*)

CITRONELLAL (*ACANTHOMYOPS CLAVIGER*)

2-HEPTANONE (*IRIDOMYRMEX PRUINOSUS*)

FOUR ALARM PHEROMONES, given off by the workers of the ant species indicated, have so far been identified. Disturbing stimuli trigger the release of these substances from various glandular reservoirs.

substances. The French biologist J. Le-Magnen has reported that the odor of Exaltolide, the synthetic lactone of 14-hydroxytetradecanoic acid, is perceived clearly only by sexually mature females and is perceived most sharply at about the time of ovulation. Males and young girls were found to be relatively insensitive, but a male subject became more sensitive following an injection of estrogen. Exaltolide is used commercially as a perfume fixative. LeMagnen also reported that the ability of his subjects to detect the odor of certain steroids paralleled that of their ability to smell Exaltolide. These observations hardly represent a case for the existence of human pheromones, but they do suggest that the relation of odors to human physiology can bear further examination.

It is apparent that knowledge of chemical communication is still at an early stage. Students of the subject are in the position of linguists who have learned the meaning of a few words of a nearly indecipherable language. There is almost certainly a large chemical vocabulary still to be discovered. Conceiv-

ably some pheromone "languages" will be found to have a syntax. It may be found, in other words, that pheromones can be combined in mixtures to form new meanings for the animals employing them. One would also like to know if some animals can modulate the intensity

or pulse frequency of pheromone emission to create new messages. The solution of these and other interesting problems will require new techniques in analytical organic chemistry combined with ever more perceptive studies of animal behavior.

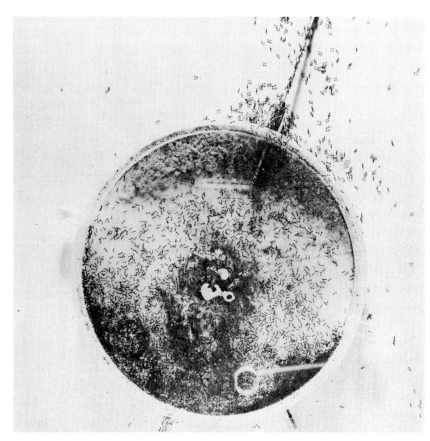

MASSIVE DOSE of trail pheromone causes the migration of a large portion of a fire ant colony from one side of a nest to another. The pheromone is administered on a stick that has been dipped in a solution extracted from the Dufour's glands of freshly killed workers.

VI

SENSORY PROCESSES

VI SENSORY PROCESSES

INTRODUCTION

Much of the electrical activity that is eventually transformed by the central nervous system into an appropriate behavioral output begins in sense organs. These range in complexity from scattered isolated cells in the skin that respond to mechanical deformation or to temperature change, to those that form highly complex, organized sensory structures in eyes, ears, or olfactory organs. The properties of sensation result from the abilities of receptor cells to change, or *transduce*, a variety of forms of stimulus energy into the common language of membrane depolarization and propagated electrical signals. The cells responsible for such transductions are, within a given category or *modality* of stimulus energy, quite similar. Most photoreceptor cells, for example, contain a series of internal membranes, stacked on top of one another; the membranes may be in the form of a stack of discs derived from cilia, as in the rods of the eyes of vertebrates, or of a honeycomb of microtubules, as in the photoreceptors of anthropods, but the basic plan is the same.

Despite the common properties implied by the basic transduction process, we know that the performances of sensory systems, even within the same category, differ dramatically. Consider, as an illustration, the receptor systems that convert mechanical energy into impulses. In some of the touch receptors of human skin, a naked nerve ending is used for this purpose. In the Pacinian corpuscles found in joints and mesenteries, the nerve ending is enclosed in a capsule; in the stretch receptors of muscles, it is wrapped around a specialized muscle fiber. In other receptor systems, the transducer may be a special cell derived from an epithelial layer. The neuromast cells of vertebrates have a crown of sensory cilia which, when subjected to a shearing force, somehow transmit the excitation to nerve fibers that connect with the base of the cell. Such receptors, which are coupled to the world beyond the neuromast cell in different ways, serve various purposes.

In the semicircular canals of the inner ear, the cilia of neuromast cells are embedded in a gelatinous cupula. The cupula, which is located in a swelling of the fluid-filled canal, acts like a swinging door; exertion of a force in one direction in the plane of the canal twists the door, and produces a shearing counterforce on the hairlike cilia so that the cells excite a higher rate of discharge in the nerve cells connected to them. Exertion of a force in the reverse direction lowers the rate of discharge. In the cochlea, the cilia of cells that are similar to the neuromast cells of the semicircular canals project to an overlying membrane, and traveling waves in the membrane that bears these cilia are transmitted from the eardrum through a series of bones in the middle ear. A shearing force similar to that which excites the cells of the semicircular canals excites the cells of the cochlea; but the coupling system is quite different, so high-frequency airborne sound waves, rather than movements of the head, are the effective natural stimuli.

The initial transduction process converts the impinging stimuli into a train of nerve impulses whose frequency is related to the energy content of the stimulus. In general, sense organs have a characteristic amplification factor: that is, there is a proportionality between the frequency of nerve discharge and the amplitude of the stimulus. The relationship is not usually a linear one; instead, the impulse frequency is proportional to the logarithm of the stimulus energy. Thus a tenfold increase in energy is required to double the discharge frequency of the sensory neuron. In effect, this lessens the absolute sensitivity of the nervous system as the energy level is increased. This relationship has its perceptual counterpart: we can readily detect a whisper in a quiet room, but that same amount of sound energy goes unnoticed if it is accompanied by a loud background noise. We say that the contrast threshold of the sensory system is a constant fraction of the total energy content of the stimulus.

The sensory process only begins with the absorption of stimulus energy and the conversion of that energy into streams of nerve impulses. Every system of receptors has its own processing machinery in the central nervous system. In higher animals, major divisions of the brain are allocated to the processing of information in each of the major sensory modalities—the so-called "special senses." These central processing stations are actually complex networks of nerve cells that filter incoming signals from the receptors and extract information that is of special importance to the organism. This aspect of sensory physiology, relatively neglected until recently, is emerging as a critical focus for research on the nervous system. It used to be thought that the sensory systems synthesized a picture of the impinging stimulus, in which those fundamental physical relationships detectable by the receptors were accurately preserved. We now have evidence, on the contrary, that the sensory parts of the nervous system work profound changes upon these relationships, emphasizing certain features, rejecting others, and altering still others. Connected networks of neurons at higher levels in the nervous system appear to be arranged to detect those aspects of the stimulus that have particular importance to the organism; hence, we call them *filters*.

Both the transducing and filtering functions of sense organs are especially well illustrated by the visual system. The general properties of the transduction process, and of light-sensitive sensory cells, are laid out by William H. Miller, Floyd Ratliff, and H. K. Hartline, in "How Cells Receive Stimuli." Although the article was published first in 1961 as a part of the special *Scientific American* issue on "The Living Cell," the authors' account of generator potentials in various kinds of receptors could hardly be improved upon today. Hartline and his collaborators at Rockefeller University have been intensively engaged for many years in the study of impulse generation and interaction between receptors in the eye of the horseshoe crab *Limulus*. Ironically, the eye of *Limulus* first attracted Hartline's attention because it appeared to be entirely lacking in interactions between the receptor units, and hence uniquely suited to an analysis of the problem of impulse initiation in a simple system of primary receptors. Hartline *was* successful in analyzing the impulse-generation process there, but perhaps an even more significant contribution was made when he and his colleagues realized that interaction between the primary sensory units occurred after all. Thereafter, in work described in the last two pages of their article, the Rockefeller group began to analyze the interaction in a careful, systematic fashion. The result was a full description of one of the most basic processing networks we know of in sensory systems: lateral inhibition. Networks of this kind, as the article describes, have inhibitory connections between nearby receptor units. They result in the enhancement of any difference in intensity that occurs along a grid of receptors, and as a consequence they achieve a perceptual enhancement of contrast. The

same enhancement occurs in the retina and central visual system of vertebrates, in the auditory system, and quite probably in a host of other sensory
"filters" as well.

The next two articles deal with a different kind of transduction, that of
light into electrical energy, as it takes place in the most complex sense organs.
In "Eye and Camera" (1950), George Wald describes the optics and the photochemistry of eyes and pursues the similarity between them and man-made
optical devices. Though this article was first published in 1950, its discussion
of the basic properties of vertebrate vision is as accurate today as it was then,
and its comparisons between visual systems and optical devices are especially
instructive. It is perhaps surprising that in his article Wald only touched upon
the biochemistry of visual pigments, about which he and his colleagues had
already gathered much of the basic information. The rod cells of the retina
invest more than 40 percent of their dry weight in the visual pigment, rhodopsin, which consists of a protein coupled to a molecule, which Wald discovered to be a derivative of vitamin A. This complex maximally absorbs
light of a wavelength of 500 millimicrons, which is blue-green; the dark-
adapted eye is therefore most sensitive to light of this wavelength. When the
complex absorbs light, the derivative of vitamin A isomerizes, and then it
dissociates from the protein component. The reunion of the two requires
that retinal tissue be able to reisomerize the vitamin A to its proper configuration. This explains why an eye that has just been exposed to bright light must
remain in the dark for several minutes before it can perceive dimly lit objects.

A similar chemical mechanism exists in the cones of the retina, which are
responsible for vision in bright light. The cone pigments differ from those of
rods, however, in that their proteins combine much more readily with the
vitamin A derivative than do the proteins of rod pigments. This enables the
cones to maintain their sensitivity in the very bright light the eye must operate in during the day. Although the biochemistry of these visual pigments
was studied in the test tube by Wald and others, analysis of them in the living
eye, where parallel physiological measurements of sensitivity could be made,
seemed to pose insurmountable difficulties. W. A. H. Rushton made such
analysis possible by the development of a remarkable technique of reflection
photometry. The method, and its application to human color vision, are reported in "Visual Pigments in Man" (1962). Rushton describes the measurements used to determine the relation between the concentration of rhodopsin
and rod sensitivity. He also provides direct evidence for the existence of separate visual pigments in man, and correlates their absence with the familiar
types of color blindness.

The discovery of the three cone pigments and their relationship to human
color vision provides us with a recent (not entirely happy) glimpse of the
human side of scientific discovery. Since the time of von Helmholtz and
Young, it had been proposed that to discriminate between the colors (wavelengths) of stimuli independently of their intensity, an eye would require
more than one, and perhaps at least three, fundamental types of receptors.
Later it was conclusively shown that the cones were entirely responsible for
color vision, and therefore it became clear that to prove the "trichromacy"
hypothesis correct, someone would have to find three types of cones, or pigments, or processes assignable to them in a color-sensitive retina—preferably
a human one. Extracting mammalian retinas chemically to obtain cone pigments was tried, of course; but these substances are present in low concentrations, and because they are swamped by the more abundant rod pigment,
the technique was abandoned. Two groups, one led by E. F. MacNichol at
Johns Hopkins and one involving Paul Brown and others in Wald's laboratory
at Harvard, began the difficult task of measuring the absorption spectra of
pigment in single, isolated cone cells directly, by the technique of micro-

spectrophotometry. At the same time, Wald was adapting human subjects to colored lights in an attempt to bleach away and remove certain primary-color-detecting retinal pigments, in order to measure the sensitivity of surviving processes which he believed depended on single pigments. He had been led to this approach by the earlier work, in England, of W. S. Stiles, who had used two-color contrast thresholds to demonstrate what looked very much like primary-color-detecting processes. Finally, several groups using retinas from different animals were trying to penetrate single cone cells directly with microelectrodes, in order to measure the spectral sensitivities of their electrical responses to colored lights.

Meanwhile Rushton, who had already perfected his technique of reflection photometry by investigating the kinetics of bleaching and regeneration of rod pigments in the peripheral human retina, was bringing the method to bear on the much more difficult task of measuring the cone pigments on the fovea. His was the most imaginative and challenging approach of all. The cone pigments were apparently dilute and hard to detect, and furthermore, the measuring light beam would be able to measure them all at once, not one at a time. For these reasons, Rushton was forced, like Wald in his experiments on the color thresholds of human subjects, to bleach away certain pigments with colored lights, and then measure the surviving pigment. But there was a difference: Rushton was measuring the pigments directly, not processes from which the absorption spectra of pigments could be inferred. Working with human vision, moreover, gave him the advantage of isolating pigments by the indirect method of choosing subjects deficient in others, namely, people who were color-blind.

In the long run, each approach succeeded. The Johns Hopkins group obtained the first successful spectra from cone cells in a color-detecting retina, that of the goldfish. Their absorption spectra fell into three categories and correlated well with electrophysiological sensitivity data. They turned to the primate retina and again found three classes of cone absorption-spectra. Their publication of the result of their work with primates, in *Science* in 1965, was virtually simultaneous with publication of the work of the Harvard group that revealed the existence of three receptor types in human cones. These also matched up quite well with the primary-color-detection processes that Wald derived by using the method adapted from Stiles.

But before any of these findings were announced, Rushton had already reported the absorption spectra he had obtained for the green- and red-sensitive cone pigments by reflection photometry, and had demonstrated the absence of predicted spectra in each of the two common clinical kinds of color-blindness. His results were sharply criticized by Wald, who pointed out that Rushton's absorption spectrum for green-sensitive pigment agreed rather poorly with the corresponding primary-color-detection process in comparison with the Harvard group's spectra, taken directly from cones. Yet in retrospect it is difficult to know how much weight to attach to this criticism. Even the most negative view of the errors in Rushton's measurement leave unchallenged his assertion that primary pigments exist with red- and green-absorbing properties, that they can be separately bleached and measured, and that their selective absence in people who are color-blind links them inextricably to the perceptual process of hue discrimination.

Recent history, however, has not tended to credit Rushton with priority for the proof that separate cone pigments are responsible for color discrimination. The 1967 Nobel Prize in physiology or medicine was widely regarded as a "vision prize." Wald and Hartline received richly deserved shares in it, which could have been justified on the basis of work having nothing to do with color vision at all. But Rushton did not receive an award. No outsider knows —though many try to guess—the selection methods of the Nobel committees;

so we cannot know how they regarded the adequacy of Rushton's proof, or how much they may have been influenced by the increasingly bitter scientific feud between Rushton and his old friend Wald. But to many, Rushton's absence from the Nobel roster stands as a perplexing omission.

The last two articles in this section return to the issue raised by Miller, Ratliff, and Hartline at the end of their description of lateral inhibition in the eye of *Limulus*. What kinds of connections in the sensory nervous system enable it to extract, from an array of receptors each producing a sequence of impulses, the information most crucial to the organism?

In "The Visual Cortex of the Brain" (1963), David H. Hubel describes some remarkable findings on the organization of a nervous center that processes information from a complex sense organ. The analysis has proceeded step by step from the neurons in the retina of the eye to those in the ultimate receiving area in the visual cortex. Each set of cells receives information from another set nearer the receptors, and in turn passes excitation on to still another layer of cells. At each successive stage, more complex stimuli are required to activate the cells; at each level of synaptic integration, a highly specific set of connections appears to be made by the preceding layer of neurons. Thus optic nerve fibers will discharge impulses when simple spots of light are shone onto the region of retina from which they come, whereas simple cortical cells will discharge impulses only in response to bars or lines with a specific angular orientation. Complex cortical cells may have responses that are similar, but they respond to moving lines, and so must be connected to simple cells over a wide area. Hubel shows that the response of cortical cells to particular linear orientations has an anatomical correlate; cells with similar requirements are aligned in vertical columns. The specificity of these cells seems to be a direct outcome of their receiving connections from a particular set of cells from a stage nearer the retina—a function for which the structure is obviously appropriate. Such connections, or the developmental machinery producing them, must be the result of a long process of natural selection that has suited the nature of the extracted information to the needs of the animal. The cortical connections ensure that certain properties of a stimulus will be attended to and many others disregarded; this finding suggests that studies on the central organization of sensory systems will provide a portrait of the subject's own perceptual world. This exciting prospect can be phrased in the reverse way: an animal's behavior may predict the behavior of his sensory cortex.

A major emphasis of Hubel's studies is clearly upon the developmental processes that give rise to these crucially important connections between one stage and the next higher one. It is not surprising that Hubel and his colleague Torsten Wiesel turned to an investigation of the consequences of altered development upon cortical organization. They have recently found, for example, that newborn kittens, before they have opened their eyes, possess a "ready-made" pattern of cortical connections resembling that found in adult cats. If, however, one eye is occluded, or made to diverge from the other in its gaze by an operation, the input to cortical cells that had previously been binocularly supplied is now delivered by one eye alone. It is as though the central neurons reject connections that are not matched, in the timing of their activity, to others. It is just this sort of problem in the development of the nervous system to which Sperry and other workers have been addressing themselves; in the introduction to Section III, results of Marcus Jacobsen were described in which optic nerve fibers revealed a genetically specified tendency to connect with particular loci in the brain. This is consistent with the results of Hubel and Wiesel, who find an apparently normal organization of the visual cortex in kittens without visual experience.

Yet the matching of the separate binocular inputs to cortical visual neurons is remarkably precise. In "The Neurophysiology of Binocular Vision" (1972),

John Pettigrew describes experiments on the receptive-field organization of "disparity detectors"—neurons in the visual cortex in which the binocular fields are slightly displaced from one another, to register the cues necessary for stereoscopic depth perception. The connections of pathways to such cells from the two eyes must be very accurately matched indeed. The question is whether genetic information alone is adequate to the task, or whether instead the environment helps during early development. To put the latter alternative in another way, does the pattern of impulse traffic along the visual pathways itself participate in instructing the nervous system how they should be connected? That connections can be *unmade* if the proper stimuli are withheld, as demonstrated by Hubel and Wiesel, suggests a role for experience. Yet the apparent adequacy of the genetic instructions would seem to deny it.

It is not surprising that Pettigrew, Horace Barlow, and Colin Blakemore—all of whom worked together on the disparity problems described in Pettigrew's article—have, like Hubel and Wiesel, turned to the developing visual system to ask these questions. Blakemore and G. Cooper in Cambridge, England, and a group working independently at Stanford, have shown that the environment *does* influence the development of the line-detection system in the visual cortex described by Hubel. If kittens are reared in artificial environments in which all the lines are vertical, they are deficient in horizontal line detectors, and if all the lines in the environment are horizontal, the kittens are deficient in vertical line detectors. It is, of course, crucial to know whether the deficiency is the result of selective elimination of cortical neurons that fail to be activated by their particular stimuli, or the result, instead, of the conversion of cortical cells of one type into the other. At this moment, the evidence suggests that it is conversion: the density of active cortical neurons is normal, although they tend to be all of one type. Robert Shlaer at The University of Chicago has similarly demonstrated that if the gaze of one eye is deviated with a prism, by one or two degrees, the degree of "disparity" found in cortical neurons is equivalently shifted. Results of this kind strongly suggest that the environment directs the formation of visual connections.

Other recent experiments suggest that the openness of the developing nervous system to this kind of environmental influence is restricted to a critical period occurring early in postnatal development, and lasting in the cat for only two or three weeks. After this period the nervous system appears to be "hard-wired" and not subject to environmental influence. But at fifteen days of postnatal life, the kitten appears astonishingly sensitive to the shape of the world around it. Only five hours of exposure at this age to a world of vertical lines will, if the rest of the kitten's time is spent in darkness, significantly bias the distribution of its cortical receptive fields. In his own experiments, Pettigrew has gone even further: he finds that *during an experiment* the receptive field preference of a single cortical neuron in the kitten can be altered by the orientation of the line stimuli used to evoke activity in it.

These very recent results have even raised some doubt about the earlier claims that genetic "wiring" is adequate to equip the newborn kitten with a normal system of line detection in its visual cortex. Pettigrew and the Berkeley group have pointed out that because stimuli used during an experiment can influence orientation selectivity, the results of the early studies by Hubel and Wiesel on neonatal kittens might have involved an experimental artifact. A brisk controversy over that point is now going on and is unresolved at this writing. This much is clear: although the nervous system clearly has a substantial capacity for genetically specified connections, we have underestimated its sensitivity to environmental influence. The new revolution in visual physiology thrusts a new question forward for students of the developing nervous system: how does the environmentally induced impulse *activity* of a particular sensory nerve fiber affect its connection to a postsynaptic neuron?

How Cells Receive Stimuli

by William H. Miller, Floyd Ratliff and H. K. Hartline
September 1961

In complex organisms certain cells are highly specialized to detect changes in the environment. The properties of such cells have been elucidated by studies of the visual receptors of the horseshoe crab

The survival of every living thing depends ultimately on its ability to respond to the world around it and to regulate its own internal environment. In most multicellular animals this response and regulation is made possible by specialized receptor cells that are sensitive to a wide variety of physical, chemical and mechanical stimuli.

In many animals, including man, these receptors provide information that far exceeds that furnished by the traditional five senses (sight, hearing, smell, taste and touch). Sense organs of which we are less aware include equally important receptors that monitor the internal environment. Receptor organs in the muscles, called muscle spindles, provide a continuous measure of muscle stretch, and other receptors sense the movement of joints. Without such receptors it would be difficult to move or talk. Receptor cells in the hypothalamus, a part of the brain, are sensitive to the temperature of the blood; pressure-sensitive cells in the carotid sinus measure the blood pressure. Still other internal receptors monitor carbon dioxide in special regions of the large arteries. Pain receptors, widely distributed throughout the body, respond to noxious stimuli of almost any nature that are likely to cause tissue damage.

Receptor cells not only have diverse functions and structures but also connect in various ways with the nerve fibers channeling into the central nervous system. Some receptor cells give rise directly to nerve fibers of their own; others make contact with nerve fibers originating elsewhere. All receptors, however, share a common function: the generation of nerve impulses. This does not imply that impulses necessarily occur in the receptor cells themselves. For example, in the eyes of vertebrates no one has yet been able to detect impulses in the photoreceptor cells: the rods and cones. Nevertheless, the rods and cones, when struck by light, set up the physicochemical conditions that trigger impulses in nerve cells lying behind them. Typical nerve impulses are readily detected in the optic nerve itself, which is composed of fibers of ganglion cells separated from the rods and cones by at least one intervening group of nerve cells.

Eventually physiologists hope to unravel the detailed train of events by which a receptor cell gives rise to a discharge of nerve impulses following mechanical deformation, absorption of light or heat, or stimulation by a particular molecule. In no case have all the events been traced out. In our discussion we will begin with the one final event common to all sensory reception—the generation of nerve impulses. We will then examine in some detail the events occurring in one particular receptor: the photoreceptor of *Limulus*, the horseshoe crab. Finally, we will describe some characteristics of the output of receptors acting singly and in concert with others.

The nerve fiber, or axon, is a thread-like extension of the nerve-cell body. The entire surface membrane of the cell, including that of the axon, is electrically polarized; the inside of the cell is some 70 millivolts negative with respect to the outside. This potential difference is called the membrane potential. In response to a suitable triggering event the membrane potential is momentarily and locally altered, giving rise to a nerve impulse, which is then propagated the whole length of the axon [see "How Cells Communicate," page 135].

In any particular nerve fiber the impulses are always of essentially the same magnitude and form and they travel with the same speed. This has been known for some 30 years, since the pioneering studies of E. D. Adrian at the University of Cambridge. He and his colleagues found that varying the intensity of the stimulus applied to a receptor cell affects not the size of the impulses but the frequency with which they are discharged; the greater the intensity, the greater the frequency of nerve impulses generated by the receptor. Thus all sensory messages—concerning light, sound, muscle position and so on—are conveyed in the same code of individual nerve impulses. The animal is able to decode the various messages because each type of receptor communicates to the higher nerve centers only through its own private set of nerve channels.

Adrian and others have investigated the problem of how the receptor cell triggers sensory nerve impulses. Adrian suggested that the receptor must somehow diminish the resting membrane potential of its nerve fiber; that is, it must locally depolarize the axon membrane. The existence of local potentials in the eye has been known since 1865, and much later similar potentials were recorded in other sense organs. But the relationship of these gross electrical changes to the discharge of nerve impulses was not clear. For some simple eyes, however, the polarity of the local potential changes in the receptors is such that they appear to depolarize the sensory nerve fibers. This led Ragnar Granit of the Royal Caroline Institute in Stockholm to propose that they be called "generator" potentials. The present view is that stimulation of the receptor cell gives rise to a sustained local depolarization of the sensory nerve fiber, which thereupon generates a train of impulses.

Some of the first direct evidence for generator potentials at the cellular level was produced in 1935 by one of the authors of this article (Hartline), then working at the Johnson Research Foundation of the University of Pennsylvania. He found what appeared to be a generator potential when he recorded the activity of a single optic nerve fiber and its receptor in the compound eye of *Limulus*. Superimposed on the potential was a train of nerve impulses [*see illustration on page 248*].

In 1950 Bernhard Katz of University College London obtained unmistakable evidence for a generator potential in a somewhat simpler receptor: the vertebrate muscle spindle. When the spindle was stretched, a small, steady depolarization could be recorded in the nerve fiber coming from the spindle. As viewed on the oscilloscope, it appeared that the base line of the recorded signal had been shifted slightly upward. Superimposed on the shifted signal, or local potential, was a series of "spikes" representing individual nerve impulses. The stronger or the more rapid the stretch, the greater the magnitude of the potential shift and the greater the frequency of the impulses [*see illustration on page 246*]. Analysis of many such records showed that in the steady state the frequency of nerve impulses depends directly on the magnitude of the altered potential. If a local anesthetic is applied to the spindle, the impulses are abolished but the potential shift remains. Katz concluded that this potential shift is an essential link between the stretching of the spindle and the discharge of nerve impulses; indeed, that it is the generator potential. Moreover, the potential can be detected only very close to the spindle, showing that it is conducted passively—which is to say poorly—along the nerve fiber.

Important confirmation of the role of the generator potential was provided by the work of Stephen W. Kuffler and

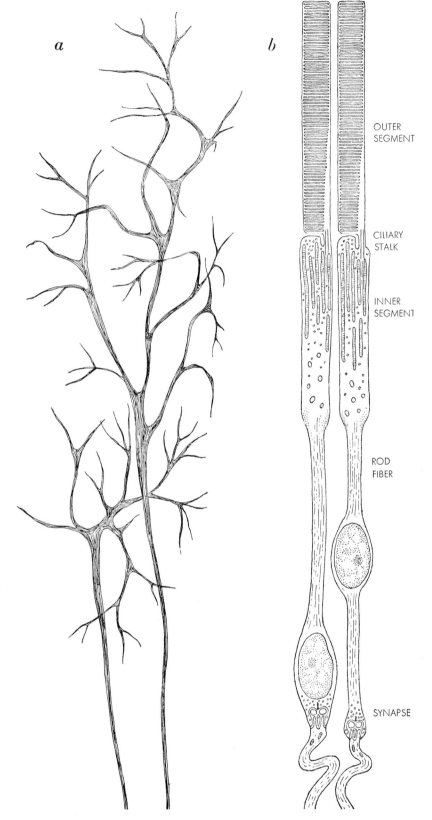

OUTER SEGMENT

CILIARY STALK

INNER SEGMENT

ROD FIBER

SYNAPSE

RECEPTOR CELLS, typical of those found in vertebrates, respond to a variety of stimuli: heat, light, chemicals and mechanical deformation. The "pit" on the head of the pit viper contains a network of free nerve endings (*a*) that are sensitive to heat and help the viper locate its prey. Rods (*b*) are light-sensitive cells in the retina of the eye; photosensitive pigment is in the laminar structure at top of drawing. Taste buds (*c*) are chemoreceptor cells embedded in the tongue. The cochlea, a spiral tube in the inner ear, contains thousands of

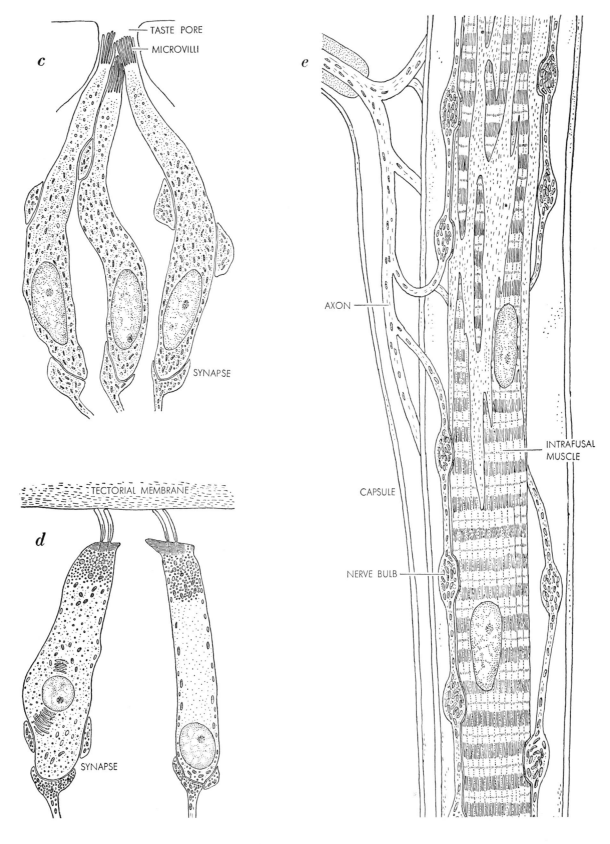

c

TASTE PORE
MICROVILLI
SYNAPSE

TECTORIAL MEMBRANE

d

SYNAPSE

e

AXON

INTRAFUSAL
MUSCLE

CAPSULE

NERVE BULB

sensitive cells (*d*) in the so-called organs of Corti. When the hairlike bristles of these cells are mechanically deformed by sound vibrations, impulses are generated in the auditory nerve fibers leading to brain. Muscle spindle (*e*) contains a number of nerve endings that respond sensitively to stretching of muscle fibers surrounding them. These illustrations of receptor cells are based on the work of the following investigators: Theodore H. Bullock of the University of California at Los Angeles (*a*), Fritiof Sjöstrand of the same institution (formerly of the Karolinska Institute, Stockholm) (*b*), A. J. de Lorenzo of the Johns Hopkins School of Medicine, Baltimore, Md. (*c*), Salvatore Iurato of the University of Milan (*d*) and Bernhard Katz of University College London (*e*).

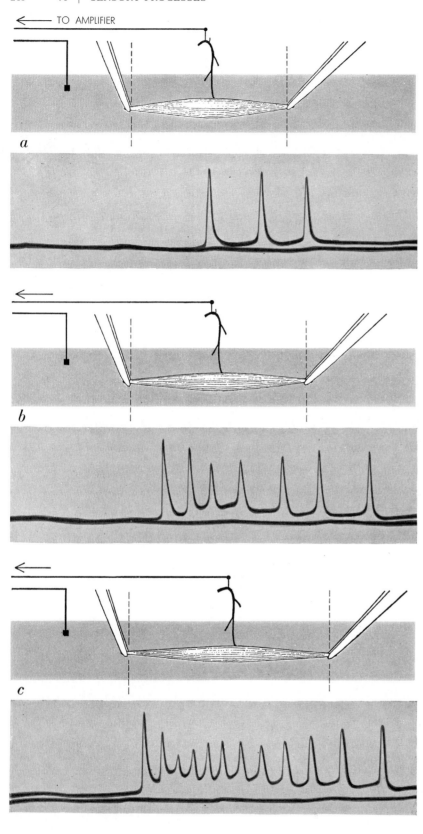

TO AMPLIFIER

a

b

c

MUSCLE SPINDLE responds to stretch by firing nerve impulses at a rate proportional to the degree and speed of stretching. These recordings made by Bernhard Katz of University College London were the first to show that stretching causes depolarization of the nerve near the spindle (*base line shifted upward in the traces*) and that this depolarization is the precondition for the firing of nerve impulses. The shift is called the generator potential.

Carlos Eyzaguirre, then at Johns Hopkins University, using the so-called Alexandrowicz stretch-receptor cells in crustaceans. These are large single receptor cells with dendrites (short fibers) that are embedded in specialized receptor muscles. Kuffler was able to insert a microelectrode within the cell and record its membrane potential as well as the nerve impulses in its axon. He found that when he distorted the cell's dendrites by stretching the receptor muscle, the cell body became depolarized and the depolarization spread passively to the site of impulse generation, which is probably in the axon close to where it emerges from the cell body. When this generator potential reached a critical level, the cell fired a train of nerve impulses; the greater the depolarization of the axon above this critical level, the higher the frequency of the discharge.

There is now abundant evidence that a receptor cell triggers a train of nerve impulses by locally depolarizing the adjacent nerve fiber—either its own fiber or one provided by another cell. With few exceptions, a fiber of a nerve trunk will not respond repetitively if one passes a sustained depolarizing current through it; it responds only briefly with one to several impulses and then accommodates to the stimulus and responds no more. Evidently that part of the sensory nerve fiber close to the receptor must be specialized so that it does not speedily accommodate to the generator potential. It is nonetheless true that a certain amount of accommodation, or adaptation, almost always takes place when a receptor cell is exposed to a sustained stimulus. In any event, the initiation of nerve impulses in the axons of receptor cells by means of a generator potential appears to be a general phenomenon.

The question still remains: How does the external stimulus produce the generator potential? In most of the receptors studied there is no evidence whatever on this point. Only in the photoreceptor do we have precise knowledge of the first step in the excitation of the sense cell. Yet the study of the photoreceptor is beset by special difficulties. In most eyes the receptors are small and densely packed, and their associated neural structures are complex and highly organized. A fortunate exception is the compound eye of *Limulus*, which provided early evidence for the generator potential. In this eye the receptor cells are large and the neural organization is relatively simple.

The coarsely faceted compound eye

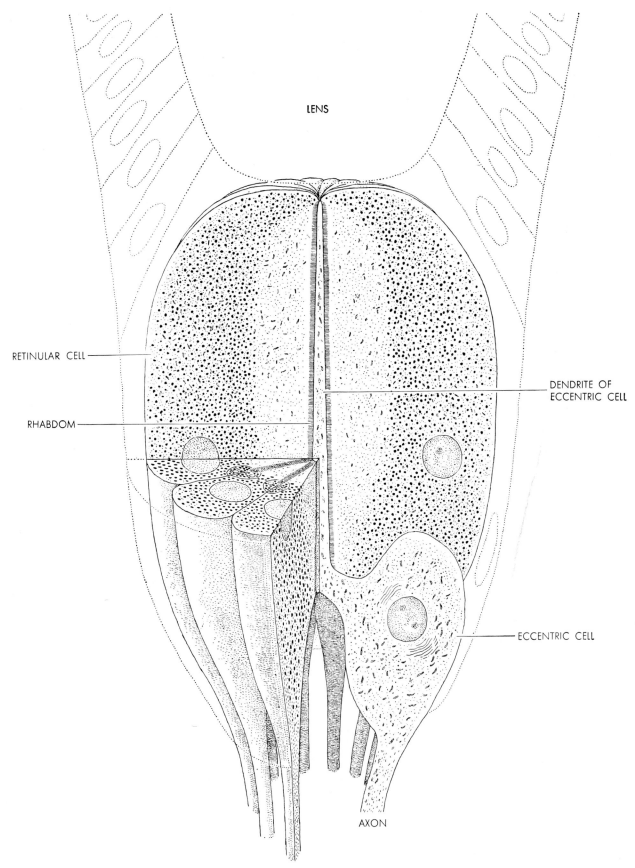

LENS

RETINULAR CELL

RHABDOM

DENDRITE OF
ECCENTRIC CELL

ECCENTRIC CELL

AXON

OMMATIDIUM OF LIMULUS is a remarkable structure roughly the size of a pencil lead. About 1,000 form the crab's compound eye. The ommatidium consists of about 12 wedge-shaped retinular cells clustered around a central fiber, which is the dendrite (sensitive process) of a nerve cell, the eccentric cell shown at lower right. When light strikes the ommatidium (*at the top*), the eccentric cell gives rise to nerve impulses (*see illustration on next page*). Photosensitive pigment rhodopsin is believed to be in the rhabdom.

of *Limulus* has some 1,000 ommatidia ("little eyes"), each of which contains about a dozen cells. The cells in each ommatidium have a regular arrangement. The retinular cells—the receptors—are arranged radially like the segments of a tangerine around the dendrite of an associated neuron: a single eccentric cell within each ommatidium [*see illustration on preceding page*].

Hartline, H. G. Wagner and E. F. MacNichol, Jr., working at Johns Hopkins University, found by the use of microelectrodes that the eccentric cell gives rise to the nerve impulses that can be recorded farther down in the nerve strand leaving the ommatidium. The microelectrode also records the generator potential of the ommatidium. Because of the anatomical complexity

of the ommatidium, the site of origin of the generator potential has not been identified with certainty. Nor has activity yet been detected in the axons of the retinular cells. As in the vertebrate and invertebrate stretch receptors, local anesthetics extinguish the nerve impulses without destroying the generator potential. Moreover, as in the stretch receptors, there is a proportional relationship between the degree of depolarization and the frequency of nerve impulses.

Recently M. G. F. Fuortes of the National Institute of Neurological Diseases and Blindness has shown that illumination increases the conductance of the eccentric cell. He postulates that the increase is produced by a chemical transmitter substance that is released by the action of light and acts on the eccentric cell's dendrite. Presumably the increased conductance of the dendrite results in a depolarization that spreads passively to the site of impulse generation, where it acts as the generator potential.

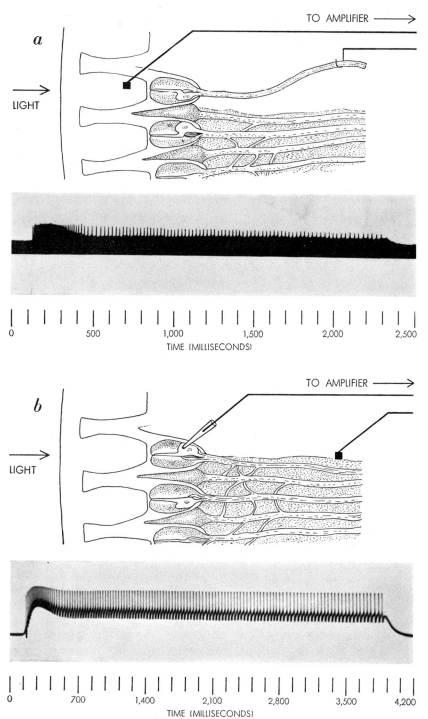

RECORDINGS FROM OMMATIDIA show trains of nerve impulses evoked by light. The upper recording from a nerve bundle (*a*) was made by one of the authors, H. K. Hartline, some 25 years ago. Shift of base line underlying nerve spikes is the generator potential. Lower recording, made with microelectrode (*b*), shows generator potential more clearly.

In photosensory cells—alone among all receptors—there exists direct experimental evidence of the initial molecular events in the receptor process. It has been known for about a century that visual receptor cells in both vertebrates and invertebrates have specially differentiated organelles containing a photosensitive pigment. In vertebrates this reddish pigment, called rhodopsin, can be clearly seen in the outer segments of rods. The absorption spectrum of human rhodopsin corresponds closely to the light-sensitivity curve for human vision under conditions of dim illumination, when only the rods of the retina are operative. This is strong evidence that rhodopsin brings about the first active event in rod vision: the absorption of light by the photoreceptor structure. (There is evidence for similar pigments in the outer segments of cones, but they have proved more difficult to isolate and study.)

The visual pigments are known to be complex proteins, but the light-absorbing part of the pigment, called the chromophore, has been found to be a relatively simple substance: vitamin A aldehyde. Because it contains a number of double chemical bonds in its make-up, vitamin A aldehyde can exist in various molecular configurations known as "*cis*" and "*trans*" isomers. We know from the work of Ruth Hubbard, George Wald and their colleagues at Harvard University that the absorption of light changes the chromophore from

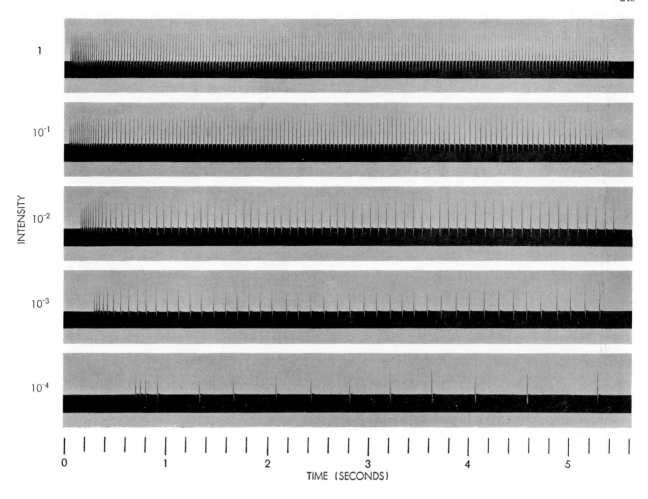

NERVE IMPULSES TRIGGERED BY LIGHT are directly related to intensity of steady light falling on the *Limulus* eye. Recordings were made from the optic nerve fiber arising from one ommatidium. At high light intensity (*top*) the nerve fires about 30 times per second. As intensity is reduced by factors of 10, firing is reduced in uniform steps, falling to a low of two or three impulses per second.

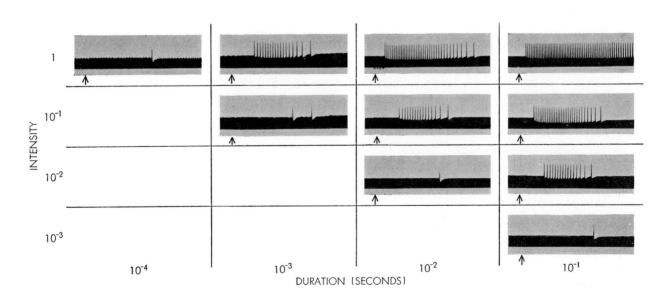

DURATION AND INTENSITY OF LIGHT have equivalent effect on the *Limulus* eye. Evidently the receptor responds to the total amount of energy received in a brief flash (*arrows*) regardless of how the energy is "packaged" in duration and intensity. Thus a brief intense flash (*top left*) evokes about the same response as a flash a 1,000th as bright lasting 1,000 times longer (*bottom right*).

11-*cis* vitamin A aldehyde to the *trans* configuration. This photochemical reaction is the first step that leads, through a chain of chemical and physical events as yet unknown, to the initiation of the generator potential of the receptor cell and finally to the discharge of impulses in the optic nerve. This is the only case in which the specific molecular mechanism is known whereby a receptor cell detects environmental conditions.

Supporting evidence that rhodopsin governs the response to a light stimulus can be found by comparing the absorption spectrum of *Limulus* rhodopsin with the sensitivity of the *Limulus* eye at various wavelengths. In 1935 Clarence H. Graham and Hartline measured the intensity of flashes at several wavelengths required to produce a fixed number of impulses in the *Limulus* optic nerve. When a sensitivity curve obtained from this experiment is superimposed on the absorption curve found by Hubbard and Wald for *Limulus* rhodopsin, the two match almost perfectly. At a wavelength of about 520 angstrom units, where rhodopsin absorbs light most strongly, the *Limulus* eye generates the highest number of impulses for a given quantity of light energy received. It turns out that the wavelength sensitivity of the *Limulus* eye is close to that of the human eye in dim light when rod vision dominates.

Many other familiar sensory experiences are manifestations of the properties of individual sense cells. Perhaps the most elementary experience is our ability to perceive when a stimulus has been increased in intensity. Under such circumstances we can be sure that the sensory fibers conveying information to the brain are firing more rapidly as the stimulus is increased. We are also familiar with the experience of sensory adaptation; for example, a strong odor usually seems to decrease in intensity after a time, although objective measurements would show that its intensity has remained constant.

We know from photography that shutter speed and lens opening can be interchanged to produce a constant exposure, which is the same as saying that intensity and duration of illumination can be interchanged (within limits) to produce a constant photochemical effect. The same equivalence holds for the human eye exposed to short flashes of light, and the equivalence can be demonstrated in the photoreceptor of *Limulus*. About the same number of nerve impulses are produced by exposing the ommatidium to a weak light for a 10th of a second as by exposing it to light 10 times as bright for a 100th of a second [*see bottom illustration on page 249*].

We also know from watching motion pictures or television that a light flickering at a high rate appears not to be flickering at all. A neural basis for this phenomenon can be seen in the generator potentials and nerve impulses recorded when a *Limulus* ommatidium is exposed to a light flickering at various rates [*see the illustration at left*]. Flicker is detectable as fluctuations in the generator potential, which in turn gives rise to bursts of impulses. As the repetition rate increases, the rate of discharge becomes steadier and finally is indistinguishable from a response to continuous illumination. As can be seen from the records, this "flicker fusion" is directly attributable to the generator potential, which becomes smooth at the highest repetition rates.

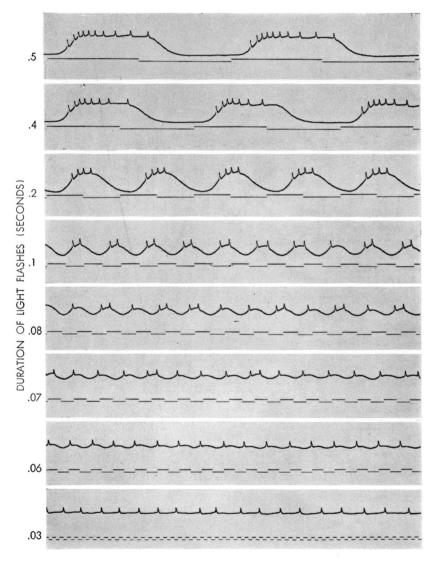

EFFECT OF FLICKERING LIGHT on the *Limulus* ommatidium provides a basis for explaining "flicker fusion": the inability to perceive a rapid flicker. The recordings show the response of the ommatidium to a light flickering at various rates; when the horizontal line is raised, the light is on. At low flicker rates the generator potential, indicated by a rise in base potential, rises and falls. As flicker rate increases, the generator potential no longer falls between flashes, and spacing between nerve impulses becomes more uniform.

The experiments described so far were carried out on single cells or single sensory units. In the eye, ear and other organs, however, receptor cells are grouped close together and usually act in concert. In fact, modern studies show that receptor cells of complex sense organs seldom act independently. In such organs the receptor cells are interconnected neurally and as a result of these connections new functional properties arise.

Although the compound eye of *Limu-*

lus is much less complex than the eyes of vertebrates, it still shows clearly the effects of neural interaction. In *Limulus* the activity of each photoreceptor unit is affected to some degree by the activity of adjacent ommatidia. The frequency of discharge of impulses in an optic nerve fiber from a particular ommatidium is decreased—that is, inhibited—when light falls on its neighbors. Since each ommatidium is a neighbor of its neighbors, mutual inhibition takes place. This inhibition is brought about by a branching array of nerve axons that make synaptic contact with each other in a feltwork of fine fibers behind the ommatidia. The inhibition probably results from a decrease in the magnitude of the generator potential at the site of origin of the nerve impulses, as a consequence of which the rate of firing is slowed down.

When two adjacent ommatidia are illuminated at the same time, each discharges fewer impulses than when it receives the same amount of light by itself [*see illustrations on this page*]. The magnitude of the inhibition exerted on each ommatidium (in the steady state) depends only on the frequency of the response of the other. The more widely separated the ommatidia, the smaller the mutual inhibitory effect. When several ommatidia are illuminated at the same time, the inhibition of each is given by the sum of the inhibitory effects from all others.

Inhibitory interaction can produce important visual effects. The more intensely illuminated retinal regions exert a stronger inhibition on the less intensely illuminated ones than the latter do on the former. As a result differences in neural activity from differently lighted retinal regions are exaggerated. In this way contrast is heightened and certain significant features of the retinal image tend to be accentuated at the expense of fidelity of representation.

This has been shown by illuminating the *Limulus* compound eye with a "step" pattern: a bright rectangle next to a dimmer one [*see illustration on page 252*]. The eye was masked so that only one ommatidium "observed" the pattern, which was moved to various positions on the retinal mosaic. At each position the steady-state frequency of discharge was measured. The result was a faithful reproduction (in terms of frequency of impulses) of the form of the pattern. Then the eye was unmasked so that all the ommatidia observed the pattern, and a recording was again made from the single ommatidium. This time

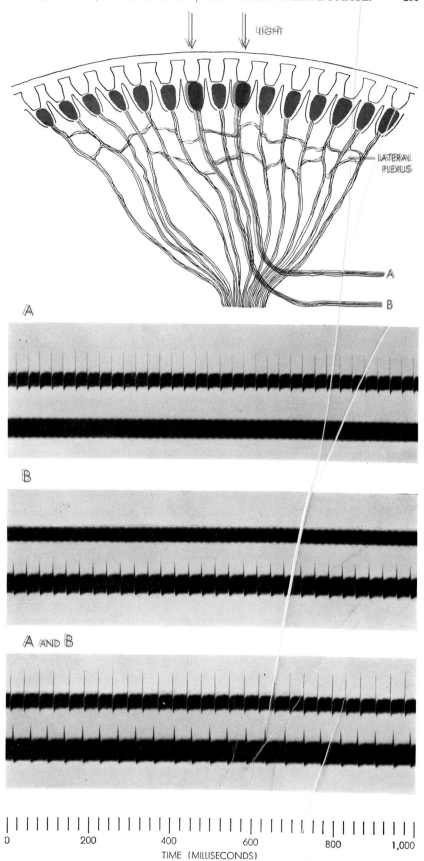

MUTUAL INHIBITION results when two neighboring ommatidia are illuminated at the same time (*top*). The inhibition is exerted by cross connections among nerve fibers. When ommatidia attached to fiber *A* and fiber *B* were illuminated separately, 34 and 30 impulses were recorded respectively in one second. Illuminated together, they fired less often.

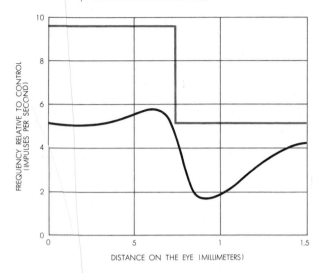

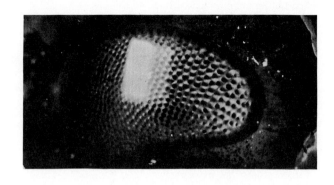

CONTRAST HEIGHTENING AT CONTOURS is demonstrated by letting "step" pattern of light, a bright area next to a darker one, fall on *Limulus* compound eye (*right*). If eye is masked so light strikes only one ommatidium, a recording of its output forms a simple step-shaped curve (*left*) as the pattern moved across the eye. If the eye is unmasked, the output of the single ommatidium is inhibited in varying degrees by the light striking its neighbors. The net effect (*lower curve*) is to heighten the contrast at light-dark boundaries.

the frequency increased on the bright side of the step and decreased near the dim side. This is expected because near the bright side of the step the neighboring ommatidia illuminated by the dim part of the step pattern have a low frequency of firing and therefore do not exert much inhibition. Consequently the frequency of discharge of the receptors on the bright side of the step is higher than its equally illuminated but more distant neighbors. Similar reasoning explains the decrease in frequency on the dim side of the step. The net effect of this pattern of response is to enhance contours, an effect we can easily demonstrate in our own vision by looking at a step pattern consisting of a series of uniform gray bands graded from white to black.

Artists are quite familiar with the existence of "border contrast" and may even heighten it in their paintings. And as we all know, significant information is conveyed by contours alone, as is demonstrated by cartoons and other line drawings. Georg von Békésy of Harvard University has suggested that a similar reciprocal inhibition in the auditory system would lead to a sharpening of the sense of pitch.

There is also evidence that in many sense organs the response can be modified by neural influences exerted back onto them by higher centers of the nervous system. Thus the sensitivity of the vertebrate stretch receptor or muscle spindle is established by variations in the length of the spindle fibers, and this length is dependent both on the output of the receptor and on its interaction with higher centers. The sensitivity of the vertebrate olfactory receptors can also be altered, in all probability, by the flow of impulses from above. Similar influences, not yet well understood, also seem to be at work in the retina of the eye.

It is evident, then, that the responses of complex sense organs are determined by the fundamental properties of the individual receptor cells, by the influences they exert on one another and by control exerted on them by other organs. In this way the activity of the receptor cells is integrated into complex patterns of nervous activity that enable organisms to survive in a world of endless variety and change.

Eye and Camera

by George Wald

August 1950

The classical comparison of the two devices is still fruitful. Today it has gone beyond their optics into their basic physics and chemistry

OF all the instruments made by man, none resembles a part of his body more than a camera does the eye. Yet this is not by design. A camera is no more a copy of an eye than the wing of a bird is a copy of that of an insect. Each is the product of an independent evolution; and if this has brought the camera and the eye together, it is not because one has mimicked the other, but because both have had to meet the same problems, and frequently have done so in much the same way. This is the type of phenomenon that biologists call convergent evolution, yet peculiar in that the one evolution is organic, the other technological.

Over the centuries much has been learned about vision from the camera, but little about photography from the eye. The camera made its first appearance not as an instrument for making pictures but as the *camera obscura* or dark chamber, a device that attempted no more than to project an inverted image upon a screen. Long after the optics of the camera obscura was well understood, the workings of the eye remained mysterious.

In part this was because men found it difficult to think in simple terms about the eye. It is possible for contempt to breed familiarity, but awe does not help one to understand anything. Men have often approached light and the eye in a spirit close to awe, probably because they were always aware that vision provides their closest link with the external world. Stubborn misconceptions held back their understanding of the eye for many centuries. Two notions were particularly troublesome. One was that radiation shines out of the eye; the other, that an inverted image on the retina is somehow incompatible with seeing right side up.

I am sure that many people are still not clear on either matter. I note, for example, that the X-ray vision of the comic-strip hero Superman, while regarded with skepticism by many adults, is not rejected on the ground that there are no X-rays about us with which to see. Clearly Superman's eyes supply the X-rays, and by directing them here and there he not only can see through opaque objects, but can on occasion shatter a brick wall or melt gold. As for the inverted image on the retina, most people who learn of it concede that it presents a problem, but comfort themselves with the thought that the brain somehow compensates for it. But of course there is no problem, and hence no compensation. We learn early in infancy to associate certain spatial relations in the outside world with certain patterns of nervous activity stimulated through the eyes. The spatial arrangements of the nervous activity itself are altogether irrelevant.

It was not until the 17th century that the gross optics of image formation in the eye was clearly expressed. This was accomplished by Johannes Kepler in 1611, and again by René Descartes in 1664. By the end of the century the first treatise on optics in English, written by William Molyneux of Dublin, contained several clear and simple diagrams comparing the projection of a real inverted image in a "pinhole" camera, in a camera obscura equipped with a lens and in an eye.

Today every schoolboy knows that the eye is like a camera. In both instruments a lens projects an inverted image of the surroundings upon a light-sensitive surface: the film in the camera and the retina in the eye. In both the opening of the lens is regulated by an iris. In both the inside of the chamber is lined with a coating of black material which absorbs stray light that would otherwise be reflected back and forth and obscure the image. Almost every schoolboy also knows a difference between the camera and the eye. A camera is focused by moving the lens toward or away from the film; in the eye the distance between the lens and the retina is fixed, and focusing is accomplished by changing the thickness of the lens.

The usual fate of such comparisons is that on closer examination they are exposed as trivial. In this case, however, just the opposite has occurred. The more we have come to know about the mechanism of vision, the more pointed and fruitful has become its comparison with photography. By now it is clear that the relationship between the eye and the camera goes far beyond simple optics, and has come to involve much of the

FORMATION OF AN IMAGE on the retina of the human eye was diagrammed by Rene Descartes in 1664. This diagram is from Descartes' *Dioptrics*.

essential physics and chemistry of both devices.

Bright and Dim Light

A photographer making an exposure in dim light opens the iris of his camera. The pupil of the eye also opens in dim light, to an extent governed by the activity of the retina. Both adjustments have the obvious effect of admitting more light through the lens. This is accomplished at some cost to the quality of the image, for the open lens usually defines the image less sharply, and has less depth of focus.

When further pressed for light, the photographer changes to a more sensitive film. This ordinarily involves a further loss in the sharpness of the picture. With any single type of emulsion the more sensitive film is coarser in grain, and thus the image cast upon it is resolved less accurately.

The retina of the eye is grainy just as is photographic film. In film the grain is composed of crystals of silver bromide embedded in gelatin. In the retina it is made up of the receptor cells, lying side by side to form a mosaic of light-sensitive elements.

There are two kinds of receptors in the retinas of man and most vertebrates: rods and cones. Each is composed of an inner segment much like an ordinary nerve cell, and a rod- or cone-shaped outer segment, the special portion of the cell that is sensitive to light. The cones are the organs of vision in bright light, and also of color vision. The rods provide a special apparatus for vision in dim light, and their excitation yields only neutral gray sensations. This is why at night all cats are gray.

The change from cone to rod vision, like that from slow to fast film, involves a change from a fine- to a coarse-grained mosaic. It is not that the cones are smaller than the rods, but that the cones act individually while the rods act in large clumps. Each cone is usually connected with the brain by a single fiber of the optic nerve. In contrast large clusters of rods are connected by single optic nerve fibers. The capacity of rods for image vision is correspondingly coarse. It is not only true that at night all cats are gray, but it is difficult to be sure that they are cats.

Vision in very dim light, such as starlight or most moonlight, involves only the rods. The relatively insensitive cones are not stimulated at all. At moderately low intensities of light, about 1,000 times greater than the lowest intensity to which the eye responds, the cones begin to function. Their entrance is marked by dilute sensations of color. Over an intermediate range of intensities rods and cones function together, but as the brightness increases, the cones come to dominate vision. We do not know that

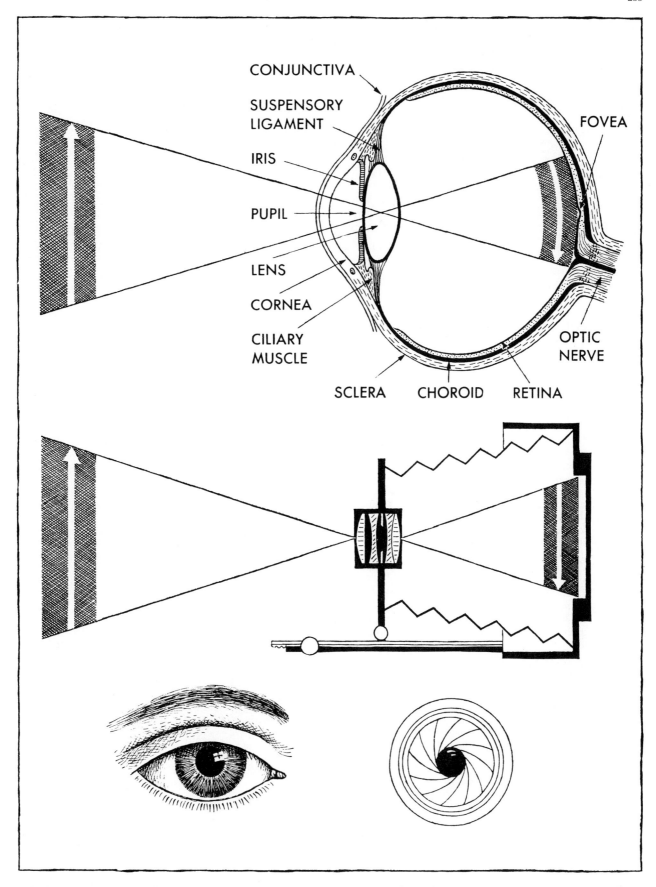

CONJUNCTIVA

SUSPENSORY LIGAMENT

IRIS

PUPIL

LENS

CORNEA

CILIARY MUSCLE

FOVEA

OPTIC NERVE

SCLERA   CHOROID   RETINA

OPTICAL SIMILARITIES of eye and camera are apparent in their cross sections. Both utilize a lens to focus an inverted image on a light-sensitive surface. Both possess an iris to adjust to various intensities of light. The single lens of the eye, however, cannot bring light of all colors to a focus at the same point. The compound lens of the camera is better corrected for color because it is composed of two kinds of glass.

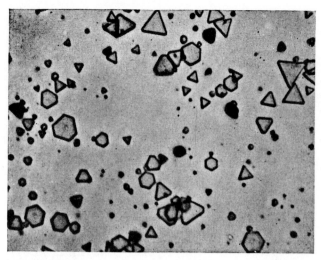 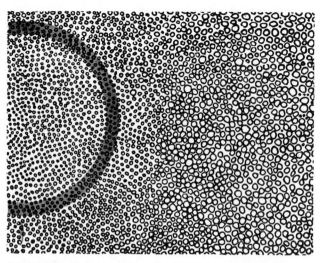

GRAIN of the photographic emulsion, magnified 2,500 times, is made up of silver-bromide crystals in gelatin.

"GRAIN" of the human retina is made up of cones and rods (*dots at far right*). Semicircle indicates fovea.

the rods actually stop functioning at even the highest intensities, but in bright light their relative contribution to vision falls to so low a level as to be almost negligible.

To this general transfer of vision from rods to cones certain cold-blooded animals add a special anatomical device. The light-sensitive outer segments of the rods and cones are carried at the ends of fine stalks called myoids, which can shorten and lengthen. In dim light the rod myoids contract while the cone myoids relax. The entire field of rods is thus pulled forward toward the light, while the cones are pushed into the background. In bright light the reverse occurs: the cones are pulled forward and the rods pushed back. One could scarcely imagine a closer approach to the change from fast to slow film in a camera.

The rods and cones share with the grains of the photographic plate another deeply significant property. It has long been known that in a film exposed to light each grain of silver bromide given enough developer blackens either completely or not at all, and that a grain is made susceptible to development by the absorption of one or at most a few quanta of light. It appears to be equally true that a cone or rod is excited by light to yield either its maximal response or none at all. This is certainly true of the nerve fibers to which the rods and cones are connected, and we now know that to produce this effect in a rod—and possibly also in a cone—only one quantum of light need be absorbed.

It is a basic tenet of photochemistry that one quantum of light is absorbed by, and in general can activate, only one molecule or atom. We must attempt to understand how such a small beginning can bring about such a large result as the development of a photographic grain or the discharge of a retinal receptor. In the photographic process the answer to this question seems to be that the ab-

sorption of a quantum of light causes the oxidation of a silver ion to an atom of metallic silver, which then serves as a catalytic center for the development of the entire grain. It is possible that a similar mechanism operates in a rod or a cone. The absorption of a quantum of light by a light-sensitive molecule in either structure might convert it into a biological catalyst, or an enzyme, which could then promote the further reactions that discharge the receptor cell. One wonders whether such a mechanism could possibly be rapid enough. A rod or a cone responds to light within a small fraction of a second; the mechanism would therefore have to complete its work within this small interval.

One of the strangest characteristics of the eye in dim light follows from some of these various phenomena. In focusing the eye is guided by its evaluation of the sharpness of the image on the retina. As the image deteriorates with the opening of the pupil in dim light, and as the retinal capacity to resolve the image falls with the shift from cones to rods, the ability to focus declines also. In very dim light the eye virtually ceases to adjust its focus at all. It has come to resemble a very cheap camera, a fixed-focus instrument.

In all that concerns its function, therefore, the eye is one device in bright light and another in dim. At low intensities all its resources are concentrated upon sensitivity, at whatever sacrifice of form; it is predominantly an instrument for seeing light, not pattern. In bright light all this changes. By narrowing the pupil, shifting from rods to cones, and other stratagems still to be described, the eye sacrifices light in order to achieve the utmost in pattern vision.

Images

In the course of evolution animals have used almost every known device

for forming or evaluating an image. There is one notable exception: no animal has yet developed an eye based upon the use of a concave mirror. An eye made like a pinhole camera, however, is found in Nautilus, a cephalopod mollusk related to the octopus and squid. The compound eye of insects and crabs forms an image which is an upright patchwork of responses of individual "eyes" or ommatidia, each of which records only a spot of light or shade. The eye of the tiny arthropod Copilia possesses a large and beautiful lens but only one light receptor attached to a thin strand of muscle. It is said that the muscle moves the receptor rapidly back and forth in the focal plane of the lens, scanning the image in much the same way as it is scanned by the light-sensitive tube of a television camera.

Each of these eyes, like the lens eye of vertebrates, represents some close compromise of advantages and limitations. The pinhole eye is in focus at all distances, yet to form clear images it must use a small hole admitting very little light. The compound eye works well at distances of a few millimeters, yet it is relatively coarse in pattern resolution. The vertebrate eye is a long-range, high-acuity instrument useless in the short distances at which the insect eye resolves the greatest detail.

These properties of the vertebrate eye are of course shared by the camera. The use of a lens to project an image, however, has created for both devices a special group of problems. All simple lenses are subject to serious errors in image formation: the lens aberrations.

Spherical aberration is found in all lenses bounded by spherical surfaces. The marginal portions of the lens bring rays of light to a shorter focus than the central region. The image of a point in space is therefore not a point, but a little "blur circle." The cost of a camera is largely determined by the extent to which this aberration is corrected by modifying the lens.

The human eye is astonishingly well corrected—often slightly overcorrected—for spherical aberration. This is accomplished in two ways. The cornea, which is the principal refracting surface of the eye, has a flatter curvature at its margin than at its center. This compensates in part for the tendency of a spherical surface to refract light more strongly at its margin. More important still, the lens is denser and hence refracts light more strongly at its core than in its outer layers.

A second major lens error, however, remains almost uncorrected in the human eye. This is chromatic aberration, or color error. All single lenses made of one material refract rays of short wavelength more strongly than those of longer wavelength, and so bring blue light to a shorter focus than red. The result is that the image of a point of white light is not a white point, but a blur circle fringed with color. Since this seriously disturbs the image, even the lenses of inexpensive cameras are corrected for chromatic aberration.

It has been known since the time of Isaac Newton, however, that the human eye has a large chromatic aberration. Its lens system seems to be entirely uncorrected for this defect. Indeed, living organisms are probably unable to manufacture two transparent materials of such widely different refraction and dispersion as the crown and flint glasses from which color-corrected lenses are constructed.

The large color error of the human eye could make serious difficulties for image vision. Actually the error is moderate between the red end of the spectrum and the blue-green, but it increases rapidly at shorter wavelengths: the blue, violet and ultraviolet. These latter parts of the spectrum present the most serious problem. It is a problem for both the eye and the camera, but one for which the eye must find a special solution.

The first device that opposes the color error of the human eye is the yellow lens. The human lens is not only a lens but a color filter. It passes what we ordinarily consider to be the visible spectrum, but sharply cuts off the far edge of the violet, in the region of wavelength 400 millimicrons. It is this action of the lens, and not any intrinsic lack of sensitivity of the rods and cones, that keeps us from seeing in the near ultraviolet. Indeed, persons who have lost their lenses in the operation for cataract and have had them replaced by clear glass lenses, have excellent vision in the ultraviolet. They are able to read an optician's chart from top to bottom in ultraviolet light which leaves ordinary people in complete darkness.

The lens therefore solves the problem of the near ultraviolet, the region of the spectrum in which the color error is greatest, simply by eliminating the region from human vision. This boon is distributed over one's lifetime, for the lens becomes a deeper yellow and makes more of the ordinary violet and blue invisible as one grows older. I have heard it said that for this reason aging artists tend to use less blue and violet in their paintings.

The lens filters out the ultraviolet for the eye as a whole. The remaining devices which counteract chromatic aberration are concentrated upon vision in bright light, upon cone vision. This is good economy, for the rods provide such a coarse-grained receptive surface that they would be unable in any case to evaluate a sharp image on the retina.

As one goes from dim to bright light, from rod to cone vision, the sensitivity of the eye shifts toward the red end of the spectrum. This phenomenon was described in 1825 by the Czech physiologist Johannes Purkinje. He had noticed that with the first light of dawn blue objects tend to look relatively bright compared with red, but that they come to look relatively dim as the morning advances. The basis of this change is a large difference in spectral sensitivity between rods and cones. Rods have their maximal sensitivity in the blue-green at about 500 millimicrons; the entire spectral sensitivity of the cones is transposed toward the red, the maximum lying in the yellow-green at about 562 millimicrons. The point of this difference for our present argument is that as one goes from dim light, in which pattern vision is poor in any case, to bright light, in which it becomes acute, the sensitivity of the eye moves away from the region of the spectrum in which the chromatic aberration is large toward the part of the spectrum in which it is least.

The color correction of the eye is completed by a third dispensation. Toward the center of the human retina there is a small, shallow depression called the fovea, which contains only cones. While the retina as a whole sweeps through a visual angle of some 240 degrees, the fovea subtends an angle of only about 1.7 degrees. The fovea is considerably smaller than the head of a pin, yet with this tiny patch of retina the eye accomplishes all its most detailed vision.

The fovea also includes the fixation point of the eye. To look directly at something is to turn one's eye so that its image falls upon the fovea. Beyond the boundary of the fovea rods appear, and they become more and more numerous as the distance from the fovea increases. The apparatus for vision in bright light is thus concentrated toward the center of the retina, that for dim light toward its periphery. In very dim light, too dim to excite the cones, the fovea is blind. One can see objects then only by looking at them slightly askance

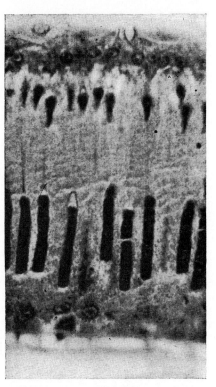

CONES of the catfish *Ameiurus* are pulled toward the surface of the retina (*top*) in bright light. The rods remain in a layer below the surface.

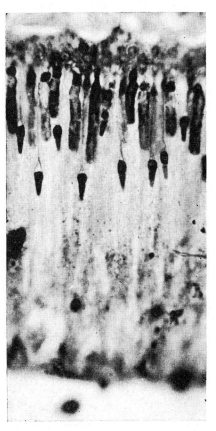

RODS advance and cones retreat in dim light. This retinal feature is not possessed by mammals. It is peculiar to some of the cold-blooded animals.

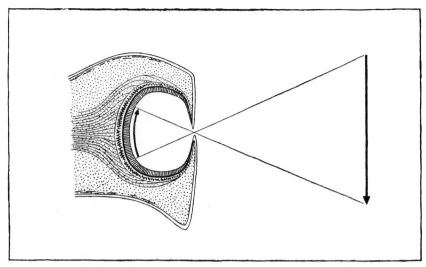

PINHOLE-CAMERA EYE is found in Nautilus, the spiral-shelled mollusk which is related to the octopus and the squid. This eye has the advantage of being in focus at all distances from the object that is viewed. It has the serious disadvantage, however, of admitting very little light to the retina.

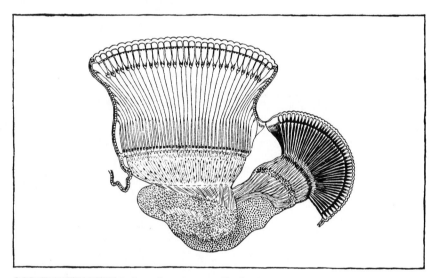

COMPOUND EYE is found in insects. Each element contributes only a small patch of light or shade to make up the whole mosaic image. This double compound eye is found in the mayfly *Chloeon*. The segment at the top provides detailed vision; the segment at the right, coarse, wide-angled vision.

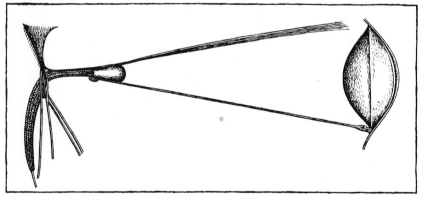

SCANNING EYE is found in the arthropod Copilia. It possesses a large lens (*right*) but only one receptor element (*left*). Attached to the receptor are the optic nerve and a strand of muscle. The latter is reported to move the receptor back and forth so that it scans the image formed by the lens.

to catch their images on areas rich in rods.

In man, apes and monkeys, alone of all known mammals, the fovea and the region of retina just around it is colored yellow. This area is called the yellow patch, or *macula lutea*. Its pigmentation lies as a yellow screen over the light receptors of the central retina, subtending a visual angle some five to 10 degrees in diameter.

Several years ago in our laboratory at Harvard University we measured the color transmission of this pigment in the living human eye by comparing the spectral sensitivities of cones in the yellow patch with those in a colorless peripheral area. The yellow pigment was also extracted from a small number of human maculae, and was found to be xanthophyll, a carotenoid pigment that occurs also in all green leaves. This pigment in the yellow patch takes up the absorption of light in the violet and blue regions of the spectrum just where absorption by the lens falls to very low values. In this way the yellow patch removes for the central retina the remaining regions of the spectrum for which the color error is high.

So the human eye, unable to correct its color error otherwise, throws away those portions of the spectrum that would make the most trouble. The yellow lens removes the near ultraviolet for the eye as a whole, the macular pigment eliminates most of the violet and blue for the central retina, and the shift from rods to cones displaces vision in bright light bodily toward the red. By these three devices the apparatus of most acute vision avoids the entire range of the spectrum in which the chromatic aberration is large.

Photography with Living Eyes

In 1876 Franz Boll of the University of Rome discovered in the rods of the frog retina a brilliant red pigment. This bleached in the light and was resynthesized in the dark, and so fulfilled the elementary requirements of a visual pigment. He called this substance visual red; later it was renamed visual purple or rhodopsin. This pigment marks the point of attack by light on the rods: the absorption of light by rhodopsin initiates the train of reactions that end in rod vision.

Boll had scarcely announced his discovery when Willy Kühne, professor of physiology at Heidelberg, took up the study of rhodopsin, and in one extraordinary year learned almost everything about it that was known until recently. In his first paper on retinal chemistry Kühne said: "Bound together with the pigment epithelium, the retina behaves not merely like a photographic plate, but like an entire photographic workshop, in which the workman continually renews the plate by laying on new light-sensitive material, while simultaneously erasing the old image."

Kühne saw at once that with this pigment which was bleached by light it might be possible to take a picture with the living eye. He set about devising methods for carrying out such a process, and succeeded after many discouraging failures. He called the process optography and its products optograms.

One of Kühne's early optograms was made as follows. An albino rabbit was fastened with its head facing a barred window. From this position the rabbit could see only a gray and clouded sky. The animal's head was covered for several minutes with a cloth to adapt its eyes to the dark, that is to let rhodopsin accumulate in its rods. Then the animal was exposed for three minutes to the light. It was immediately decapitated, the eye removed and cut open along the equator, and the rear half of the eyeball containing the retina laid in a solution of alum for fixation. The next day Kühne saw, printed upon the retina in bleached and unaltered rhodopsin, a picture of the window with the clear pattern of its bars.

I remember reading as a boy a detective story in which at one point the detective enters a dimly lighted room, on the floor of which a corpse is lying. Working carefully in the semidarkness, the detective raises one eyelid of the victim and snaps a picture of the open eye. Upon developing this in his darkroom he finds that he has an optogram of the last scene viewed by the victim, including of course an excellent likeness of the murderer. So far as I know Kühne's optograms mark the closest approach to fulfilling this legend.

The legend itself has nonetheless flourished for more than 60 years, and all of my readers have probably seen or heard some version of it. It began with Kühne's first intimation that the eye resembles a photographic workshop, even before he had succeeded in producing his first primitive optogram, and it spread rapidly over the entire world. In the paper that announces his first success in optography, Kühne refers to this story with some bitterness. He says: "I disregard all the journalistic potentialities of this subject, and willingly surrender it in advance to all the claims of fancy-free coroners on both sides of the ocean, for it certainly is not pleasant to deal with a serious problem in such company. Much that I could say about this had better be suppressed, and turned rather to the hope that no one will expect from me any corroboration of announcements that have not been authorized with my name."

Despite these admirable sentiments we find Kühne shortly afterward engaged in a curious adventure. In the nearby town of Bruchsal on November 16, 1880, a young man was beheaded by

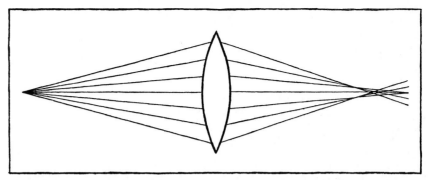

SPHERICAL ABERRATION occurs when light is refracted by a lens with spherical surfaces. The light which passes through the edge of the lens is brought to a shorter focus than that which passes through the center. The result of this is that the image of a point is not a point but a "blur circle."

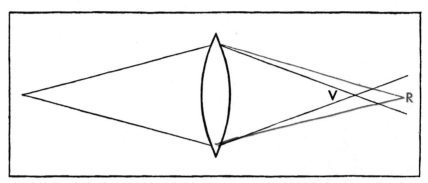

CHROMATIC ABERRATION occurs when light of various colors is refracted by a lens made of one material. The light of shorter wavelength is refracted more than that of longer wavelength, *i.e.*, violet is brought to a shorter focus than red. The image of a white point is a colored blur circle.

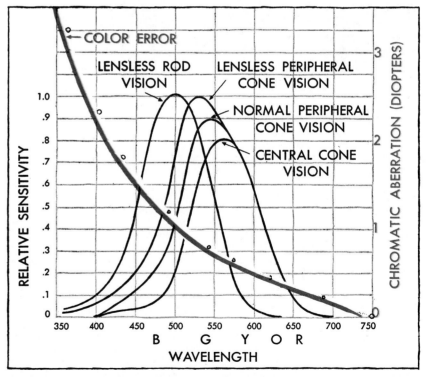

CHROMATIC ABERRATION of the human eye is corrected by various stratagems which withdraw the cones from the region of maximum aberration, *i.e.*, the shorter wavelengths. The horizontal coordinate of this diagram is wavelength in millimicrons; the colors are indicated by initial letters.

guillotine. Kühne had made arrangements to receive the corpse. He had prepared a dimly lighted room screened with red and yellow glass to keep any rhodopsin left in the eyes from bleaching further. Ten minutes after the knife had fallen he obtained the whole retina from the left eye, and had the satisfaction of seeing and showing to several colleagues a sharply demarcated optogram printed upon its surface. Kühne's drawing of it is reproduced at the bottom of page 262. To my knowledge it is the only human optogram on record.

Kühne went to great pains to determine what this optogram represented. He says: "A search for the object which served as source for this optogram remained fruitless, in spite of a thorough inventory of all the surroundings and reports from many witnesses. The delinquent had spent the night awake by the light of a tallow candle; he had slept

human eye as did the original subject of the picture.

How the human eye resolves colors is not known. Normal human color vision seems to be compounded of three kinds of responses; we therefore speak of it as trichromatic or three-color vision. The three kinds of response call for at least three kinds of cone differing from one another in their sensitivity to the various regions of the spectrum. We can only guess at what regulates these differences. The simplest assumption is that the human cones contain three different light-sensitive pigments, but this is still a matter of surmise.

There exist retinas, however, in which one can approach the problem of color vision more directly. The eyes of certain turtles and of certain birds such as chickens and pigeons contain a great predominance of cones. Since cones are the organs of vision in bright light as well as

In a paper published in 1907 the German ophthalmologist Siegfried Garten remarked that he was led by such retinal color filters to invent a system of color photography based upon the same principle. This might have been the first instance in which an eye had directly inspired a development in photography. Unfortunately, however, in 1906 the French chemist Louis Lumière, apparently without benefit of chicken retinas, had brought out his autochrome process for color photography based upon exactly this principle.

To make his autochrome plates Lumiere used suspensions of starch grains from rice, which he dyed red, green and blue. These were mixed in roughly equal proportions, and the mixture was strewn over the surface of an ordinary photographic plate. The granules were then squashed flat and the interstices were filled with particles of carbon. Each dyed granule served as a color filter for the patch of silver-bromide emulsion that lay just under it.

Just as the autochrome plate can accomplish color photography with a single light-sensitive substance, so the cones of the chicken retina should require no more than one light-sensitive pigment. We extracted such a pigment from the chicken retina in 1937. It is violet in color, and has therefore been named iodopsin from *ion*, the Greek word for violet. All three pigments of the colored oil globules have also been isolated and crystallized. Like the pigment of the human macula, they are all carotenoids: a greenish-yellow carotene; the golden mixture of xanthophylls found in chicken egg yolk; and red astaxanthin, the pigment of the boiled lobster.

Controversy thrives on ignorance, and we have had many years of disputation regarding the number of kinds of cone concerned in human color vision. Many investigators prefer three, some four, and at least one of my English colleagues seven. I myself incline toward three. It is a good number, and sufficient unto the day.

The appearance of three colors of oil globule in the cones of birds and turtles might be thought to provide strong support for trichromatic theories of color vision. The trouble is that these retinas do in fact contain a fourth class of globule which is colorless. Colorless globules have all the effect of a fourth color; there is no doubt that if we include them, bird and turtle retinas possess the basis for four-color vision.

RETINAL PHOTOGRAPH, or an optogram, was drawn in 1878 by the German investigator Willy Kühne. He had exposed the eye of a living rabbit to a barred window, killed the rabbit, removed its retina and fixed it in alum.

from four to five o'clock in the morning; and had read and written, first by candlelight until dawn, then by feeble daylight until eight o'clock. When he emerged in the open, the sun came out for an instant, according to a reliable observer, and the sky became somewhat brighter during the seven minutes prior to the bandaging of his eyes and his execution, which· followed immediately. The delinquent, however, raised his eyes only rarely."

Color

One of the triumphs of modern photography is its success in recording color. For this it is necessary not only to graft some system of color differentiation and rendition upon the photographic process; the finished product must then fulfill the very exacting requirement that it excite the same sensations of color in the

of color vision, these animals necessarily function only at high light intensities. They are permanently night-blind, due to a poverty or complete absence of rods. It is for this reason that chickens must roost at sundown.

In the cones of these animals we find a system of brilliantly colored oil globules, one in each cone. The globule is situated at the joint between the inner and outer segments of the cone, so that light must pass through it just before entering the light-sensitive element. The globules therefore lie in the cones in the position of little individual color filters.

One has only to remove the retina from a chicken or a turtle and spread it on the stage of a microscope to see that the globules are of three colors: red, orange and greenish yellow. It was suggested many years ago that they provide the basis of color differentiation in the animals that possess them.

Latent Images

Recent experiments have exposed a wholly unexpected parallel between vision and photography. Many years ago Kühne showed that rhodopsin can be extracted from the retinal rods into clear water solution. When such solutions are

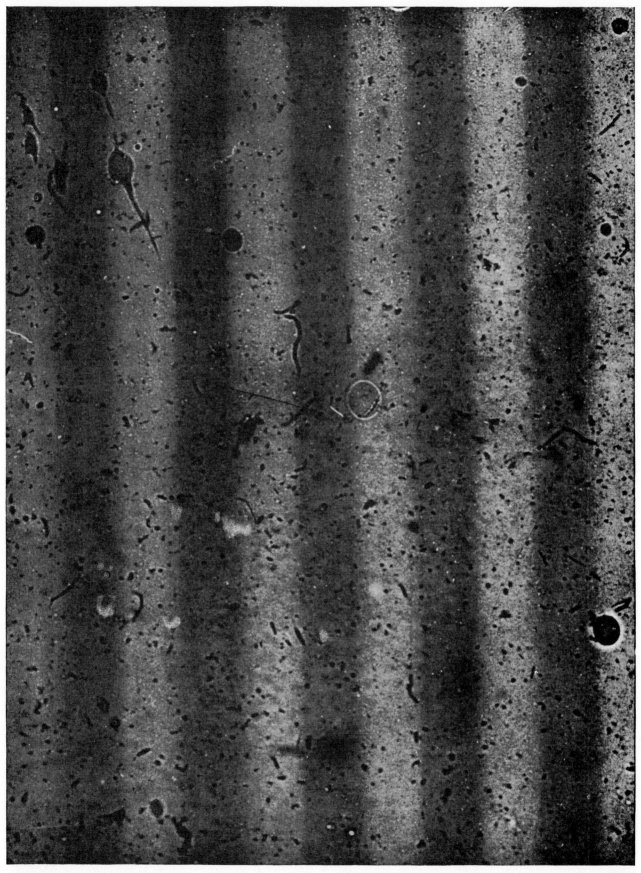

RHODOPSIN PHOTOGRAPH was made by the author and his associates Paul K. Brown and Oscar Starobin. Rhodopsin, the light-sensitive red pigment of rod vision, had been extracted from cattle retinas, mixed with gelatin and spread on celluloid. This was then dried and exposed to a pattern made up of black and white stripes. When the film was wetted in the dark with hydroxylamine the rhodopsin bleached in the same pattern.

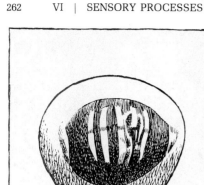

FROG OPTOGRAM showing a barred pattern was made by the German ophthalmologist Siegfried Garten. The retina is mounted on a rod.

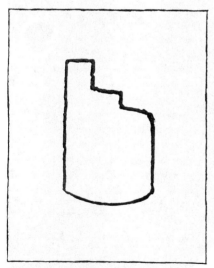

HUMAN OPTOGRAM was drawn by Kühne after he had removed the retina of a beheaded criminal. Kühne could not determine what it showed.

exposed to light, the rhodopsin bleaches just as it does in the retina.

It has been known for some time that the bleaching of rhodopsin in solution is not entirely accomplished by light. It is started by light, but then goes on in the dark for as long as an hour at room temperature. Bleaching is therefore a composite process. It is ushered in by a light reaction that converts rhodopsin to a highly unstable product; this then decomposes by ordinary chemical reactions—"dark" reactions in the sense that they do not require light.

Since great interest attaches to the initial unstable product of the light reaction, many attempts were made in our laboratory and at other laboratories to seize upon this substance and learn its properties. It has such a fleeting existence, however, that for some time nothing satisfactory was achieved.

In 1941, however, two English workers, E. E. Broda and C. F. Goodeve, succeeded in isolating the light reaction by irradiating rhodopsin solutions at about −73 degrees Celsius, roughly the temperature of dry ice. In such extreme cold, light reactions are unhindered, but ordinary dark processes cannot occur. Broda and Goodeve found that an exhaustive exposure of rhodopsin to light under these conditions produced only a very small change in its color, so small that though it could be measured one might not have been certain merely by looking at these solutions that any change had occurred at all. Yet the light reaction had been completed, and when such solutions were allowed to warm up to room temperature they bleached *in the dark*. We have recently repeated such experiments in our laboratory. With some differences which need not be discussed, the results were qualitatively as the English workers had described them.

These observations led us to re-examine certain early experiments of Kühne's. Kühne had found that if the retina of a frog or rabbit was thoroughly dried over sulfuric acid, it could be exposed even to brilliant sunlight for long periods without bleaching. Kühne concluded that dry rhodopsin is not affected by light, and this has been the common understanding of workers in the field of vision ever since.

It occurred to us, however, that dry rhodopsin, like extremely cold rhodopsin, might undergo the light reaction, though with such small change in color as to have escaped notice. To test this possibility we prepared films of rhodopsin in gelatin, which could be dried thoroughly and were of a quality that permitted making accurate measurements of their color transmission throughout the spectrum.

We found that when dry gelatin films of rhodopsin are exposed to light, the same change occurs as in very cold rhodopsin. The color is altered, but so slightly as easily to escape visual observation. In any case the change cannot be described as bleaching; if anything the color is a little intensified. Yet the light reaction is complete; if such exposed films are merely wetted with water, they bleach in the dark.

We have therefore two procedures—cooling to very low temperatures and removal of water—that clearly separate the light from the dark reactions in the bleaching of rhodopsin. Which of these reactions is responsible for stimulating rod vision? One cannot yet be certain, yet the response of the rods to light occurs so rapidly that only the light reaction seems fast enough to account for it.

What has been said, however, has a further consequence that brings it into direct relation with photography. Everyone knows that the photographic process also is divided into light and dark components. The result of exposing a film to light is usually invisible, a so-called "latent image." It is what later occurs in the darkroom, the dark reaction of development, that brings out the picture.

This now appears to be exactly what happens in vision. Here as in photography light produces an almost invisible result, a latent image, and this indeed is probably the process upon which retinal excitation depends. The visible loss of rhodopsin's color, its bleaching, is the result of subsequent dark reactions, of "development."

One can scarcely have notions like this without wanting to make a picture with a rhodopsin film; and we have been tempted into making one very crude rhodopsin photograph. Its subject is not exciting—only a row of black and white stripes—but we show it at the right for what interest it may have as the first such photograph. What is important is that it was made in typically photographic stages. The dry rhodopsin film was first exposed to light, producing a latent image. It was then developed in the dark by wetting. It then had to be fixed; and, though better ways are known, we fixed this photograph simply by redrying it. Since irradiated rhodopsin bleaches rather than blackens on development, the immediate result is a positive.

Photography with rhodopsin is only in its first crude stages, perhaps at the level that photography with silver bromide reached almost a century ago. I doubt that it has a future as a practical process. For us its primary interest is to pose certain problems in visual chemistry in a provocative form. It does, however, also add another chapter to the mingled histories of eye and camera.

Visual Pigments in Man

by W. A. H. Rushton
November 1962

One pigment records images in black and white when the light level is low; two other pigments distinguish between green and red when the level is raised. The blue-sensitive pigment is yet to be found

Everyone knows that the eye is a camera—more properly a television camera—that not only forms a picture but also transmits it in code via the optic nerves to the brain. In this article I shall not discuss how the lens forms an image on the retina; it does so in virtually the same way that the lens of a photographic camera forms an image on a piece of film, and the process needs no explanation here. Nor shall I treat of the encoding of nerve messages in the eye, still less of their decoding in the brain, because on those topics reliable information remains extremely scanty. I shall deal rather with the light-sensitive constituents of the retina of the eye—the "silver bromide" of vision—and their relation to the perception of light and color.

It is no use taking a snapshot with color film if the illumination is poor; the only hope of getting a picture is to use sensitive black-and-white film. If the light signal is only sufficient to silhouette outlines, it cannot provide additional information for the discrimination of color. Thus for a camera to be well equipped to extract the maximum information from any kind of scene it must be provided with sensitive black-and-white film for twilight and color film for full daylight. The eye is furnished with a retina having precisely this dual purpose. The saying goes, "In the twilight all cats are gray," but by day some cats are tortoise-shell.

We cannot slip off our daylight retina and wind on the twilight roll; the two films must remain in place all the time. They are not situated one behind the other but are mixed together, the grains of the two "emulsions" lying side by side. The color grains are too insensitive to contribute to the twilight picture, which is therefore formed entirely by the black-and-white grains; these, on the other hand, give only a rather faint picture, which in daylight is quite overpowered by the color grains.

Of course the actual grains in the retina are not inorganic crystals such as silver bromide but are the specialized body cells known as rods and cones. The rods and cones do, however, contain a photosensitive pigment that is laid down in a molecular array so well ordered as to be quasi-crystalline. The rods are the grains responsible for twilight vision, and their photosensitive pigment is rhodopsin, often called visual purple. The cones are the grains of daylight vision, and the photosensitive pigments they contain will be one of the topics of this article.

It was first noticed almost a century ago that if a frog's eye was dissected in dim light and if the excised retina was then brought out into diffuse daylight, the initial rose-pink color of the retina would gradually fade and become almost transparent. The fading of the retina was the more rapid the stronger the light to which it was exposed; hence the term "bleaching" is used to describe the chemical change brought about when light falls on the photosensitive constituents of the rods and cones. If a microscope is employed to observe the retina as it bleaches, one can see that the pink color resides only in the rods. The cones appear to possess no colored pigment at all.

The presence of a photosensitive pigment in the rods does not prove that this is the chemical that catches the light with which we see; the pigment may be doing something quite different. There is one rather strict test that must be satisfied if rhodopsin, the pink pigment, is the starting point of vision.

Since the pigment looks pink by transmitted light, it obviously absorbs green and transmits red (and some blue). With a spectrophotometer it is quite easy to measure the absorption of a rhodopsin solution at various wavelengths. When this is done, one obtains a bell-shaped curve with a peak close to a wavelength of 500 millimicrons, in the blue-green region of the spectrum. If rhodopsin catches the light we see in twilight, we should see best precisely those wavelengths that are best caught. In other words, the spectral absorption curve of rhodopsin should coincide with the spectral sensitivity curve of human twilight vision. Actual measurements of the twilight sensitivity of the eye at various wavelengths leave no doubt that rhodopsin is indeed the pigment that enables us to see at night [*see illustration on page 266*].

The eye is able to discriminate differences in brightness efficiently over a range in which the brightest light is a billion times more intense than the dimmest. Any instrument that can do that must have a variable "gain," or sensitivity-multiplying factor, and some means of adjusting the gain to match the level of signal to be discriminated. It is common experience that the eye adjusts its gain so smoothly that when the sun goes behind a cloud, the details of the scene appear just as distinct as before, and indeed we have so little clue to the eye's automatic compensation that when (as in photography) we want an estimate of the light intensity, it is safer to use a photoelectric meter. The change in gain of the eye is called visual adaptation.

It is plain that visual adaptation adjusts itself automatically to the prevailing brightness. To explain how this could occur Selig Hecht of Columbia University 40 years ago drew attention to the visual pigments and suggested that their color intensity seems to vary with the level of light. He hypothesized that in bright light these pigments are somewhat bleached and that in the dark they are regenerated from precursors stored in the eye or conveyed by the blood. Under steady illumination a balance will be struck between these two processes, and the equilibrium level of rhodopsin will be lower the stronger the bleaching light is. Hecht suggested that the level of rod adaptation is controlled by the level of the rhodopsin in the rods.

One difficulty in accepting this rather plausible suggestion is that until one can measure the actual rhodopsin level in the eye and correlate it with the corresponding state of visual adaptation, the idea remains speculative and very insecure. This indeed was the situation for some 30 years, but now it is possible to measure rhodopsin and cone pigments in the normal human eye by a procedure requiring only about seven seconds. As a result one can now follow the time course of bleaching and regeneration and test Hecht's suggestion.

Most people have at one time or another seen the eyes of a cat in the glare of an automobile headlight. The brilliant yellow-green eyes shining out of the darkness are a striking sight. The effect is caused simply by the reflection of light from the back of the cat's eye. What is important for our purpose is that these rays are reflected from behind the cat's retina and have therefore passed twice through the retina and the rhodopsin contained in the retina. This by itself would make the eye look pink, as it does in the case of the dissected frog retina. The cat, however, has a brilliant green backing to its retina and it is this backing that colors the returning light. To see the color of rhodopsin itself we need an animal whose retina has a white backing. If instead of a cat there were an alligator in the road, we should see the eye-shine colored pink by rhodopsin.

By using a photocell to analyze the returning light one can measure the rhodopsin no matter whether the eye is backed by green as it is in the cat, by white as in the alligator or even by black as in man. Regardless of its color, the reflectivity of the rear surface is unchanging, whereas the rhodopsin lying in front can be bleached away by strong light. It follows that if one measures not the color but the intensity of the returning light, one can find how much

of the light was absorbed by rhodopsin.

The illustration on this page shows schematically the instrument used to measure the bleaching of human eye pigments in my laboratory at the University of Cambridge. Light enters the eye through the upper half of the pupil, which has been dilated by a drug to allow more light to pass. It returns after reflection from the black rear surface, having twice traversed the retina. A small mirror intercepts the light from the lower half of the pupil and deflects it into a photomultiplier tube, which provides a measure of the light absorbed by the retinal pigments. If a powerful light is shined into the eye, the light bleaches away some of the pigment. This leaves less pigment to absorb the light traversing the retina; consequently the photocell output will be greater than before. The output can be returned to its former value by reduction of the measuring light. This is done by interposing a purple wedge in the beam of light entering the eye. The initial photocell output is restored when the amount of purple added by the wedge exactly matches the visual purple—the rhodopsin—removed by bleaching. The change in rhodopsin is thus measured simply by the change in wedge thickness that replaces it. The wedge scale is calibrated so that the reading is zero when all the rhodopsin is bleached away. Therefore the wedge setting for constant photocell output gives the rhodopsin density at that moment.

The intensity of the light reaching the photocell is only about a 20,000th of that falling on the eye, and the light striking the eye has to be so weak that it will not appreciably bleach the pigment it measures. Thus the equipment needs some rather careful compensations if measurements are to be reliable. We are not concerned here, however, with the technique of measurement but with the results in relation to the physiology of vision, and in particular with the question of the relation of rhodopsin level to visual adaptation.

The top illustration on page 268 shows the first measurements of this kind. They were made on my eye by F. W. Campbell at the University of Cambridge in 1955. The black dots show the wedge readings when a moderately bright bleaching light (one "bleaching unit") was applied to the dark-adapted eye. The pigment at first bleaches fast, then more slowly, and in five minutes it levels out, either because all the pigment is now bleached or because bleaching is just counterbalanced by the regeneration

METHOD OF MEASURING VISUAL PIGMENTS depends on the bleaching produced by light. Light enters the eye through a purple wedge, and the amount reflected is measured by a photocell. When the pigment rhodopsin, or "visual purple," is bleached from the retina, an equivalent amount of wedge is inserted in the light beam to keep the electric output the same after bleaching as before. The change in pigment is measured by the wedge displacement; a change of one unit means reflectivity of the eye has changed by a factor of 10.

EYE OF THE ALLIGATOR, which has a white reflecting layer behind the retina, illustrates how rhodopsin bleaches in the light and regenerates in the dark. The eye of the alligator above is light-adapted; the light of a stroboscopic-flash lamp, reflected from the white layer through the retina, is essentially colorless. The eyes of the alligator below are dark-adapted; the light reflected is red. The photographs were made at the New York Zoological Park with the kind assistance of Herndon G. Dowling and Stephen Spencook.

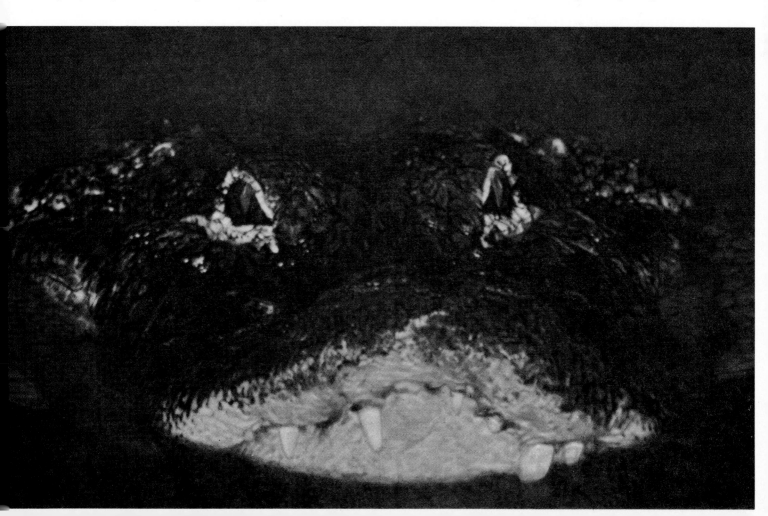

process. The latter is obviously the correct explanation, since by increasing the intensity of the bleaching light fivefold, further bleaching occurs and a lower level of equilibrium is achieved. In fact, a further increase of a hundredfold is needed to bleach the pigment entirely. The rate of pigment regeneration in the dark following total bleaching is plotted by the colored dots in the illustration. The regeneration follows an exponential curve and is about 90 per cent complete in 15 minutes.

Let us now examine Hecht's suggestion that it is the level of rhodopsin in the rods that defines the state of adaptation in twilight vision. But before doing so we must distinguish two quite different visual processes that are often designated by the word "adaptation." One process is exemplified by the quick changes in sensitivity that occur at night when the moon is fitfully obscured by passing clouds. This can be called field adaptation. When, on the other hand, we have got well adapted to bright light and then go into the dark—from sunlight into a theater, for instance—a different process occurs, which can be called adaptation of bleaching.

Now, field adaptation has nothing to do with the level of rhodopsin in the rods (or of visual pigments in the cones); the light intensity involved is only about a 100,000th of the bleaching unit referred to earlier, so that no appreciable bleaching can have occurred. Moreover, the time of adjustment to the new light level when the moon pops in and out of cloud is of the order of two seconds, rather than the 1,000 seconds required for the regeneration of rhodopsin. This rapid change of gain is in all likelihood produced entirely by the activity of nerve cells. Conceivably a feedback mechanism in the neural system maintains a constant signal strength by exchanging sensitivity for space-time discrimination. The adaptation of bleaching, on the other hand, turns out to be tightly linked to the level of rhodopsin in the rods.

The simplest way to examine this relation is to illuminate the eye with a powerful beam of light, a beam having an intensity of 100 bleaching units. After a minute or two all the rhodopsin will be bleached away and the course of pigment regeneration can be followed. The experiment is now repeated, but instead of measuring rhodopsin we determine the threshold of the eye by finding what is the weakest flash that can be detected at various intervals as the pigment regenerates. This is conveniently done by inserting a gray wedge to reduce the flash to threshold strength. The wedge displacement will now give the threshold directly on a logarithmic scale. A plot of this threshold yields the well-known dark-adaptation curve, shown in the bottom illustration on page 268.

As can be seen, the curve for the normal eye consists of two branches, the first of which corresponds to the log threshold of cones; the second, to the log threshold of rods. Only the rod threshold is related to rhodopsin, and it is a serious drawback that so much of this curve is hidden by the cone branch. Fortunately the complete rod curve can be obtained by using test subjects with a rare congenital abnormality in which rods are normal but cones entirely lack function. The dark-adaptation curve for such a subject is the black curve in the bottom illustration on page 268. It can be seen that the curve exactly follows the time course of the regeneration of rhodopsin, whether measured in the same subject or in a normal subject. It is therefore plain that the increase in light sensitivity of the rods waits precisely on the return of rhodopsin in the rods. What is far from plain, however, is what the increase in sensitivity waits for.

The change of sensitivity gain by nerve feedback in field adaptation is purposeful and efficient. The coupling of gain to the regeneration of rhodopsin in the adaptation of bleaching seems both pointless and clumsy. I have a far greater faith in nature, however, than in myself. I am sure that someone with deeper insight will eventually show that the deficiencies in dark adaptation, which to me seem unnecessary, are in fact inevitable.

The rapid and unconscious change of gain that makes absolute levels of light intensity hard to judge applies to cones as well as to rods, but in cones there is also the appreciation of color, which has its own adaptations. In judging brightness we estimate the brightness of parts with respect to the mean brightness of the whole. Thus the actual intensity of light reflected from black print in the noonday sun is far greater than that from white paper after sunset, yet the first looks black and the second white.

In color judgments wavelengths en-

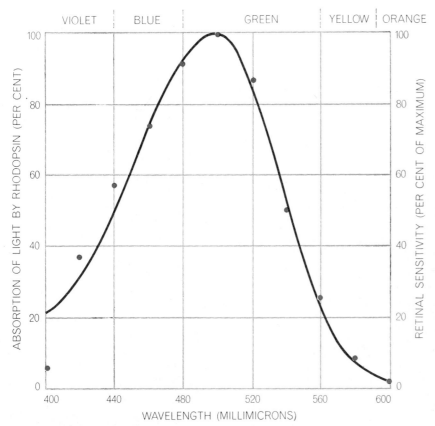

MEASUREMENTS OF RHODOPSIN show it to be the pigment responsible for twilight vision. The black curve indicates how a solution of rhodopsin, obtained from retinal rods, absorbs light of various wavelengths. Dots show sensitivity of the eye in twilight.

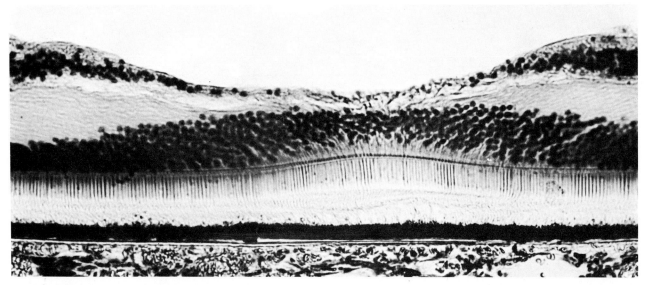

HUMAN RETINA, magnified about 370 diameters, is shown sectioned through the fovea, the tiny central region responsible for acute vision. The rods and cones, the photoreceptor cells containing the visual pigments, are the closely packed vertical stalks extending across the picture. Above the rods and cones are several layers composed chiefly of nerve cells that relay signals from the retina to the brain. At the fovea, which contains few if any rods, these layers are much thinned out to expose the light-sensitive part of the cones to incident light. This micrograph was made by C. M. H. Pedler of the Institute of Ophthalmology at University of London.

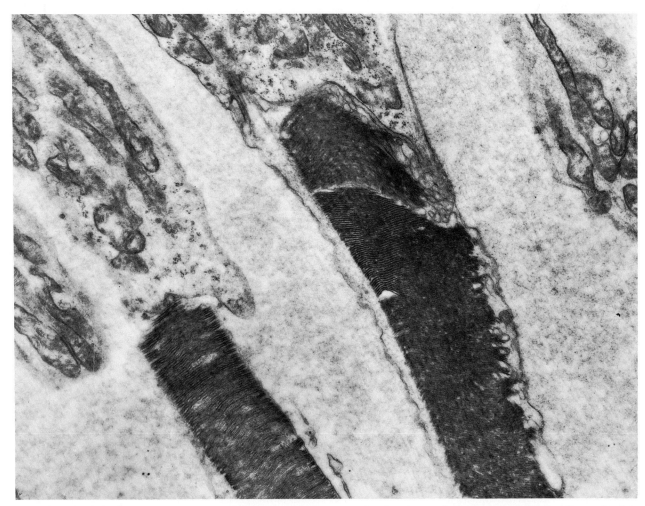

HUMAN ROD AND CONE are magnified about 20,000 diameters in this electron micrograph. The rod is on the left; the cone, on the right. The lamellated structures are the photoreceptor segments, believed to contain the visual pigments. These segments are joined at their base to the inner segments filled with mitochondria, which supply the cell with energy. The inner segments are positioned nearest the incoming light. The picture was made by Ben S. Fine of the Armed Forces Institute of Pathology in Washington.

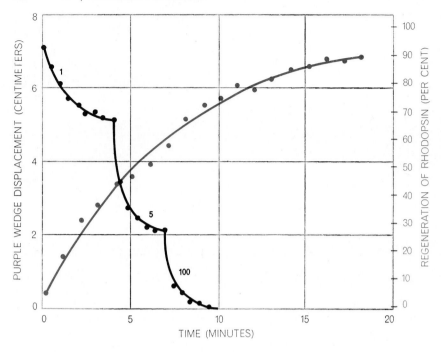

BLEACHING AND REGENERATION OF RHODOPSIN are shown in the two curves obtained by the method illustrated on page 264. The black curve records the time course of bleaching for a light of moderate intensity (*1*) and for lights five and 100 times brighter. In the dark, rhodopsin regenerates as shown by the colored curve. The measurements were made on the eye of the author by F. W. Campbell at the University of Cambridge.

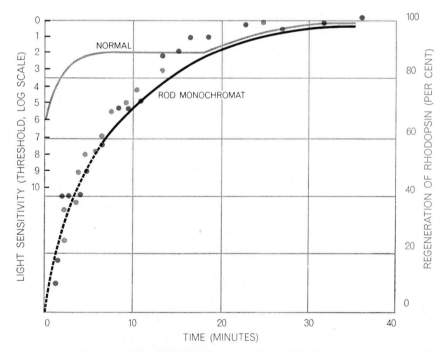

ROD AND CONE LIGHT SENSITIVITY can be distinguished by comparing a normal eye with that of a "rod monochromat," a person whose retinal cones do not function. The rhodopsin is fully bleached and the weakest detectable flash of light is measured. As the rhodopsin regenerates, the eye detects flashes that are weaker and weaker. The light sensitivity of the normal eye follows a discontinuous curve. The initial sensitivity increase is due to cones; the final increase is due to rods. In the rod monochromat the sensitivity rises more slowly but in a smooth curve. Independent measurements with the purple-wedge technique show that rhodopsin regeneration goes hand in hand with increased light sensitivity in the rod monochromat (*dark-colored dots*). In the normal eye, however, rhodopsin regeneration (*light-colored dots*) follows only the rod branch of the light-sensitivity curve.

ter in, and we estimate the color of parts of a scene in relation to the mean wavelength of the whole. The fact that our perceptions of color can be independent of wavelength to a surprising degree has been brought into great prominence by the striking demonstrations of Edwin H. Land of the Polaroid Corporation [see "Experiments in Color Vision," by Edwin H. Land; SCIENTIFIC AMERICAN Offprint 223]. Land has shown, for example, that two superimposed images of a scene, made on black-and-white film through different filters, will appear to contain a large range of color when one image is projected by red light and the other by white light. To say that the eye uses the average wavelength of such a red-and-white projection to judge the color of its parts is not meant to "explain" the Land phenomena, still less to suggest that no explanation is needed. It is merely a reminder that owing to some sort of adaptation—which Land has recently shown to be instantaneous —the eye is almost as bad at making absolute judgments of color as it is of brightness.

What the eye can do very well, however, is to make color *matches,* and these remain good even in the conditions of Land's projections. For instance, if monochromatic beams of red and green light are superimposed by projection on a screen, they can be made to match the yellow of a sodium lamp exactly, just by suitably adjusting the intensity of the red and of the green. If this red-green mixture is now substituted for the sodium yellow in one of Land's two-color projections, the colored picture resulting is exactly what it was before. Although many strange things appear in Land's pictures, one thing is clear: If red and green match yellow in one situation, they will match it in every other situation. Why, we may ask, are color matches stable under conditions where color appearance changes so greatly, and what colors can be matched by a mixture of others? A century ago James Clerk Maxwell showed that all colors could be matched by a suitable mixture of red, green and blue primaries, and indeed that any three colors could be chosen as primaries provided that no one of them could be matched by a mixture of the other two. The trichromaticity of color implies that the cones have three and only three ways of catching light. It seems reasonable, therefore, that there may be three and only three different cone pigments.

Since the rods have only one pigment,

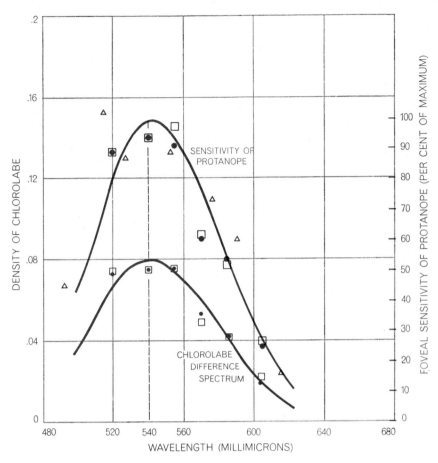

□ AFTER PARTIAL BLEACHING WITH
  BLUE-GREEN LIGHT

● AFTER PARTIAL BLEACHING WITH RED LIGHT

▣ AFTER BLEACHING WITH WHITE LIGHT

△ ACTION SPECTRUM

GREEN-CATCHING PIGMENT, called chlorolabe, can be measured in the eye of a protanope, the name given to a person who is red-blind. The pigment in the fovea of a protanope is partially bleached with red light and the change in reflectivity is measured at six wavelengths (*small dots*). The reflectivity change is then measured after partial bleaching with blue-green light (*small squares*). Since the protanope's fovea responds in the same way to both bleaches, it evidently contains only one pigment. The two sets of measurements define the difference spectrum of chlorolabe. Bleaching with white light, which shows total pigment present, shifts the foveal reflectivity upward at each wavelength (*larger squares and dots*). White-bleaching measurements coincide well with measurements of the protanope's sensitivity to white light (*colored curve*), made by F. H. G. Pitt of Imperial College. Still another way to measure bleaching, described in text, defines the "action spectrum" (*triangles*). It also supports the view that cones of the protanope contain one pigment.

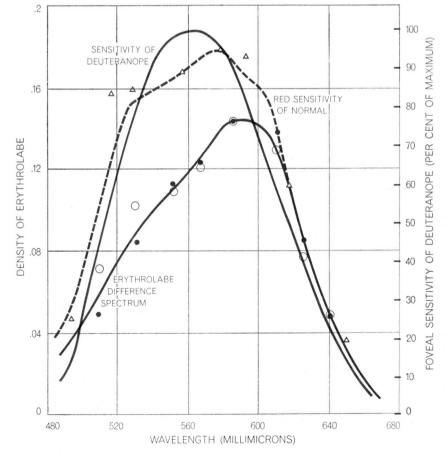

● AFTER BLEACHING WITH RED LIGHT

○ AFTER BLEACHING WITH BLUE-GREEN LIGHT

△ ACTION SPECTRUM

RED-CATCHING PIGMENT, erythrolabe, can be measured in the eye of a deuteranope, a person who is green-blind. The experiments are similar to those performed on the protanope. The black dots show the change in reflectivity of the fovea after partial bleaching with red light, the open circles after partial bleaching with blue-green light. The curve fitted to the two sets of circles is the difference spectrum of erythrolabe, the single visual pigment in the foveal cones of the deuteranope. The erythrolabe difference spectrum, however, does not coincide well with measurements by Pitt showing the deuteranope's sensitivity to white light (*colored curve*). This suggests that erythrolabe forms a colored photoproduct when bleached, which reduces foveal reflectivity below the values expected. The efficacy of bleaching as measured by the action spectrum comes closer to matching the deuteranope's visual sensitivity. It also agrees well with the sensitivity of the normal eye to red light alone (*broken curve*).

two lights of different wavelength composition will appear identical if they are scaled in intensity so that both are equally absorbed by rhodopsin. By the same token it should be possible to scale the intensity of two lights of different composition so that they will be absorbed equally by any one cone pigment. To that pigment the two lights would appear to have the same color. The scaling that will deceive the red pigment, however, will be detected by the green and blue pigments. It needs rather careful adjustment of two different color mixtures if they are to match; that is, if they are to deceive all three cone pigments at the same time. When this is achieved, the two inputs to the eye are in fact identical, and no one—not even Land—has the magic to show as different what all three cone pigments agree is the same.

Now we see why color matches are stable although color appearances change. Matches depend simply on the wavelength and intensity of light striking the three pigments and on the absorption spectra of these three chemicals. But appearances are subject to the whole complex of nervous interaction, not only between cone and cone in the retina but also between sensation and preconception in the mind. Let us therefore leave the rarefied atmosphere of color appearance and return to the solid ground of cone pigments.

If the cones contain three visual pigments, it should be possible to detect them and measure some of their properties by the method described for rhodopsin. To be sure, the human retina, like that of the frog, contains such a preponderance of rhodopsin that it is hard to measure anything else. Fortunately the fovea, that precious central square millimeter of the retina that we use for reading, contains no rods. It is also deficient in blue cones. Therefore if pigment-absorption measurements are confined to this tiny area, they should reveal the properties of just the red and green cones. One can simplify even further.

The common red-green color blindness is of two kinds: in one the color-blind individual is red-blind, in the other he is not. It turns out that the first individual lacks the red-sensitive pigment and that the second lacks the green-sensitive pigment. Therefore by measuring the fovea of the red-blind person, or protanope, we obtain information about the green-sensitive pigment only. The results of an analysis of this kind are

set forth in the top illustration on the preceding page.

It will be recollected that what we do is to adjust the wedge so that the output of the photocell is the same after bleaching as it was before. For the protanope experiment we use a gray wedge and express this displacement in terms of the corresponding change in optical density of the cone pigment. Since light passes through the pigment twice, once on entering and once on returning, measurements indicate a "double density" of pigment. Such measurements, made in lights of six wavelengths, are shown by the squares and dots in the illustration. The change in the reflectivity of the fovea, caused by bleaching, is maximal when measured with light that has a wavelength of 540 millimicrons and diminishes on each side. The small squares represent change in the reflectivity after bleaching with blue-green light; the small dots, after bleaching with very bright red light. These changes define

a curve that we call a difference spectrum. The fact that both curves coincide means that there is only one pigment present. If there had been a mixture, the more red-sensitive of the two would have shown a greater change after bleaching with red light; the other, after bleaching with blue-green light. A second series of measurements made after bleaching with a bright white light shows the total pigment present.

To discover whether or not this photosensitive pigment is indeed the basis of cone vision in the protanope we apply the test discussed earlier for rhodopsin. We simply ask: Does the spectral absorption coincide with the spectral sensitivity? The colored curve in the top illustration on the facing page shows how the cone sensitivity of the protanope does in fact correspond to the absorption measurements. We may conclude, therefore, that the protanope in daylight sees by this pigment, which is called chlorolabe, after the Greek words for "green-catching."

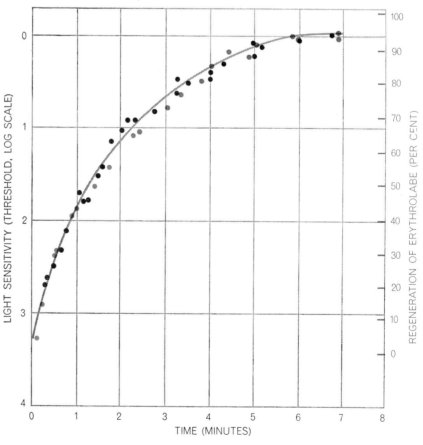

LIGHT SENSITIVITY AND REGENERATION OF ERYTHROLABE follow the same time course. The black dots show how the deuteranope's fovea becomes increasingly sensitive to brief flashes of light as the eye becomes dark-adapted. The colored dots are separate measurements made after the erythrolabe has been fully bleached. They show that the foveal pigment regenerates in seven minutes. The coincidence of the two sets of measurements implies that cones, like rods, have dark adaptation linked to pigment regeneration.

The other type of color-blind person, lacking chlorolabe, is known as a deuteranope. That he also has a single foveal pigment is established by the similar change in foveal reflectivity produced by either deep red light or blue-green light, as shown in the bottom illustration on page 269. It is plain that this pigment, which is called erythrolabe ("red-catching"), has a difference spectrum extending much further into the red than chlorolabe. If erythrolabe is the pigment that catches the light by which the deuteranope sees, he ought to be able to see further into the red end of the spectrum than the protanope can. This indeed is the case, but it is also apparent that the difference spectrum of chlorolabe does not coincide too well with the spectral sensitivity of the deuteranope, shown by the colored curve. Therefore the matter needs to be studied further.

If erythrolabe is the cone pigment of the deuteranope, lights of various wavelengths adjusted in intensity so that each appears equally bright to the deuteranope ought also to prove equivalent in the rate at which they bleach erythrolabe. Measurements of bleaching efficacy for lights of various wavelengths produce an "action spectrum," shown by triangles in the two illustrations on page 269. It can be seen in the bottom illustration that the action spectrum coincides reasonably well with the sensitivity of the deuteranope and also with the sensitivity of the red mechanism in the normal eye, shown by the broken curve. Thus there is fair agreement between sensitivity and bleaching power, and erythrolabe has a strong claim as the visual pigment of the deuteranope and of the normal red color mechanism.

It is also possible to measure the time required for the erythrolabe in the deuteranope's fovea to regenerate after bleaching. The curve in the illustration at the left resembles that for rhodopsin but rises about four times faster. It can be seen that the light sensitivity of the deuteranope, also plotted, increases precisely in step with the return of erythrolabe. So we are reasonably confident that erythrolabe is the pigment with which the deuteranope catches light.

Now we are in a position to prove that the normal fovea contains both green-sensitive chlorolabe and red-sensitive erythrolabe. The pertinent measurements are shown in the illustration below. The black dots show the bleaching produced by deep red light, and it is evident that they define a curve identical to the difference spectrum of erythrolabe, as measured in the deuteranope.

If in the deuteranope we changed the bleaching light from red to blue-green, no alteration would occur, since both lights bleach the deuteranope's single pigment equally. But when blue-green light is used to bleach the normal eye, one discovers that additional bleaching takes place, which cannot be attributed to erythrolabe. This additional bleaching is shown by the open circles in the illustration. Since no change in erythrolabe can contribute to this increment, it must represent the pure change in a second pigment in the normal eye. To see if this pigment is chlorolabe we draw on the same chart the difference spectrum of chlorolabe, as measured in the protanope, and we find that it closely follows the open circles. Thus the normal fovea is seen to contain both erythrolabe and chlorolabe.

A person with normal color vision can distinguish colors in the red-orange-yellow-green range of the spectrum because all of these colors affect the pigments erythrolabe and chlorolabe in different proportions. In this range protanopes and deuteranopes have only the one, or only the other, of these pigments; hence they have no more means of distinguishing these colors by day than a person with normal vision has by night. They can see only one color because they have only one pigment.

The reader will ask: What about the blues? Is there a "blue-catcher"—a cyanolabe—to complete the triad of cone pigments? I think there is, but it is much harder than the others to measure and there is not much at present to be said about it.

Practically all the ideas in this article have been entertained long ago by acute investigators; they have also often been disputed. What the measurement of pigments in man has done is to bring some degree of exactness and security to ideas that were enticing but speculative. The precision of measurement, however, lies not in the investigator who turns the knobs but in the subjects who sit with clamped head and fixed eye gazing steadfastly 20 minutes at a time through flashing and gloom. These are my students, some normal, some color-blind— volunteers from the classes in physiology in the University of Cambridge.

TWO PIGMENTS IN NORMAL CONES are demonstrated by bleaching the eye with deep red light, then with blue-green light and recording the change in reflectivity of the fovea at eight wavelengths. Bleaching with red light gives the results shown by black dots and coincides with the erythrolabe difference spectrum (*broken curve*) found in the deuteranope (*see bottom illustration, page 269*). When the bleaching light is blue-green, the reflectivity of the fovea increases beyond that observed when the bleach is red. The additional reflectivity is shown by open circles and conforms to the difference spectrum of chlorolabe (*solid curve*), as measured in the protanope (*see top illustration, page 269*).

25

The Visual Cortex
of the Brain

by David H. Hubel
November 1963

*A start toward understanding how it analyzes images
on the retina can be made through studies of the
responses that individual cells in the visual system of
the cat give to varying patterns of light*

An image of the outside world striking the retina of the eye activates a most intricate process that results in vision: the transformation of the retinal image into a perception. The transformation occurs partly in the retina but mostly in the brain, and it is, as one can recognize instantly by considering how modest in comparison is the achievement of a camera, a task of impressive magnitude.

The process begins with the responses of some 130 million light-sensitive receptor cells in each retina. From these cells messages are transmitted to other retinal cells and then sent on to the brain, where they must be analyzed and interpreted. To get an idea of the magnitude of the task, think what is involved in watching a moving animal, such as a horse. At a glance one takes in its size, form, color and rate of movement. From tiny differences in the two retinal images there results a three-dimensional picture. Somehow the brain manages to compare this picture with previous impressions; recognition occurs and then any appropriate action can be taken.

The organization of the visual system —a large, intricately connected population of nerve cells in the retina and brain —is still poorly understood. In recent years, however, various studies have begun to reveal something of the arrangement and function of these cells. A decade ago Stephen W. Kuffler, working with cats at the Johns Hopkins Hospital, discovered that some analysis of visual patterns takes place outside the brain, in the nerve cells of the retina. My colleague Torsten N. Wiesel and I at the Harvard Medical School, exploring the first stages of the processing that occurs in the brain of the cat, have mapped the visual pathway a little further: to what appears to be the sixth step from the retina to the cortex of the cerebrum. This kind of work falls far short of providing a full understanding of vision, but it does convey some idea of the mechanisms and circuitry of the visual system.

In broad outline the visual pathway is clearly defined [*see bottom illustration on opposite page*]. From the retina of each eye visual messages travel along the optic nerve, which consists of about a million nerve fibers. At the junction known as the chiasm about half of the nerves cross over into opposite hemispheres of the brain, the other nerves remaining on the same side. The optic nerve fibers lead to the first way stations in the brain: a pair of cell clusters called the lateral geniculate bodies. From here new fibers course back through the brain to the visual area of the cerebral cortex. It is convenient, although admittedly a gross oversimplification, to think of the pathway from retina to cortex as consisting of six types of nerve cells, of which three are in the retina, one is in the geniculate body and two are in the cortex.

Nerve cells, or neurons, transmit messages in the form of brief electrochemical impulses. These travel along the outer membrane of the cell, notably along the membrane of its long principal fiber, the axon. It is possible to obtain an electrical record of impulses of a single nerve cell by placing a fine electrode near the cell body or one of its fibers. Such measurements have shown that impulses travel along the nerves at velocities of between half a meter and 100 meters per second. The impulses in a given fiber all have about the same amplitude; the strength of the stimuli that give rise to them is reflected not in amplitude but in frequency.

At its terminus the fiber of a nerve cell makes contact with another nerve cell (or with a muscle cell or gland cell), forming the junction called the synapse. At most synapses an impulse on reaching the end of a fiber causes the release of a small amount of a specific substance, which diffuses outward to the membrane of the next cell. There the substance either excites the cell or inhibits it. In excitation the substance acts to bring the cell into a state in which it is more likely to "fire"; in inhibition the substance acts to prevent firing. For most synapses the substances that act as transmitters are unknown. Moreover, there is no sure way to determine from microscopic appearances alone whether a synapse is excitatory or inhibitory.

It is at the synapses that the modification and analysis of nerve messages take place. The kind of analysis depends partly on the nature of the synapse: on how many nerve fibers converge on a single cell and on how the excitatory and inhibitory endings distribute themselves. In most parts of the nervous system the anatomy is too intricate to reveal much about function. One way to circumvent this difficulty is to record impulses with microelectrodes in anesthetized animals, first from the fibers coming into a structure of neurons and then from the neurons themselves, or from the fibers they send onward. Comparison of the behavior of incoming and outgoing fibers provides a basis for learning what the structure does. Through such exploration of the different parts of the brain concerned with vision one can hope to build up some idea of how the entire visual system works.

That is what Wiesel and I have undertaken, mainly through studies of the visual system of the cat. In our experiments the anesthetized animal faces a wide screen 1.5 meters away, and we shine various patterns of white light on the screen with a projector. Simultaneously we penetrate the visual portion of the cortex with microelectrodes. In that way we can record the responses of individual cells to the light patterns. Sometimes it takes many hours to find the region of the retina with which a particular visual cell is linked and to work out the optimum stimuli for that cell. The reader should bear in mind the relation between each visual cell—no matter how far along the visual pathway it may be—and the retina. It requires an image on the retina to evoke a meaningful response in any visual cell, however indirect and complex the linkage may be.

The retina is a complicated structure, in both its anatomy and its physiology, and the description I shall give is highly simplified. Light coming through the lens of the eye falls on the mosaic of receptor cells in the retina. The receptor cells do not send impulses directly through the optic nerve but instead connect with a set of retinal cells called bipolar cells. These in turn connect with retinal ganglion cells, and it is the latter set of cells, the third in the visual pathway, that sends its fibers—the optic nerve fibers—to the brain.

This series of cells and synapses is no simple bucket brigade for impulses: a receptor may send nerve endings to more than one bipolar cell, and several receptors may converge on one bipolar cell. The same holds for the synapses between the bipolar cells and the retinal ganglion cells. Stimulating a single receptor by light might therefore be expected to have an influence on many bipolar or ganglion cells; conversely, it should be possible to influence one bipolar or retinal ganglion cell from a number of receptors and hence from a substantial area of the retina.

The area of receptor mosaic in the retina feeding into a single visual cell is called the receptive field of the cell. This term is applied to any cell in the visual system to refer to the area of retina with which the cell is connected—the retinal area that on stimulation produces a response from the cell.

Any of the synapses with a particular cell may be excitatory or inhibitory, so that stimulation of a particular point on the retina may either increase or decrease the cell's firing rate. Moreover, a single cell may receive several excitatory and inhibitory impulses at once, with the result that it will respond according to the net effect of these inputs. In considering the behavior of a single cell an observer should remember that it is just one of a huge popu-

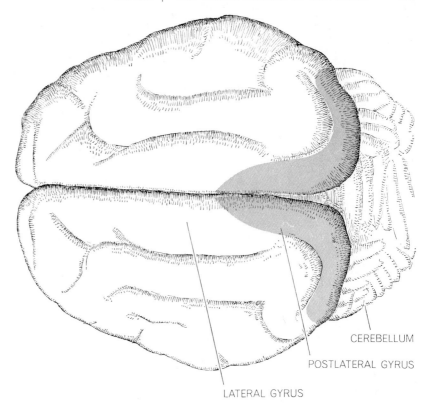

CEREBELLUM

POSTLATERAL GYRUS

LATERAL GYRUS

CORTEX OF CAT'S BRAIN is depicted as it would be seen from the top. The colored region indicates the cortical area that deals at least in a preliminary way with vision.

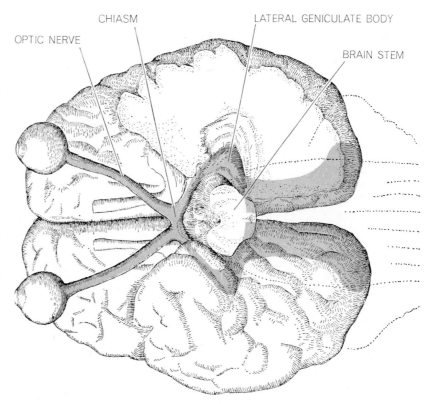

CHIASM

LATERAL GENICULATE BODY

OPTIC NERVE

BRAIN STEM

VISUAL SYSTEM appears in this representation of the human brain as viewed from below. Visual pathway from retinas to cortex via the lateral geniculate body is shown in color.

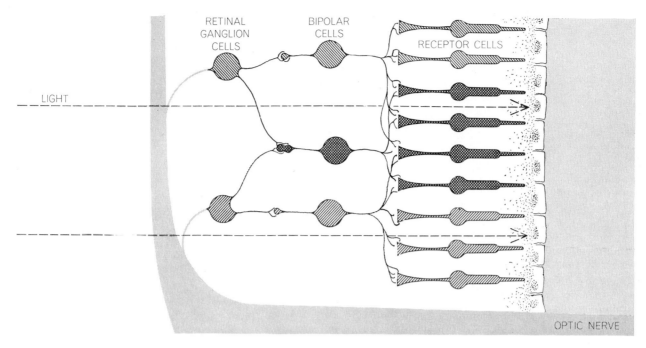

STRUCTURE OF RETINA is depicted schematically. Images fall on the receptor cells, of which there are about 130 million in each retina. Some analysis of an image occurs as the receptors transmit messages to the retinal ganglion cells via the bipolar cells. A group of receptors funnels into a particular ganglion cell, as indicated by the shading; that group forms the ganglion cell's receptive field. Inasmuch as the fields of several ganglion cells overlap, one receptor may send messages to several ganglion cells.

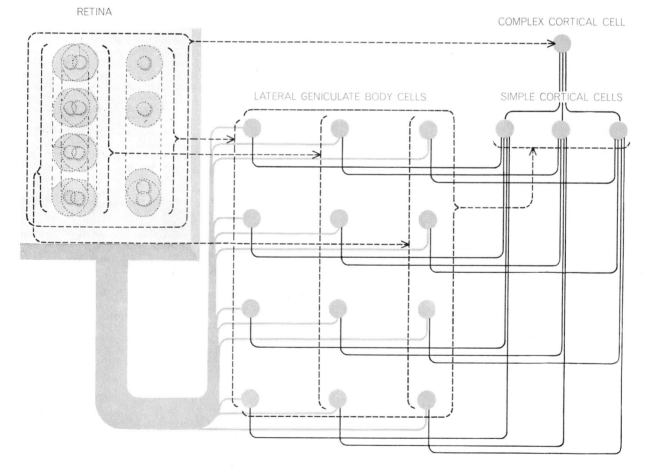

VISUAL PROCESSING BY BRAIN begins in the lateral geniculate body, which continues the analysis made by retinal cells. In the cortex "simple" cells respond strongly to line stimuli, provided that the position and orientation of the line are suitable for a particular cell. "Complex" cells respond well to line stimuli, but the position of the line is not critical and the cell continues to respond even if a properly oriented stimulus is moved, as long as it remains in the cell's receptive field. Broken lines indicate how receptive fields of all these cells overlap on the retina; solid lines, how several cells at one stage affect a single cell at the next stage.

lation of cells: a stimulus that excites one cell will undoubtedly excite many others, meanwhile inhibiting yet another array of cells and leaving others entirely unaffected.

For many years it has been known that retinal ganglion cells fire at a fairly steady rate even in the absence of any stimulation. Kuffler was the first to observe how the retinal ganglion cells of mammals are influenced by small spots of light. He found that the resting discharges of a cell were intensified or diminished by light in a small and more or less circular region of the retina. That region was of course the cell's receptive field. Depending on where in the field a spot of light fell, either of two responses could be produced. One was an "on" response, in which the cell's firing rate increased under the stimulus of light. The other was an "off" response, in which the stimulus of light decreased the cell's firing rate. Moreover, turning the light off usually evoked a burst of impulses from the cell. Kuffler called the retinal regions from which these responses could be evoked "on" regions and "off" regions.

On mapping the receptive fields of a large number of retinal ganglion cells into "on" and "off" regions, Kuffler discovered that there were two distinct cell types. In one the receptive field consisted of a small circular "on" area and a surrounding zone that gave "off" responses. Kuffler termed this an "on"-center cell. The second type, which he called "off"-center, had just the reverse form of field—an "off" center and an "on" periphery [*see top illustration on this page*]. For a given cell the effects of light varied markedly according to the place in which the light struck the receptive field. Two spots of light shone on separate parts of an "on" area produced a more vigorous "on" response than either spot alone, whereas if one spot was shone on an "on" area and the other on an "off" area, the two effects tended to neutralize each other, resulting in a very weak "on" or "off" response. In an "on"-center cell, illuminating the entire central "on" region evoked a maximum response; a smaller or larger spot of light was less effective.

Lighting up the whole retina diffusely, even though it may affect every receptor in the retina, does not affect a retinal ganglion cell nearly so strongly as a small circular spot of exactly the right size placed so as to cover precisely the receptive-field center. The main concern of these cells seems to be the contrast in illumination between one retinal region and surrounding regions.

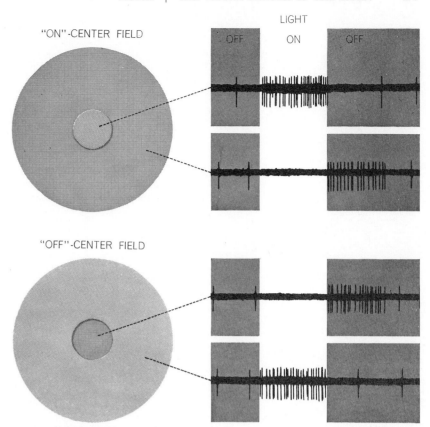

CONCENTRIC FIELDS are characteristic of retinal ganglion cells and of geniculate cells. At top an oscilloscope recording shows strong firing by an "on"-center type of cell when a spot of light strikes the field center; if the spot hits an "off" area, the firing is suppressed until the light goes off. At bottom are responses of another cell of the "off"-center type.

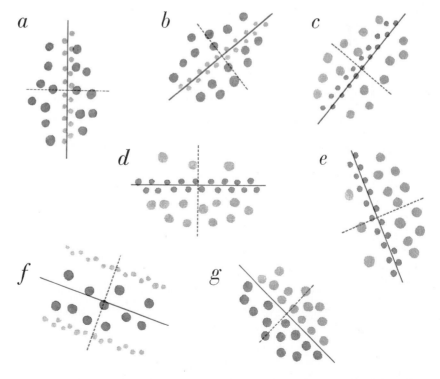

SIMPLE CORTICAL CELLS have receptive fields of various types. In all of them the "on" and "off" areas, represented by colored and gray dots respectively, are separated by straight boundaries. Orientations of fields vary, as indicated particularly at *a* and *b*. In the cat's visual system such fields are generally one millimeter or less in diameter.

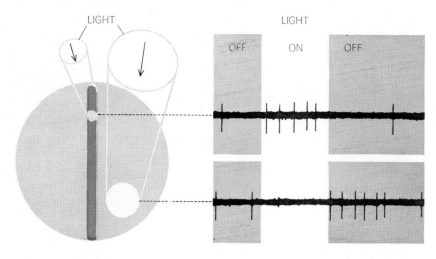

RESPONSE IS WEAK when a circular spot of light is shone on receptive field of a simple cortical cell. Such spots get a vigorous response from retinal and geniculate cells. This cell has a receptive field of type shown at *a* in bottom illustration on preceding page.

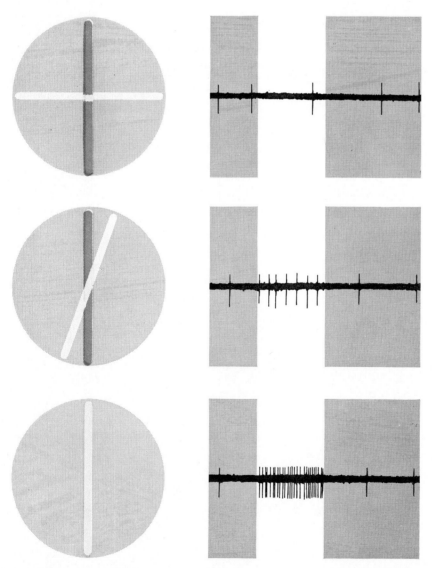

IMPORTANCE OF ORIENTATION to simple cortical cells is indicated by varying responses to a slit of light from a cell preferring a vertical orientation. Horizontal slit *(top)* produces no response, slight tilt a weak response, vertical slit a strong response.

Retinal ganglion cells differ greatly in the size of their receptive-field centers. Cells near the fovea (the part of the retina serving the center of gaze) are specialized for precise discrimination; in the monkey the field centers of these cells may be about the same size as a single cone—an area subtending a few minutes of arc at the cornea. On the other hand, some cells far out in the retinal periphery have field centers up to a millimeter or so in diameter. (In man one millimeter of retina corresponds to an arc of about three degrees in the 180-degree visual field.) Cells with such large receptive-field centers are probably specialized for work in very dim light, since they can sum up messages from a large number of receptors.

Given this knowledge of the kind of visual information brought to the brain by the optic nerve, our first problem was to learn how the messages were handled at the first central way station, the lateral geniculate body. Compared with the retina, the geniculate body is a relatively simple structure. In a sense there is only one synapse involved, since the incoming optic nerve fibers end in cells that send their fibers directly to the visual cortex. Yet in the cat many optic nerve fibers converge on each geniculate cell, and it is reasonable to expect some change in the visual messages from the optic nerve to the geniculate cells.

When we came to study the geniculate body, we found that the cells have many of the characteristics Kuffler described for retinal ganglion cells. Each geniculate cell is driven from a circumscribed retinal region (the receptive field) and has either an "on" center or an "off" center, with an opposing periphery. There are, however, differences between geniculate cells and retinal ganglion cells, the most important of which is the greatly enhanced capacity of the periphery of a geniculate cell's receptive field to cancel the effects of the center. This means that the lateral geniculate cells must be even more specialized than retinal ganglion cells in responding to spatial differences in retinal illumination rather than to the illumination itself. The lateral geniculate body, in short, has the function of increasing the disparity—already present in retinal ganglion cells—between responses to a small, centered spot and to diffuse light.

In contrast to the comparatively simple lateral geniculate body, the cerebral cortex is a structure of stupendous complexity. The cells of this great plate of

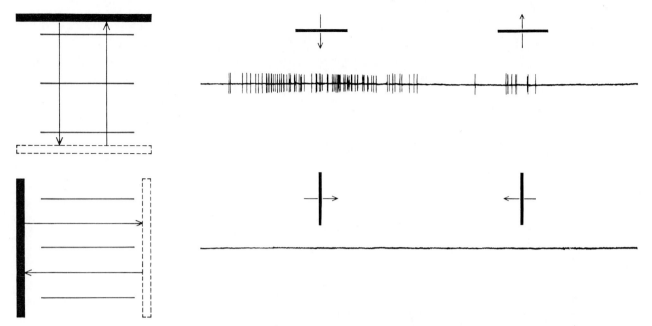

COMPLEX CORTICAL CELL responded vigorously to slow downward movement of a dark, horizontal bar. Upward movement of bar produced a weak response and horizontal movement of a vertical bar produced no response. For other shapes, orientations and movements there are other complex cells showing maximum response. Such cells may figure in perception of form and movement.

gray matter—a structure that would be about 20 square feet in area and a tenth of an inch thick if flattened out—are arranged in a number of more or less distinct layers. The millions of fibers that come in from the lateral geniculate body connect with cortical cells in the layer that is fourth from the top. From here the information is sooner or later disseminated to all layers of the cortex by rich interconnections between them. Many of the cells, particularly those of the third and fifth layers, send their fibers out of the cortex, projecting to centers deep in the brain or passing over to nearby cortical areas for further processing of the visual messages. Our problem was to learn how the information the visual cortex sends out differs from what it takes in.

Most connections between cortical cells are in a direction perpendicular to the surface; side-to-side connections are generally quite short. One might therefore predict that impulses arriving at a particular area of the cortex would exert their effects quite locally. Moreover, the retinas project to the visual cortex (via the lateral geniculate body) in a systematic topologic manner; that is, a given area of cortex gets its input ultimately from a circumscribed area of retina. These two observations suggest that a given cortical cell should have a small receptive field; it should be influenced from a circumscribed retinal region only, just as a geniculate or retinal ganglion cell is. Beyond this the anatomy provides no hint of what the cortex does

with the information it receives about an image on the retina.

In the face of the anatomical complexity of the cortex, it would have been surprising if the cells had proved to have the concentric receptive fields characteristic of cells in the retina and the lateral geniculate body. Indeed, in the cat we have observed no cortical cells with concentric receptive fields; instead there are many different cell types, with fields markedly different from anything seen in the retinal and geniculate cells.

The many varieties of cortical cells may, however, be classified by function into two large groups. One we have called "simple"; the function of these cells is to respond to line stimuli—such shapes as slits, which we define as light lines on a dark background; dark bars (dark lines on a light background), and edges (straight-line boundaries between light and dark regions). Whether or not a given cell responds depends on the orientation of the shape and its position on the cell's receptive field. A bar shone vertically on the screen may activate a given cell, whereas the same cell will fail to respond (but others will respond) if the bar is displaced to one side or moved appreciably out of the vertical. The second group of cortical cells we have called "complex"; they too respond best to bars, slits or edges, provided that, as with simple cells, the shape is suitably oriented for the particular cell under observation. Complex cells, how-

ever, are not so discriminating as to the exact position of the stimulus, provided that it is properly oriented. Moreover, unlike simple cells, they respond with sustained firing to moving lines.

From the preference of simple and complex cells for specific orientation of light stimuli, it follows that there must be a multiplicity of cell types to handle the great number of possible positions and orientations. Wiesel and I have found a large variety of cortical cell responses, even though the number of individual cells we have studied runs only into the hundreds compared with the millions that exist. Among simple cells, the retinal region over which a cell can be influenced—the receptive field—is, like the fields of retinal and geniculate cells, divided into "on" and "off" areas. In simple cells, however, these areas are far from being circularly symmetrical. In a typical example the receptive field consists of a very long and narrow "on" area, which is adjoined on each side by larger "off" regions. The magnitude of an "on" response depends, as with retinal and geniculate cells, on how much either type of region is covered by the stimulating light. A long, narrow slit that just fills the elongated "on" region produces a powerful "on" response. Stimulation with the slit in a different orientation produces a much weaker effect, because the slit is now no longer illuminating all the "on" region but instead includes some of the antagonistic "off" region. A slit at right angles to the optimum orientation for a

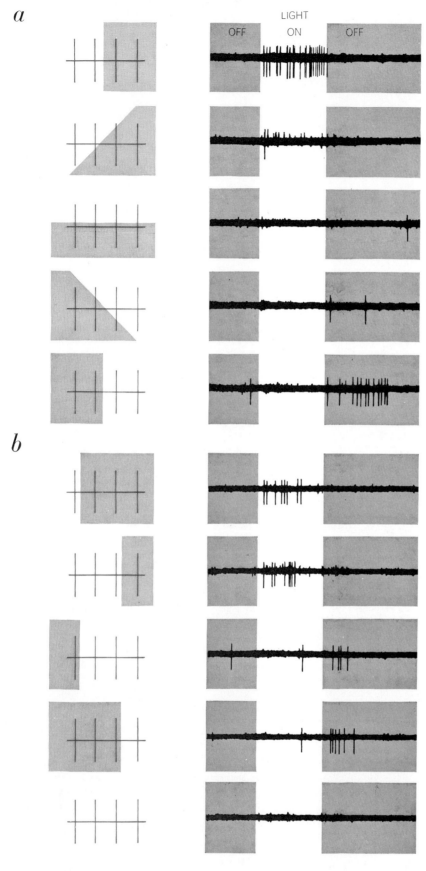

SINGLE COMPLEX CELL showed varying responses to an edge projected on the cell's receptive field in the retina. In group *a* the stimulus was presented in differing orientations. In group *b* all the edges were vertical and all but the last evoked responses regardless of where in the receptive field the light struck. When a large rectangle of light covered entire receptive field, however, as shown at bottom, cell failed to respond.

cell of this type is usually completely ineffective.

In the simple cortical cells the process of pitting these two antagonistic parts of a receptive field against each other is carried still further than it is in the lateral geniculate body. As a rule a large spot of light—or what amounts to the same thing, diffuse light covering the whole retina—evokes no response at all in simple cortical cells. Here the "on" and "off" effects apparently balance out with great precision.

Some other common types of simple receptive fields include an "on" center with a large "off" area to one side and a small one to the other; an "on" and an "off" area side by side; a narrow "off" center with "on" sides; a wide "on" center with narrow "off" sides. All these fields have in common that the border or borders separating "on" and "off" regions are straight and parallel rather than circular [see bottom illustration on page 275]. The most efficient stimuli—slits, edges or dark bars—all involve straight lines. Each cell responds best to a particular orientation of line; other orientations produce less vigorous responses, and usually the orientation perpendicular to the optimum evokes no response at all. A particular cell's optimum, which we term the receptive-field orientation, is thus a property built into the cell by its connections. In general the receptive-field orientation differs from one cell to the next, and it may be vertical, horizontal or oblique. We have no evidence that any one orientation, such as vertical or horizontal, is more common than any other.

How can one explain this specificity of simple cortical cells? We are inclined to think they receive their input directly from the incoming lateral geniculate fibers. We suppose a typical simple cell has for its input a large number of lateral geniculate cells whose "on" centers are arranged along a straight line; a spot of light shone anywhere along that line will activate some of the geniculate cells and lead to activation of the cortical cell. A light shone over the entire area will activate all the geniculate cells and have a tremendous final impact on the cortical cell [see bottom illustration on page 274].

One can now begin to grasp the significance of the great number of cells in the visual cortex. Each cell seems to have its own specific duties; it takes care of one restricted part of the retina, responds best to one particular shape of stimulus and to one particular orientation. To look at the problem from the opposite direction, for each stimulus—each area of the retina stimulated, each type of line (edge, slit or bar) and each orientation of stimulus—there is a particular set of simple cortical cells that will respond; changing any of the stimulus arrangements will cause a whole new population of cells to respond. The number of populations responding successively as the eye watches a slowly rotating propeller is scarcely imaginable.

Such a profound rearrangement and analysis of the incoming messages might seem enough of a task for a single structure, but it turns out to be only part of what happens in the cortex. The next major transformation involves the cortical cells that occupy what is probably the sixth step in the visual pathway: the complex cells, which are also present in this cortical region and to some extent intermixed with the simple cells.

Complex cells are like simple ones in several ways. A cell responds to a stimulus only within a restricted region of retina: the receptive field. It responds best to the line stimuli (slits, edges or dark bars) and the stimulus must be oriented to suit the cell. But complex fields, unlike the simple ones, cannot be mapped into antagonistic "on" and "off" regions.

A typical complex cell we studied happened to fire to a vertical edge, and it gave "on" or "off" responses depending on whether light was to the left or to the right. Other orientations were almost completely without effect [see illustration on opposite page]. These responses are just what could be expected from a simple cell with a receptive field consisting of an excitatory area separated from an inhibitory one by a vertical boundary. In this case, however, the cell had an additional property that could not be explained by such an arrangement. A vertical edge evoked responses anywhere within the receptive field, "on" responses with light to the left, "off" responses with light to the right. Such behavior cannot be understood in terms of antagonistic "on" and "off" subdivisions of the receptive field, and when we explored the field with small spots we found no such regions. Instead the spot either produced responses at both "on" and "off" or evoked no responses at all.

Complex cells, then, respond like simple cells to one particular aspect of the stimulus, namely its orientation. But when the stimulus is moved, without changing the orientation, a complex cell differs from its simple counterpart chiefly in responding with sustained firing. The firing continues as the stimulus is moved over a substantial retinal area, usually the entire receptive field of the cell, whereas a simple cell will respond to movement only as the stimulus crosses a very narrow boundary separating "on" and "off" regions.

It is difficult to explain this behavior by any scheme in which geniculate cells project directly to complex cells. On the other hand, the findings can be explained fairly well by the supposition that a complex cell receives its input from a large number of simple cells. This supposition requires only that the simple cells have the same field orientation and be all of the same general type. A complex cell responding to vertical edges, for example, would thus receive fibers from simple cells that have vertically oriented receptive fields. All such a scheme needs to have added is the requirement that the retinal positions of these simple fields be arranged throughout the area occupied by the complex field.

The main difficulty with such a scheme is that it presupposes an enormous degree of cortical organization. What a vast network of connections must be needed if a single complex cell is to receive fibers from just the right simple cells, all with the appropriate field arrangements, tilts and positions! Yet there is unexpected and compelling evidence that such a system of connections exists. It comes from a study of what can be called the functional architecture of the cortex. By penetrating with a microelectrode through the cortex in many directions, perhaps many times in a single tiny region of the brain, we learned that the cells are arranged not in a haphazard manner but with a high degree of order. The physiological results show that functionally the cortex is subdivided like a beehive into tiny columns, or segments [see illustration on next page], each of which extends from the surface to the white matter lower in the brain. A column is de-

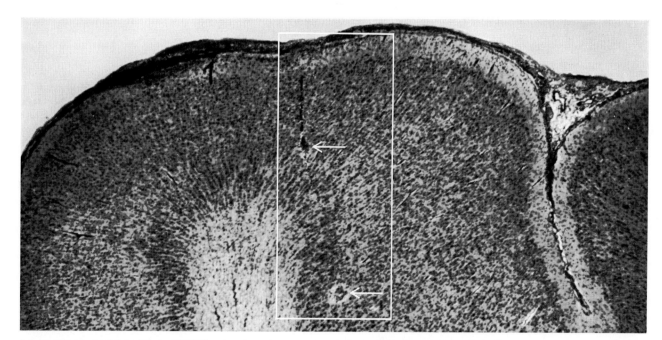

SECTION OF CAT'S VISUAL CORTEX shows track of microelectrode penetration and, at arrows, two points along the track where lesions were made so that it would be possible to ascertain later where the tip of the electrode was at certain times. This section of cortex is from a single gyrus, or fold of the brain; it was six millimeters wide and is shown here enlarged 30 diameters.

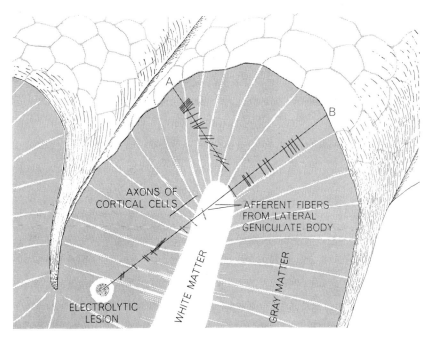

FUNCTIONAL ARRANGEMENT of cells in visual cortex resembled columns, although columnar structure is not apparent under a microscope. Lines *A* and *B* show paths of two microelectrode penetrations; colored lines show receptive-field orientations encountered. Cells in a single column had same orientation; change of orientation showed new column.

fined not by any anatomically obvious wall—no columns are visible under the microscope—but by the fact that the thousands of cells it contains all have the same receptive-field orientation. The evidence for this is that in a typical microelectrode penetration through the cortex the cells—recorded in sequence as the electrode is pushed ahead—all have the same field orientation, provided that the penetration is made in a direction perpendicular to the surface of the cortical segment. If the penetration is oblique, as we pass from column to column we record several cells with one field orientation, then a new sequence of cells with a new orientation, and then still another.

The columns are irregular in cross-sectional shape, and on the average they are about half a millimeter across. In respects other than receptive-field orientation the cells in a particular column tend to differ; some are simple, others complex; some respond to slits, others prefer dark bars or edges.

R eturning to the proposed scheme for explaining the properties of complex cells, one sees that gathered together in a single column are the very cells one should expect to be interconnected: cells whose fields have the same orientation and the same general retinal position, although not the same position. Furthermore, it is known from the anatomy that there are rich interconnections between neighboring cells, and the preponderance of these connections in a vertical direction fits well with the long, narrow, more or less cylindrical shape of the columns. This means that a column may be looked on as an independent functional unit of cortex, in which simple cells receive connections from lateral geniculate cells and send projections to complex cells.

It is possible to get an inkling of the part these different cell types play in vision by considering what must be happening in the brain when one looks at a form, such as, to take a relatively simple example, a black square on a white background. Suppose the eyes fix on some arbitrary point to the left of the square. On the reasonably safe assumption that the human visual cortex works something like the cat's and the monkey's, it can be predicted that the near edge of the square will activate a particular group of simple cells, namely cells that prefer edges with light to the left and dark to the right and whose fields are oriented vertically and are so placed on the retina that the boundary between "on" and "off" regions falls exactly along the image of the near edge of the square. Other populations of cells will obviously be called into action by the other three edges of the square. All the cell populations will change if the eye strays from the point fixed on, or if the square is moved while the eye remains stationary, or if the square is rotated.

In the same way each edge will activate a population of complex cells, again cells that prefer edges in a specific orientation. But a given complex cell, unlike a simple cell, will continue to be activated when the eye moves or when the form moves, if the movement is not so large that the edge passes entirely outside the receptive field of the cell, and if there is no rotation. This means that the populations of complex cells affected by the whole square will be to some extent independent of the exact position of the image of the square on the retina.

Each of the cortical columns contains thousands of cells, some with simple fields and some with complex. Evidently the visual cortex analyzes an enormous amount of information, with each small region of visual field represented over and over again in column after column, first for one receptive-field orientation and then for another.

I n sum, the visual cortex appears to have a rich assortment of functions. It rearranges the input from the lateral geniculate body in a way that makes lines and contours the most important stimuli. What appears to be a first step in perceptual generalization results from the response of cortical cells to the orientation of a stimulus, apart from its exact retinal position. Movement is also an important stimulus factor; its rate and direction must both be specified if a cell is to be effectively driven.

One cannot expect to "explain" vision, however, from a knowledge of the behavior of a single set of cells, geniculate or cortical, any more than one could understand a wood-pulp mill from an examination of the machine that cuts the logs into chips. We are now studying how still "higher" structures build on the information they receive from these cortical cells, rearranging it to produce an even greater complexity of response.

In all of this work we have been particularly encouraged to find that the areas we study can be understood in terms of comparatively simple concepts such as the nerve impulse, convergence of many nerves on a single cell, excitation and inhibition. Moreover, if the connections suggested by these studies are remotely close to reality, one can conclude that at least some parts of the brain can be followed relatively easily, without necessarily requiring higher mathematics, computers or a knowledge of network theories.

26

The Neurophysiology
of Binocular Vision

by John D. Pettigrew
August 1972

*The ability of certain mammals, including man, to
visually locate objects in the third dimension is traced
to the selective activity of single binocular nerve cells
in the visual cortex of the brain*

Man, along with cats, predatory birds and most other primates, is endowed with binocular vision. That is to say, both of his eyes look in the same direction and their visual fields (each about 170 degrees) overlap to a considerable extent. In contrast, many animals, such as rabbits, pigeons and chameleons, have their eyes placed so as to look in different directions, thereby providing a more panoramic field of view. Two questions come to mind: First, why do we have binocular vision instead of panoramic vision? Second, how is it that our single impression of the outside world results from the two different views we have of it by virtue of the separation of our eyes?

In answer to the first question, it is now known that binocular vision provides a powerful and accurate means of locating objects in space, a visual aptitude called stereopsis, or solid vision. Of course, it is possible to judge distance from the visual image of one eye by using indirect cues such as the angle subtended by an object of known size, the effort used in focusing the lens of the eye or the effect of motion parallax (in which the relative motions of near and far objects differ). These cues cannot be used in all situations, however, and they are not as accurate or as immediate as the powerful sensation of stereopsis, which is perhaps most familiar in the context of stereoscopic slide-viewers, three-dimensional motion pictures and so on. Some 2 percent of the population cannot enjoy stereopsis because of undefined anomalies of binocular vision. It is the aim of this article to give an account of recent work that shows how the brain achieves the very first stages of binocular depth discrimination.

Although it was not until the 19th century that the advantages of binoc-ular vision were clearly demonstrated, man has pondered the arrangement of his eyes from earliest times. Of more concern to early investigators was not the first question, "Why binocular vision?" but rather "How does my single unified impression of the world result from the two views I have of it?" This second question is almost as difficult to answer today as it was when it was first asked by the ancient Greeks. The problem of "fusing" two slightly differing views of the world, however, is closely akin to the problem of using the slight differences between the views to achieve stereopsis. Thus a better understanding of the events in the nervous system underlying stereopsis should also throw some light on the problem of binocular fusion.

Galen taught in the second century that the fluid-filled ventricles of the brain were the seat of union, with a flow of visual spirit outward to both eyes. Galen's teachings were influential until the Renaissance, when scholars realized that the transfer of information is from the world to the eye, rather than in the reverse direction. René Descartes proposed in the 17th century that fibers from each eye might converge on the pineal gland for unification [see illustration on page 283]. His scheme, although incorrect, clearly indicates the now established principle that fibers from roughly corresponding regions of each eye converge on a single site in the brain. It was Isaac Newton who in 1704 first proposed that where the optic nerves cross in the optic chiasm there is an exchange of fibers. An early drawing of this concept, called partial decussation, shows how the fibers come together to carry information from corresponding parts of each eye [see illustration on page 284].

Newton's proposal, unlike that of Ga-len or Descartes, has been extensively verified. The number of uncrossed fibers in the optic chiasm depends on the amount of overlap of the two visual fields, and this number tended to increase as animals evolved with eyes occupying a more frontal position [see illustration on page 285]. The rabbit, with only a tiny binocular portion in its visual field, has a very small number of ipsilateral, or uncrossed, fibers in the optic chiasm, whereas each of its cerebral hemispheres is heavily dominated by contralateral, or crossed, fibers from the opposite side. As the amount of binocular overlap increases from animal to animal, so does the number of ipsilateral fibers. In man there is almost complete overlap and 50 percent of the fibers of the optic nerve are uncrossed.

Although partial decussation provides the opportunity for the optic nerve fibers to come together in the brain, for a long time there was controversy over whether this coming together does in fact occur. For instance, at the first way station for the optic-nerve fibers in the brain, the lateral geniculate nucleus, the inputs from the two eyes are carefully segregated into layers. The more binocular overlap the animal has, the more obvious the layering is. The segregation is confirmed by physiological recordings that show that a neuron, or nerve cell, in a given layer can be excited by light stimuli falling on one eye only. The segregation is reinforced by inhibitory connections between corresponding neurons in adjacent layers.

At the level of the visual cortex of the brain, however, single neurons do receive excitatory inputs from both eyes. David H. Hubel and Torsten N. Wiesel of the Harvard Medical School demonstrated this effect for the first time in

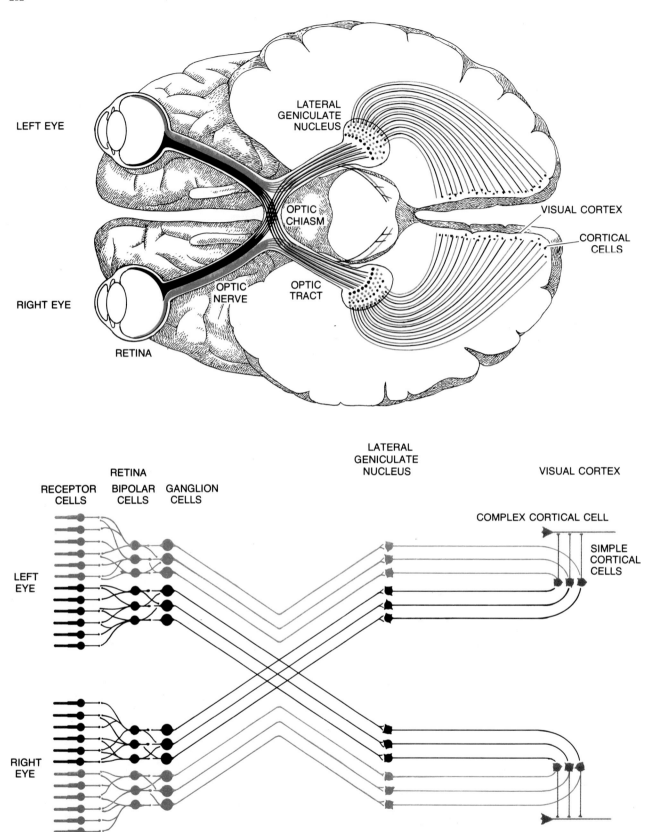

LEFT EYE

LATERAL GENICULATE NUCLEUS

VISUAL CORTEX

CORTICAL CELLS

OPTIC CHIASM

OPTIC NERVE

OPTIC TRACT

RIGHT EYE

RETINA

LATERAL GENICULATE NUCLEUS

VISUAL CORTEX

RETINA

RECEPTOR CELLS

BIPOLAR CELLS

GANGLION CELLS

COMPLEX CORTICAL CELL

SIMPLE CORTICAL CELLS

LEFT EYE

RIGHT EYE

ANATOMY OF BINOCULAR VISION is represented in the drawing at top, which shows the human brain as viewed from below. The visual pathway from retina to cortex consists essentially of six types of neurons, or nerve cells, of which three are in the retina, one is in the lateral geniculate nucleus and two are in the cortex (*see schematic diagram at bottom*). Roughly half of the fibers of the optic nerve from each eye remain uncrossed at the optic chiasm. These ipsilateral, or uncrossed, fibers (*color*) from the outer part of the retina of one eye join with the contralateral, or crossed, fibers (*black*) from the inner part of the retina of the other eye, and the two types of fiber travel together along the optic tract to the lateral geniculate nucleus, where fibers from each eye are segregated into layers. The fibers that emerge from this body carrying an input from both eyes converge on single neurons in the cortex.

1959 by recording from single neurons in the visual cortex of the cat [see the article "The Visual Cortex of the Brain," by David H. Hubel, beginning on page 272]. For almost every nerve cell studied two areas could be defined where light stimuli evoked a response, one associated with each eye. The proportion of binocularly activated neurons found has increased as the technique of presenting stimuli has become refined. Recent work by P. O. Bishop, Geoffrey H. Henry and John S. Coombs of the Australian National University shows that all the cells in the striate cortex of the cat receive an excitatory input from both eyes. Since each neuron in the striate cortex is simultaneously "looking" in two directions (one direction through each eye), the striate cortex can be regarded as the "cyclopean eye" of the binocular animal. The term cyclopean eye, derived from the mythological Cyclops, who had one eye in the center of his forehead, was first used by Ewald Hering and Hermann von Helmholtz in the 19th century to describe the way the visual cortex resolves the different directions of a given object as seen by each eye. Besides assessing visual direction, the cyclopean eye has an ability not possessed by either eye alone: it can use binocular parallax to ascertain the distance of an object.

Imagine looking down on an upturned bucket [see illustration on page 286]. If one directs each eye's fovea (the central high-resolution area of the retina) toward the cross at the center of the bucket, each eye will receive a slightly different view. The difference between the two views is called binocular parallax. Since each retinal image of the bottom of the bucket (small circles) is equidistant from the fovea, these two images must lie on exactly corresponding retinal points and are said to have zero disparity. Because of the horizontal separation between the eyes, the images of the rim of the bucket (large circles) are displaced horizontally with respect to the small circles and the foveas. These images lie on disparate retinal points. The retinal disparity between two such images is measured as an angle that corresponds to the difference between the angular separations of the two images from some known point such as the fovea (in the case of absolute disparity) or the smaller circle (in the case of relative disparity). If the disparity between the retinal images of the large circle is not too great, one sees not two large circles but a single large circle floating in depth behind the small one.

The basis of this powerful depth sensation of stereopsis was first demonstrated by Sir Charles Wheatstone (of

Wheatstone-bridge fame) in 1838. By providing very precise localization of objects in visual space, stereopsis can be regarded as the *raison d'être* of binocular vision. Whenever in evolution the need for the protection of panoramic vision was lessened (by the animal's taking to the trees as in the case of the primates, or by the animal's becoming predatory as in the case of the cats), then binocular vision developed to make it possible to use a depth cue more direct and accurate than the depth cues available to one eye alone. In addition, stereopsis enables a predator to penetrate the camouflage used by its prey, because monocular form perception is not a necessary prerequisite for stereoscopic vision. For example, an insect disguised as a leaf may be invisible monocularly but stand out in a different depth plane from real leaves when it is viewed stereoscopically. One can readily demonstrate this effect for oneself with the aid of random-dot stereograms devised by Bela Julesz of the Bell Telephone Laboratories [see "Texture and Visual Perception," by Bela Julesz; SCIENTIFIC AMERICAN Offprint 318]. Here a given pattern, such as a square, may be invisible to monocular inspection but stand out vividly when viewed stereoscopically.

The sole basis of stereopsis is the horizontal disparity between the two retinal

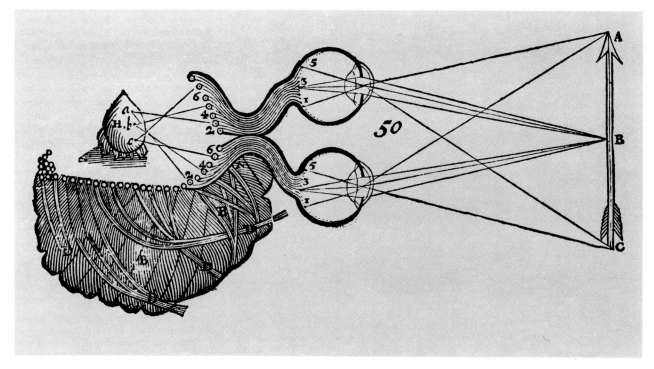

CONVERGENCE of nerve fibers from corresponding regions of each eye on a single site in the brain was proposed more than 300 years ago by René Descartes. In this early drawing of Descartes's scheme, reproduced from his study *Traite de l'Homme*, the optic nerve fibers from each eye are shown converging on the pineal gland (*H*), where they are rearranged, with those from corresponding retinal regions merging together. It is now known that fibers from both eyes do in fact converge, but not on the pineal.

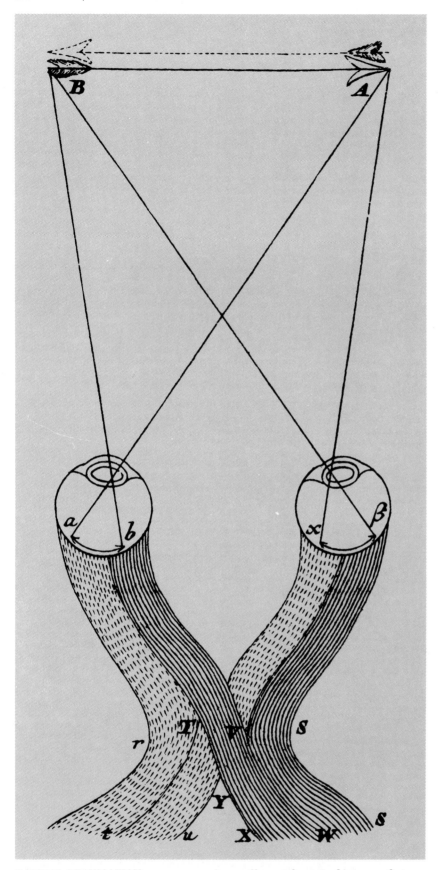

PARTIAL DECUSSATION, or crossover, of nerve fibers at the optic chiasm was first proposed by Isaac Newton in 1704. This drawing, made by a contemporary of Newton's, shows how fibers carrying information from corresponding parts of the eye come together on the same side of the brain in the interests of binocular vision. Newton's scheme was verified in the 19th century by the first ophthalmologists, Heinrich Müller and Bernhard von Gudden.

images. Of course, the brain must have a means by which it can first select those parts of the two images that belong to each other in the sense that they are images of the same feature in space. The horizontal disparities between the paired parts give the cue to depth. Julesz' experiments show that the binocular assessment of depth does not require the prior recognition of form, suggesting that the disparity information is processed by the brain fairly early in visual perception. Encouraged by this suggestion, neurophysiologists have examined the properties of binocularly activated neurons in the visual cortex, and over the past five years they have learned more about the neural mechanisms of binocular depth discrimination.

The experimental arrangement for studying binocularly activated neurons in the visual cortex is shown in the illustration on page 287. A cat that has been anesthetized and given a muscular paralyzing agent to prevent eye movements faces a screen onto which a variety of visual stimuli can be projected. A microelectrode inserted into the visual cortex samples activity from single nerve cells, and this activity is amplified so that it can be displayed on an oscilloscope, recorded and fed into a loudspeaker, in order that the experimenter can readily follow the response. Unlike neurons in the retina or the lateral geniculate nucleus, cortical neurons need exquisitely defined stimuli if they are to fire. Each cell requires that a line of a particular orientation (for instance a white slit on a dark background, a dark bar on a light background or a dark-light border) be placed on a narrowly defined region of each retina. Neurons with the same requirements for orientation of the stimulus are grouped together in columns that run from the surface of the brain to the brain's white matter. For some neurons within the column absolute position of the line is very important. Plotting with small spots of light suggests that such neurons receive a fairly direct input from the fibers coming into the cortex. These are "simple" neurons. Other neurons, called "complex," probably signal the output of the column; they behave as if they receive an input from a large number of simple neurons.

The speed and the direction of the moving stimulus are also important. Each of these stimulus requirements is more or less critical and each varies from neuron to neuron. As one slowly advances the electrode to pick up neurons, one must be continually moving a complicated pattern in front of the animal's eyes, in order to activate neurons that would otherwise be missed because of their lack of activity in the absence of the specific stimulus. (I sometimes wear a knitted sweater with a regular design as I move about in front of the cat.) Once one has found the specific stimulus for a given neuron, it is possible to define a region in the visual field of each eye where that stimulus will cause excitation of the cell. This region, called the response field, is plotted for each eye on a screen in front of the animal. The eye on the same side of the brain as the neuron in question is called ipsilateral; the eye on the opposite side is called contralateral. The study of a number of neurons in succession gives an array of ipsilateral and contralateral response fields [*see illustration on page 289*]. The arrays for each eye are separated on the screen because of the slightly divergent position the eyes assume in paralysis. Normally the eyes would be lined up so that the arrays could overlap and a single object might stimulate both response fields for a given neuron.

The highly specific stimulus requirements of cortical neurons could provide a means of identifying the parts of the two images corresponding to a single feature in object space. Because the number of identical features in a small part of the image in one eye is likely to be low, similar features lying in roughly corresponding regions in each eye can be assumed to belong to the same object. For example, a black line with a particular orientation and direction of movement in one image would be associated with a similar line at the most nearly corresponding position in the other image, because both are likely to be images of the same object. Since binocular cortical neurons have properties suited to the detection of the pair of retinal images produced by a given object, it was of great interest to see if they could also detect disparity between the pairs of images.

When one takes a close look at the position of each response field compared with the position of its partner in the opposite eye, it is immediately obvious that it is not possible to superimpose every response field on its partner simultaneously because of the greater scatter in the fields of the ipsilateral eye compared with those of the contralateral eye. The response fields therefore do not lie in corresponding regions of each retina and may be said to show disparity. I had

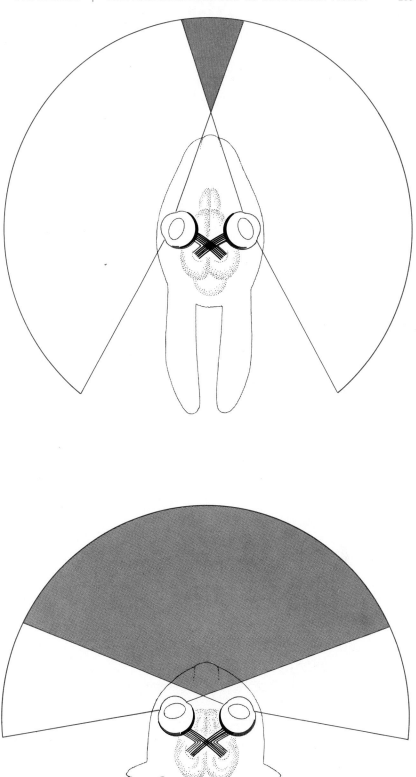

AMOUNT OF BINOCULAR OVERLAP of an animal's two visual fields is proportional to the percentage of uncrossed fibers in the optic chiasm. As animals evolved with eyes occupying a more frontal position this percentage tended to increase. The rabbit, for example, has only a tiny amount of binocular overlap and accordingly has a very small number of uncrossed fibers (*top*). The cat, in contrast, has a much larger binocular overlap and a correspondingly higher percentage of uncrossed fibers (*bottom*). In man there is almost complete binocular overlap and 50 percent of the fibers are uncrossed. The uncrossed nerve fibers carry information from the outer part of retina, the region responsible for binocular vision.

noticed this phenomenon in 1965 while working with Bishop at the University of Sydney and had considered the possibility that the variation in the position of one eye's response field with respect to the position of the corresponding response field of the opposite eye might play a role in the detection of retinal-image disparity and therefore in binocular depth discrimination. At that time, however, there were two major difficul-

ties involved in the interpretation of the phenomenon.

The first difficulty was residual eye movement, which is present in small amounts even after the standard muscular paralyzing agents are applied. Since determination of the two response fields for one neuron can take hours (because one has to find the best stimulus orientation, speed of movement, exact position on the screen and so forth), one has

to be sure that the eyes do not move in that time. Eye movement would produce spurious response-field disparities.

The second problem concerns the specificity of a neuron to binocular stimulation in the situation where a single stimulus is presented simultaneously to both eyes. It could be argued that the response-field disparities observed are not significant functionally since the neuron might tolerate large amounts of

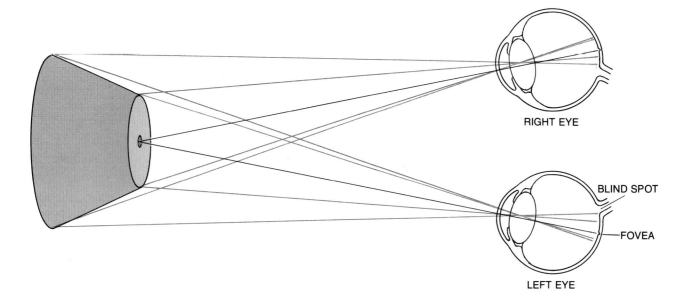

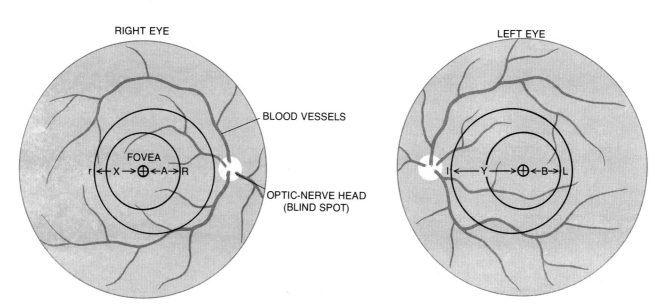

BINOCULAR PARALLAX is the term used to describe the disparity produced between two retinal images when one views a three-dimensional object. In this case a subject has been instructed to direct each eye toward the cross at the center of an upturned bucket (*top*). The drawings at bottom show how the fundus of each eye would look in an indirect ophthalmoscope. The fovea, the central high-resolution area of each retina, would appear as a shallow depression on which the cross would be imaged. The optic-nerve head, or blind spot, where the fibers of the optic nerve leave the retina and the blood vessels enter, would appear as a white disk.

Since the small circles corresponding to the retinal images of the bottom of the bucket are each equidistant from the fovea (that is, distance A is equal to distance B), these two images must lie on exactly corresponding retinal points (R and L); these images are said to have zero disparity. The large circles corresponding to the images of the rim of the bucket, on the other hand, are displaced horizontally with respect to the small circles and the foveas (that is, distance X is not equal to distance Y); these images are said to lie on disparate retinal points (r and l). The amount of retinal disparity, usually expressed as an angle, is the difference between X and Y.

overlap of its receptive fields without changing its response. If the tolerated overlap were of the same order of magnitude as the variation in response-field disparity from neuron to neuron, then the latter variation would be of no use.

Both of these problems were worked out in the succeeding years by me in collaboration with Bishop and Tosaku Nikara at Sydney, and with Horace B. Barlow and Colin Blakemore at the University of California at Berkeley. The problem of eye movement was solved by resorting to a number of measures simultaneously. A particularly potent mixture of neuromuscular blocking drugs was developed to reduce eye movement to a minimum without toxicity to the cat's heart and blood vessels. The sympathetic nerves to the orbit of the eye were cut to eliminate movements due to the involuntary muscles near the eye. Any residual drift was carefully monitored by plotting the projection of some small blood vessel inside the eye onto a screen. Any tiny amount of movement between response-field plots could then be corrected. In the Berkeley experiments we carefully attached the margins of each eye to rigidly held rings, which kept the eyes fixed for the duration of the long measurements.

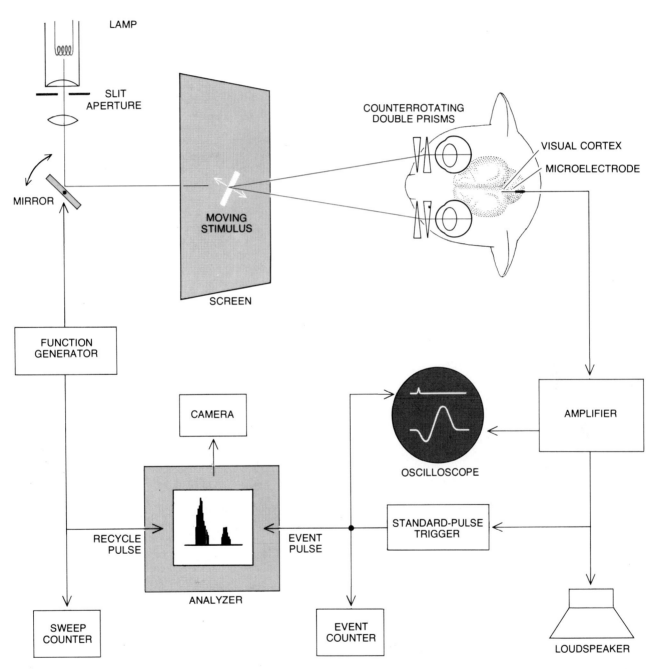

EXPERIMENTAL SETUP used by the author and his colleagues to study binocularly activated neurons in the visual cortex of the cat is depicted here. A cat that has been immobilized by a number of measures faces a screen onto which a moving line stimulus of any orientation, direction and speed can be projected from behind. A microelectrode inserted into the visual cortex samples activity from single neurons, and this activity is amplified so that it can be displayed on an oscilloscope, recorded and fed into a loudspeaker. Once a particular orientation and direction are discovered that will make the neuron fire, the stimulus is moved back and forth repeatedly while the neuron's response pattern is worked out. Counterrotating double prisms of variable power placed before the eyes enable the experimenter to determine the effect of changing retinal disparity as the stimulus moves in a fixed plane in front of the cat.

a

ELECTRODE

VISUAL CORTEX

b

VISUAL CORTEX

ELECTRODE

COLUMN

WHITE
MATTER

GRAY MATTER

c

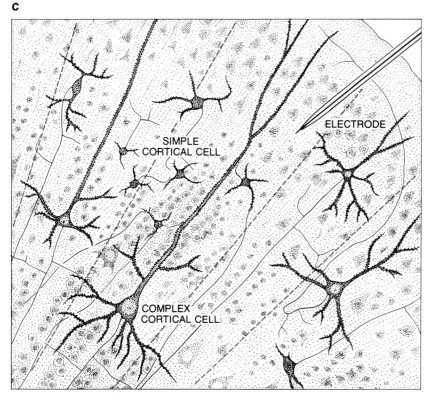

ELECTRODE

SIMPLE
CORTICAL CELL

COMPLEX
CORTICAL CELL

PENETRATION of the microelectrode through the cat's visual cortex is represented in this sequence of successively enlarged views. The top view of the entire brain (*a*) shows that the cortical region associated with vision (*color*) corresponds to a single gyrus, or fold, located toward the rear of each of the brain's two hemispheres. The cross section of this gyrus (*b*) reveals that the visual cortex is functionally divided into an array of tiny columns (*broken colored lines*) that run from the surface of the brain to the interior white matter. The magnified cross section (*c*) shows that each of these columns (which are not visible under the microscope) consists of a group of neurons with the same requirements for orientation of the stimulus. "Simple" neurons are those that appear to receive a fairly direct input from the fibers coming into the cortex. "Complex" neurons, which behave as if they receive an input from a large number of simple neurons, probably signal the output of the column.

The question "How specific is the response of a single neuron for the retinal disparity produced by a stimulus presented to both eyes?" was answered with the help of the following technique. The specific stimulus was swept forward and backward over both of the response fields of a cortical neuron when these had been lined up on the screen by the use of a double prism of variable power. The double prism was used to vary the visual direction of one eye (and therefore any response fields of that eye) in finely graded steps. Since the eyes are fixed, this maneuver changes the disparity between the retinal images of the moving stimulus, and if the change in prism power for one eye is in the horizontal plane, then the effect is identical with the one produced by a change in the distance of the stimulus along a line through the opposite eye and its response field.

We found that the binocular response of a particular cortical neuron in the left visual cortex to a moving slit of light varies considerably as the disparity is changed [*see illustration on page 291*] The responses were averaged for stimulation of each eye alone and for binocular stimulation when the two response fields have different amounts of overlap on the tangent screen where the light stimulus is moving forward and backward. The cortical neuron responded only to an oblique slit, and there was no response when the orientation was rotated more than 10 degrees in either direction. There was a response only as the slit moved across the screen from left to right and not in the reverse direction. Moreover, the response elicited from the left (ipsilateral) eye alone was weak.

The relative strength of the responses elicited from each eye varies from one neuron to another, and the particular neuron shown in the illustration is a case of extreme contralateral dominance. The inhibitory contribution from the ipsilateral eye is far from weak, however, as can be seen by the reduction of the binocular response when the response fields are stimulated in the "wrong" spatiotemporal relationship, as for example when the prism setting is such that the response fields are side by side instead of being superimposed.

Our findings demonstrated that the binocular response of the neuron is in fact very sensitive to slight changes in the overlap of its response fields in the plane of the stimulus. Since the response fields themselves are quite small (less than half a degree), this means a high

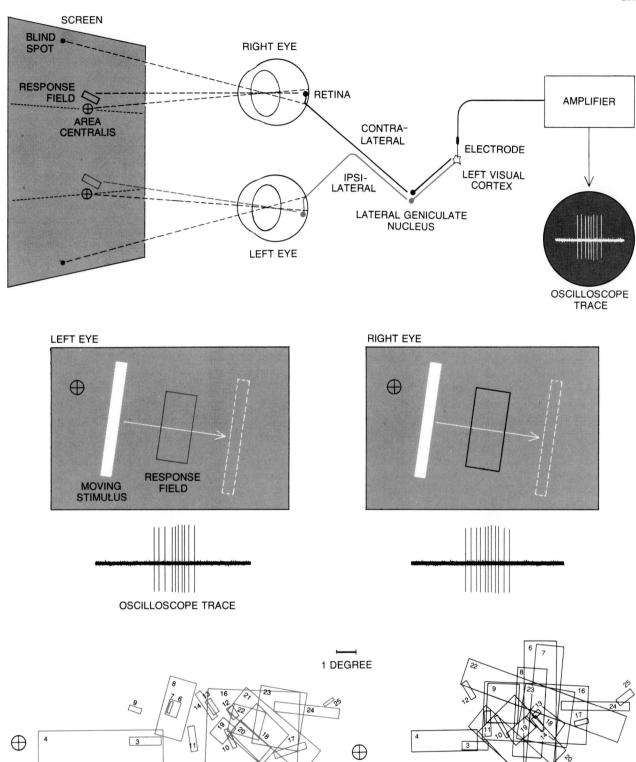

RESPONSE FIELDS FOR LEFT (IPSILATERAL) EYE

RESPONSE FIELDS FOR RIGHT (CONTRALATERAL) EYE

PAIRED RESPONSE FIELDS, one for each eye, can be plotted for a given neuron in the visual cortex of the cat (top). The response field of the neuron is defined as the region in the visual field of each eye where a specific stimulus will cause excitation of the cell (middle). In this illustration the response fields are separated on the screen because of the slightly divergent position the eyes assume in paralysis. Normally the fields would tend to overlap each other. By moving the microelectrode carefully through the cortex it is possible to record successively from a large number of different neurons; when the paired response fields of these neurons are plotted (bottom), those for the ipsilateral eye (the eye on the same side of the brain as the neuron in question) are more scattered than those for the contralateral eye (the eye on the opposite side). This means that it is not possible to superimpose all response fields on the same plane at the same time and that therefore different neurons would be optimally stimulated by objects in different planes.

level of disparity specificity. This particular neuron could indicate, by a marked decrease in firing rate, a disparity change as small as two minutes of arc, a feat approaching human performance. (The human threshold disparity is about 10 seconds of arc, or approximately 10 times better.)

It is perhaps not too surprising, in view of the very small size of the response fields, that the two retinal images of a binocularly presented stimulus must be very precisely located in order to produce a good response from the neuron. More surprising is the almost total suppression of the strong response from an appropriately located image in one eye if the image is inappropriately located in the other. This inhibition persists when the image is moved (for example by inserting the prism or by changing the distance of the stimulus) more than one degree of arc in either direction from the optimal position with respect to its correctly located partner in the other eye. In other words, binocular inhibition extends for more than one degree of retinal disparity on each side of the optimal disparity for a given neuron. The significance of this conclusion can be seen when one considers how nearby neurons behave with respect to one another in binocular vision; for those binocular neurons concerned with central vision the total range of optimal disparities is also a couple of degrees.

Let us now look at another disparity-specific binocular neuron recorded from the same column of tissue in the cortex as the one just described. Its stimulus requirements were quite similar (a slowly moving slit of light with the same orientation) except that the optimal disparity was 1.7 degrees more convergent because of the different position of its ipsilateral response field. Thus an oblique slit, in spite of the fact that it stimulates the contralateral response fields of both neurons, will under binocular viewing conditions excite one of them and inhibit the other, depending critically on its distance from the cat.

This binocular inhibition, operating over the same range as the range of disparity from one neuron to another, may be part of the explanation for the phenomenon of binocular fusion. A binocularly viewed target can be seen as being single in spite of the fact that it appears to lie in two different directions when the views from each eye alone are compared. In the upturned-bucket example, if the disparity between the retinal images of the larger circles is not too great, then one sees not two large circles but

a single (fused) large circle floating in depth. It is reasonable to suppose that the failure to see a second large circle is due to the binocular inhibition of those neurons that were activated monocularly by such a circle. The narrowing down of the amount of activity among different neurons narrows down in turn the number of stimulus possibilities from which the brain has to choose. In this case groups of binocular neurons associated with the same contour but with different retinal disparities are narrowed to one group and therefore a particular disparity.

Both of the neurons described above belong to Hubel and Wiesel's class of simple neurons, that is, neurons that respond only to stimuli on narrowly defined areas of the retina. It was particularly interesting to examine the binocular properties of complex neurons, since they are thought to receive an input from a number of simple neurons and therefore to respond over a wider area of retina. Would they also respond over a wider range of disparity?

Two types of disparity-specific complex neuron were found. In one group there was a high degree of specificity in spite of the large size of the response field. One binocular complex neuron had response fields six degrees across but could still detect changes of disparity as accurately as most simple neurons (which have fields less than one degree across). This astonishing precision means that the neuron would signal with a change in firing rate that a stimulus moving anywhere over a six-degree area had produced a change of just a few minutes of arc in retinal disparity. With the eyes in a constant position a disparity-specific complex neuron "looks" at a thin sheet suspended in space and fires if a stimulus with the correct orientation and speed of movement appears anywhere on the sheet (but not in front of or behind it).

Disparity-specific complex neurons behave as if they receive an input from a number of simple neurons with different absolute response-field positions in each eye but with the same relative position, so that they all have the same optimal disparity. In fact, we noticed groups of such neurons in the Berkeley experiments, and Hubel and Wiesel have recently shown that binocular neurons with the same disparity specificity in the monkey's cortex appear to be grouped in cortical columns similar to the columns for orientation specificity. We therefore have another example of the

cortical column as a system for extracting information about one specific type of stimulus while generalizing for others. A disparity-specific complex neuron can accordingly respond to a vertical edge moving over a wide region of the retina but over a very narrow depth in space. Directional specificity is lost but orientation and disparity information are retained.

There is some evidence for another type of binocular cortical column where all the neurons have response fields in the same position for the contralateral eye but have scattered fields for the ipsilateral eye. Blakemore calls these structures "constant direction" columns, because the neurons associated with them appear to respond at different disparities but to stimuli that are in the same direction from the contralateral eye. The output cell from such a column would presumably generalize for disparity but would be specific for the orientation and direction of the object.

Other complex binocular neurons responded over a wide range of disparity as well as of visual field. Since these neurons are active over the same range in which one observes binocular inhibition, they may be the source of the inhibition for simple neurons.

Once small residual eye movements had been accounted for and disparity specificity had been demonstrated, we were able to go ahead and compare the response-field pairs of a large number of different binocular neurons. In that way we could assess the total range of disparity variation. This was of particular interest because of a large body of observations obtained in psychophysical experiments on humans showing the range of disparity over which there is binocular fusion and the range of disparity over which stereopsis operates, both for central (foveal) vision and as one moves into the lower-resolution, peripheral visual field. The measurements were tedious because of the great length of time it takes to characterize a disparity-specific cortical neuron. In a typical experiment it took us three days to accumulate the 21 disparity-specific neurons whose response fields are shown in the illustration on the preceding page. All the neurons were recorded from the left striate cortex, and inspection of their response fields reveals a greater scatter in the fields of the left eye than in the fields of the right. This general observation that the ipsilateral receptive fields show more horizontal scatter is of interest in view of the fact that the ipsilateral

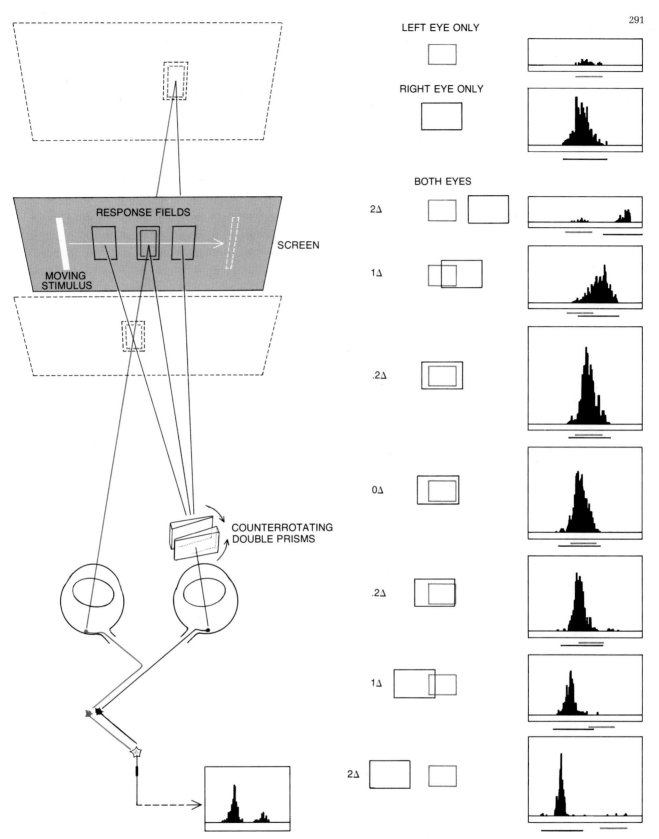

SPECIFICITY of a given binocular neuron for varying degrees of retinal disparity can be studied with the aid of counterrotating double prisms of variable power. The slight shifts of the two response fields with respect to each other in the plane of the moving stimulus are equivalent to setting that plane nearer to or farther from the animal (*left*). In this particular example, a case of extreme contralateral dominance, the response to a moving slit of light elicited from the left (ipsilateral) eye alone was weak; the inhibitory contribution from the ipsilateral eye is far from weak, however, as can be seen from the reduction of the binocular response when the prism setting is such that the response fields are side by side instead of being superimposed (*right*). The response for this particular neuron falls off rapidly for disparities on the order of tenths of a degree; hence there would be good discrimination among neurons whose optimal disparities cover a range of several degrees. A prism setting of one diopter (Δ) is equal to a retinal disparity of one centimeter at a distance of one meter; expressed as an angle, a disparity of one Δ equals .57 degree.

fibers are the ones that have arisen most recently in the evolution of binocular vision.

One can get a measure of the range of disparity in the response fields by shifting each pair of fields horizontally so that all the left-eye fields are superimposed. It is clear that if there were no disparity between different pairs of response fields, then that would lead to superimposition of the right-eye fields also. The degree to which the fields do not superimpose can be measured, and in this case there was a range of six degrees of horizontal disparity and two degrees of vertical disparity. It is not immediately obvious why the neurons should cover a range of vertical disparity, since only the horizontal component can be used for stereopsis. The psychophysical studies show, however, that although the visual system cannot make use of vertical disparity, allowance must be made for such disparities so that the system can still operate when they are introduced. Vertical disparities arise at close viewing distances (where the image of a given object may be significantly larger on one retina) and also in the course of eye movements (where the two eyes may not remain perfectly aligned vertically).

The total range of disparity surveyed by a given cortical area varies according to retinal eccentricity. Binocular neurons concerned with the area centralis (the high-resolution part of the cat's retina that corresponds to the human fovea) cover a disparity range of two degrees compared with six degrees for those neurons dealing with the visual field about 10 degrees away from the midline. The small total range for the area centralis not only allows fine discrimination within that range but also requires fine control of eye movements so that the target being examined can be kept within the range. The range of disparities for central vision appears to be even narrower in humans and monkeys, where there is exquisite control of convergent and divergent eye movements so that the midpoint of the range can be varied. The fineness of the range is attested by the double vision that occurs if there is the slightest imbalance in the muscular system.

The preliminary results described here provide some insight into the initial operations performed by the visual cortex in extracting the information about disparity between small elements of the two retinal images. Much remains to be determined about how these first steps are utilized by the brain to yield our

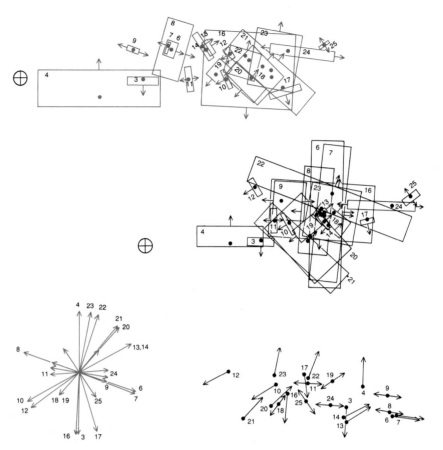

RANGE OF OPTIMAL RETINAL DISPARITIES for the 21 binocular neurons whose response field are plotted in the illustration on page 289 (and reproduced at the top of this illustration) can be calculated by shifting each pair of response fields horizontally so that the binocular centers (*black dots*) of all the left eye's field are superimposed (*bottom*). The scatter of the right eye's response fields then gives the range of disparities. For those fields located away from the central area there is a range of six degrees of horizontal disparity and two degrees of vertical disparity. For neurons with fields closer to the center of the retina the range is smaller and hence the neurons are capable of finer discrimination.

complete three-dimensional view of the world. Here are two examples of the kind of problem that remains to be solved: (1) Since convergent and divergent eye movements themselves produce changes in retinal disparity, how are these movements taken into account so that an absolute depth sense results that does not change with eye position? (2)

How is a synthesis achieved from the disparity information about the myriad contours of a visual scene? The answers to these and many more perplexing questions about the brain may be best answered by the combination of the approaches of psychophysics and neurophysiology that has proved fruitful thus far.

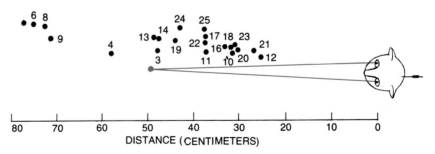

DEPTH DISCRIMINATION attributable to binocular vision is represented by this plot of the points in space at which a correctly oriented contour would optimally stimulate the 21 binocular cortical neurons whose response fields are shown in the illustration at top of this page, provided that the cat's eyes are fixed on a point 50 centimeters in front (*colored dot*).

VII

THE CONTROL
OF BEHAVIOR

VII THE CONTROL OF BEHAVIOR

INTRODUCTION

The result of most complex nervous and endocrine activity is an overt response by an animal. Some responses are clearly related to a stimulus; some appear to be caused by an internal state. Behavior takes the form of a coordinated set of movements that change the animal's relation to its environment; it may range from reflexes as simple as withdrawal of a limb, through such cyclically repeated locomotor patterns as the flight of an insect, to highly complex ritualized activities like some of those in courtship. Much behavior appears to be entirely innate; that is, much of it appears to be dependent upon a set of genetically determined central nervous connections. Other actions are obviously learned, and many others contain elements that are both innate and learned.

The study of behavior is comprehensive, drawing upon many disciplines other than biology. Since this book is concerned with the elaboration of cellular elements into interacting systems, the emphasis of the following five selections is upon the physiological mechanisms underlying behavior. Analysis of these mechanisms is extremely difficult, because in most situations so many underlying elements are involved that a meaningful cause-and-effect relationship is hard to extract. Fortunately, some systems exhibit either an extreme specialization or a simplification of cellular elements that makes them promising as subjects for study. Others attract attention because of their ubiquity, or because of their great complexity and importance to man.

The first article in the section, "How We Control the Contraction of Our Muscles" (1972), by P. A. Merton, is an excellent example of one kind of approach to the central nervous control of behavior. The behavior Merton selected for study is not a single, ritualized act; rather, it is a system of reflexes that controls static body position or executes volitional movements. Merton addresses the question of what kind of relationship exists between receptors that sense movement and motor systems that initiate movement. The analysis, then, amounts to a study of feedback in a control system, and Merton combines the approach of a biologist and that of a systems engineer. He shows that the sensory feedback from load-moving skeletal muscles acts as a kind of amplifier (or *servo assistance*) for the movement: since the signal from these receptors is proportional to the load being moved, receptors in effect cancel loads and ensure that a centrally issued command for movement will reach the same final position independently of variations in the loads imposed. It is interesting that receptors of an analogous type have evolved independently in invertebrate animals, and that these perform a function that is precisely identical.

The following three articles represent a quite different approach, because the authors of these articles have studied relatively accessible, easily observed nervous systems in the belief that these provide experimental material especially well suited to analysis of the control of particular behaviors. The general principles are discussed in "Small Systems of Nerve Cells" (1967). Many invertebrates present unique advantages to neurobiologists because of the reduced number of their nerve cells, and because the nerve cells of some invertebrates are large and identifiable—that is, the same unique cell can be recorded again and again in preparation after preparation. For these reasons too, stereotyped behaviors can be selected for study or, alternatively, phenomena of general importance—like the studies on lateral inhibition and resistance reflexes mentioned in the article—can be analyzed precisely.

Since the article on small nerve-cell systems was first published, a number of changes have been made in the zoological garden of neurobiologically useful invertebrates. The sea-hare *Aplysia*, discussed in that article, is still popular, and studies of the mollusc with the five-celled eye, *Hermissenda*, have been conducted in several laboratories. A number of other marine (and a few terrestrial) gastropod mollusca have more recently been brought into the service of physiology, and circuits of identified neurons are being published at a bewildering rate. A by-product of the increasing research has been the creation of a lively cottage industry to meet the new scientific demand for exotic, semitropical marine organisms. The business is so lively, in fact, that some neurobiologists worry about the stability of the supply, to say nothing of the ecology of the locales where the collecting is done. Among the arthropods, studies of insects have been increasing for two reasons: first, they present interesting stereotyped behavior patterns that are well suited to physiological analysis; second, insects have actual or at least potential use in genetic experiments designed to demonstrate the mechanisms responsible for the assembly of the neural circuits. Like the animals mentioned in the article on small systems, insects have a relatively small number of sensory and motor nerve cells, and an experimenter can insert electrodes into an insect to monitor the activity of particular nerve cells with only minor impairment to the behavior of the insect. The following two articles by Wilson and Roeder represent differing applications of this approach.

In his account of "The Flight-Control System of the Locust" (1968), the late Donald M. Wilson focuses on a simple, cyclic piece of locomotor behavior and attempts to learn how the central nervous system controls it. The work he describes was perhaps more influential than any other set of experiments in changing biologists' attitudes about how reflexes participate in more complex acts. Previously walking, swimming, flight, and similar behavior in which motion is repeated in cycles had been thought of in terms of a "chain-reflex" model. According to this view, the execution of each part of the whole cycle of a movement activates receptors in the muscles, skin, and joints, that act as triggers for the next subunit of the behavior. Wilson's experiments tested this proposal directly with insect flight, and showed conclusively that the chain-reflex idea was wrong. Instead of providing signals that evoke the next element of movement, the stretch receptors of the locust's wing provide a signal that is averaged over many wingbeat cycles and tends to compensate for unexpected variations in load or lift of the wings. In fact, the receptors have a function not unlike that proposed by Merton for the muscle spindles, which also serve to compensate load. Many neurobiologists now share Wilson's view that in insects most of the neural networks for motor control generate patterns of signals to the muscles through connections programmed by the genes, and *not* in response to sequential information sent to the central nervous system from the movement. Wilson's final proof of the centrality of the motor pattern of flight is that it can be produced by an isolated central nervous system stimulated only by a random pattern of electrical pulses.

Deafferentation—removal of all sources of sensory input—is the ultimate test for whether a particular behavior depends upon purely central patterning. It has worked in several experiments on invertebrates, including some of those described in "Small Systems of Nerve Cells"—where it is shown that complex patterns may be released by the stimulation of single central neurons. But the test has proven difficult or impossible in studies on the locomotion of vertebrates, despite more than thirty years of effort. Early experiments conducted on fish and amphibians in the 1930's and 1940's showed that swimming or walking movements survived sectioning of most of the sensory pathways, but when the last pathways were cut, the animals became immobile. The results do not argue against a central pattern generator and in favor of reflex control, since the experimental animals may simply have been deprived of so much nonspecific excitation by the deafferentation that a pattern of movement could not emerge. Recently, such tests have been more decisive. In several lower vertebrates, normal locomotor patterns have been observed after all or nearly all spinal sensory inputs have been cut; it is highly unlikely that sensory feedback through the surviving sources has any relation to the way the limbs move. Central generation of patterns of movement in mammals has been demonstrated too. A group of investigators in the Soviet Union, at Moscow State University, have developed an experiment in which a brain-sectioned cat will walk on a treadmill if electrical stimulation is supplied to a particular area in the midbrain. Deafferented single limbs retain their normal cycles of movement on the treadmill, and eventually it may be possible to apply the ultimate test for centrality. At the moment, neurobiologists can only speculate that the statements made by Wilson about the central control of locomotor rhythms in insect flight will be applicable also to higher vertebrates.

In the second article, "Moths and Ultrasound" (1965), Kenneth D. Roeder discusses another navigation system; this one is composed of just four sensory cells and is employed in evading a predator. The predator is a bat, which emits ultrasonic cries and uses the echoes of these cries to determine the location of insect prey. Some of the prey species, however, have evolved accoustical receptors sensitive to the same frequency range employed by the bats. Moreover, the sensory cells of moths have the capability to locate the sources of ultrasonic signals; the complex interactions between a moth's uses of its sensory cells and variations in the position of its wings during the flight cycle has been analyzed in detail. Especially interesting is the fact that different frequencies of sensory discharge, which depend upon differences in sound intensity, may produce alternate forms of behavior. Faint sounds cause the moth to turn directly away from the sound source, and loud sounds evoke erratic, evasive flight. In most sensory systems, changes in the intensity of a stimulus merely change the intensity of the response; but in the sensory system of the moth, a curious "switch," which apparently changes the *character* of the response instead, is present.

The evolutionary struggle between moths and bats has its own fascinations. It is an example of the kinds of selectional races waged between prey and predator organisms of all sorts; but it is of special interest because it involves acoustical communication. Sonar is one of several sensory systems in which an organism uses acoustical communication in an *exploratory* way, evaluating the returned messages in terms of the movements that produced them. Other exploratory sensory systems are the electrical signals used for orientation by certain fish, and the exploratory manipulations of the hand in man and other primates. By having evolved a mechanism for intercepting the signals, the moths gain a measure of immunity.

The last selection considers a kind of behavioral performance that is uniquely human: language. Like other behaviors, this one has special brain areas devoted to it. But to study the central nervous control of language we cannot employ the usual strategy of analyzing comparable performance in animals related to man. Electrical stimulation of equivalent areas in monkey brains do not produce vocalization; in fact, the evolutionary development of the human cerebral cortex involves a selective expansion of those areas we know to be concerned with language. Thus there is every reason to believe that language is the most recently evolved human ability—one for which there is simply no adequate animal model.

In "Language and the Brain" (1972), Norman Geschwind describes the results of an alternative strategy. Careful study of clinical cases in which language is impaired tells much about the localities within the brain that control language. It is, in the first place, lateralized—restricted, in 95 percent of all humans, to the left hemisphere of the cortex. Separate areas seem to control the primary encoding of speech and in its subsequent translation into appropriate output, whether written or spoken. Recent work described by Geschwind has led to a better understanding of the communication between the cortical hemispheres that controls language, and helps to explain some perplexing deficits in reading, speaking, and writing ability.

But the big questions in language are still unanswered. Many modern linguists believe "grammar"—the *structure* of written or spoken speech—is innate; that is, it depends upon the properties of those special language areas of the brain, whose function is to encode thoughts into the speech stream. If so, grammar is potentially analyzable in the same terms as other brain functions, with the limitation that in humans we are not free to interfere in the brain experimentally. What strategy can we devise to go further with the problem?

27

How We Control the Contraction of Our Muscles

by P. A. Merton
May 1972

*Voluntary muscular movements are driven by a
servomechanism similar in many respects to the
automatic feedback system employed to control
power-assisted steering in an automobile*

Psychophysics is the branch of experimental science that deals with the relation between conscious mental events and physical events within and without the body. Most psychophysics is sensory psychophysics, which deals with the relation between a physical stimulus and the resulting sensation experienced by the subject. The object of sensory-psychophysical experiments is to gain understanding of the physiological mechanisms that lie between the stimulus and the sensation, and to be able to draw inferences about what goes on inside a sense organ, a nerve or the brain. Measurements of subjective sensory thresholds in any sensory mode (tactile, visual, auditory or whatever), perceptions of color matches and judgments of the pitch of a note or the direction of a sound are examples of sensory-psychophysical observations. Sensory psychophysics is an old and highly respectable subject. In the hands of such investigators as Thomas Young, Jan Purkinje, Hermann von Helmholtz, James Clerk Maxwell, Lord Rayleigh and their modern successors it has told us a great deal about vision, hearing and other senses. Young's celebrated three-color theory of color vision, published in 1802, was formulated entirely on psychophysical evidence and is the basis of modern color photography and color television.

The other branch of psychophysics, motor psychophysics, does not have these credentials. It deals with the reciprocal problem, the relation between a conscious effort of will and the resulting physical movement of the body. It is just as important to know how we move as how we feel, but on the motor side much less has been achieved, partly, I suspect, because physiologists for metaphysical reasons feel that conscious volition is a faintly disreputable thing for them to have dealings with.

In sensory psychophysics it is easy to find illustrative examples of sensory phenomena that have an analytical character, that is, examples that provide some insight into sensory mechanisms, but on the motor side it is not so easy. I can think of one striking instance. A motor psychophysical fact of immense everyday importance is the individuality of a person's signature. Whenever Mr. X makes the appropriate volitional effort and signs his name, it always comes out the same (or enough so to be recognizable) and different from what anyone can write if he tries to write the same name. This is not an analytical observation; it is just a mysterious physiological fact, which we take for granted because we are so familiar with it. What does tell us something, however, is the further observation that if Mr. X takes a piece of chalk and signs his name in large letters on a blackboard, it again comes out the same. The muscles used are different but the individuality remains. From this observation we learn something about the organization of the motor system.

In this article evidence from both branches of psychophysics is taken into account, but the main object is to redress the balance in favor of the motor side. In more concrete terms we ask: What has been learned by making observations on voluntary movements in man about the physiological mechanisms that make our muscles do what we expect of them? Not, of course, very much. The title of this article is somewhat pretentious, as titles will be. There are a few definite phenomena to describe. With them we reach a new point of view, from which I hope we can see a general line of advance. I shall stick to simple movements and not come close to explaining the individuality of handwriting. (That subject was introduced partly to advertise the fact that sensory physiologists do not have all the glamor problems.) It will be useful to start by drawing an analogy between the human body and an automobile.

In the old days the steering wheel of a motorcar was directly connected to the road wheels by a series of levers and linkages, and the brake pedal similarly applied pressure directly to the brake shoes. On coming to a hill a gearshift could be moved to engage a suitable pair of gears to climb the hill with.

Today, in order to enable the driver, no matter how frail, to control a massive vehicle with the flick of a wrist or ankle, sophisticated mechanisms have been developed to assist with steering, braking and gear-shifting. All these mechanisms have devices (sensors, we may call them) that measure some physical variable (for example brake pressure or engine revolutions) and use the "feedback" information from them to control the mechanism that assists the driver. Let us concentrate on the mechanism that assists with steering. In its essentials it works as follows. Each position of the steering wheel corresponds to a certain angle of the front road wheels that the driver would like them to assume with respect to the fore-and-aft axis of the chassis. A sensor at the bottom of the steering column detects the difference between this "demanded" position and the actual position of the road wheels. Signals from

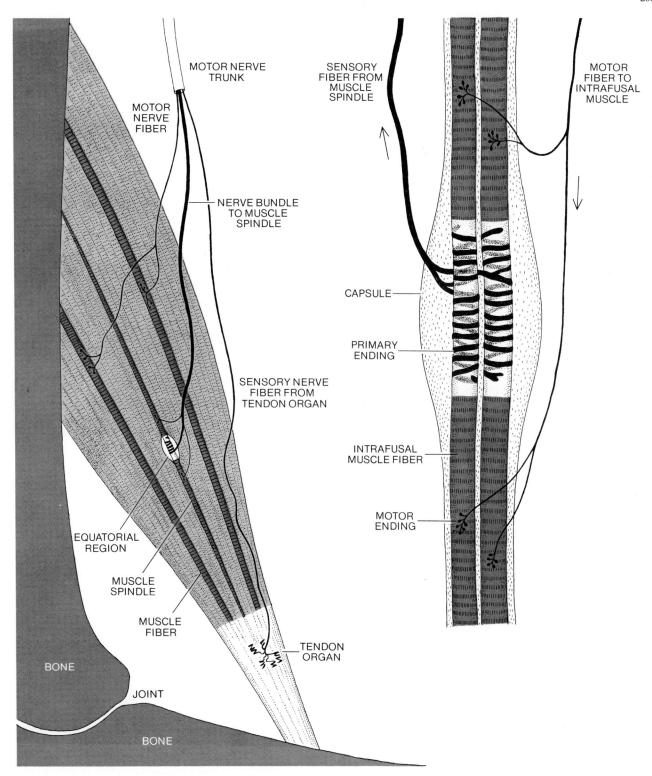

MOTOR NERVE TRUNK

MOTOR NERVE FIBER

NERVE BUNDLE TO MUSCLE SPINDLE

SENSORY NERVE FIBER FROM TENDON ORGAN

EQUATORIAL REGION

MUSCLE SPINDLE

MUSCLE FIBER

TENDON ORGAN

BONE

JOINT

BONE

SENSORY FIBER FROM MUSCLE SPINDLE

MOTOR FIBER TO INTRAFUSAL MUSCLE

CAPSULE

PRIMARY ENDING

INTRAFUSAL MUSCLE FIBER

MOTOR ENDING

ARRANGEMENT OF SENSE ORGANS in a typical muscle is indicated in these simplified diagrams. The proportions in the diagram at left are highly distorted. A real muscle fiber is only about a tenth of a millimeter in diameter, but it is often several centimeters long. A muscle spindle is somewhat thinner; it consists of even finer specialized structures called intrafusal muscle fibers. Only two ordinary muscle fibers and one spindle are depicted in detail; a real muscle may contain tens of thousands of muscle fibers and hundreds of spindles. The diagram at right gives an enlarged view of the equatorial region of a muscle spindle. Wrapped around the intrafusal muscle fibers are the terminations of the sensory nerve fiber; the function of these sense endings is to respond to mechanical deformation by causing nerve impulses to be sent up the sensory

nerve. In the equatorial region the cross striations, which are an indication of the presence of a contractile mechanism within the fiber, are absent. Hence when the intrafusal fibers contract, this region is extended and excites the sensory endings, just as if the region had been extended by stretching the entire muscle and the spindle within it. In this diagram only two intrafusal fibers are shown; a real spindle often has half a dozen or more. Moreover, intrafusal fibers come in two distinct varieties, only one of which is shown here. Another complication is the fact that there are three distinct kinds of motor nerve to the intrafusal fibers. In a real spindle the equatorial region is also much longer than depicted here. Photomicrographs showing the innervation of a tendon organ and the equatorial region of a muscle spindle appear on page 302.

the sensor, called the misalignment detector, are used to turn on a small servomotor (from the Latin *servus*, meaning slave), which turns the road wheels in such a direction as to cancel the misalignment. Thus the road wheels are made to point in the direction the driver wants, without his having to exert himself. As he turns the steering wheel the road wheels follow automatically.

Such is power-assisted steering. An engineer calls it a follow-up servomechanism. An important point to note is that, the function of the device being to help the driver automatically, he does not want to be bothered with the details of its operation; in particular he would only be distracted from his task of keeping his eyes on the road to see where to steer if signals from the sensor were relayed to him. They give information that is relevant only to the functioning of what ought to be a completely subservient mechanism, and they should remain private to that mechanism.

Power-assisted steering relieves the driver of physical effort only; other such devices relieve him of mental effort too. The automatic transmission, for example, does away with the need to decide when to change gear, as well as the need to perform the change. In an aircraft the automatic pilot does everything and leaves the human pilot completely free.

In the human body there are numerous automatic feedback mechanisms of this kind controlling physiological functions without any mental effort on our part. For instance, the blood pressure

and the output of the heart are controlled so as to suit the current needs of the body; we are quite unaware of the functioning of these systems and of the signals from the pressure sensors in the walls of the arteries and elsewhere that are a part of them.

Such mechanisms are commonplace physiology; they are in the textbooks for medical students and nurses. When we come to muscle, however, the situation is different. To return to our analogy, in the case of the automobile we know what we want to control—direction, speed or retardation—and the problem is to design servomechanisms to help the driver, with appropriate sensors in each instance. The signals from the sensors are just part of the engineering technology, and so we do not display them on dials on the dashboard. They would only put the driver off. In the human machine we have muscles to control. How do we do it? Do the orders to contract go directly from the brain? Presumably not, since on examination it appears that muscles, like the automobile, are equipped with sensors of their own, of whose signals the owner of the muscles, like the owner of the automobile, remains unaware. Presumably, like the sensors in the automobile, they are taking part in automatic mechanisms that assist the subject in controlling his muscles. What are they helping to control? Muscle tension perhaps? It could be; some of them measure tension. Length? Others of them respond to changes in muscle length. A combina-

tion of tension and length? Sometimes tension and sometimes length? Now we see the nature of the problem. It is the inverse of the automobile designer's. We are presented with the sensors and we have to discover what the mechanism they are part of was designed to do. What precisely do we ask of our muscles that they need these confidential sensors to make them do it? It is by no means obvious.

Having thus briefly sketched the picture, let me now go into the physiology in more detail. It falls into two sections. The first presents the evidence that muscles incorporate sensory receptors of whose signals we are not consciously aware; the second discusses what is known of the mechanisms in which they take part.

In the 18th century the great Swiss physiologist Albrecht von Haller established for the first time that the internal organs of the body, such as the heart, the stomach and the brain, are in general insensitive to the kind of stimuli that are so readily felt by the skin: pricking, pinching, cutting, burning and so forth. It is this fact that enables surgeons to perform operations on, say, the brain substance with only local anesthesia around the incision. In his studies of muscle Haller found that stretching a muscle by pulling gently on the tendon exposed in a wound in a human subject did not cause sensations of either movement or tension. (Pulling hard, however, is painful.) Reflecting on Haller's observations, one can perceive that the viscera and the muscles are really in different categories. It is not at all surprising, when one comes to think of it, that the liver should be insensitive to cutting with a knife or burning with a cigarette; such stimuli would be so rare without the animal's getting an earlier and more effective warning from the abdominal skin that to develop a system to report them would give the animal a negligible evolutionary advantage, whereas sensitivity to mechanical contacts, which the skin preeminently possesses, would very likely be a positive disadvantage. Imagine what life would be like if throughout it one were as vividly aware of the beat of one's heart as the surgeon who puts a finger on it exposed during an operation! With muscle, however, it is quite otherwise. It might be useful for us to be conscious of how extended our muscles are at any moment, since that determines the position of our limbs, and also to know their rate of shortening or elongation and the tension in them. If we are to be-

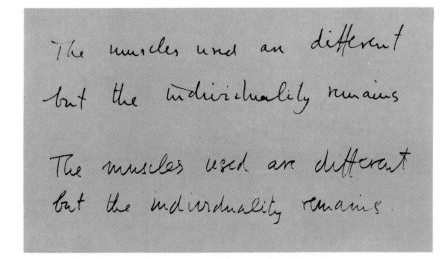

STRIKING EXAMPLE of a simple experimental observation that provides some insight into the organization of the motor psychophysical system is represented by these two hand-written versions of a sentence taken from the text of this article. The sentence was written large on a wall with a felt-tipped pen (*top*) and small on a piece of paper with a fine mapping pen (*bottom*). The writing on the wall is about 10 times larger. The large writing was done by movements of the wrist, elbow and shoulder, whereas the small writing used muscles in the hand itself. Nevertheless, the character of the writing is the same in both cases.

lieve Haller, this is just what they do not tell us. For this and other reasons that will shortly emerge we find that, whereas the insentience of the viscera has long been received as physiological dogma with a status comparable to the circulation of the blood, the insentience of muscle has often been called in question and probably cannot be regarded as universally accepted even today.

For me the question was both raised and answered the day I read the arguments of Helmholtz, published in 1867 in his *Handbook of Physiological Optics*. Helmholtz reached the same conclusion as Haller by experiments with the eye, which have the merit that anyone can repeat them and convince himself of the facts. Helmholtz starts with the familiar observation that if one takes hold of the skin at the outer corner of the eyelids and jerks it sideways, the eye itself is moved and what one sees with that eye appears to jump about. On the other hand, we know that if one moves one's eyes voluntarily, the scene one is looking at does not appear to jump. Helmholtz argues as follows. In both cases the image of the external world moves over the retina as the eye moves. When one moves one's eyes actively, by voluntary effort, one allows for the eye movement and does not interpret the movement of the image on the retina as signifying a movement of the external world. When the eyes are moved passively by an external pull, however, one interprets what one sees as if the eye had remained still. The movement of the image on the retina is assumed to be due to a movement of the external world and not to a movement of the eye. Hence we only know in which direction our eyes are pointing when we move them voluntarily, and this must be because we make an unconscious estimate of the effort put into moving them. (We have a "sense of effort.") Sense organs in the eye muscles (or elsewhere around the eye, if there are any) do not tell us which way our eyes are pointing, because when the eyes are moved passively, we do not seem to know they have moved.

This argument, as it stands, is not conclusive, because when the eyelids are pulled, the sense organs in the eye muscles or elsewhere might not be excited in the same manner as when the eye is turned normally by the contraction of its muscles. The apparent movement of the external world during a passive movement of the eye might therefore be due to a misjudgment of the eye's direction rather than to a complete ignorance of its movement.

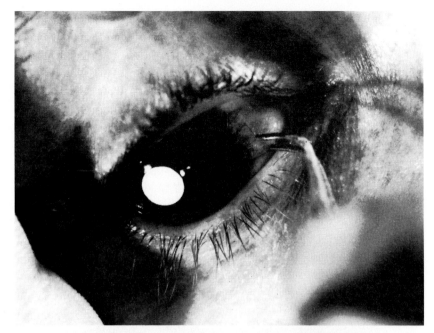

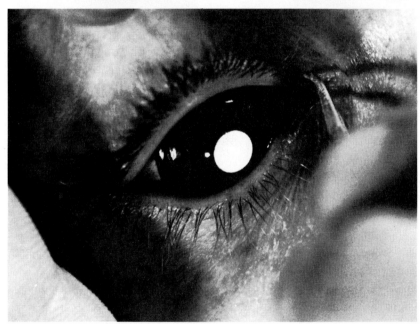

INSENTIENCE OF EYE MUSCLES was demonstrated a few years ago by means of an ingenious experiment devised by G. S. Brindley, now at the Maudsley Hospital in London. In these photographs, made in the course of the experiment, Brindley is manipulating the author's eye with forceps to test whether, after blinding it with a black cap, there was any awareness of passive movements. There was not. The white spot on the cap is to give an indication of eye position. The eye and the lids were treated with local anesthetic.

This objection, as Helmholtz argues, can be answered by considering afterimages. If one stares fixedly at a bright light for 15 to 30 seconds (please, not the sun!), then on looking elsewhere an afterimage of the bright light is perceived and persists for a minute or so. When an object is fixated steadily, the afterimage likewise stays still, but when the gaze is shifted, the afterimage also moves. This, of course, refers to active voluntary eye movements. In passive movements quite the opposite is found. No matter how hard one pulls on the eyelids the afterimage appears to remain completely stationary. In order to be certain of this phenomenon it is necessary to view the afterimage against a featureless background, such as a sheet of plain paper held close to the eye; otherwise the concomitant apparent jerking around of external objects may make the judgment difficult. Hence during passive movements we interpret what we see precisely as if the eye had not moved at all. It is not a matter of a quantitative misjudgment. The reader is encouraged to repeat for himself these crucial observations and reflect on the compelling conclusions Helmholtz drew from them.

A few years ago my friend G. S. Brindley (now at the Maudsley Hospital in London), who has a genius for settling or eliminating argument by incisive experiment, proposed that we confirm Helmholtz directly by blinding an eye with a black cap on the cornea (the eye's transparent front surface) and then moving the eye around with forceps to see if the subject could feel the movement. (Pain was prevented by instilling gener-

ous quantities of local-anesthetic eye drops.) The test proved that subjects are quite unaware of large passive rotations of the eye in its socket of 30 degrees or more; they do not know the eye is being manipulated at all unless the forceps happen to touch the eyelid. Another important point was that if the subject was invited to voluntarily move his eyeball while the forceps were gripping it, he was unable to tell whether the experimenter holding the forceps was allowing the movement to take place or was preventing the eye from moving.

The unequivocal conclusion of all these experiments is that we have no sense organs in the eye muscles or near them that tell us which way our eyes are pointing. We normally know which way

we are looking, but only because an internal "sense of effort" gives us an estimate of how much we have exerted our eye muscles. If voluntary movements are artificially impeded, or if passive movements are imposed, we absolutely do not know what is going on—unless we can see and reason back from the visual illusions we receive.

So much for the eyes. In the limbs the same facts are less easily demonstrated. To use Haller's method with patients whose tendons have been exposed under local anesthetic in the course of an orthopedic operation is one possibility, but it does not satisfy the powerful compulsion that all investigators in sensory physiology have to try it for themselves. A paper on visual illusions in which the author had not experienced the phenomena himself is almost unthinkable, and rightly so. What better way could he have of satisfying himself that they were correctly reported? Hence it is desirable to find a method for studying muscular sensibility in ordinary limb muscles of healthy subjects. The difficulty, of course, is to devise a way of stretching a muscle without the subject's knowing what is being done, since he can feel pressure on the skin or the movement of a joint. Local anesthesia of an extremity provides an answer. Investigators have variously injected local anesthetic around the joint at the base of the big toe or at the base of a finger, or have anesthetized the entire hand by cutting off the blood supply with a pneumatic tourniquet around the wrist for about 90 minutes. Movement of an anesthetized digit then stretches the muscles that move it, which lie above the anesthetized region. My collaborators and I use the top joint of the thumb, which has the advantage that only one muscle (lying well up in the forearm) flexes it, whereas the joints of the fingers are operated by more than one muscle, some in the hand and some in the forearm. Thus when the thumb is anesthetized by a tourniquet at the wrist, voluntary movements of the top joint are unimpaired in strength. We have also used injection of local anesthetic around the base of the thumb.

The uniform result of numerous experiments is that, with an adequate depth of anesthesia, the subject (whose eyes are shut) cannot tell in what position the experimenter is holding the top of his thumb, or whether he is bending it backward and forward. This is true only provided that the movement is not rapid and that the thumb is not forcibly extended or flexed at the limits of its range of movement. It is also the case

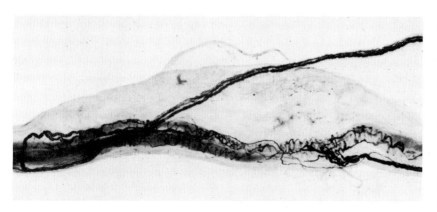

TENDON ORGAN contains sense endings that signal to the nervous system the tension in the part of the muscle in which they lie. A typical location of a tendon organ is shown in the diagram on page 299. The single sensory nerve fiber that services the tendon organ has been made to appear black in this photograph by means of a special silver stain. The nerve fiber divides many times, terminating in very fine branches with knobs at the ends. These structures, in some unknown way, sense the deformation produced by tension and cause nerve impulses to be sent up the sensory fiber at a rate that is determined by the tension. This tendon organ was dissected out of the leg muscle of a cat; it is about half a millimeter long. Surrounding one end are the remains of muscle fibers. Both photographs on this page were made by Colin Smith, Michael Stacey and David Barker of the University of Durham.

EQUATORIAL REGION of a muscle spindle dissected from the leg muscle of a rabbit appears in this photomicrograph; the part shown is about a millimeter long. Again the nerve fibers and nerve endings have been stained with a silver stain, making it possible to distinguish clearly the equatorial capsule, the intrafusal muscle fibers and the sensory endings wrapped around them. The nerve ending to the right is a primary ending; its sensory nerve fiber enters from lower right. The other ending is a secondary ending; its nerve fiber enters from upper right. The finer nerve fibers are part of the motor nervous system.

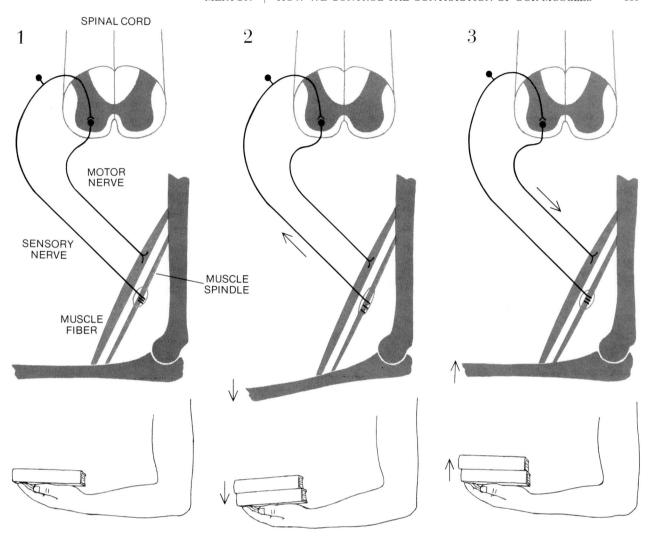

STRETCH REFLEX is mediated by the nervous mechanism depicted in this highly schematic illustration. A muscle is under the influence of the stretch reflex when it is engaged in a steady contraction of a voluntary nature, as when a person's elbow is flexed steadily against a load (*1*). A sudden unexpected increase in the load (*2*) stretches the muscle, causing the sense ending on the muscle spindle to send nerve impulses to the spinal cord (*upward arrow*), where they impinge on a motor nerve cell at a synapse and excite it. As a result motor impulses are sent back down to the muscle (*downward arrow*), where they cause it to contract (*3*). More complicated nervous pathways than the one shown may also be involved in the stretch reflex. Any real muscle is, of course, supplied with many motor nerve fibers and spindles. In addition the synaptic connections to even a single motor nerve cell are multiple.

that if the subject attempts to flex his thumb, he cannot tell whether he has been successful, or whether the experimenter has prevented it from moving. Thus with skin and joint sensation eliminated the thumb behaves just like the eye. Muscle *is* insentient.

I have already argued that one would not on general grounds expect the liver, say, to have sensibility like the skin's. Indeed, if one looks at the liver through a microscope, it has none of the elaborate apparatus of sensibility seen in the skin—no network of branching nerve fibers ending in a variety of characteristic sensitive structures: the sense organs. The same goes for other viscera. Muscles are not so obliging. They are supposed to be insentient, but when we look inside them, they turn out to be full

of sense organs, and very fine sense organs at that. The principal kind, the muscle spindles, are the most elaborate sensory structures in the body outside of the eyes and ears. This deep paradox (for which the reader has already been prepared) is at the back of everything in this article. All the essential facts that create it have been known since 1894, when Sir Charles Sherrington proved conclusively that there were nerve fibers going to the muscle spindles that belonged to the body's system of sensory nerves, and hence established that the muscle spindles were sense organs. Unfortunately in those distant days Sherrington was insensitive to the class distinction between the information on the road sign that tells the driver to turn right and the information from the sensors in his power-assisted steering gear

that enables him to do so effortlessly (between, one might say, the different types of information required by the legislature and the executive). He allowed himself to be persuaded that Helmholtz had been wrong and that his own discovery showed that muscles were sentient after all.

Sherrington had thus taken the view that in effect there was no paradox, and his influence was so immense that it was 60 years before the true situation was at last clearly perceived. By this time the paradox had much less impact, since physiologists had discovered many of the facts about the muscle spindle needed for its resolution. Before going on to these facts I should finish the present story.

In the past few years the paradox has been given a further twist. Several groups of workers on both sides of the Atlantic, whose members are too numerous to name individually, have found that signals from muscle sense organs find their way to the cerebral cortex. It seems that they get to the cortex but we remain unconscious of them. This is very surprising. No one imagines for a moment that we do not make use of all the information our eyes send to the cerebral cortex to build up the picture of the outside world we consciously perceive, and I am sure that a few years ago any ordinary physiologist would have been prepared to extend this point of view to sensory information of any kind that could be shown to get to the cortex.

The evidence for what I have just said is not complete. The animal most resembling man in which signals from muscle sense organs have been shown to reach the cortex is the baboon. It seems unlikely that they do not reach the cortex in man, and equally unlikely that a baboon should be conscious of the signals from its muscles when a man is not. A strong hint also comes from the cat. John E. Swett and C. M. Bourassa of the Upstate Medical Center of the State University of New York showed that muscle sense organs send signals to the cat's cerebral cortex, but unlike signals from the skin (or for that matter from the eyes or ears) they cannot be used to set up a conditioned reflex. Without explaining what is meant by this fact in detail one can say that it strongly suggests the cat is not conscious of the signals from its muscles.

The first part of this article was intended to introduce the reader to the idea that muscle organs function at a subconscious level in a purely subservient role. Like the perfect servant, they work so unobtrusively that we are unconscious of them, but the findings about cortical projection begin to strain the analogy. The eccentric 18th-century scientist Henry Cavendish reportedly dismissed any servant he caught sight of. He wrote down what food he wanted and it was put out for him. It would have been going too far to expect the butler to wait on him at table without betraying his presence, but that is what the muscle sense organs seem to manage to do!

Scarcely less remarkable than the mere existence of the muscle spindles is the fact that they (the most important of the two kinds of muscle sense organ) are themselves contractile. This is a unique property among sensory structures. That was perfectly clear to Sherrington in 1894, but it still remains one of the most challenging observations in the physiology of the motor system; even if the interpretations to be put forward later in this article are on the right lines, it is most improbable that they are more than one facet of the truth.

Muscle spindles (they are called spindles because they are long and thin and have pointed ends) consist of a bundle of modified muscle fibers, the intrafusal muscle fibers (from the Latin *fusus*, meaning spindle), with the sensory nerve fibers wrapped around a short specialized region somewhere near the middle of their length. The stimulus that excites a muscle spindle is the stretching of this specialized sensory region. Now, as I have said, the muscle spindles are contractile. They are not, however, equally contractile along their entire length; the contractile apparatus fades out in the sensory region, and the middle of the sensory region, where the sense endings connected to the largest nerve fibers lie, probably does not contract at all. When the spindle contracts, these sense endings (known as the primary endings) are stretched by the contraction of the remainder of the spindle and discharge nerve impulses.

The next point to observe is that the

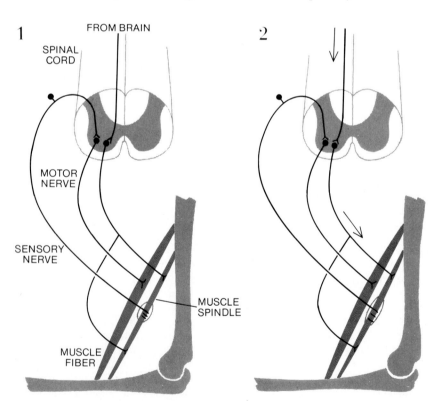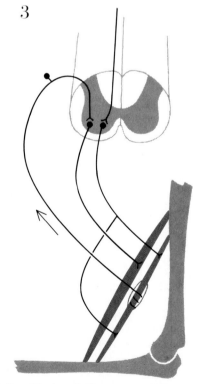

SERVOMECHANISM involved in the control of voluntary muscular contractions is shown here. The basic diagram (*1*) is the same as it is in the illustration of the stretch reflex, but with provision made for signals from the brain to cause the muscle spindle to contract by way of a special motor nerve fiber. When a signal is transmitted along this special fiber (*2*), the spindle contracts, exciting the spindle sensory ending, just as if the spindle had been stretched. Consequently a contraction of the main muscle is excited by way of the stretch-reflex pathway (*3, 4*). In a real muscle this picture is further complicated by the existence of a direct pathway

muscle spindles lie among the ordinary muscle fibers (the much larger red stringy structures, visible to the unaided eye, that actually do the work) and share their attachments to bone or tendon. Hence they change length as the main muscle fibers change length. If a contraction of a muscle spindle, which excites its primary ending, is succeeded by an equal contraction of the main muscle, the stretch will be taken off the sensory region and the ending will be silenced. The spindle primary, in fact, is sensitive to the difference in length between the spindle and the main muscle fibers; it is a misalignment detector. It discharges if contraction of the spindle is not matched by contraction of the main muscle, or, vice versa, if extension of the main muscle is not accompanied by relaxation of the spindle. There is no obligation for the muscle spindles and the main muscle to contract and relax together, because the motor nerve fibers that run to them and carry the nerve impulses from the central nervous system that cause them to contract are largely separate. The spindles could therefore be activated while the main muscle remained passive, and vice versa.

Having seen the circumstances under which nerve impulses are discharged by the spindle primary endings,

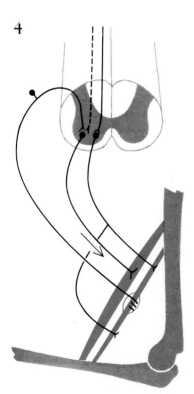

4

(*broken line in diagram 4*) from the brain to the main motor nerve cells. In the power-steering analogy this pathway corresponds to a direct connection between the steering wheel and the road wheels of an automobile.

the next question is: What do these impulses do when they reach the central nervous system? Their best-established function is to excite an automatic contraction—the stretch reflex—in the muscle from which they come. This they do, at least in part, by impinging directly on those nerve cells in the spinal cord that give rise to the motor nerve fibers to the muscle in question.

The most familiar manifestation of the stretch reflex is the knee jerk, widely used in medicine to test the state of the nervous pathways concerned. A physician strikes the tendon below the knee-cap with a rubber hammer, and in a healthy subject the muscles that straighten the knee briefly contract involuntarily. The effect of striking the tendon is slightly and suddenly to stretch these muscles, and so to excite their muscle spindles. The tendon itself has no part in the sensory mechanism. The tendon jerk is quite transient, but under suitable circumstances a slower, sustained extension of a muscle will result in a sustained reflex contraction. If the reaction in a patient who is otherwise relaxed is exaggerated, the limb is said to be "spastic," that is, affected by spasm.

Human muscles in general can be shown to be under the influence of the stretch reflex when they are engaged in steady contractions of a voluntary nature. The main evidence for this is that if a subject is invited, say, to flex his elbow steadily against a load, it is found that a sudden unexpected increase in the load, which causes his elbow to extend, calls up a larger contraction of his biceps muscle, and conversely a decrease in load causes a relaxation. Electrical recording methods reveal that these reactions begin so soon (within about a twentieth of a second) that they must be automatic, reflex responses.

It has been realized for half a century that the stretch reflex confers valuable self-regulating properties on a muscle, causing it automatically to adjust to changes in load, without any need for the orders that the brain sends down to be altered. Everyone believes the reason the horse does not sag at the knees when Douglas Fairbanks leaps from the castle parapet onto its back is that the horse's leg muscles immediately respond to the extra strain by way of their stretch reflexes. If this interpretation is correct, we have one answer to the question: What does the horse expect of its muscles? In this situation it expects them not only to exert enough force to support its body weight but also to adjust automatically to extra weight. Clearly what the horse really wants is for the

length of the muscles to be kept roughly constant so that posture is maintained. The stretch reflex can achieve this result for the horse because it is based on a sensor—the muscle spindle—that measures length, or, to be more exact, differences in length.

What happens when it is desired that the muscles should execute a movement, not merely maintain a stationary posture or some other steady contraction? The obvious trick is to cause the spindles to contract at the desired rate so that the sensory endings on the spindles will be excited if the main muscle does not itself keep up with the spindles, that is, does not contract at the desired rate. In this way the advantages of automatic compensation for changes of load by means of the stretch reflex could be retained during active shortening. Contraction of the spindles would in effect drive the main muscle by means of the stretch reflex, turning on more contraction if an unexpected obstruction were met with, or if the rate of shortening for any other reason fell behind, and, vice versa, damping down contraction automatically if the load unexpectedly diminished, or if for some other reason the movement undesirably accelerated. Within the past year C. D. Marsden, H. B. Morton and I have obtained direct evidence that this kind of rapid, reflex compensation does in fact occur during voluntary movements in man.

In this mode of operation the stretch reflex, as the reader will have perceived, functions as a follow-up servomechanism, closely analogous to power-assisted steering in an automobile. Contraction of the spindle corresponds to turning of the steering wheel, shortening of the main muscle to turning of the road wheels, with the spindle sensory ending acting as the misalignment detector. The subject can demand of his muscles either a certain limb position or a certain rate of change of limb position, and within limits (limits not yet known in quantitative terms) his demands will be automatically met by his muscle servo.

That, in brief outline, is as far as we have gone in understanding how, when we make a voluntary effort, the muscle sense organs act at a subconscious level to ensure that our muscles do what we expect of them. Many facts have had to be left out and without doubt many more remain to be discovered. To attempt any account at this stage requires a certain presumption. I can only hope that when the whole truth emerges, it will prove to be an extension and not a contradiction of the story I have told here.

28

Small Systems of Nerve Cells

by Donald Kennedy
May 1967

In some invertebrate animals complete behavioral functions may be controlled by a very few cells. This makes it possible to trace out the interactions of the cells and so to investigate nervous integration

The nervous system of a man comprises between 10 billion and 100 billion cells, and the "lower" mammals men study in an effort to understand their own brains may have two or three billion nerve cells. Even specialized parts of vertebrate nervous systems have an awesome number of elements: the retina of the eye has about 130 million receptor cells and sends more than a million nerve fibers to the brain; a single segment of spinal cord controls the few muscles it operates through several thousand motoneurons, or motor nerve cells, which in turn receive instructions from a larger number of sensory elements.

These vast populations of cells present a formidable challenge to biologists trying to understand how the nervous system works. Since the system is made up of cells, one would like to understand it in terms of cellular activities, and by examining the activity of single nerve cells investigators have been able to learn a great deal about the nature of the nerve impulse and about the generation and transmission of the patterns of impulses that constitute nervous signals. The ultimate object must be, however, to understand not only the activities of single cells but also the rules of their interaction. Since one cannot expect to understand even the most restricted systems by predicting the possible interactions of an inadequately sampled population of cells, mammalian physiologists have devised ingenious ways of circumventing the superabundance of elements, involving particularly biochemical studies of regions of the brain and sophisticated computer analyses of brain waves. The trouble is that most of these methods treat cells as anonymous members of a population rather than as interacting individuals.

Some biologists are taking a different approach, one that retains the individual nerve cell as the focus of attention and yet attempts to deal with groups of them. This approach is made possible by the availability of animals that have fewer nerve cells than any vertebrate and that nonetheless display reasonably complex behavior. The claw-bearing limb of the shore crab, for example, shows impressive coordination and range of movement, made possible by six movable joints with pairs of muscles acting in opposition to each other. As a mechanical device it is in most ways the equal of a mammalian limb, yet the crab operates all this machinery with about two dozen motor nerve cells. A mammal of comparable size would employ several thousand for the analogous purpose.

Nor is such parsimony confined to the motor apparatus: in contrast to the several billion cells of the entire mammalian nervous system, the crab has only half a million or so. The real utility of such systems to the investigator becomes apparent only when one concentrates on the nerve cells belonging to one functional unit such as a reflex pathway, a special sense organ or a particular pattern of behavior. Indeed, the nervous systems of some of the higher invertebrate animals are so economically built that for certain functions one may hope to specify the activity of every individual cell.

This achievement would be a hollow one insofar as learning about mammalian nervous systems is concerned if the additional complexity of more "advanced" systems depended heavily on new capabilities of the individual cells comprising them. All the evidence, however, indicates that the performance limits of the single nerve cell are already reached in relatively simple animals.

The central nervous elements found in the lobster or the sluglike "sea hare" nearly equal those of mammals in size, quantity of input and structural complexity. The marvelous performance of the mammalian brain is, it appears, not so dependent on the individual capabilities of its cells as it is on their greater number and the resulting opportunities for permutation. Therefore an understanding of the connections underlying behavior in a simple system can lead to useful conclusions about the organization of much more complicated ones.

The difficulty lies in the choice of appropriate experimental objects. Ideally one needs a nervous system that produces a reasonably complex repertory of behavior and has only a few cells, each of which can be recognized and located time after time. In certain animals specialized giant cells offer this ready identifiability. The giant axons, or nerve fibers, of the squid and the lobster are an example, and biophysicists have long exploited them for experiments on the properties of nerve-cell membranes. A student of integrative processes in the nervous system needs more; he wants to specify individual properties for each cell in an entire functional assembly.

Two different kinds of nervous system seem particularly promising for this purpose: that of mollusks and that of arthropods. The most thoroughly studied mollusk is the sea hare *Aplysia*, a snail that has only a vestigial shell and leads a somewhat more mobile life than its relatives. Its nervous system is concentrated in a few large ganglia connected by nerve trunks. Several features of the cells are advantageous. The cell bodies are unusually large, some with a diameter of .8 millimeter and the rest well sorted in size below this maximum. They contain

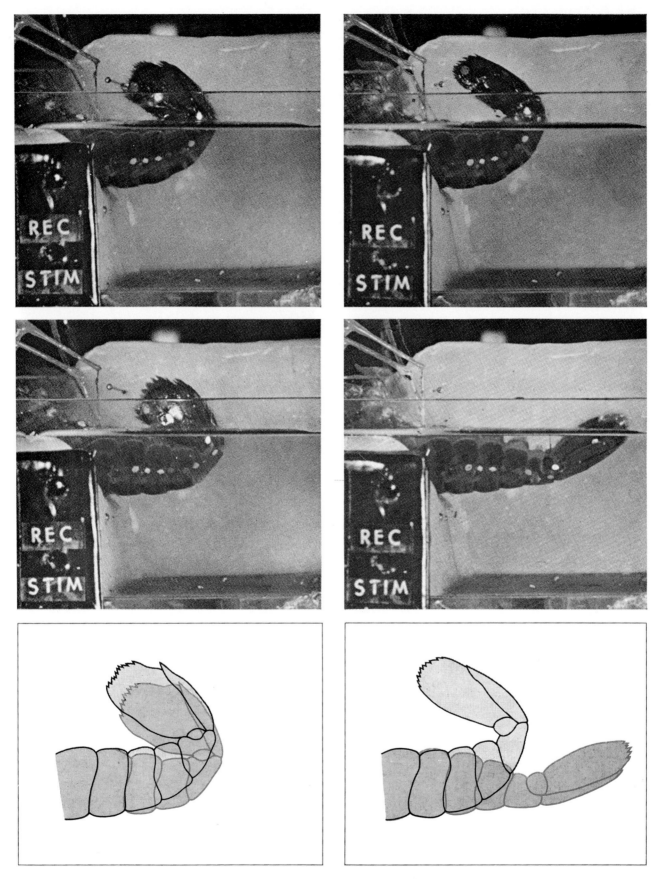

CRAYFISH ABDOMEN responds in complex and specific ways to stimulation of different single cells in the central nervous system. Frames from a motion-picture film made by Benjamin Dane in the author's laboratory record the effects of two different "command fibers"; the drawings specify the initial (gray) and the final (red) positions of the abdomen. One fiber evoked activity primarily in the forward segments (left); the other produced extension in all segments (right). Spots painted on the abdomen facilitated precise measurement of segment angles. Visible effects were confirmed by recording the activity of motor nerve fibers in the first segment.

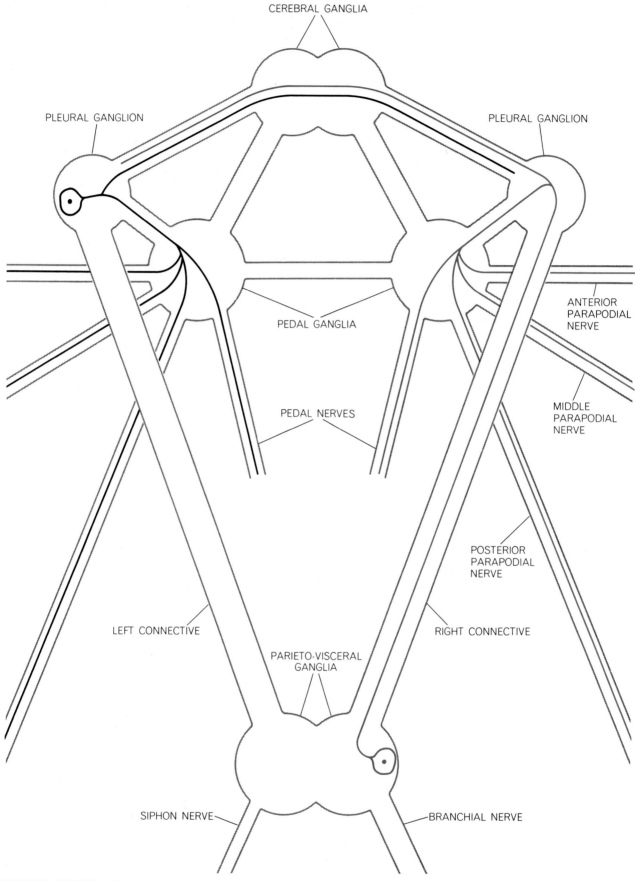

CEREBRAL GANGLIA

PLEURAL GANGLION

PLEURAL GANGLION

ANTERIOR
PARAPODIAL
NERVE

PEDAL GANGLIA

MIDDLE
PARAPODIAL
NERVE

PEDAL NERVES

POSTERIOR
PARAPODIAL
NERVE

LEFT CONNECTIVE

RIGHT CONNECTIVE

PARIETO-VISCERAL
GANGLIA

SIPHON NERVE

BRANCHIAL NERVE

NERVOUS SYSTEM of the "sea hare" *Aplysia,* a mollusk, lends itself to investigation because its cells are few, large and identifiable. Here the entire system is diagrammed and the paths of two nerve cells are shown in color. One of them (*red*) is cell No. 1 in the ganglion illustrated on the opposite page. The routes of a large number of *Aplysia* cells were worked out by L. Tauc of the Centre National de Recherche Scientifique in Paris and G. M. Hughes of Bristol University by recording from individual nerve cell bodies.

a variety of yellow and orange pigments in different proportions. Particular cells are consistent in position from one animal to the next. Several dozen cells can therefore be reliably recognized in different individuals by their size, color and position [*see illustrations on this page*].

Aplysia was first intensively studied in the late 1940's by A. Arvanitaki-Chalazonitis in her laboratory in Monaco. It has since attracted a number of investigators, notably L. Tauc of the Centre National de Recherche Scientifique in Paris, Eric R. Kandel of New York University and Felix Strumwasser of the California Institute of Technology. Tauc and his collaborators, notably G. M. Hughes of Bristol University, have constructed ingenious physiological maps of several *Aplysia* cells [*see illustration on opposite page*]. They accomplished this by inserting a glass microelectrode into a cell body and placing wire electrodes on all the nerve trunks connecting with the ganglion in which the cell body was located. If a particular nerve contained an axon of the cell in which the microelectrode was located, the microelectrode recorded an impulse when that nerve was stimulated with a brief electric shock. The branches of the axons may be arranged in an extremely complex way, but each cell is characterized by a constant arrangement of branches.

Other workers have demonstrated that the connections made in turn by such branches with other nerve cells are also constant. Microelectrodes were inserted in several identified cells, and one microelectrode was used to stimulate the cell it had penetrated while the others recorded impulse activity. A particular cell produced either of two kinds of effect in a nerve cell to which it was connected: an excitation, sometimes strong enough to evoke an impulse in the second cell, or an inhibition, which opposed the discharge of impulses. Strumwasser and Kandel have demonstrated that a single cell can directly excite some cells and inhibit others, and that a certain set of identifiable cells is always excited by a given cell and another set always inhibited.

The ganglia of *Aplysia* have been used in the investigation of two other important problems in nerve physiology: the cellular modifications that take place during "conditioning" and the origin of "discharge rhythms." To investigate the first, Kandel and Tauc recorded from an identifiable cell with a microelectrode while stimulating two nerve trunks containing nerve fibers that excited the recorded cell. A shock delivered to one of

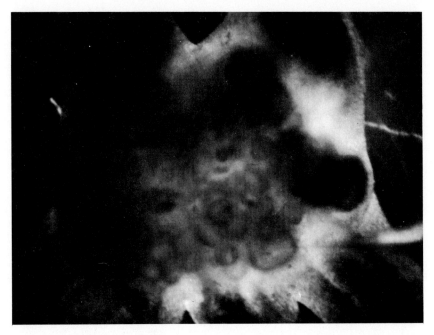

PARIETO-VISCERAL GANGLION of *Aplysia*, photographed unstained by Felix Strumwasser of the California Institute of Technology, measures about 2.5 millimeters across.

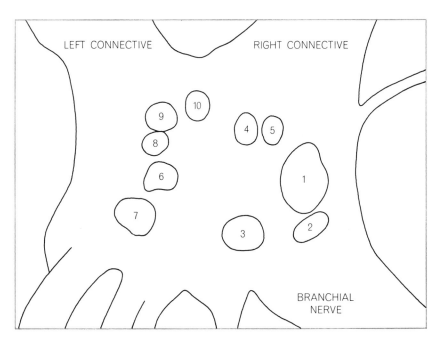

INDIVIDUAL CELLS that are consistently identifiable in the ganglion are diagrammed. Not all are clearly seen in the photograph, in which a microelectrode points to cell No. 3.

these nerves produced strong excitation, indicating that it was the more effective pathway; the other produced only a weak effect. If the shocks to the two pathways were repeatedly paired so that the weak followed the strong at a fixed, short interval, the response to the weak pathway became "conditioned" to the strong excitation; the response increased dramatically and stayed elevated for 15 minutes or more after the conditioning period. If the same number of shocks

was delivered to both pathways for the same period but at random intervals, conditioning did not occur; the response to the weak pathway was not affected. Such systems promise that electrophysiological studies on learning may at last be brought down to the cellular level.

Nerve cells often discharge rhythmically, in bursts that last for a few seconds and are separated by longer intervals. Strumwasser discovered a particularly dramatic instance of rhythmicity

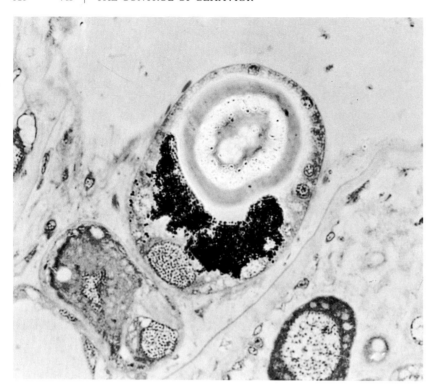

EYE OF *HERMISSENDA* is enlarged about 900 diameters in a photomicrograph made by the late John Barth at Stanford University. The longitudinal section shows the lens (*concentric rings*) and the retina below it. The granular gray structure just below and to the left of the black pigment screening the retina is the nucleus of one of the five receptor cells.

in a certain *Aplysia* cell whose activity—recorded over long periods of time in an isolated ganglion—exhibits a "circadian," or approximately 24-hour, rhythm. Its most active period is a few hours before what would be sunset on a normal light schedule. This inherent periodicity, like that of other biological clocks, is maintained even if such environmental variables as light and temperature are held constant. Strumwasser is now investigating the chemical changes that are correlated with the discharge cycle.

The large size and small number of nerve cells in certain mollusks makes them convenient systems in which to study a number of sensory phenomena. The mollusk *Hermissenda*, for example, has provided information on the nerve-cell interactions that underlie the processing of visual information in a simple "camera" eye: an eye in which a single lens focuses light on a retina of receptor cells. Vertebrate camera eyes have too many cells for such analysis, and so investigators have relied largely on simpler compound eyes, in which a number of independent elements, each with its own lens and sensory cells, send fibers to the central nervous system. Studies of such compound eyes, particularly in the horseshoe crab *Limulus*, revealed the

existence of one especially important process in visual systems: H. K. Hartline and his associates at the Rockefeller Institute found that when a given visual element is illuminated, it not only sends a train of impulses to the central nervous system along its own nerve fiber; it also inhibits the discharge of neighboring elements in the eye. This "lateral inhibition" decreases with distance from the visual element, and it functions to raise the level of contrast at boundaries of stimulus intensity [see "How Cells Receive Stimuli," by William H. Miller, Floyd Ratliff and H. K. Hartline, the article beginning on page 243].

Is lateral inhibition also an essential component of image-formation in other kinds of eye? What is the minimum number of light-receptor cells necessary to form a useful spatial representation of the visual field? What rules are followed in connecting them? These are among the questions investigated in our laboratory at Stanford University, first by the late John Barth and more recently by Michael Dennis, in the course of a study of the eye of *Hermissenda*. The eye measures less than .1 millimeter in its long dimension and consists of a lens, a cup of black pigment to ensure that light enters only from the right direction, and an underlying retina of receptor cells.

The remarkable feature of this miniature camera eye is that its retina consists of only five receptor cells, each one large enough to be penetrated with microelectrodes.

Barth found that cells exposed to light might respond in one of three ways: with an accelerated discharge, with a slower rate of firing followed by an "off" discharge (signaling the cessation of illumination) or with some complex mixture of the two. Dennis' experiments show clearly that this differentiation is not the result of separate classes of receptors. Both the excitatory response and the inhibitory one show similar peaks of sensitivity across the visible spectrum, indicating that they depend on the same light-absorbing pigment. When the activity of a pair (or a trio) of cells is recorded simultaneously, each impulse in one cell is followed by inhibition in the other (or the other two), indicating the presence of cross-connections among them [see *upper illustration on opposite page*]. Since this situation always holds, we conclude that the inhibitory network connects every cell with all four others.

We originally doubted that a mosaic of only five cells could actually form images as larger camera eyes do, but Dennis has demonstrated that the mosaic does indeed have the ability to detect the position of small light sources in the visual field. In one experiment a spot of light five degrees in diameter was moved from right to left and back again on a screen facing the eye [see *lower illustration on opposite page*]. Two of the five receptor cells were impaled with microelectrodes; one, whose activity is shown in the lower trace of each record, responded more strongly when the spot was moved to the left, and held its neighbor, whose activity is shown in the upper trace, under effective inhibitory check. When the light was at the right, the discharge ratio had been such that both cells were firing at almost the same frequency, and it returned to this former value when the spot was moved back to the right. Clearly the relative intensities impinging on the two cells must have been different for the two positions of the light. With the light at the left, the lower cell was more strongly illuminated and consequently fed a stronger inhibition to its neighbor; with the light at the right, the intensities were presumably nearly balanced. In addition, it may be that specific cells display individual personalities even under perfectly homogeneous illumination; those with comparatively little light-sensitive pigment or with relatively strong inhibitory input, for example, would be particularly likely to respond in a predominantly inhibitory fashion and so would be characterized as "off" cells.

As the records show, there is a strong tendency for pairs of cells to fire at the same instant, even when their frequencies of discharge are quite different. This behavior cannot be accounted for on the basis of the inhibitory interaction alone, and it turns out that there is an additional kind of interaction of cells. It is of an excitatory nature, is very brief and is probably mediated by direct electrical connections between two cells. It promotes simultaneous discharge by acting as a trigger for the initiation of impulses in neighboring cells that are nearly ready to fire anyway. The resulting tendency to synchronize may be of value to the region of the brain that receives the visual messages.

These results indicate the technical advantage of small systems of nerve cells to the physiologist. In a larger sense

they illustrate how a few elements connected in simple ways can serve an organism remarkably well. It appears that with five receptor cells, a modest optical system and two types of interaction *Hermissenda* can build a crude image-forming system capable of enhancing contrast at boundaries and—at least theoretically—of measuring the speed and direction of moving objects.

The nerve cells of some crustaceans and insects are also easy to recognize individually, and there are relatively few of them. (They are not so spectacularly large as those in *Aplysia,* and instead of recording from the cells with microelectrodes most investigators dissect single axons from the connective nerves that run between central ganglia.) In such nervous systems a very few cells sometimes control a specific, anatomically restricted process. A network of this kind is found, for example, in the hearts of crabs and lobsters, where the beat is

triggered and spread by an assembly of only nine to 11 cells embedded in the heart muscle. Some of these cells are "pacemakers," which initiate impulses spontaneously at regular intervals; others are "followers" activated by the pacemakers. Connections among the follower cells marshal their responses into a burst of activity that grows and then subsides until the next pacemaker signals arrive. Activity from the followers also has a subtle feedback effect on the pacemaker frequency. With this system Theodore H. Bullock and his collaborators at the University of California at Los Angeles have pioneered in examining restricted ensembles of nerve cells to establish principles of nervous integration.

Can such analyses be expanded to deal with larger groups of nerve cells that control entire systems of muscles, or even control behavior complex enough to orient an animal in its environment or move the animal through it? They can, provided that the controlling cells are in-

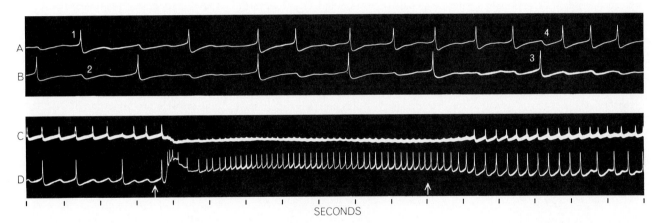

"ON" AND "OFF" responses are seen in records made simultaneously in two *Hermissenda* receptor cells. An impulse from either cell *A* or *B* (*1, 3*) produces an inhibitory hyperpolarization in the other cell (*2, 4*). The illumination of cell *D* (*interval between arrows*) makes it fire faster, inhibiting cell *C*. When the light is turned off, cell *C* is released from inhibition and discharges again.

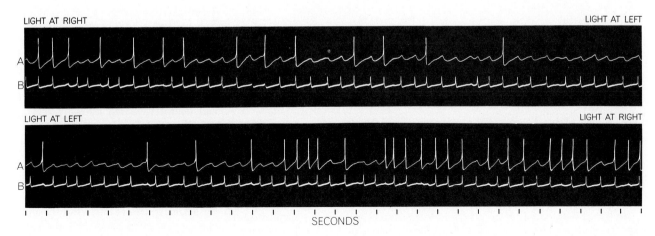

POSITION IS DETECTED by the differential responses of two receptor cells in *Hermissenda*. The discharge rates of cells *A* and *B* were about the same when a light was placed to the right of the field of view. As the light was moved to the left cell *B* discharged relatively more rapidly, effectively inhibiting cell *A*. When the light was moved back again, the former discharge ratio was reestablished.

dividually unique and are identifiable by the experimenter. The feasibility of such studies has been demonstrated primarily by C. A. G. Wiersma and his co-workers at the California Institute of Technology. In a series of investigations spanning the past three decades Wiersma has concerned himself with crustacean muscles and the motor-nerve axons that innervate them. The motor axons—which are remarkably few in number—can be distinguished from one another by their different electrical effects on the muscle and by the kinds of contraction they evoke. Some produce quick twitches by generating large, abrupt depolarizations (reductions in membrane potential that lead to excitation) in the muscle fibers; others make the muscle contract in a more sustained way by producing small and gradually augmenting depolarizations; still others prevent contraction. The number of nerve fibers serving any given muscle is small enough so that one can distinguish the impulses of each one in an electrical record from the entire nerve and correlate these impulses with events in the muscle.

Wiersma has exposed an even more remarkable differentiation of elements in the central nervous system of the crayfish. Single interneurons—nerve cells that collect information from a number of sensory fibers—can be isolated for electrical recording from a part of the central nervous system in the abdomen;

elsewhere such a cell runs its normal course, making connections by branching in each of several different ganglia. Wiersma has prepared maps that give the distribution over the animal's surface of the sensory receptors that will excite each such cell. He has shown that each interneuron is uniquely connected with a set of sensory-nerve fibers, so that a cell with a specific map is always found in the same anatomical location in the central nervous system.

A particular group of touch-sensitive hairs on the back of the fourth abdominal segment, for example, might connect with a dozen different central interneurons. Each interneuron, however, responds to some unique combination of that group of hair receptors and other groups. One interneuron, for instance, might be excited by the fourth-segment group alone; another might be activated by the corresponding group on segment No. 5 as well as by the hairs on segment No. 4; another by those on Nos. 6, 5 and 4; another by those on all segments on one side or on both sides, and so forth. Precise duplication of function is apparently absent; each element encodes a unique spatial combination of sensory inputs [see illustration below].

This specificity suggests that the position and connection pattern of each central nerve cell is precisely determined in the course of differentiation of the nervous system. The reliability of this

mechanism has been demonstrated impressively by Melvin J. Cohen and his colleagues at the University of Oregon, who have analyzed the organization of central ganglia in the cockroach. Cohen has located the cell bodies of specific motoneurons by taking advantage of a striking response shown by such cells when their axons are cut: If a motor nerve is severed at the periphery, even very near the muscle, each cell body supplying an axon in that nerve quickly develops a dense ring around its nucleus. This response, which can be detected with suitable stains, occurs within 12 hours [see illustration on page 314]. The new material comprising the ring has been identified as ribonucleic acid (RNA). Presumably it is required for the protein synthesis associated with regeneration; in any event, the ring provides an unambiguous label for associating a specific central cell with the peripheral destination of its axon.

By cutting individual motor nerves and locating their cell bodies in this way, Cohen has constructed a map that gives the positions for most of the motoneurons in a ganglion. The maps of ganglia from different individuals appear to be almost identical; indeed, specific cells are in nearly perfect register when sections of corresponding ganglia from several animals are superposed. Not only are the cell bodies of motoneurons that serve particular muscles precisely arrayed; they also appear to have rather specific biochemical personalities. Some motoneurons show the ring reaction especially strongly and others show it quite weakly, and these differences are consistently associated with particular cells—as judged by the position of the cell body and the peripheral destination of its axon.

In our own laboratory we have analyzed how the central nervous system of the crayfish deploys a limited array of identifiable nerve cells to control the posture of the abdomen. This structure consists of five segments with joints between them. Its shape is continuously and delicately varied by the action of thin sheets of muscle operating in antagonistic pairs—extensor and flexor—at each of the five joints. Both sets of muscles are symmetrical on each side of the abdomen. The extensors of each half-segment receive six nerve fibers, as do the flexors; the nerve fibers, like the muscles, are repeated almost identically in each segment.

Each individual motor nerve cell can be identified by the size of its impulses

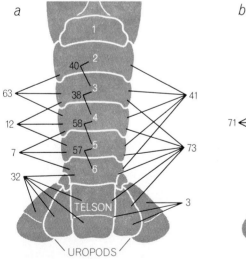
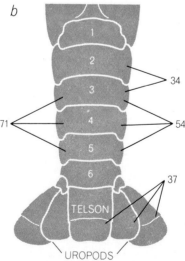

SPECIFIC GROUPS of touch-sensitive hairs on the segments of the crayfish abdomen excite specific interneurons located within the central nervous system. The receptive field of each interneuron listed here (*black numbers*) was mapped by C. A. G. Wiersma's group at Cal Tech. Each responds to stimulation of hair receptors on a unique combination of segments (*white numbers*), as shown by the pointers. The system is bilaterally symmetrical, with an interneuron No. 40, for instance, on each side. Neurons mapped at *a* respond to stimuli delivered to the same side of the animal. Of those mapped at *b*, those listed at the right respond to stimuli on the side opposite them; No. 71 responds to stimuli on either side.

in an electrical record obtained from the nerve bundles that run to flexor or extensor muscles. As one might expect, particular reflexes activate the nerves and muscles in rather stereotyped ways. If, for example, one forcibly flexes the abdomen of a crayfish while holding its thorax clamped and then releases it, the segments extend to approximately their former position. Howard Fields of our laboratory found that this action depends on a pair of receptors that span the dorsal joints in the abdomen. Flexion lengthens the muscle strands associated with these receptor cells, and the cells then discharge impulses that travel toward the central nervous system and activate motoneurons supplying the extensor muscles, which are thereby caused to contract against the imposed load. Such "resistance reflexes" are known in a variety of other systems, including the limbs of mammals and of crabs; in the crayfish they can be studied in a simplified situation, with a single receptor cell and six well-characterized motor cells constituting the entire neural equipment for the reflex loop.

Since only about 120 motoneurons are involved in the regulation of the entire abdomen's position, and since we were able to identify each of them, William H. Evoy and I decided to analyze the central control system for abdominal posture. While recording the motor discharge in several segments at once, we isolated and then stimulated single interneurons located within the central nervous system.

As we had anticipated, most cells had no effect on motor discharge, but we encountered some that regularly released intense, fully coordinated motor-output patterns when we stimulated them with a series of electric shocks. In every case the output was reciprocal: flexors were excited while extensors were inhibited, or vice versa, and the response was similar in several segments. Although equally complex behavior can be produced in many animals by localized stimulation of central nervous structures, it is likely that in most such cases many cells—perhaps thousands—are simultaneously activated by the comparatively gross stimulating procedure that is employed. In the crayfish abdomen, however, we have been able to demonstrate directly that a complete behavioral output can be the result of activity in a single central interneuron. Motor effects produced by stimulation of such central neurons in crayfish had been described earlier by Wiersma and K. Ikeda, who coined the term "command fiber" for those inter-

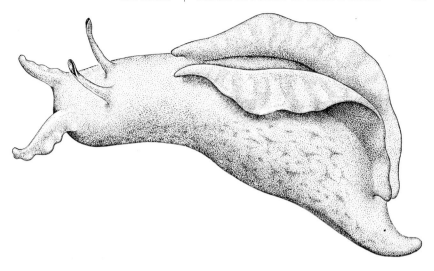

SEA HARE *APLYSIA* is a mollusk with a vestigial shell under the mantle between the winglike parapodia, which are used for locomotion. The animal is about 10 inches long.

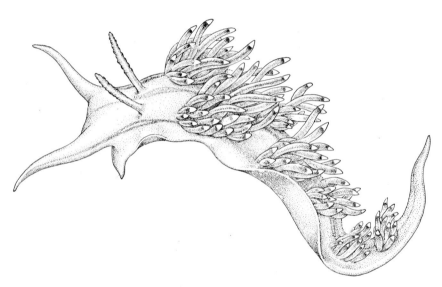

HERMISSENDA, another mollusk, is only one to three inches long. The "camera" eyes are embedded just behind the rhinophores, the two upright stalks near the left end of the animal.

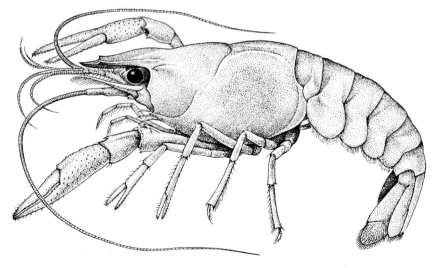

CRAYFISH *Procambarus*, the invertebrate in which complex motor effects produced by single cells are studied by Wiersma and by the author, is from three to five inches long.

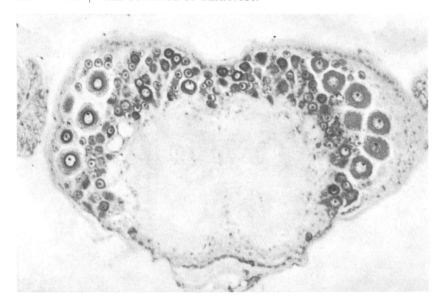

DENSE RINGS of ribonucleic acid surround the nuclei of the cells on one side (*left*) of a stained section of a cockroach ganglion about a millimeter across. Two days before the photograph was made by Melvin J. Cohen of the University of Oregon, all peripheral nerves from one side of this ganglion had been cut. The rings appear in the cells on that side.

neuron axons that produce stereotyped, complex motor effects.

When we compared, in single animals, a series of central nerve cells that produce extension, we found that each produced a special distribution of motor output in different segments. One strongly influenced the last two segments but had a weak effect on segments farther forward; another yielded an almost even balance of output; another activated the forward segments more strongly, and so on. This result led naturally to the inference that individual command elements might each code for a unique combination of segmental actions and thereby produce specific movements or positions. In collaboration with Benjamin Dane, who is now at Tufts University, we filmed the maneuvers of the abdomen that occurred when command fibers were stimulated, while simultaneously recording the discharge of motoneurons supplying muscles in one of the segments. This experiment confirmed the neurophysiological predictions: the films show that two command elements with a generally similar action affect the segments in different ratios and so produce unique abdominal postures [*see illustration on page 307*].

These results indicate that the coordination of at least some behavior is based on the intrinsic "wiring" of a nervous center and can be released by single-cell triggers. Such a conclusion would have been more surprising a decade ago than it is now. It was once

thought that complex sequential behavior such as locomotion depended for its organization on a series of instructions fed back to the central nervous system from peripheral sensory cells. Each new act, according to this view, was initiated by proprioceptive sense organs that were excited by the animal's own movement and signaled that one response had been completed; the whole sequence was likened to a chain of reflexes successively activated by the continuous inflow of sensory signals from the periphery. In recent years biologists have become more convinced that the connections among interneurons at the nervous center, selected by evolution and precisely made during the developmental process, contain most of the organization inherent in such behavior, and that sense organs play a role that is more permissive than instructive.

One of the most convincing demonstrations of the importance of these central connections was provided by the experiments of Donald M. Wilson of the University of California at Berkeley on the flight of locusts. This behavior pattern consists of a sequence of actions in the muscles that elevate, depress and twist the wings. Since these muscles are served by only one or two motor nerve cells, one can obtain a record of the activity of each cell by inserting fine wires into the muscles of a relatively intact animal. In this way Wilson and Torkel Weis-Fogh were able to define the impulse sequences, along various motor nerves, characteristic of normal flight.

The pattern can be triggered in several ways. When the insect is stimulated by having air blown on its head, the motor activity characteristic of flight is frequently released, but Wilson has shown that the output pattern has no relation to the timing of the input. Indeed, when he stimulated one of the central connectives electrically at random intervals, the flight motor output still had its normal frequency and pattern.

Associated with the base of each wing of the locust there is a sensory-nerve cell that responds with single impulses or a short burst when the wing comes to the top of its upstroke. It had been supposed that these receptors might be providing phase information about the wingbeat cycle and perhaps initiating activity in the motor nerves controlling the downstroke. Wilson eliminated these receptors and found that although the frequency of flight motor output dropped somewhat, its repetitive character and the sequence of events within a single cycle were unimpaired. This showed that the stretch receptors, although they participate in a reflex that controls frequency, do not provide any information about the phase relations within a single cycle. Since the entire pattern of behavior can be produced by a central nervous system isolated from peripheral structures—as long as enough excitation of some kind is supplied—it must be concluded that all the information for the flight sequence is stored in a set of central connections.

At this point one can only assume that the sequential events of the flight pattern, like the motions that result in postural changes in the crayfish abdomen, are released by activity in central "command" elements. Some single interneurons are known to release similar complex, sequential behavior in the crayfish. One, for example, produces a sequence of flexions that ascends slowly from segment to segment until it reaches the forward end of the abdomen and then begins again at the back. Another causes an intricate series of movements in the appendages of the tail. We wonder how such central command neurons are activated during voluntary movements, whether they are organized entirely in parallel or in part as a hierarchy of related elements, and what kinds of sensory stimulus excite them. While it would be premature to assume that all these questions can be answered or that we can apply the results to other systems, it is the peculiar advantage of small networks of nerve cells that such a prospect does not seem hopeless.

29

The Flight-Control System
of the Locust

by Donald M. Wilson
May 1968

Groups of nerve cells controlling such activities as locomotion are regulated not only by simple reflex mechanisms but also by behavior patterns apparently coded genetically in the central nervous system

Physicists can properly be concerned with atoms and subatomic particles as being important in themselves, but biologists often study simple or primitive structures with the long-range hope of understanding the workings of the most complex organisms, including man. Studies of viruses and bacteria made it possible to understand the basic molecular mechanisms that we believe control the heredity of all living things. Adopting a similar approach, investigators concerned with the mechanisms of behavior have turned their attention to the nervous systems of lower animals, and to isolated parts of such systems, in the hope of discovering the physiological mechanisms by which behavior is controlled.

One way to approach the study of behavior mechanisms is to ask: Where does the information come from that is needed to coordinate the observable activities of the nervous system? We know that certain behavior patterns are inherited.

This means that some of the informational input must be directly coded in the genetic material and therefore has an origin that is remote in time. Nonetheless, probably all behavior patterns depend to some degree on information supplied directly by the environment by way of the sense organs. Behavior that is largely triggered and coordinated by the nervous input of the moment is commonly called reflex behavior. Much of neurophysiological research has been directed at the analysis of reflex behavior mechanisms. Recent work makes it clear, however, that whole programs for the control of patterns of animal activity can be stored within the central nervous system [see the article "Small Systems of Nerve Cells," by Donald Kennedy, beginning on page 306]. Apparently these inherited nervous programs do not require much special input information for their expression.

I should point out here that whereas there is now general agreement among

biologists that many aspects of animal behavior are under genetic control, it is not easy to show in particular cases that a kind of behavior is inherited and not learned. I believe, however, that this is a reasonable assumption for the cases to be discussed in this article, namely flight and walking by arthropods (insects, crustaceans and other animals with an external skeleton).

The studies I shall describe were begun as part of an effort to demonstrate how several reflexes could be coordinated into an entire behavior pattern. Until recently it was thought by most students of simple behavior such as locomotion that much of the patterning of the nervous command that sets the muscles into rhythmic movement flowed rather directly from information in the immediately preceding sensory input. Each phase of movement was assumed to be triggered by a particular pattern of input from various receptors. According to this hypothesis, known as the peripheral-

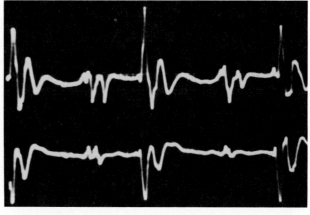

LOCUST WING position and wing-muscle action potentials were recorded in synchronous photographs. The flash that illuminates the locust (*left*) is triggered by the first muscle potential (*at left on oscilloscope trace*). The wing motion is traced by spots of white paint on each wing tip that reflect room light through the open shutter. The trace at the top shows three "doublet" firings of downstroke muscles controlling the forewing; the bottom trace shows similar firings for the hindwing. The smaller potentials visible between the large doublets are from elevator muscles more remote from the electrodes. The oscilloscope traces span 100 milliseconds.

control hypothesis, locomotion might begin because of a signal from external sense organs such as the eye or from brain centers, but thereafter a cyclic reflex process kept it properly timed.

This cyclic reflex could be imagined to operate as follows. An initiating input causes motor nerve impulses to travel to certain muscles, and the muscles cause a movement. The movement is sensed by position or movement receptors within the body (proprioceptors), which send impulses back to the central nervous system. This proprioceptive feedback initiates activity in another set of muscles, perhaps muscles that are antagonists of the first set. The sequence of motor outputs and feedbacks is connected so that it is closed on itself and cyclic activity results. Clearly such a system depends on a well-planned (probably inherited) set of connections among the many parts involved; thus both the central nervous system and its peripheral extensions (the

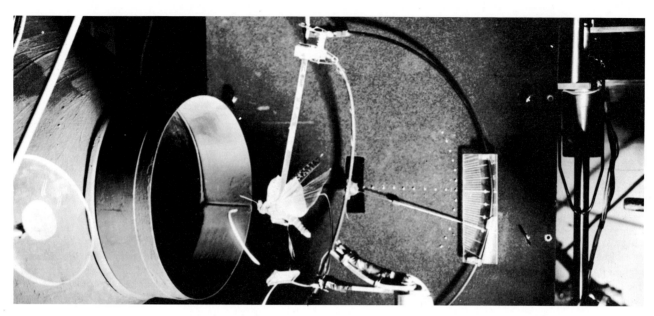

NERVE AND MUSCLE impulses were recorded during flight with this equipment. The locust is flying, suspended at the end of a pendulum, at the mouth of a wind tunnel. The scale at the right registers the insect's angle of pitch. The wires lead to amplifiers.

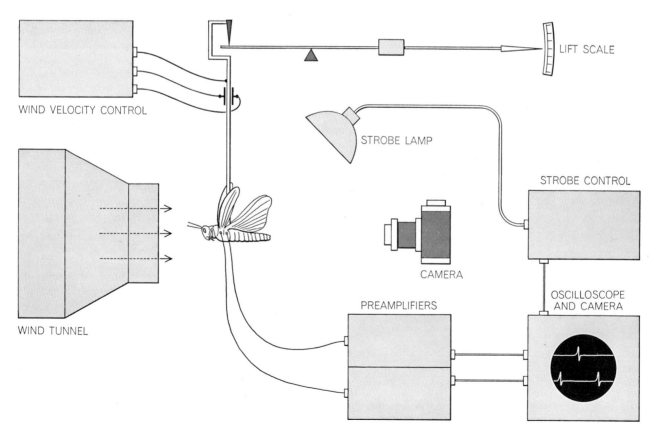

EXPERIMENTAL SETUP is diagrammed. The motion of the pendulum controls the wind-tunnel blower, so that the insect can fly at its desired wind speed. Muscle or nerve impulses are displayed on the oscilloscope, which in turn controls the stroboscopic flash lamp.

muscular and sensory structures) are crucial to the basic operation of the system.

An alternative hypothesis, known as the central-control hypothesis, suggests that the output pattern of motor nerve impulses controlling locomotion can be generated by the central nervous system alone, without proprioceptive feedback. This hypothesis has received much support from studies of embryological development. Only a few zoophysiologists have favored it, however, because the existence of proprioceptive reflexes had been clearly demonstrated. It seemed that if such reflexes exist, they must operate.

Proprioceptive reflexes certainly play an important role in the maintenance of posture. I suspect that this may be their basic and primitive function. In many animals—insects and man included—proprioceptive reflexes help to maintain a given body position against the force of gravity. A simple example was described by Gernot Wendler of the Max Planck Institute for the Physiology of Behavior in Germany. The stick insect, named for its appearance, stands so that its opposed legs form a flattened "M" [*see illustration at right*]. Sensory hairs are bent in proportion to the angle of the leg joints. The hairs send messages to the central nervous ganglia, concentrated groups of nerve cells and their fibrous branches that act as relay and coordinating centers. If too many impulses from the hairs are received, motor nerve cells are excited that cause muscles to contract, thereby moving the joint in the direction that decreases the sensory discharge. Thus the feedback is negative, and it results in the equilibration of a certain position.

If a weight is placed on the back of the insect, one would expect the greater force to bend the leg joints. Instead the proprioceptive feedback loop adjusts muscle tension to compensate for the extra load. The body position remains approximately constant, unless the weight is more than the muscles can bear. If the hair organs are destroyed, the feedback loop is opened and the body sags in relation to the weight, as one would expect in an uncontrolled system.

If the leg reflexes of arthropods are studied under dynamic rather than static conditions, one finds also that they are similar to the reflexes of vertebrates. When a leg of an animal is pushed and pulled rhythmically, the muscles respond reflexively with an output at the same frequency. At high frequencies of movement the reflex system cannot keep up and the output force developed by the muscles lags behind the input move-

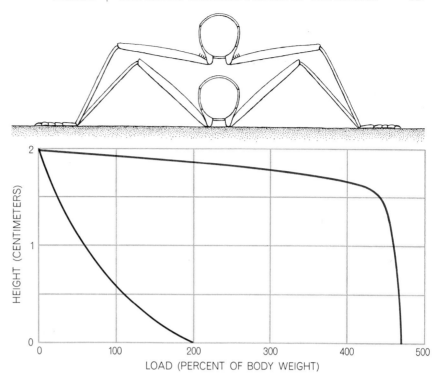

STICK INSECT keeps its body height nearly constant as the load on its leg muscles changes. Hairs at the first leg joint sense the angle of the joint and a reflex loop maintains the angle until the animal is overloaded (*colored curve*). If the sensory hairs are damaged, the reflex loop is opened and the body sags quickly as more weight is added to it (*black curve*).

ment. The peripheral-control hypothesis postulates an interaction of similar reflexes in each of the animal's legs. If one leg is commanded to lift, the others must bear more weight and postural reflexes presumably produce the increased muscle force that is needed. If any part oscillates, other parts oscillate too, perhaps in other phase relations. Although no such total system has been analyzed, one can imagine a sequence of reflex relationships that could coordinate all the legs into a smooth gait.

Against this background I shall describe the work on the nervous control of flight in locusts I began in the laboratory of Torkel Weis-Fogh at the University of Copenhagen in 1959. Weis-Fogh and his associates had already investigated many aspects of the mechanisms of insect flight, including the sense organs and their role in the initiation and maintenance of flight [see "The Flight of Locusts," by Torkel Weis-Fogh; Scientific American, March, 1956]. These studies, and the general climate of opinion among physiologists, tended to support a peripheral-control hypothesis based on reflexes. To test this hypothesis I set out to analyze the details of the reflex mechanisms.

An important consideration in the early phases of the work was how to study nervous activities in a small, rapidly moving animal. This was accomplished by having locusts fly in front of a wind tunnel while they were suspended on a pendulum that served as the arm of an extremely sensitive double-throw switch. The switch operated relays that controlled the blower of the tunnel, so that whenever the insect flew forward, the wind velocity increased and vice versa. Thus the insect chose its preferred wind speed, but it stood approximately still in space. Other devices measured aerodynamic lift and body and wing positions; wires that terminated in the muscles or on nerves conducted electrical impulses to amplifying and recording apparatus.

Early in the program of research we found that fewer than 20 motor nerve cells control the muscles of each wing, and that we could record from any of the motor units controlled by these cells during normal flight. We drew up a table showing when each motor unit was activated for various sets of aerodynamic conditions. The results of this rather tedious work were not very exciting but did provide a necessary base for further investigation. Moreover, I think we can say that these results constitute one of the first and most complete descriptions of the activity of a whole animal analyzed in terms of the activities of single motor nerve cells. In brief, we found that the output pattern consists of nearly synchronous impulses in two small populations of cooperating motor units, with activity alternating between antagonistic sets of muscles, the muscles that elevate the wings and the muscles that depress them. Each muscle unit normally receives one or two impulses per wingbeat or no impulse at all. The variation in the number of excitatory impulses sent to the different muscles serves to control flight power and direction.

We also found it possible to record from the sensory nerves that innervate, or carry signals to, the wings. These nerves conduct proprioceptive signals from receptors in the wing veins and in the wing hinge. The receptors in the

wing veins register the upward force, or lift, on the wing; the receptors in the wing hinge indicate wing position and movement in relation to the body. These sensory inputs occur at particular phases of the wing stroke. The lift receptors usually discharge during the middle of the downstroke; each wing-hinge proprioceptor is a stretch receptor that discharges one, two or several impulses toward the end of the upstroke [*see illustration on page 320*].

Everything I have described so far about the motor output and sensory input of an insect in flight is consistent with the peripheral-control hypothesis. Motor impulses cause the movements

the receptors register. According to the hypothesis the sensory feedback should trigger a new round of output. Does this actually happen?

A useful test of feedback-loop function is to open the loop. This we did simply by cutting or damaging the sense organs or sensory nerves that provide the feedback. Cutting the sensory nerve carrying the information about lift forces caused little change in the basic pattern of motor output, although it did affect the insect's ability to make certain maneuvers. On the other hand, burning the stretch receptors that measure wing position and angular velocity always resulted in a drastic reduction in wingbeat frequency. These proprioceptors provide the only input we could discover that had such an effect. Most important of all, we found that, even when we eliminated all sources of sensory feedback, the wings could be kept beating in a normal phase pattern, although at a somewhat reduced frequency, simply by stimulating the central nervous system with random electrical impulses.

From these studies we must conclude that the flight-control system of the locust is not adequately explained by the peripheral-control hypothesis and patterned feedback. Instead we find that the coordinated action of locust flight muscles depends on a pattern-generating system that is built into the central nervous system and can be turned on by an unpatterned input. This is a significant finding because it suggests that the networks within the nerve ganglia are endowed through genetic and developmental processes with the information needed to produce an important pattern of behavior and that proprioceptive reflexes are not major contributors of coordinating information.

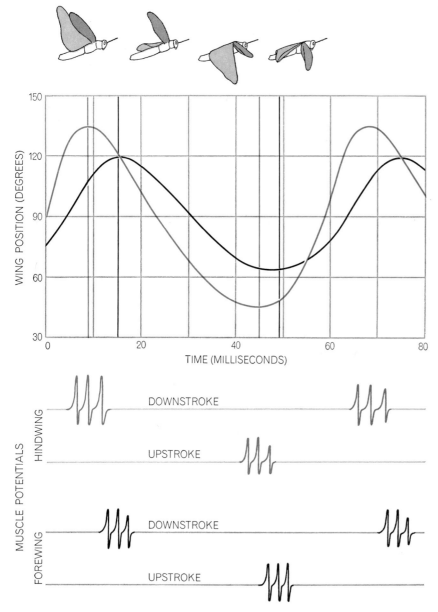

MUSCLE-POTENTIAL RECORDS are summarized in relation to wing positions in a flying locust. The curves (*top*) show the angular position (90 degrees is horizontal) of the hindwings (*color*) and forewings (*black*). The four simulated traces at the bottom show how the downstroke and upstroke muscles respectively fire at the high and low point for each wing.

Erik Gettrup and I were particularly curious to learn how the wing-hinge proprioceptor, a stretch receptor, helped to control the frequency of wingbeat. When Gettrup analyzed the response of this receptor to various wing movements, he found that to some degree it signaled to the central nervous system information on wing position, wingbeat amplitude and wingbeat frequency. We then cut out the four stretch receptors so that the wingbeat frequency was reduced to about half the normal frequency and artificially stimulated the stumps of the stretch receptors in an attempt to restore normal function. Under these conditions we found that electrical stimulation of the stumps could raise the frequency of wingbeat no matter what input pattern we used. Although the normal input

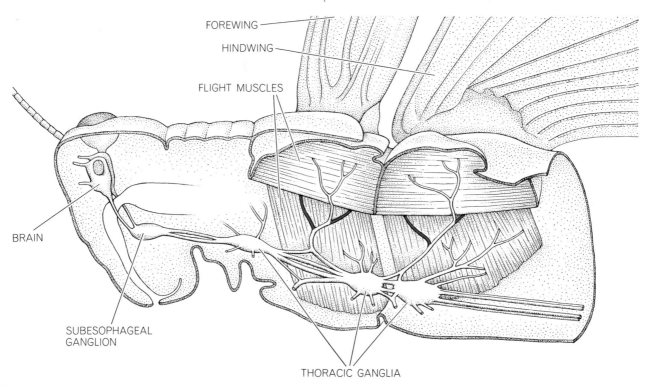

FOREWING

HINDWING

FLIGHT MUSCLES

BRAIN

SUBESOPHAGEAL
GANGLION

THORACIC GANGLIA

NERVES AND MUSCLES controlling flight in the locust are shown in simplified form. The central nervous system includes the brain and the various ganglia. From the thoracic ganglia, motor nerves lead to the wing's upstroke muscles (*vertical fibers*) and the downstroke muscles (*horizontal fibers*) above them. There are also sensory nerves (*color*) that sense wing position and aerodynamic forces.

from the stretch receptor arrives at a definite, regular time with respect to the wingbeat cycle, in our artificially stimulated preparations the effect was the same no matter what the phase of the input was.

We also found that the response to the input took quite a long time to develop. When the stimulator was turned on, the motor output frequency would increase gradually over about 20 to 40 wingbeat cycles. Hence it appears that the ganglion averages the input from the four stretch receptors (and other inputs too) over a rather long time interval compared with the wingbeat period, and that this averaged level of excitation controls wingbeat frequency. In establishing this average of the input most of the detailed information about wing position, frequency and amplitude is lost or discarded, with the result that no reflection of the detailed input pattern is found in the motor output pattern. We must therefore conclude that the input turns on a central pattern generator and regulates its average level of activity, but that it does not determine the main features of the pattern it produces. These features are apparently genetically programmed into the central network.

If that is so, why does the locust even have a stretch reflex to control wingbeat frequency? If the entire ordered pattern needed to activate flight muscles can be coded within the ganglion, why not also include the code for wingbeat frequency? The answer to this dual question can probably be found in mechanical considerations. The wings, muscles and skeleton of the flight system of the locust form a mechanically resonant system— a system with a preferred frequency at which conversion of muscular work to aerodynamic power is most efficient. This frequency is a function of the insect's size. It seems likely that even insects with the same genetic makeup may reach different sizes because of different environmental conditions during egg production and development. Hence each adult insect must be able to measure its own size, as it were, to find the best wingbeat frequency. This measurement may be provided by the stretch reflex, automatically regulating the wingbeat frequency to the mechanically resonant one.

What kind of pattern-generating nerve network is contained in the ganglia? We do not know as yet. Nonetheless, a plausible model can be suggested. The arguments leading to this model are not rigorous and the evidence in its favor is not overwhelming, but it is always useful to have a working hypothesis as a guide in planning future experiments. Also, it seems worthwhile to present a hypothesis of how a simply structured network might produce a special temporally patterned output when it is excited by an unpatterned input.

When neurophysiologists find a system in which there is alternating action between two sets of antagonistic muscles, they tend to visualize a controlling nerve network in which there is reciprocal inhibition between the two sets of nerve cells [*see top illustration on page 321*]. Such a network can turn on one or both sets at first, but one soon dominates and the other is silenced. When the dominant set finally slows down from fatigue, the inhibiting signal it sends to the silent set also decreases, with the result that the silent set turns on. It then inhibits the first set. This reciprocating action is analogous to the action of an electronic flip-flop circuit; timing cues are not needed in the input. The information required for the generation of the output pattern is contained largely in the structure of the network and not in the input, which only sets the average level of activity. A nerve network that acts in this way can consist of as few as two cells or be made up of two populations of cells in which there is some mechanism to keep the cooperating units working together.

In the locust flight-control system several tens of motor nerve cells work together in each of the two main sets. The individual nerve cells within each set

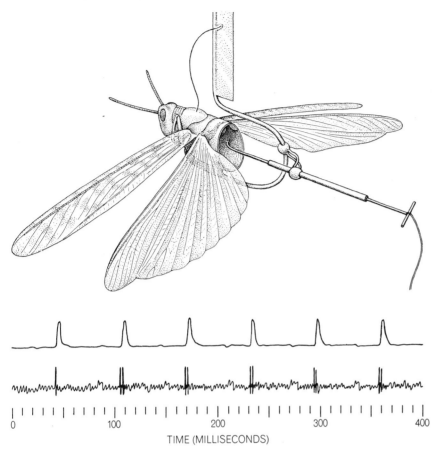

TIME (MILLISECONDS)

SENSORY DISCHARGES in nerves from the wing and wing hinge are recorded with wires manipulated into the largely eviscerated thoracic cavity of a locust. The top record is of downstroke muscle potentials, which are repeating at the wingbeat frequency. The bottom record is of a sensory (stretch) receptor from one wing, firing one or two times per wingbeat.

The tarantula can lose several of its legs and still walk. Suppose the first and third pairs of legs are amputated. If the spider's legs were coordinated by means of a simple preprogrammed circuit like the one controlling the locust's flight muscles, one would expect the spider to move the remaining two legs on one side in step with each other and out of step with the legs on the other side. A four-legged spider that did this would fall over. In actuality the spider adjusts relations between the remaining legs to achieve the diagonal rhythm. Other combinations of amputations give rise to other adaptations that also maintain the mechanically more stable diagonal rhythm.

Thus it appears that the pattern of coordination does depend on input from the legs. One can advance a possible explanation. Each leg is either driven by a purely central nervous oscillator or each leg and its portion of ganglion forms an oscillating reflex feedback loop. Suppose the several oscillators are negatively coupled. A pair of matched negatively coupled oscillators will operate out of phase. If the nearest leg oscillators are negatively coupled more strongly than the ones farther apart, the normal diagonal rhythm will result. For example, if left leg 1 has a strong tendency to alternate with right leg 1 and right leg 1 alternates with right leg 2, then left leg 1 must operate synchronously with right leg 2, to which it is more weakly connected. Now if some of the oscillators are turned off by amputating legs, so that either the postulated oscillatory feedback loop is broken or the postulated central oscillator receives insufficient excitatory input, new patterns of leg movements will appear that will always exhibit a diagonal rhythm.

The real nature of the oscillators involved in the leg rhythms is not known. These results and the postulated model nonetheless illustrate how sensory feedback could be used by the nervous system in such a way that the animal could adjust to genetically unpredictable conditions of the body or environment without recourse to learning mechanisms. Could this be the role of reflexes in general? We have seen that in the locust flight system much information for pattern generation is centrally stored—presumably having been provided genetically—and that the reflexes do seem to supply only information that could not have been known genetically.

A way to describe the two general models of muscle-control systems has been suggested by Graham Hoyle of the University of Oregon. He calls the cen-

seem to share some excitatory interconnections. These not only provide the coupling that keeps the set working efficiently but also have a further effect of some importance. Strong positive coupling between nerve cells can result in positive feedback "runaway"; the network, once it is activated, produces a heavy burst of near-maximum activity until it is fatigued [see middle illustration on opposite page] and then turns off altogether until it recovers. A network of this kind can also produce sustained oscillations consisting of successive bursts of activity alternating with periods of silence without any patterned input. Thus either reciprocal inhibition or mutual excitation can give rise to the general type of burst pattern seen in locust motor units. Both mechanisms have been demonstrated in various behavior-control systems. It is likely that both are working in the locust and that these two mechanisms, as well as others, converge to produce a pattern of greater stability than might otherwise be achieved.

In summary, the model suggests that each group of cooperating nerve cells is

mutually excitatory, so that the units of each group tend to fire together and produce bursts of activity even when their input is steady. In addition the two sets are connected to each other by inhibitory linkages that set the two populations into alternation. Notice that in these hypothetical networks the temporal pattern of activity is due to the network structure, not to the pattern of the input. Even the silent network stores most of the information needed to produce the output pattern.

The locust flight-control system consists in large part of a particular kind of circuit built into the central ganglia. Some other locomotory systems seem much more influenced by reflex inputs. As an example of such a system I shall describe briefly the walking pattern of the tarantula spider. There is much variation in the relative timing of the eight legs of this animal, but on the average the legs exhibit what is called a diagonal rhythm. Opposite legs of one segment alternate and adjacent legs on one side alternate so that diagonal pairs of legs are in step [see illustration on page 322].

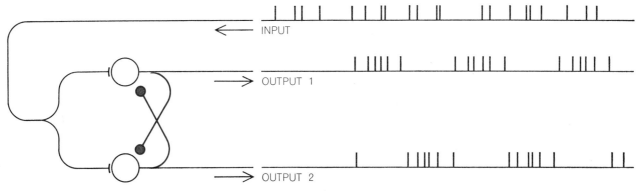

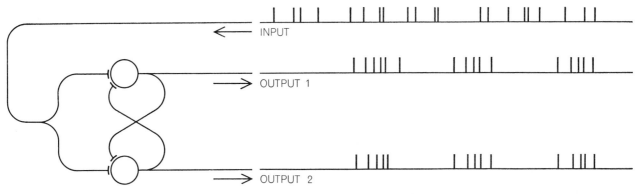

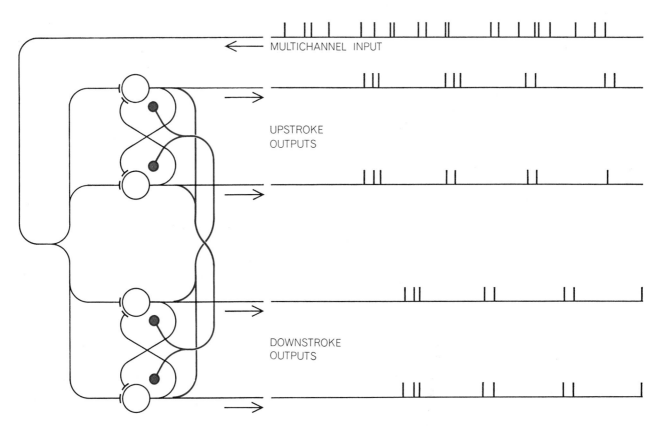

CROSS INHIBITION is one kind of interaction between nerve cells. The cells are connected in such a way that impulses from one inhibit the other (*color*). This can cause a pattern of alternating bursts, each cell firing (inhibiting the other) until fatigued. The hypothetical network shows how an unpatterned input can be transformed into a patterned output by structurally coded information.

IN CROSS EXCITATION the output from each cell excites the other. This makes for approximate synchrony. There may also be a positive feedback "runaway" until fatigue causes deceleration or a pause; once rested, the network begins another accelerating burst.

HYPOTHETICAL NETWORK of nerve cells in the locust might involve two cell populations, an upstroke group (*top*) and a downstroke group (*bottom*). Cells *within* a group excite one another but there are inhibitory connections *between* groups (*color*). The inhibition keeps the activity of one group out of phase with that of the other, so that upstroke and downstroke muscles alternate.

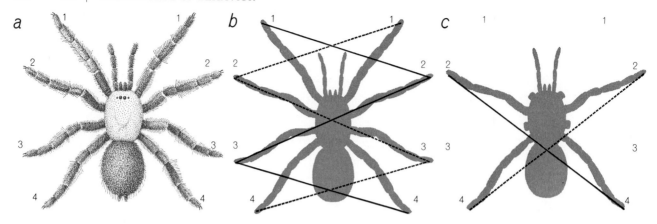

LEGS OF TARANTULA (*a*) move in diagonally arranged groups of four (*b*). The diagonal pattern persists if some legs are removed (*c*), suggesting that proprioceptive reflexes set the pattern or that changed input alters inherent central nervous system activity.

trally stored system of pattern generation a motor-tape system. In such a system a preprogrammed "motor score" plays in a stereotyped manner whenever it is excited. For a system in which reflexes significantly modify the behavioral sequences, Hoyle suggests the term "sensory tape" system. I prefer the term "sensory template." Such systems have a preprogrammed input requirement that can be achieved by various outputs. If the output at any moment results in an input that does not match the template, the mismatch results in a changed output pattern until the difference disappears. Such a goal-oriented feedback system can adapt to unexpected environments or to bodily damage. Spider and insect walking patterns show a kind of plasticity in which new movement patterns can compensate for the loss of limbs. In the past this kind of plasticity has often been interpreted as evidence for the reflex control of locomotory pattern. The locust flight-control system, on the other hand, certainly has a motor score that is not organized as a set of reflexes. Can a motor-score system show the plastic behavior usually associated with reflex, or sensory-template, systems?

For several years we thought that the locust flight-control system was relatively unplastic. We knew about the control of wingbeat frequency by the stretch reflex and about other reflexes, for example a reflex that tends to keep the body's angle of pitch constant. When I had made recordings of nerve impulses to some important control muscles before and after cutting out other muscles or whole wings, however, I had not found differences in the motor output. I therefore concluded that the flight-control system was not capable of a wide range of adaptive behavior. On this basis one would predict that a damaged locust could not fly, or could fly only in circles.

The locust has four wings. Recently I cut whole wings from several locusts, threw the locusts into the air and found to my surprise that they flew quite well. The flying locust shows just as much ability to adapt to the loss of a limb as a walking insect or spider does. For the crippled locust to fly it must significantly change its motor output pattern. Why did the locust not show this change in the experiments in which I made recordings of nerve impulses to its muscles?

In all the laboratory experiments the locust was approximately fixed in space. If the insect made a motor error, it might sense the error proprioceptively, but it could not receive a feedback signal from the environment around it indicating that it was off course. The free-flying locust has at least two extraproprioceptive sources of feedback about its locomotory progress: signals from its visual system and signals from directionally sensitive hairs on its head that respond to the flow of air. Either or both of these extraproprioceptive sources can tell the locust that it is turning in flight. In the free-flying locust the signals are involved in a negative feedback control that tends to keep the animal flying straight in spite of functional or anatomical errors in the insect's basic motor system. When animals are studied in the laboratory under conditions that do allow motor output errors or anatomical damage to produce turning motions, then compensatory changes in the motor output pattern are observed provided only that the appropriate sensory structures are intact.

The locust flight-control system consists in part of a built-in motor score, but it also shows the adaptability expected of a reflex, or sensory-template, type of control. From these observations on plasticity in the locust flight system one can see that there may be no such thing as a pure motor-tape or a pure sensory-template system. Many behavior systems probably have some features of each.

What we are striving for in studies such as the ones I have reported here is a way of understanding the functioning of networks of nerve cells that control animal behavior. Neurophysiologists have already acquired wide knowledge about single nerve cells—how their impulses code messages and how the synapses transmit and integrate the messages. Much is also known about the electrical behavior and chemistry of large masses of nerve cells in the brains of animals. The intermediate level, involving networks of tens or hundreds of nerve cells, remains little explored. This is an area in which many neurobiologists will probably be working in the next few years. I suspect that it is an area in which important problems are ripe for solution.

I shall conclude with a few remarks on the unraveling of the mechanisms of genetically coded behavior. As I see it, there are two major stages in the readout of genetically coded behavioral information. The first stage is the general process of development of bodily form, including the detailed form of the networks of the central nervous system and the form of peripheral body parts, such as muscles and sense organs that are involved in the reflexes. This stage of the genetic readout is not limited to neurobiology, of course. It is a stage that will probably be analyzed largely at the molecular level. The second stage involves a problem that is primarily neurobiological: How can information that is coded in the grosser level of nervous-system structure, in the shapes of whole nerve cells and networks of nerve cells, be translated into temporal sequences of behavior?

Moths and Ultrasound

by Kenneth D. Roeder
April 1965

*Certain moths can hear the ultrasonic cries by which
bats locate their prey. The news is sent from ear to
central nervous system by only two fibers. These can
be tapped and the message decoded*

If an animal is to survive, it must be able to perceive and react to predators or prey. What nerve mechanisms are used when one animal reacts to the presence of another? Those animals that have a central nervous system perceive the outer world through an array of sense organs connected with the brain by many thousands of nerve fibers. Their reactions are expressed as critically timed sequences of nerve impulses traveling along motor nerve fibers to specific muscles. Exactly how the nervous system converts a particular pattern of sensory input into a specific pattern of motor output remains a subject of investigation in many branches of zoology, physiology and psychology.

Even with the best available techniques one can simultaneously follow the traffic of nerve impulses in only five or perhaps 10 of the many thousands of separate nerve fibers connecting a mammalian sense organ with the brain. Trying to learn how information is encoded and reported among all the fibers by following the activity of so few is akin to basing a public opinion poll on one or two interviews. (Following the activity of all the fibers would of course be like sampling public opinion by having the members of the population give their different answers in chorus.) Advances in technique may eventually make it possible to follow the traffic in thousands of fibers; in the meantime much can be learned by studying animals with less profusely innervated sense organs.

With several colleagues and students at Tufts University I have for some time been trying to decode the sensory patterns connecting the ear and central nervous system of certain nocturnal moths that have only two sense cells in each ear. Much of the behavior of these simple invertebrates is built in, not learned, and therefore is quite stereotyped and stable under experimental conditions. Working with these moths offers another advantage: because they depend on their ears to detect their principal predators, insect-eating bats, we are able to discern in a few cells the nervous mechanisms on which the moth's survival depends.

Insectivorous bats are able to find their prey while flying in complete darkness by emitting a series of ultrasonic cries and locating the direction and distance of sources of echoes. So highly sophisticated is this sonar that it enables the bats to find and capture flying insects smaller than mosquitoes. Some night-flying moths—notably members of the families Noctuidae, Geometridae and Arctiidae—have ears that can detect the bats' ultrasonic cries. When they hear the approach of a bat, these moths take evasive action, abandoning their usual cruising flight to go into sharp dives or erratic loops or to fly at top speed directly away from the source of ultrasound. Asher E. Treat of the College of the City of New York has demonstrated that moths taking evasive action on a bat's approach have a significantly higher chance of survival than those that continue on course.

A moth's ears are located on the sides of the rear part of its thorax and are directed outward and backward into the constriction that separates the thorax and the abdomen [see top illustration on page 325]. Each ear is externally visible as a small cavity, and within the cavity is a transparent eardrum. Behind the eardrum is the tympanic air sac; a fine strand of tissue containing the sensory apparatus extends across the air sac from the center of the eardrum to a skeletal support. Two acoustic cells, known as *A* cells, are located within this Utrand. Each *A* cell sends a fine sensory strand outward to the eardrum and a nerve fiber inward to the skeletal support. The two *A* fibers pass close to a large nonacoustic cell, the *B* cell, and are joined by its nerve fiber. The three fibers continue as the tympanic nerve into the central nervous system of the moth. From the two *A* fibers, then, it is possible—and well within our technical means—to obtain all the information about ultrasound that is transmitted from the moth's ear to its central nervous system.

Nerve impulses in single nerve fibers can be detected as "action potentials," or self-propagating electrical transients, that have a magnitude of a few millivolts and at any one point on the fiber last less than a millisecond. In the moth's *A* fibers action potentials travel from the sense cells to the central nervous system in less than two milliseconds. Action potentials are normally an all-or-nothing phenomenon; once initiated by the sense cell, they travel to the end of the nerve fiber. They can be detected on the outside of the fiber by means of fine electrodes, and they are displayed as "spikes" on the screen of an oscilloscope.

Tympanic-nerve signals are demonstrated in the following way. A moth, for example the adult insect of one of the common cutworms or armyworms, is immobilized on the stage of a microscope. Some of its muscles are dissected away to expose the tympanic nerves at a point outside the central nervous system. Fine silver hooks are placed under one or both nerves, and the pattern of passing action potentials is observed on the oscilloscope. With moths thus prepared we have spent much time in impromptu outdoor laboratories, where the cries of passing bats provided the necessary stimuli.

In order to make precise measurements we needed a controllable source of ultrasonic pulses for purposes of comparison. Such pulses can be generated by electronic gear to approximate natural bat cries in frequency and duration. The natural cries are frequency-modulated: their frequency drops from about 70 kilocycles per second at the beginning of each cry to some 35 kilocycles at the end. Their duration ranges from one to 10 milliseconds, and they are repeated from 10 to 100 times a second. Our artificial stimulus is a facsimile of an average series of bat cries; it is not frequency-modulated, but such modulation is not detected by the moth's ear. Our sound pulses can be accurately graded in intensity by decibel steps; in the sonic range a decibel is roughly equivalent to the barely noticeable difference to human ears in the intensity of two sounds.

By using electronic apparatus to elicit and follow the responses of the A cells we have been able to define the amount of acoustic information avail-

MOTH EVADED BAT by soaring upward just as the bat closed in to capture it. The bat entered the field at right; the path of its flight is the broad white streak across the photograph. The smaller white streak shows the flight of the moth. A tree is in background. The shutter of the camera was left open as contest began. Illumination came from continuous light source below field.

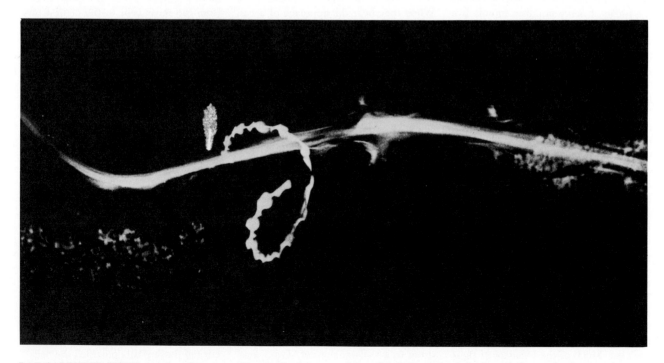

BAT CAPTURED MOTH at point where two white streaks intersect. Small streak shows the flight pattern of the moth. Broad streak shows the flight path of the bat. Both streak photographs were made by Frederic Webster of the Sensory Systems Laboratories.

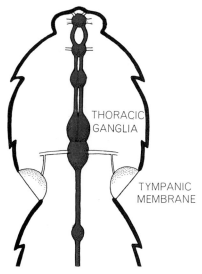

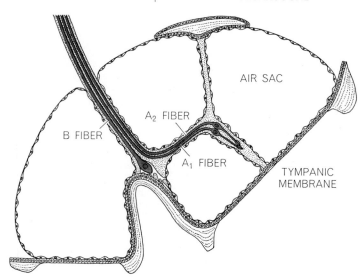

NERVES FROM EAR to central nervous system of moth are shown at two magnifications. Drawing at left indicates position of the tympanic organs on each side of the moth and the tympanic nerves connecting them with the thoracic ganglia. Central nervous system is colored. Drawing at right shows two nerve fibers of the acoustic cells joined by a nonacoustic fiber to form the tympanic nerve.

able to the moth by way of its tympanic nerve. It appears that the tympanic organ is not particularly sensitive; to elicit any response from the A cell requires ultrasound roughly 100 times more intense than sound that can just be heard by human ears. The ear of a moth can nonetheless pick up at distances of more than 100 feet ultrasonic bat cries we cannot hear at all. The reason it cannot detect frequency modulation is simply that it cannot discriminate one frequency from another; it is tone-deaf. It can, however, detect frequencies from 10 kilocycles to well over 100 kilocycles per second, which covers the range of bat cries. Its greatest talents are the detection of pulsed sound—short bursts of sound with intervening silence—and the discrimination of differences in the loudness of sound pulses.

When the ear of a moth is stimulated by the cry of a bat, real or artificial, spikes indicating the activity of the A cell appear on the oscilloscope in various configurations. As the stimulus increases in intensity several changes are apparent. First, the number of A spikes increases. Second, the time interval between the spikes decreases. Third, the spikes that had first appeared only on the record of one A fiber (the "A_1" fiber, which is about 20 decibels more sensitive than the A_2 fiber) now appear on the records of both fibers. Fourth, the greater the intensity of the stimulus, the sooner the A cell generates a spike in response.

The moth's ears transmit to the oscilloscope the same configuration of spikes they transmit normally to the central nervous system, and therein lies our interest. Which of the changes in auditory response to an increasingly in-

tense stimulus actually serve the moth as criteria for determining its behavior under natural conditions? Before we face up to this question let us speculate on the possible significance of these criteria from the viewpoint of the moth. For the moth to rely on the first kind of information—the number of A spikes—might lead it into a fatal error: the long, faint cry of a bat at a distance could be confused with the short, intense cry of a bat closing for the kill. This error could be avoided if the moth used the second kind of information—the interval between spikes—for estimating the loudness of the bat's cry. The third kind of information—the activity of the A_2 fiber—might serve to change an "early warning" message to a "take cover" message. The fourth kind of information—the length of time it takes for a spike to be generated—might provide the moth with

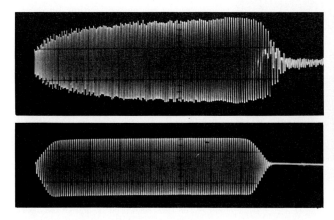

OSCILLOSCOPE TRACES of a real bat cry (top) and a pulse of sound generated electronically (bottom) are compared. The two ultrasonic pulses are of equal duration (length), 2.5 milliseconds, but differ in that the artificial pulse has a uniform frequency.

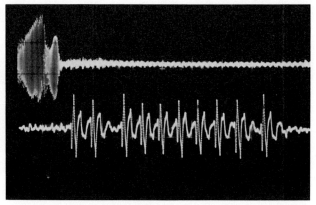

BAT CRY AND MOTH RESPONSE were traced on same oscilloscope from tape recording by Webster. The bat cry, detected by microphone, yielded the pattern at left in top trace. Reaction of the moth's acoustic cells produced the row of spikes at bottom.

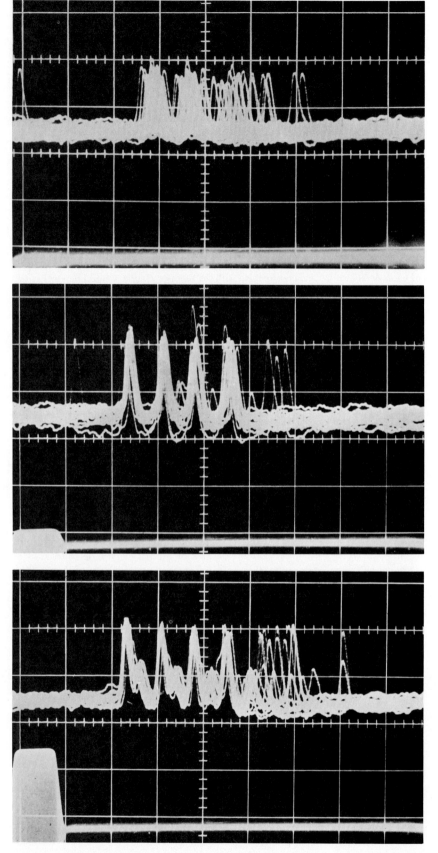

CHANGES ARE REPORTED by moth's tympanic nerve to the oscilloscope as pulses used to simulate bat cries gain intensity. Pulses (*lower trace in each frame*) were at five decibels (*top frame*), 20 (*middle*) and 35 (*bottom*). An increased number of tall spikes appear as intensity of stimulus rises. The time interval between spikes decreases slightly. Smaller spikes from the less sensitive nerve fiber appear at the higher intensities, and the higher the intensity of the stimulus, the sooner (*left on horizontal axis*) the first spike appears.

the means for locating a cruising bat; for example, if the sound was louder in the moth's left ear than in its right, then A spikes would reach the left side of the central nervous system a fraction of a millisecond sooner than the right side.

Speculations of this sort are profitable only if they suggest experiments to prove or disprove them. Our tympanic-nerve studies led to field experiments designed to find out what moths do when they are exposed to batlike sounds from a loudspeaker. In the first such study moths were tracked by streak photography, a technique in which the shutter of a camera is left open as the subject passes by. As free-flying moths approached the area on which our camera was trained they were exposed to a series of ultrasonic pulses.

More than 1,000 tracks were recorded in this way. The moths were of many species; since they were free and going about their natural affairs most of them could not be captured and identified. This was an unavoidable disadvantage; earlier observations of moths captured, identified and then released in an enclosure revealed nothing. The moths were apparently "flying scared" from the beginning, and the ultrasound did not affect their behavior. Hence all comers were tracked in the field.

Because moths of some families lack ears, a certain percentage of the moths failed to react to the loudspeaker. The variety of maneuvers among the moths that did react was quite unpredictable and bewildering [*see illustrations at top of next page*]. Since the evasive behavior presumably evolved for the purpose of bewildering bats, it is hardly surprising that another mammal should find it confusing! The moths that flew close to the loudspeaker and encountered high-intensity ultrasound would maneuver toward the ground either by dropping passively with their wings closed, by power dives, by vertical and horizontal turns and loops or by various combinations of these evasive movements.

One important finding of this field work was that moths cruising at some distance from the loudspeaker would turn and fly at high speed directly away from it. This happened only if the sound the moths encountered was of low intensity. Moths closer to the loudspeaker could be induced to flee only if the signal was made weaker. Moths at about the height of the loudspeaker flew away in the horizontal plane; those above the loudspeaker were observed to turn directly upward

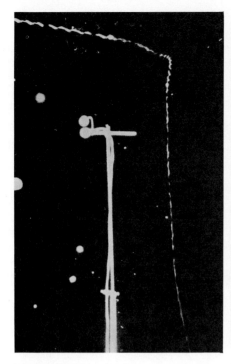

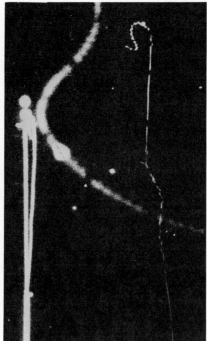

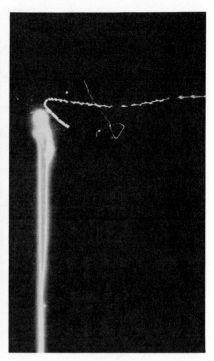

POWER DIVE is taken by moth on hearing simulated bat cry from loudspeaker mounted on thin tower (*left of moth's flight path*).

PASSIVE DROP was executed by another moth, which simply folded its wings. Blur at left and dots were made by other insects.

TURNING AWAY, an evasive action involving directional change, is illustrated. These streak photographs were made by author.

or at other sharp angles. To make such directional responses with only four sensory cells is quite a feat. A horizontal response could be explained on the basis that one ear of the moth detected the sound a bit earlier than the other. It is harder to account for a vertical response, although experiments I shall describe provide a hint.

Our second series of field experiments was conducted in another outdoor laboratory—my backyard. They were designed to determine which of the criteria of intensity encoded in the pattern of A-fiber spikes play an important part in determining evasive behavior. The percentage of moths showing "no re-

action," "diving," "looping" and "turning away" was noted when a 50-kilocycle signal was pulsed at different rates and when it was produced as a continuous tone. The continuous tone delivers more A impulses in a given fraction of a second and therefore should be a more effective stimulus if the number of A impulses is important. On the other hand, because the A cells, like many other sensory cells, become progressively less sensitive with continued stimulation, the interspike interval lengthens rapidly as continuous-tone stimulation proceeds. When the sound is pulsed, the interspike interval remains short because the A cells have had time to regain their sensitivity during the

brief "off" periods. If the spike-generation time—which is associated with difference in the time at which the A spike arrives at the nerve centers for each ear—plays an important part in evasive behavior, then continuous tones should be less effective. The difference in arrival time would be detected only once at the beginning of the stimulus; with pulsed sound it would be reiterated with each pulse.

The second series of experiments occupied many lovely but mosquito-ridden summer nights in my garden and provided many thousands of observations. Tabulation of the figures showed that continuous ultrasonic tones were much less effective in producing evasive

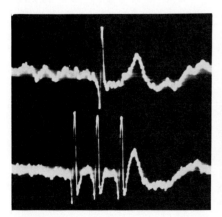

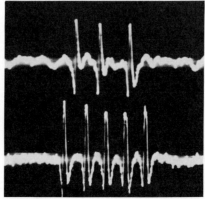

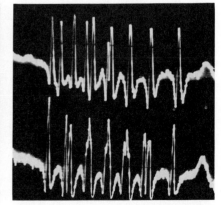

RESPONSE BY BOTH EARS of a moth to an approaching bat was recorded on the oscilloscope and photographed by the author. In trace at left the tympanic nerve from one ear transmits only one spike (*upper curve*) while the nerve from the other ear sends three. As the bat advances, the ratio becomes three to five (*middle*), then 10 to 10 (*right*), suggesting that the bat has flown overhead.

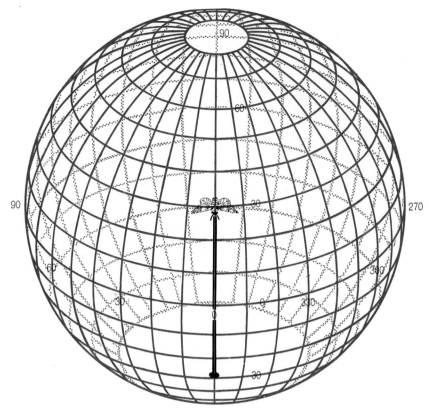

SPHERE OF SENSITIVITY, the range in which a moth with wings in a given position can hear ultrasound coming from various angles, was the subject of a study by Roger Payne of Tufts University and Joshua Wallman, a Harvard undergraduate. Moths with wings in given positions were mounted on a tower in an echo-free chamber. Data were compiled on the moths' sensitivity to ultrasound presented from 36 latitude lines 10 degrees apart.

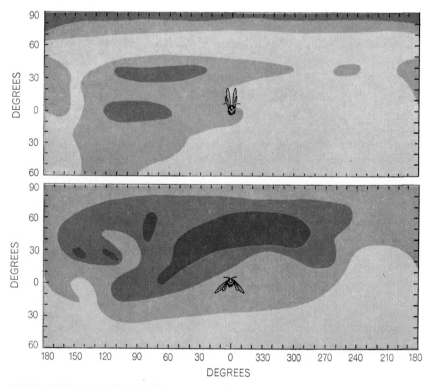

MERCATORIAL PROJECTIONS represent auditory environment of a moth with wings at end of upstroke (*top*) and near end of downstroke (*bottom*). Vertical scale shows rotation of loudspeaker around moth's body in vertical plane; horizontal scale shows rotation in horizontal plane. At top the loudspeaker is above moth; at far right and left, behind it. In Mercatorial projections, distortions are greatest at poles. The lighter the shading at a given angle of incidence, the more sensitive the moth to sound from that angle.

behavior than pulses. The number of nonreacting moths increased threefold, diving occurred only at higher sound intensities and turning away was essentially absent. Only looping seemed to increase slightly.

Ultrasound pulsed between 10 and 30 times a second proved to be more effective than ultrasound pulsed at higher or lower rates. This suggests that diving, and possibly other forms of nondirectional evasive behavior, are triggered in the moth's central nervous system not so much by the number of A impulses delivered over a given period as by short intervals (less than 2.5 milliseconds) between consecutive A impulses. Turning away from the sound source when it is operating at low intensity levels seems to be set off by the reiterated difference in arrival time of the first A impulse in the right and left tympanic nerves.

These conclusions were broad but left unanswered the question: How can a moth equipped only with four A cells orient itself with respect to a sound source in planes that are both vertical and horizontal to its body axis? The search for an answer was undertaken by Roger Payne of Tufts University, assisted by Joshua Wallman, a Harvard undergraduate. They set out to plot the directional capacities of the tympanic organ by moving a loudspeaker at various angles with respect to a captive moth's body axis and registering (through the A_1 fiber) the organ's relative sensitivity to ultrasonic pulses coming from various directions. They took precautions to control acoustic shadows and reflections by mounting the moth and the recording electrodes on a thin steel tower in the center of an echo-free chamber; the effect of the moth's wings on the reception of sound was tested by systematically changing their position during the course of many experiments. A small loudspeaker emitted ultrasonic pulses 10 times a second at a distance of one meter. These sounds were presented to the moths from 36 latitude lines 10 degrees apart.

The response of the A fibers to the ultrasonic pulses was continuously recorded as the loudspeaker was moved. At the same time the intensity of ultrasound emitted by the loudspeaker was regulated so that at any angle it gave rise to the same response. Thus the intensity of the sound pulses was a measure of the moth's acoustic sensitivity. A pen recorder continuously graphed the changing intensity of the ultrasonic pulses against the angle from which they were presented to the moth. Each chart provided a profile of sensitivity in a certain plane, and the data from it were assembled with those from others to provide a "sphere of sensitivity" for the moth at a given wing position.

This ingenious method made it possible to assemble a large amount of data in a short time. In the case of one moth it was possible to obtain the data for nine spheres of sensitivity (about 5,000 readings), each at a different wing position, before the tympanic nerve of the moth finally stopped transmitting impulses. Two of these spheres, taken from one moth at different wing positions, are presented as Mercatorial projections in the bottom illustration on the preceding page.

It is likely that much of the information contained in the fine detail of such projections is disregarded by a moth flapping its way through the night. Certain general patterns do seem related, however, to the moth's ability to escape a marauding bat. For instance, when the moth's wings are in the upper half of their beat, its acoustic sensitivity is 100 times less at a given point on its side facing away from the source of the sound than at the corresponding point on the side facing toward the source. When flight movements bring the wings below the horizontal plane, sound coming from each side above the moth is in acoustic shadow, and the left-right acoustic asymmetry largely disappears. Moths commonly flap their wings from 30 to 40 times a second. Therefore left-right acoustic asymmetry must alternate with up-down asymmetry at this frequency. A left-right difference in the A-fiber discharge when the wings are up might give the moth a rough horizontal bearing on the position of a bat with respect to its own line of flight. The absence of a left-right difference and the presence of a similar fluctuation in both left and right tympanic nerves at wingbeat frequency might inform the moth that the bat was above it. If neither variation occurred at the regular wingbeat frequency, it would mean that the bat was below or behind the moth.

This analysis uses terms of precise directionality that idealize the natural situation. A moth certainly does not zoom along on an even keel and a straight course like an airliner. Its flapping progress—even when no threat is imminent—is marked by minor yawing and pitching; its overall course is rare-

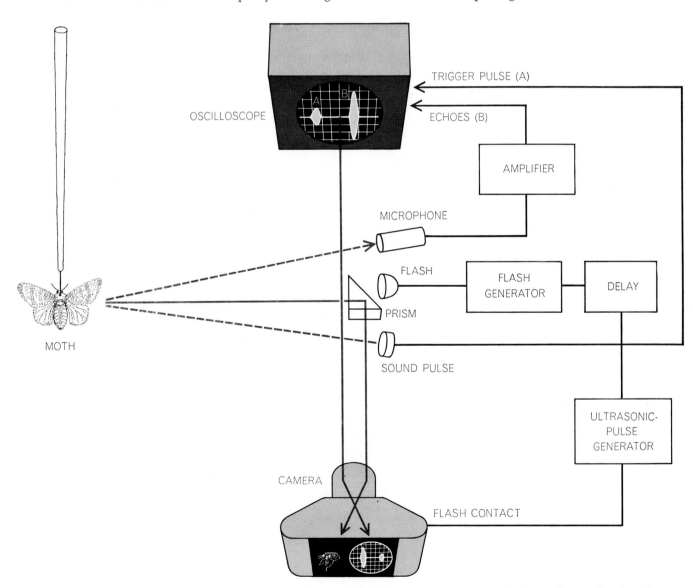

ARTIFICIAL BAT, the electronic device depicted schematically at right, was built by the author to determine at what position with respect to a bat a moth casts its greatest echo. As a moth supported by a wire flapped its wings in stationary flight, a film was made by means of a prism of its motions and of an oscilloscope that showed the pulse generated by the loudspeaker and the echo picked up by the microphone. Each frame of film thus resembled the composite picture of moth and two pulses shown inverted at bottom.

ly straight and commonly consists of large loops and figure eights. Even so, the localization experiments of Payne and Wallman suggest the ways in which a moth receives information that enables it to orient itself in three dimensions with respect to the source of an ultrasonic pulse.

The ability of a moth to perceive and react to a bat is not greatly superior or inferior to the ability of a bat to perceive and react to a moth. Proof of this lies in the evolutionary equality of their natural contest and in the observation of a number of bat-moth confrontations. Donald R. Griffin of Harvard University and Frederic Webster of the Sensory Systems Laboratories have studied in detail the almost unbelievable ability of bats to locate, track and intercept

small flying targets, all on the basis of a string of echoes thrown back from ultrasonic cries. Speaking acoustically, what does a moth "look like" to a bat? Does the prey cast different echoes under different circumstances?

To answer this question I set up a crude artificial bat to pick up echoes from a live moth. The moth was attached to a wire support and induced to flap its wings in stationary flight. A movie camera was pointed at a prism so that half of each frame of film showed an image of the moth and the other half the screen of an oscilloscope. Mounted closely around the prism and directed at the moth from one meter away were a stroboscopic-flash lamp, an ultrasonic loudspeaker and a microphone. Each time the camera shutter opened and exposed a frame of film a

short ultrasonic pulse was sent out by the loudspeaker and the oscilloscope began its sweep. The flash lamp was controlled through a delay circuit to go off the instant the ultrasonic pulse hit the moth, whose visible attitude was thereby frozen on the film. Meanwhile the echo thrown back by the moth while it was in this attitude was picked up by the microphone and finally displayed as a pulse of a certain height on the oscilloscope. All this took place before the camera shutter closed and the film moved on to the next frame. Thus each frame shows the optical and acoustic profiles of the moth from approximately the same angle and at the same instant of its flight. The camera was run at speeds close to the wingbeat frequency of the moth, so that the resulting film presents a regular series of wing positions and the echoes cast by them.

Films made of the same moth flying at different angles to the camera and the sound source show that by far the strongest echo is returned when the moth's wings are at right angles to the recording array [see illustrations at left]. The echo from a moth with its wings in this position is perhaps 100 times stronger than one from a moth with its wings at other angles. Apparently if a bat and a moth were flying horizontal courses at the same altitude, the moth would be in greatest danger of detection if it crossed the path of the approaching bat at right angles. From the bat's viewpoint at this instant the moth must appear to flicker acoustically at its wingbeat frequency. Since the rate at which the bat emits its ultrasonic cries is independent of the moth's wingbeat frequency, the actual sequence of echoes the bat receives must be complicated by the interaction of the two frequencies. Perhaps this enables the bat to discriminate a flapping target, likely to be prey, from inert objects floating in its acoustic field.

The moth has one advantage over the bat: it can detect the bat at a greater range than the bat can detect it. The bat, however, has the advantage of greater speed. This creates a nice problem for a moth that has picked up a bat's cries. If a moth immediately turns and flies directly away from a source of ultrasound, it has a good chance of disappearing from the sonar system of a still-distant bat. If the bat has also detected the moth, and is near enough to receive a continuous signal from its target, turning away on a straight course is a bad tactic because the moth is not likely to outdistance its pursuer. It is then to the moth's advantage to

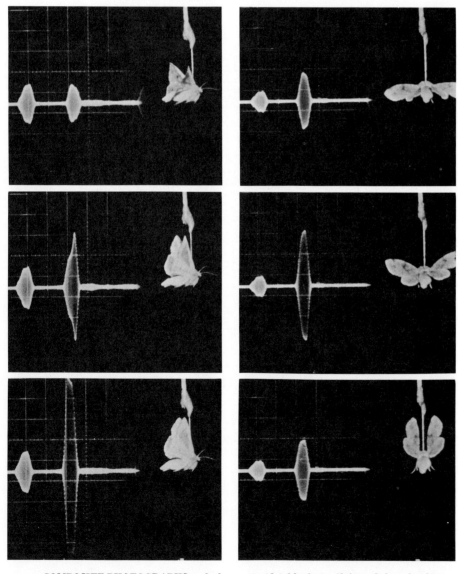

COMPOSITE PHOTOGRAPHS each show an artificial bat's cry (left) and the echo thrown back (middle) by a moth (right). The series of photographs at left is of a moth in stationary flight at right angles to the artificial bat. Those at right are of a moth oriented in flight parallel to the bat. The echo produced in the series of photographs at left is much the larger.

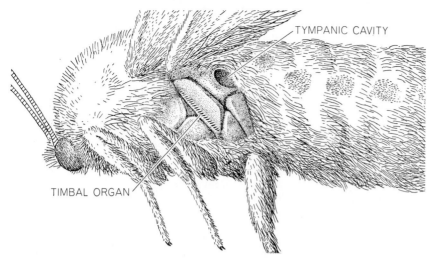

TYMPANIC CAVITY

TIMBAL ORGAN

NOISEMAKING ORGAN possessed by many moths of the family Arctiidae and of other families is a row of fine parallel ridges of cuticle that bend and unbend when a leg muscle contracts and relaxes. This produces a rapid sequence of high-pitched clicks.

go into tight turns, loops and dives, some of which may even take it toward the bat.

In this contest of hide-and-seek it seems much to a moth's advantage to remain as quiet as possible. The sensitive ears of a bat would soon locate a noisy target. It is therefore surprising to find that many members of the moth family Arctiidae (which includes the moths whose caterpillars are known as woolly bears) are capable of generating trains of ultrasonic clicks. David Blest and David Pye of University College London have demonstrated the working of the organ that arctiids use for this purpose.

In noisemaking arctiids the basal joint of the third pair of legs (which roughly corresponds to the hip) bulges outward and overlies an air-filled cavity. The stiff cuticle of this region has a series of fine parallel ridges [see illustration above]. Each ridge serves as a timbal that works rather like the familiar toy incorporating a thin strip of spring steel that clicks when it is pressed by the thumb. When one of the moth's leg muscles contracts and relaxes in rapid sequence, it bends and unbends the overlying cuticle, causing the row of timbals to produce rapid sequences of high-pitched clicks. Blest and Pye found that such moths would click when they were handled or poked, that the clicks occurred in short bursts of 1,000 or more per second and that each click contained ultrasonic frequencies within the range of hearing of bats.

My colleagues and I found that certain arctiids common in New England could also be induced to click if they were exposed to a string of ultrasonic pulses while they were suspended in stationary flight. In free flight these moths showed the evasive tactics I have already described. The clicking seems almost equivalent to telling the bat, "Here I am, come and get me." Since such altruism is not characteristic of the relation between predators and prey, there must be another answer.

Dorothy C. Dunning, a graduate student at Tufts, is at present trying to find it. She has already shown that partly tamed bats, trained to catch mealworms that are tossed into the air by a mechanical device, will commonly swerve away from their target if they hear tape-recorded arctiid clicks just before the moment of contact. Other ultrasounds, such as tape-recorded bat cries and "white" noise (noise of all frequencies), have relatively little effect on the bats' feeding behavior; the tossed mealworms are caught in midair and eaten. Thus the clicks made by arctiids seem to be heeded by bats as a warning rather than as an invitation. But a warning against what?

One of the pleasant things about scientific investigation is that the last logbook entry always ends with a question. In fact, the questions proliferate more rapidly than the answers and often carry one along unexpected paths. I suggested at the beginning of this article that it is my intention to trace the nervous mechanisms involved in the evasive behavior of moths. By defining the information conveyed by the acoustic cells I have only solved the least complex half of that broad problem. As I embark on the second half of the investigation, I hope it will lead up as many diverting side alleys as the study of the moth's acoustic system has.

31

Language and the Brain

by Norman Geschwind
April 1972

*Aphasias are speech disorders caused by brain damage.
The relations between these disorders and specific kinds
of brain damage suggest a model of how the language
areas of the human brain are organized*

Virtually everything we know of how the functions of language are organized in the human brain has been learned from abnormal conditions or under abnormal circumstances: brain damage, brain surgery, electrical stimulation of brains exposed during surgery and the effects of drugs on the brain. Of these the most fruitful has been the study of language disorders, followed by postmortem analysis of the brain, in patients who have suffered brain damage. From these studies has emerged a model of how the language areas of the brain are interconnected and what each area does.

A disturbance of language resulting from damage to the brain is called aphasia. Such disorders are not rare. Aphasia is a common aftereffect of the obstruction or rupture of blood vessels in the brain, which is the third leading cause of death in the U.S. Although loss of speech from damage to the brain had been described occasionally before the 19th century, the medical study of such cases was begun by a remarkable Frenchman, Paul Broca, who in 1861 published the first of a series of papers on language and the brain. Broca was the first to point out that damage to a specific portion of the brain results in disturbance of language output. The portion he identified, lying in the third frontal gyrus of the cerebral cortex, is now called Broca's area [*see illustration on page 334*].

Broca's area lies immediately in front of the portion of the motor cortex that controls the muscles of the face, the jaw, the tongue, the palate and the larynx, in other words, the muscles involved in speech production. The region is often called the "motor face area." It might therefore seem that loss of speech from damage to Broca's area is the result of paralysis of these muscles. This explana-tion, however, is not the correct one. Direct damage to the area that controls these muscles often produces only mild weakness of the lower facial muscles on the side opposite the damage and no permanent weakness of the jaw, the tongue, the palate or the vocal cords. The reason is that most of these muscles can be controlled by either side of the brain. Damage to the motor face area on one side of the brain can be compensated by the control center on the opposite side. Broca named the lesion-produced language disorder "aphemia," but this term was soon replaced by "aphasia," which was suggested by Armand Trousseau.

In 1865 Broca made a second major contribution to the study of language and the brain. He reported that damage to specific areas of the left half of the brain led to disorder of spoken language but that destruction of corresponding areas in the right side of the brain left language abilities intact. Broca based his conclusion on eight consecutive cases of aphasia, and in the century since his report his observation has been amply confirmed. Only rarely does damage to the right hemisphere of the brain lead to language disorder; out of 100 people with permanent language disorder caused by brain lesions approximately 97 will have damage on the left side. This unilateral control of certain functions is called cerebral dominance. As far as we know man is the only mammal in which learned behavior is controlled by one half of the brain. Fernando Nottebohm of Rockefeller University has found unilateral neural control of birdsong. It is an interesting fact that a person with aphasia of the Broca type who can utter at most only one or two slurred words may be able to sing a melody rapidly, correctly and even with elegance. This is another proof that aphasia is not the result of muscle paralysis.

In the decade following Broca's first report on brain lesions and language there was a profusion of papers on aphasias of the Broca type. In fact, there was a tendency to believe all aphasias were the result of damage to Broca's area. At this point another great pioneer of the brain appeared on the scene. Unlike Broca, who already had a reputation at the time of his first paper on aphasia, Carl Wernicke was an unknown with no previous publications; he was only 26 years old and a junior assistant in the neurological service in Breslau. In spite of his youth and obscurity his paper on aphasia, published in 1874, gained immediate attention. Wernicke described damage at a site in the left hemisphere outside Broca's area that results in a language disorder differing from Broca's aphasia.

In Broca's aphasia speech is slow and labored. Articulation is crude. Characteristically, small grammatical words and the endings of nouns and verbs are

LOCATION OF SOME LESIONS in the brain can be determined by injecting into the bloodstream a radioactive isotope of mercury, which is taken up by damaged brain tissue. The damaged region is identified by scanning the head for areas of high radioactivity. The top scan on the opposite page was made from the back of the head; the white area on the left shows that the damage is in the left hemisphere. The bottom scan is of the left side of the head and shows that the uptake of mercury was predominantly in the first temporal gyrus, indicating damage to Wernicke's speech area by occlusion of blood vessels. David Patten and Martin Albert of the Boston Veterans Administration Hospital supplied the scans.

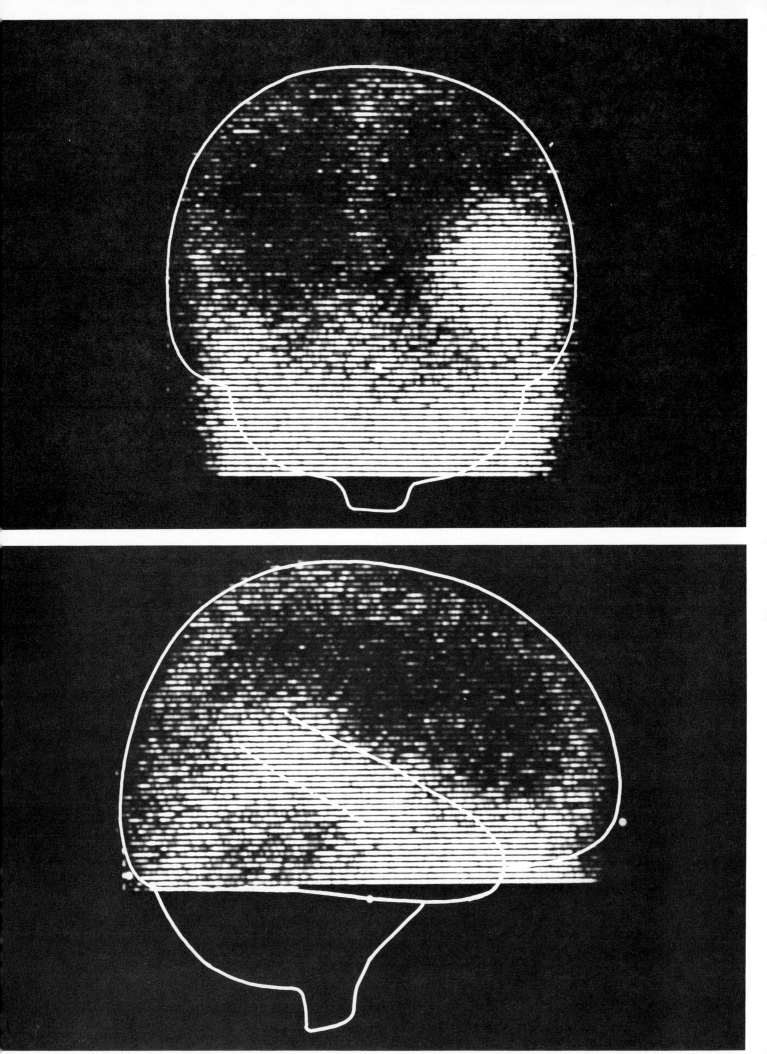

omitted, so that the speech has a telegraphic style. Asked to describe a trip he has taken, the patient may say "New York." When urged to produce a sentence, he may do no better than "Go... New York." This difficulty is not simply a desire to economize effort, as some have suggested. Even when the patient does his best to cooperate in repeating words, he has difficulty with certain grammatical words and phrases. "If he were here, I would go" is more difficult than "The general commands the army." The hardest phrase for such patients to repeat is "No ifs, ands or buts."

The aphasia described by Wernicke is quite different. The patient may speak very rapidly, preserving rhythm, grammar and articulation. The speech, if not listened to closely, may almost sound normal. For example, the patient may say: "Before I was in the one here, I was over in the other one. My sister had the department in the other one." It is abnormal in that it is remarkably devoid of content. The patient fails to use the correct word and substitutes for it by

circumlocutory phrases ("what you use to cut with" for "knife") and empty words ("thing"). He also suffers from paraphasia, which is of two kinds. Verbal paraphasia is the substitution of one word or phrase for another, sometimes related in meaning ("knife" for "fork") and sometimes unrelated ("hammer" for "paper"). Literal or phonemic paraphasia is the substitution of incorrect sounds in otherwise correct words ("kench" for "wrench"). If there are several incorrect sounds in a word, it becomes a neologism, for example "pluver" or "flieber."

Wernicke also noted another difference between these aphasic patients and those with Broca's aphasia. A person with Broca's aphasia may have an essentially normal comprehension of language. Indeed, Broca had argued that no single lesion in the brain could cause a loss of comprehension. He was wrong. A lesion in Wernicke's area can produce a severe loss of understanding, even though hearing of nonverbal sounds and music may be fully normal.

Perhaps the most important contribu-

tion made by Wernicke was his model of how the language areas in the brain are connected. Wernicke modestly stated that his ideas were based on the teachings of Theodor Meynert, a Viennese neuroanatomist who had attempted to correlate the nervous system's structure with its function. Since Broca's area was adjacent to the cortical region of the brain that controlled the muscles of speech, it was reasonable to assume, Wernicke argued, that Broca's area incorporated the programs for complex coordination of these muscles. In addition Wernicke's area lay adjacent to the cortical region that received auditory stimuli [see *illustration below*]. Wernicke made the natural assumption that Broca's area and Wernicke's area must be connected. We now know that the two areas are indeed connected, by a bundle of nerve fibers known as the arcuate fasciculus. One can hypothesize that in the repetition of a heard word the auditory patterns are relayed from Wernicke's area to Broca's area.

Comprehension of written language

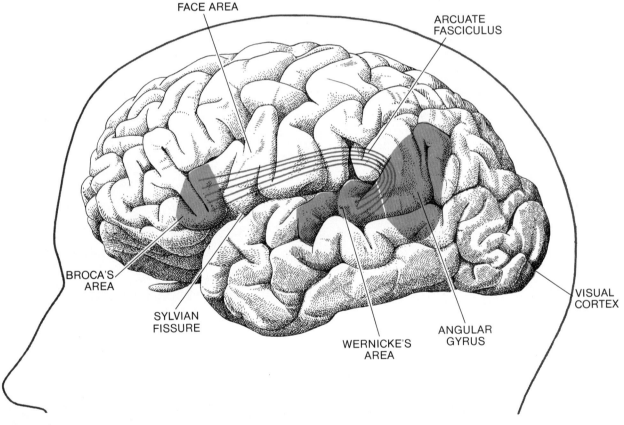

PRIMARY LANGUAGE AREAS of the human brain are thought to be located in the left hemisphere, because only rarely does damage to the right hemisphere cause language disorders. Broca's area, which is adjacent to the region of the motor cortex that controls the movement of the muscles of the lips, the jaw, the tongue, the soft palate and the vocal cords, apparently incorporates programs for the coordination of these muscles in speech. Damage to Broca's area results in slow and labored speech, but comprehension of language remains intact. Wernicke's area lies between Heschl's gyrus, which is the primary receiver of auditory stimuli, and the angular gyrus, which acts as a way station between the auditory and the visual regions. When Wernicke's area is damaged, speech is fluent but has little content and comprehension is usually lost. Wernicke and Broca areas are joined by a nerve bundle called the arcuate fasciculus. When it is damaged, speech is fluent but abnormal, and patient can comprehend words but cannot repeat them.

would require connections from the visual regions to the speech regions. This function is served by the angular gyrus, a cortical region just behind Wernicke's area. It acts in some way to convert a visual stimulus into the appropriate auditory form.

We can now deduce from the model what happens in the brain during the production of language. When a word is heard, the output from the primary auditory area of the cortex is received by Wernicke's area. If the word is to be spoken, the pattern is transmitted from Wernicke's area to Broca's area, where the articulatory form is aroused and passed on to the motor area that controls the movement of the muscles of speech. If the spoken word is to be spelled, the auditory pattern is passed to the angular gyrus, where it elicits the visual pattern. When a word is read, the output from the primary visual areas passes to the angular gyrus, which in turn arouses the corresponding auditory form of the word in Wernicke's area. It should be noted that in most people comprehension of a written word involves arousal of the auditory form in Wernicke's area. Wernicke argued that this was the result of the way most people learn written language. He thought, however, that in people who were born deaf, but had learned to read, Wernicke's area would not be in the circuit.

According to this model, if Wernicke's area is damaged, the person would have difficulty comprehending both spoken and written language. He should be unable to speak, repeat and write correctly. The fact that in such cases speech is fluent and well articulated suggests that Broca's area is intact but receiving inadequate information. If the damage were in Broca's area, the effect of the lesion would be to disrupt articulation. Speech would be slow and labored but comprehension should remain intact.

This model may appear to be rather simple, but it has shown itself to be remarkably fruitful. It is possible to use it to predict the sites of brain lesions on the basis of the type of language disorder. Moreover, it gave rise to some definite predictions that lesions in certain sites should produce types of aphasia not previously described. For example, if a lesion disconnected Wernicke's area from Broca's area while leaving the two areas intact, a special type of aphasia should be the result. Since Broca's area is preserved, speech should be fluent but abnormal. On the other hand, comprehension should be intact because Wernicke's area is still functioning. Rep-

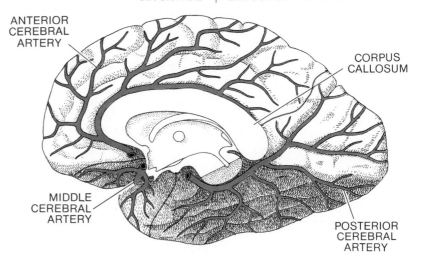

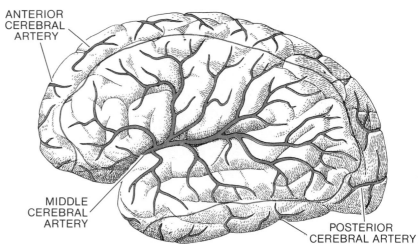

CEREBRAL AREAS are nourished by several arteries, each supplying blood to a specific region. The speech and auditory region is nourished by the middle cerebral artery. The visual areas at the rear are supplied by the posterior cerebral artery. In patients who suffer from inadequate oxygen supply to the brain the damage is often not within the area of a single blood vessel but rather in the "border zones" (*colored lines*). These are the regions between the areas served by the major arteries where the blood supply is marginal.

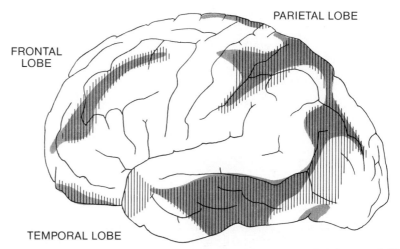

ISOLATION OF SPEECH AREA by a large C-shaped lesion produced a remarkable syndrome in a woman who suffered from severe carbon monoxide poisoning. She could repeat words and learn new songs but could not comprehend the meaning of words. Postmortem examination of her brain revealed that in the regions surrounding the speech areas of the left hemisphere, either the cortex (*colored areas*) or the underlying white matter (*hatched areas*) was destroyed but that the cortical structures related to the production of language (Broca's area and Wernicke's area) and the connections between them were left intact.

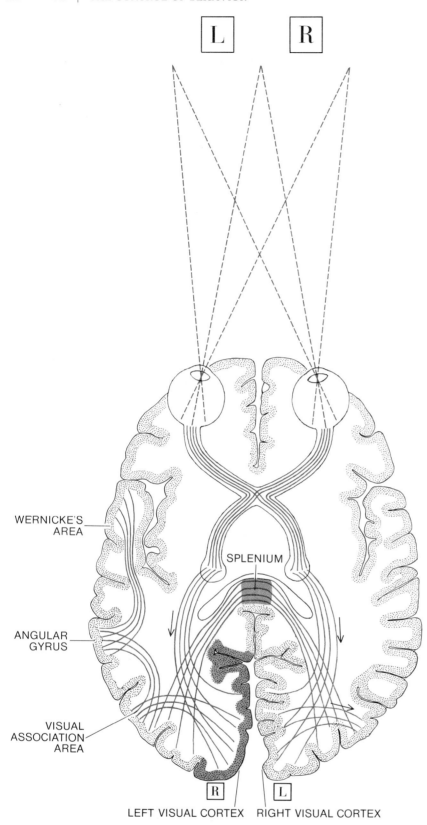

WERNICKE'S
AREA

SPLENIUM

ANGULAR
GYRUS

VISUAL
ASSOCIATION
AREA

LEFT VISUAL CORTEX   RIGHT VISUAL CORTEX

CLASSIC CASE of a man who lost the ability to read even though he had normal visual acuity and could copy written words was described in 1892 by Joseph Jules Dejerine. Postmortem analysis of the man's brain showed that the left visual cortex and the splenium (*dark colored areas*) were destroyed as a result of an occlusion of the left posterior cerebral artery. The splenium is the section of the corpus callosum that transfers visual information between the two hemispheres. The man's left visual cortex was inoperative, making him blind in his right visual field. Words in his left visual field were properly received by the right visual cortex, but could not cross over to the language areas in the left hemisphere because of the damaged splenium. Thus words seen by the man remained as meaningless patterns.

etition of spoken language, however, should be grossly impaired. This syndrome has in fact been found. It is termed conduction aphasia.

The basic pattern of speech localization in the brain has been supported by the work of many investigators. A. R. Luria of the U.S.S.R. studied a large number of patients who suffered brain wounds during World War II [see "The Functional Organization of the Brain," by A. R. Luria; SCIENTIFIC AMERICAN Offprint 526]. When the wound site lay over Wernicke's or Broca's area, Luria found that the result was almost always severe and permanent aphasia. When the wounds were in other areas, aphasia was less frequent and less severe.

A remarkable case of aphasia has provided striking confirmation of Wernicke's model. The case, described by Fred Quadfasel, Jose Segarra and myself, involved a woman who had suffered from accidental carbon monoxide poisoning. During the nine years we studied her she was totally helpless and required complete nursing care. She never uttered speech spontaneously and showed no evidence of comprehending words. She could, however, repeat perfectly sentences that had just been said to her. In addition she would complete certain phrases. For example, if she heard "Roses are red," she would say "Roses are red, violets are blue, sugar is sweet and so are you." Even more surprising was her ability to learn songs. A song that had been written after her illness would be played to her and after a few repetitions she would begin to sing along with it. Eventually she would begin to sing as soon as the song started. If the song was stopped after a few bars, she would continue singing the song through to the end, making no errors in either words or melody.

On the basis of Wernicke's model we predicted that the lesions caused by the carbon monoxide poisoning lay outside the speech and auditory regions, and that both Broca's area and Wernicke's area were intact. Postmortem examination revealed a remarkable lesion that isolated the speech area from the rest of the cortex. The lesion fitted the prediction. Broca's area, Wernicke's area and the connection between them were intact. Also intact were the auditory pathways and the motor pathways to the speech organs. Around the speech area, however, either the cortex or the underlying white matter was destroyed [see *bottom illustration on preceding page*]. The woman could not comprehend speech because the words did not arouse associations in other portions of the cortex. She could repeat speech correctly because the internal connections of the speech region were intact. Presumably well-learned word sequences stored in Broca's area could be triggered by the beginning phrases. This syndrome is called isolation of the speech area.

Two important extensions of the Wernicke model were advanced by a French neurologist, Joseph Jules Dejerine. In 1891 he described a disorder called alexia with agraphia: the loss of the ability to read and write. The patient could, however, speak and understand spoken language. Postmortem examination showed that there was a lesion in the angular gyrus of the left hemisphere, the area of the brain that acts as a way station between the visual and the auditory region. A lesion here would separate the visual and auditory language areas. Although words and letters would be seen correctly, they would be meaningless visual patterns, since the visual pattern must first be converted to the auditory form before the word can be comprehended. Conversely, the auditory pattern for a word must be transformed into the visual pattern before the word can be spelled. Patients suffering from alexia with agraphia cannot recognize words spelled aloud to them nor can they themselves spell aloud a spoken word.

Dejerine's second contribution was showing the importance of information transfer between the hemispheres. His patient was an intelligent businessman who had awakened one morning to discover that he could no longer read. It was found that the man was blind in the right half of the visual field. Since the right half of the field is projected to the left cerebral hemisphere, it was obvious that the man suffered damage to the visual pathways on the left side of the brain [*see illustration on opposite page*]. He could speak and comprehend spoken language and could write, but he could not read even though he had normal visual acuity. In fact, although he could not comprehend written words, he could copy them correctly. Postmortem examination of the man's brain by Dejerine revealed two lesions that were the result of the occlusion of the left posterior cerebral artery. The visual cortex of the left hemisphere was totally destroyed. Also destroyed was a portion of the corpus callosum: the mass of nerve fibers that interconnect the two cerebral hemispheres. That portion was the splenium, which carries the visual information between the hemispheres. The destruction of the splenium prevented stimuli from the visual cortex of the right hemisphere

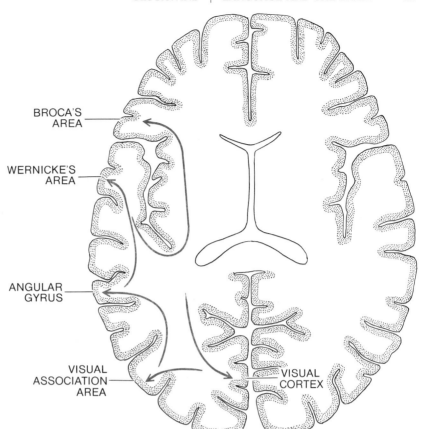

SAYING THE NAME of a seen object, according to Wernicke's model, involves the transfer of the visual pattern to the angular gyrus, which contains the "rules" for arousing the auditory form of the pattern in Wernicke's area. From here the auditory form is transmitted by way of the arcuate fasciculus to Broca's area. There the articulatory form is aroused, is passed on to the face area of the motor cortex and the word then is spoken.

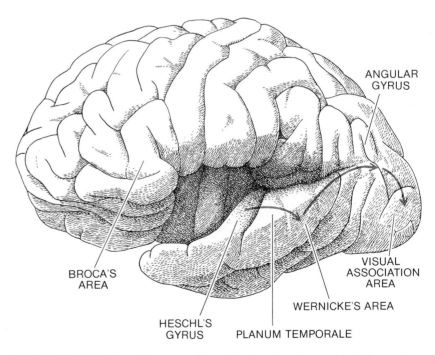

UNDERSTANDING the spoken name of an object involves the transfer of the auditory stimuli from Heschl's gyrus (the primary auditory cortex) to Wernicke's area and then to the angular gyrus, which arouses the comparable visual pattern in the visual association cortex. Here the Sylvian fissure has been spread apart to show the pathway more clearly.

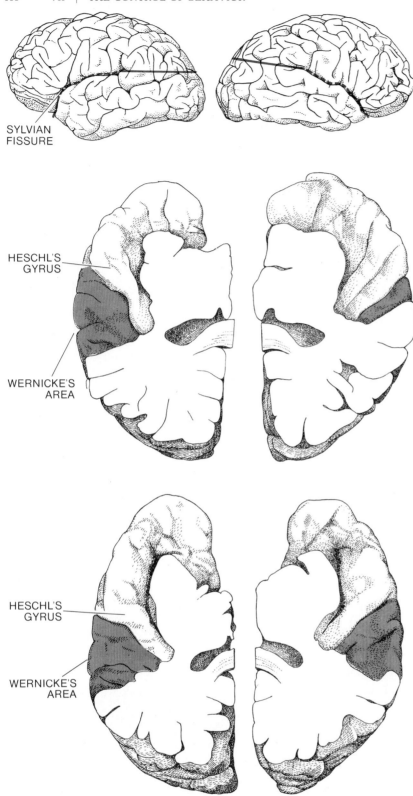

SYLVIAN FISSURE

HESCHL'S GYRUS

WERNICKE'S AREA

HESCHL'S GYRUS

WERNICKE'S AREA

ANATOMICAL DIFFERENCES between the two hemispheres of the human brain are found on the upper surface of the temporal lobe, which cannot be seen in an intact brain because it lies within the Sylvian fissure. Typically the Sylvian fissure in the left hemisphere appears to be pushed down compared with the Sylvian fissure on the right side (*top illustration*). In order to expose the surface of the temporal lobe a knife is moved along the fissure (*broken line*) and then through the brain, cutting away the top portion (*solid line*). The region studied was the planum temporale (*colored areas*), an extension of Wernicke's area. The middle illustration shows a brain with a larger left planum; the bottom illustration shows left and right planums of about the same size. In a study of 100 normal human brains planum temporale was larger on the left side in 65 percent of the cases, equal on both sides in 24 percent of the cases and larger on the right side in 11 percent.

from reaching the angular gyrus of the left hemisphere. According to Wernicke's model, it is the left angular gyrus that converts the visual pattern of a word into the auditory pattern; without such conversion a seen word cannot be comprehended. Other workers have since shown that when a person is blind in the right half of the visual field but is still capable of reading, the portion of the corpus callosum that transfers visual information between the hemispheres is not damaged.

In 1937 the first case in which surgical section of the corpus callosum stopped the transfer of information between the hemispheres was reported by John Trescher and Frank Ford. The patient had the rear portion of his corpus callosum severed during an operation to remove a brain tumor. According to Wernicke's model, this should have resulted in the loss of reading ability in the left half of the visual field. Trescher and Ford found that the patient could read normally when words appeared in his right visual field but could not read at all in his left visual field.

Hugo Liepmann, who was one of Wernicke's assistants in Breslau, made an extensive study of syndromes of the corpus callosum, and descriptions of these disorders were a standard part of German neurology before World War I. Much of this work was neglected, and only recently has its full importance been appreciated. Liepmann's analysis of corpus callosum syndromes was based on Wernicke's model. In cases such as those described by Liepmann the front four-fifths of the corpus callosum is destroyed by occlusion of the cerebral artery that nourishes it. Since the splenium is preserved the patient can read in either visual field. Such a lesion, however, gives rise to three characteristic disorders. The patient writes correctly with his right hand but incorrectly with the left. He carries out commands with his right arm but not with the left; although the left hemisphere can understand the command, it cannot transmit the message to the right hemisphere. Finally, the patient cannot name objects held in his left hand because the somesthetic sensations cannot reach the verbal centers in the left hemisphere.

The problem of cerebral dominance in humans has intrigued investigators since Broca first discovered it. Many early neurologists claimed that there were anatomical differences between the hemispheres, but in the past few decades there has been a tendency to assume that the left and right hemispheres are

	BROCA'S APHASIA				WERNICKE'S APHASIA
MEANING	KANA		KANJI		
	PATIENT'S	CORRECT	PATIENT'S	CORRECT	
INK	キンス (KINSU)	インキ (INKI)	墨 (SUMI)	墨	参 答 微 (LONG TIME) 久 (SOLDIER)
UNIVERSITY	ヌイ (TAI)	ダイガク (DAIGAKU)	大學	大学 (GREAT LEARNING)	
TOKYO	トツ (TOU)	トウキヨウ (TOKYO)	東京	東京 (EAST CAPITAL)	

JAPANESE APHASICS display some characteristics rarely found in Western patients because of the unique writing system used in Japan. There are two separate forms of such writing. One is Kana, which is syllabic. The other is Kanji, which is ideographic. Kana words are articulated syllable by syllable and are not easily identified at a glance, whereas each Kanji character simultaneously represents both a sound and a meaning. A patient with Broca's aphasia, studied by Tsuneo Imura and his colleagues at the Nihon University College of Medicine, was able to write a dictated word correctly in Kanji but not in Kana (*top left*). When the patient was asked to write the word "ink," even though there is no Kanji character for the word, his first effort was the Kanji character "sumi," which means india ink. When required to write in Kana, the symbols he produced were correct but the word was wrong. Another patient who had Wernicke's aphasia wrote Kanji quickly and without hesitation. He was completely unaware that he was producing meaningless ideograms, as are patients who exhibit paraphasias in speech. Only two of characters had meaning (*top right*).

symmetrical. It has been thought that cerebral dominance is based on undetected subtle physiological differences not reflected in gross structure. Walter Levitsky and I decided to look again into the possibility that the human brain is anatomically asymmetrical. We studied 100 normal human brains, and we were surprised to find that striking asymmetries were readily visible. The area we studied was the upper surface of the temporal lobe, which is not seen in the intact brain because it lies within the depths of the Sylvian fissure. The asymmetrical area we found and measured was the planum temporale, an extension of Wernicke's area [*see illustration on opposite page*]. This region was larger on the left side of the brain in 65 percent of the cases, equal in 24 percent and larger on the right side in 11 percent. In absolute terms the left planum was nine millimeters longer on the average than the right planum. In relative terms the left planum was one-third longer than the right. Statistically all the differences were highly significant. Juhn A. Wada of the University of British Columbia subsequently reported a study that confirmed our results. In addition Wada studied a series of brains from infants who had died soon after birth and found that the planum asymmetry was present. It seems likely that the asymmetries of the brain are genetically determined.

It is sometimes asserted that the anatomical approach neglects the plasticity of the nervous system and makes the likelihood of therapy for language disorders rather hopeless. This is not the case. Even the earliest investigators of aphasia were aware that some patients developed symptoms that were much milder than expected. Other patients recovered completely from a lesion that normally would have produced permanent aphasia. There is recovery or partial recovery of language functions in some cases, as Luria's large-scale study of the war wounded has shown. Of all the patients with wounds in the primary speech area of the left hemisphere, 97.2 percent were aphasic when Luria first examined them. A follow-up examination found that 93.3 percent were still aphasic, although in most cases they were aphasic to a lesser degree.

How does one account for the apparent recovery of language function in some cases? Some partial answers are available. Children have been known to make a much better recovery than adults with the same type of lesion. This suggests that at least in childhood the right hemisphere has some capacity to take over speech functions. Some cases of adult recovery are patients who had suffered brain damage in childhood. A number of patients who have undergone surgical removal of portions of the speech area for the control of epileptic seizures often show milder language disorders than had been expected. This probably is owing to the fact that the patients had suffered from left temporal epilepsy involving the left side of the brain from childhood and had been using the right hemisphere for language functions to a considerable degree.

Left-handed people also show on the average milder disorders than expected when the speech regions are damaged, even though for most left-handers the left hemisphere is dominant for speech just as it is for right-handers. It is an interesting fact that right-handers with a strong family history of left-handedness show better speech recovery than people without left-handed inheritance.

Effective and safe methods for studying cerebral dominance and localization of language function in the intact, normal human brain have begun to appear. Doreen Kimura of the University of Western Ontario has adapted the technique of dichotic listening to investigate the auditory asymmetries of the brain. More recently several investigators have found increased electrical activity over the speech areas of the left hemisphere during the production or perception of speech. Refinement of these techniques could lead to a better understanding of how the normal human brain is organized for language. A deeper understanding of the neural mechanisms of speech should lead in turn to more precise methods of dealing with disorders of man's most characteristic attribute, language.

BIBLIOGRAPHIES

I AN INTRODUCTION TO CELLULAR COMMUNICATION

1. Cellular Communication

STRUCTURALISM. Jean Piaget. Basic Books, Inc., 1970.

PAPERS IN CELLULAR NEUROPHYSIOLOGY. Edited by I. Cooke and M. Lipkin, Jr. Holt, Rinehart, and Winston, 1972.

II LEVELS OF CELLULAR COMPLEXITY

2. The Living Cell

THE CELL: BIOCHEMISTRY, PHYSIOLOGY, MORPHOLOGY. Vol. I: METHODS; PROBLEMS OF CELL BIOLOGY. VOL. II: CELLS AND THEIR COMPONENT PARTS. Edited by Jean Brachet and Alfred E. Mirsky. Academic Press Inc., 1959; 1961.

CELL GROWTH AND CELL FUNCTION. Torbjörn O. Caspersson. W. W. Norton and Company, Inc., 1950.

THE ULTRASTRUCTURE OF CELLS. Marcel Bessis, Sandoz Monographs, 1960.

3. Viruses and Genes

THE CONCEPT OF VIRUS. A. Lwoff in The Journal of General Microbiology, Vol. 17, No. 2, pages 239–253; October, 1957.

GENETIC CONTROL OF VIRAL FUNCTIONS. François Jacob in The Harvey Lectures, Series LIV, 1958–1959, pages 1–39. Academic Press Inc., 1960.

MICROBIAL GENETICS. Tenth Symposium of the Society for General Microbiology. Cambridge University Press, 1960.

PHYSIOLOGICAL ASPECTS OF BACTERIOPHAGE GENETICS. S. Brenner in Advances in Virus Research, Vol. 6, pages 137–158; 1959.

A SYMPOSIUM ON THE CHEMICAL BASIS OF HEREDITY. Edited by William D. McElroy. Johns Hopkins Press, 1957.

VIRUSES AS INFECTIVE GENETIC MATERIAL. S. E. Luria in Immunity and Virus Infection, edited by Victor A. Najjar, pages 188–195. John Wiley and Sons, Inc., 1959.

THE VIRUSES: BIOCHEMICAL, BIOLOGICAL, AND BIOPHYSICAL PROPERTIES. Edited by F. M. Burnet and W. M. Stanley. Academic Press Inc., 1956.

4. The Smallest Living Cells

BIOLOGY OF THE PLEUROPNEUMONIA-LIKE ORGANISMS. Annals of the New York Academy of Sciences, Vol. 79, Art. 10, pages 304–758; January 15, 1960.

NUTRITION, METABOLISM, AND PATHOGENICITY OF MYCOPLASMAS. H. E. Adler and Moshé Shifrine in Annual Review of Microbiology, Vol. 14, pages 141–160; 1960.

THE PLEUROPNEUMONIA GROUP OF ORGANISMS: A REVIEW, TOGETHER WITH SOME NEW OBSERVATIONS. D. G. ff. Edward in The Journal of General Microbiology, Vol. 10, No. 1, pages 27–64; February, 1954.

PLEUROPNEUMONIA-LIKE ORGANISMS (PPLO)—MYCOPLASMATACEAE. E. Klieneberger-Nobel. Academic Press Inc., 1962.

5. The Blue-Green Algae

PHYSIOLOGY AND BIOCHEMISTRY OF ALGAE. Edited by R. A. Lewin. Academic Press Inc., 1962.

THE RELATIONSHIP BETWEEN BLUE-GREEN ALGAE AND BACTERIA. P. Echlin and I. Morris in *Biological Reviews*, Vol. 40, No. 2, pages 143–187; May, 1965.

THE STRUCTURE AND REPRODUCTION OF THE ALGAE: VOLUME II. F. E. Fritsch. Cambridge University Press, 1965.

6. Differentiation in Social Amoebae

THE CELLUAR SLIME MOLDS. John Tyler Bonner, Princeton University Press, 1959.

EVIDENCE FOR THE SORTING OUT OF CELLS IN THE DEVELOPMENT OF THE CELLULAR SLIME MOLDS. John Tyler Bonner in *Proceedings of the National Academy of Sciences*, Vol. 45, No. 3, pages 379–384; March, 1959.

III   CELLULAR DIFFERENTIATION AND DEVELOPMENT

7. "The Organizer"

ANALYSIS OF DEVELOPMENT. Edited by Benjamin H. Willier, Paul A. Weiss and Viktor Hamburger. W. B. Saunders Company, 1955.

ASPECTS OF SYNTHESIS AND ORDER IN GROWTH. Edited by Dorothea Rudnick. Princeton University Press, 1954.

DYNAMICS OF GROWTH PROCESSES. Edited by Edgar J. Boell. Princeton University Press, 1954.

EMBRYONIC TRANSPLANTATION AND THE DEVELOPMENT OF THE NERVOUS SYSTEM. Ross G. Harrison in *The Harvey Lectures*, pages 199–222; 1907-08.

EMBRYONIC DEVELOPMENT AND INDUCTION. Hans Spemann. Yale University Press, 1938.

8. Transplanted Nuclei and Cell Differentiation

NUCLEOCYTOPLASMIC INTERACTIONS IN EGGS AND EMBRYOS. Robert Briggs and Thomas J. King in *The Cell: Biochemistry, Physiology, Morphology, Vol. I*, edited by Jean Brachet and Alfred E. Mirsky. Academic Press, 1959.

NUCLEAR TRANSPLANTATION IN AMPHIBIA AND THE IMPORTANCE OF STABLE NUCLEAR CHANGES IN PROMOTING CELLULAR DIFFERENTIATION. J. B. Gurdon in *The Quarterly Review of Biology*, Vol. 38, No. 1, pages 54–78; March, 1963.

INTERACTING SYSTEMS IN DEVELOPMENT. James D. Ebert. Holt, Rinehart, and Winston, 1965.

NUCLEAR TRANSPLANTATION IN AMPHIBIA. Thomas J. King in *Methods in Cell Physiology, Vol. II*, edited by David M. Prescott. Academic Press, 1966.

THE CYTOPLASMIC CONTROL OF NUCLEAR ACTIVITY IN ANIMAL DEVELOPMENT. J. B. Gurdon and H. R. Woodland in *Biological Reviews of the Cambridge Philosophical Society*, Vol. 43, No. 2, pages 233–267; May, 1968.

9. Phases in Cell Differentiation

EARLY PANCREAS ORGANOGENESIS: MORPHOGENESIS, TISSUE INTERACTIONS, AND MASS EFFECTS. Norman K. Wessells and Julia H. Cohen in *Developmental Biology*, Vol. 15, No. 3, pages 237–270; March, 1967.

MULTIPHASIC REGULATION IN CYTODIFFERENTIATION. William J. Rutter, William R. Clark, John D. Kemp, William S. Bradshaw, Thomas G. Sanders and William D. Ball in *Epithelial-Mesenchymal Interactions*, edited by Raul Fleischmajer and Rupert E. Billingham. The Williams and Wilkins Company, 1968.

ULTRASTRUCTURAL STUDIES OF EARLY MORPHOGENESIS AND CYTODIFFERENTIATION IN THE EMBRYONIC MAMMALIAN PANCREAS. Norman K. Wessells and Jean Evans in *Developmental Biology*, Vol. 17, No. 4, pages 413–446; April, 1968.

REGULATION OF SPECIFIC PROTEIN SYNTHESIS IN CYTODIFFERENTIATION. W. J. Rutter, J. D. Kemp, W. S. Bradshaw, W. R. Clark, R. A. Ronzio and T. G. Sanders in *Journal of Cellular Physiology*, Vol. 72, No. 2, Part II, pages 1–18; October, 1968.

10. How Cells Associate

GUIDING PRINCIPLES IN CELL LOCOMOTION AND CELL AGGREGATION. Paul Weiss in *Experimental Cell Research*, Supplement 8, pages 260–281; 1961.

IMMUNOLOGICAL RECOGNITION OF SELF. F. M. Burnet in *Science*, Vol. 133, No. 3449, pages 307–311; February 3, 1961.

ROTATION-MEDIATED HISTOGENIC AGGREGATION OF DISSOCIATED CELLS. A. Moscona in *Experimental Cell Research*, Vol. 22, pages 455–475; 1961.

11. The Growth of Nerve Circuits

DEVELOPMENTAL BASIS OF BEHAVIOR. R. W. Sperry in *Behavior and Evolution,* edited by Anne Roe and George Gaylord Simpson, pages 128–138. Yale University Press, 1958.

MECHANISMS OF NEURAL MATURATION. R. W. Sperry in *Handbook of Experimental Psychology,* edited by S. S. Stevens, pages 236–275. John Wiley and Sons, Inc., 1951.

PHYSIOLOGICAL PLASTICITY AND BRAIN CIRCUIT THEORY. R. W. Sperry in *Biological and Biochemical Bases of Behavior,* edited by Harry H. Harlow and Clinton N. Woolsey, pages 401–421. University of Wisconsin Press, 1958.

THE PROBLEM OF CENTRAL NERVOUS REORGANIZATION AFTER NERVE REGENERATION AND MUSCLE TRANSPOSITION. R. W. Sperry in *The Quarterly Review of Biology,* Vol. 20, No. 4, pages 311–369; December, 1945.

REGULATIVE FACTORS IN THE ORDERLY GROWTH OF NEURAL CIRCUITS. R. W. Sperry in *Growth Symposium, Vol. X,* pages 63–87; 1951.

IV THE CELLULAR BASIS OF INTEGRATION

12. How Cells Communicate

BIOPHYSICAL ASPECTS OF NEUROMUSCULAR TRANSMISSION. J. del Castillo and B. Katz in *Progress in Biophysics and Biophysical Chemistry,* Vol. 6, pages 121–170; 1956.

IONIC MOVEMENT AND ELECTRICAL ACTIVITY IN GIANT NERVE FIBRES. A. L. Hodgkin in *Proceedings of the Royal Society,* Series B, Vol. 148, No. 930, pages 1–37; January 1, 1958.

MICROPHYSIOLOGY OF THE NEUROMUSCULAR JUNCTION, A PHYSIOLOGICAL "QUANTUM OF ACTION" AT THE MYONEURAL JUNCTION. Bernhard Katz in *Bulletin of the Johns Hopkins Hospital,* Vol. 102, No. 6, pages 275–312; June, 1958.

THE PHYSIOLOGY OF NERVE CELLS. John Carew Eccles. The Johns Hopkins Press, 1957.

13. How Living Cells Change Shape

THE ASSEMBLY OF MICROTUBULES AND THEIR ROLE IN THE DEVELOPMENT OF CELL FORM. Lewis G. Tilney in *The Emergence of Order in Developing Systems: The 27th Symposium of the Society for Developmental Biology,* edited by Michael Locke. Academic Press, 1968.

INTRA- AND EXTRACELLULAR CONTROL OF EPITHELIAL MORPHOGENESIS. Merton R. Bernfield and Norman K. Wessells in *Changing Syntheses in Development: The 29th Symposium of the Society for Developmental Biology,* edited by Meredith N. Runner. Academic Press, 1971.

MICROFILAMENTS IN CELLULAR AND DEVELOPMENTAL PROCESSES. N. K. Wessells, B. S. Spooner, J. F. Ash, M. O. Bradley, M. A. Luduena, E. L. Taylor, J. T. Wrenn and K. M. Yamada in *Science,* Vol. 171, No. 3967, pages 135–143; January 15, 1971.

14. The Mechanism of Muscular Contraction

A DISCUSSION OF THE PHYSICAL AND CHEMICAL BASIS OF MUSCULAR CONTRACTION. Organized by A. F. Huxley and H. E. Huxley in *Proceedings of the Royal Society,* Series B, Vol. 160, No. 981, pages 433–542; October 27, 1964.

ELECTRON MICROSCOPE STUDIES ON THE STRUCTURE OF NATURAL AND SYNTHETIC PROTEIN FILAMENTS FROM STRIATED MUSCLE. H. E. Huxley in *Journal of Molecular Biology,* Vol. 7, No. 3, pages 281–308; September, 1963.

FILAMENT LENGTHS IN STRIATED MUSCLE. Sally G. Page and H. E. Huxley in *The Journal of Cell Biology,* Vol. 19, No. 2, pages 369–390; November, 1963.

THE STRUCTURE OF F-ACTIN AND OF ACTIN FILAMENTS ISOLATED FROM MUSCLE. Jean Hanson and J. Lowy in *Journal of Molecular Biology,* Vol. 6, No. 1, pages 46–60; January, 1963.

X-RAY DIFFRACTION FROM LIVING STRIATED MUSCLE DURING CONTRACTION. G. F. Elliott, J. Lowy and B. M. Millman in *Nature,* Vol. 206, No. 4991, pages 1357–1358; June 26, 1965.

15. The Nerve Axon

THE CONDUCTION OF THE NERVOUS IMPULSE. A. L. Hodgkin. Liverpool University Press, 1964.

16. The Synapse

EXCITATION AND INHIBITION IN SINGLE NERVE CELLS. Stephen W. Kuffler in *The Harvey Lectures, Series 54.* Academic Press, 1960.

PHYSIOLOGY OF NERVE CELLS. John C. Eccles. Johns Hopkins Press, 1957.

THE PHYSIOLOGY OF SYNAPSES. John Carew Eccles. Academic Press, 1964.

THE TRANSMISSION OF IMPULSES FROM NERVE TO MUSCLE, AND THE SUBCELLULAR UNIT OF SYNAPTIC ACTION. B. Katz in *Proceedings of the Royal Society,* Vol. 155, No. 961, Series B, pages 455–477; April, 1962.

V   CHEMICAL COMMUNICATION

17. Chromosome Puffs

CHROMOSOMES AND CYTODIFFERENTIATION. Joseph G. Gall in *Cytodifferentiation and Macromolecular Synthesis,* edited by Michael Locke. Academic Press, 1963.

NUCLEIC ACIDS AND CELL MORPHOLOGY IN DIPTERAN SALIVARY GLANDS. Hewson Swift in *The Molecular Control of Cellular Activity,* edited by John M. Allen. McGraw-Hill Book Company, 1962.

RIESENCHROMOSOMEN. Wolfgang Beermann in *Protoplasmatologia,* Vol. VI/D. Springer-Verlag, 1962.

UNTERSUCHUNGEN AN RIESENCHROMOSOMEN ÜBER DIE WIRKUNGSWEISE DER GENE. Ulrich Clever in *Materia Medica Nordmark,* Vol. 15, No. 10, pages 438–452; July, 1962.

18. Hormones and Genes

EFFECT OF ACTINOMYCIN AND INSULIN ON THE METABOLISM OF ISOLATED RAT DIAPHRAGM. Ira G. Wool and Arthur N. Moyer in *Biochimica et Biophysica Acta,* Vol. 91, No. 2, pages 248–256; October 16, 1964.

ON THE MECHANISM OF ACTION OF ALDOSTERONE ON SODIUM TRANSPORT: THE ROLE OF RNA SYNTHESIS. George A. Porter, Rita Bogoroch and Isidore S. Edelman in *Proceedings of the National Academy of Sciences,* Vol. 52, No. 6, pages 1326–1333; December, 1964.

PREVENTION OF HORMONE ACTION BY LOCAL APPLICATION OF ACTINOMYCIN D. G. P. Talwar and Sheldon J. Segal in *Proceedings of the National Academy of Sciences,* Vol. 50, No. 1, pages 226–230; July 15, 1963.

SELECTIVE ALTERATIONS OF MAMMALIAN MESSENGER-RNA SYNTHESIS: EVIDENCE FOR DIFFERENTIAL ACTION OF HORMONES ON GENE TRANSCRIPTION. Chev Kidson and K. S. Kirby in *Nature,* Vol. 203, No. 4945, pages 599–603; August 8, 1964.

TRANSFER RIBONUCLEIC ACIDS. E. N. Carlsen, G. J. Trelle and O. A. Schjeide in *Nature,* Vol. 202, No. 4936, pages 984–986; June 6, 1964.

19. Cyclic AMP

CYCLIC ADENOSINE MONOPHOSPHATE IN BACTERIA. Ira Pastan and Robert Perlman in *Science,* Vol. 169, No. 3943, pages 339–344; July 24, 1970.

CYCLIC AMP. G. Alan Robison, Reginald W. Butcher and Earl W. Sutherland. Academic Press, 1971.

CYCLIC AMP AND CELL FUNCTION. Edited by G. Alan Robison, Gabriel G. Nahas and Lubos Triner in *Annals of the New York Academy of Sciences,* Vol. 185; December 3, 1971.

20. Intercellular Communication

PERMEABILITY OF MEMBRANE JUNCTIONS. Werner R. Loewenstein in *Annals of the New York Academy of Sciences,* Vol. 137, Art. 2, pages 441–472; July 14, 1966.

CONTACT AND SHORT-RANGE INTERACTIONS AFFECTING GROWTH OF ANIMAL CELLS IN CULTURE. Michael Stoker in *Current Topics in Developmental Biology: Vol. II,* edited by A. A. Moscona and Alberto Monroy. Academic Press, 1967.

COMMUNICATIONS THROUGH CELL JUNCTIONS: IMPLICATIONS IN GROWTH AND DIFFERENTIATION. Werner R. Loewenstein in *The Emergence of Order in Developing Systems: Developmental Biology, Supplement 2,* edited by Michael Locke. Academic Press, 1968.

LOW-RESISTANCE JUNCTIONS BETWEEN CELLS IN EMBRYOS AND TISSUE CULTURE. Edwin J. Furshpan and David D. Potter in *Current Topics in Developmental Biology: Vol. III,* edited by A. A. Moscona and Alberto Monroy. Academic Press, 1968.

21. Pheromones

OLFACTORY STIMULI IN MAMMALIAN REPRODUCTION. A. S. Parkes and H. M. Bruce in *Science*, Vol. 134, No. 3485, pages 1049–1054; October, 1961.

PHEROMONES (ECTOHORMONES) IN INSECTS. Peter Karlson and Adolf Butenandt in *Annual Review of Entomology*, Vol. 4, pages 39–58; 1959.

THE SOCIAL BIOLOGY OF ANTS. Edward O. Wilson in *Annual Review of Entomology*, Vol. 8, pages 345–368; 1963.

VI SENSORY PROCESSES

22. How Cells Receive Stimuli

INITIATION OF IMPULSES AT RECEPTORS. J. A. B. Gray in *Handbook of Physiology, Vol. I, Section I: Neurophysiology*, pages 123–145. American Physiological Society, 1959.

THE NEURAL MECHANISMS OF VISION. H. K. Hartline in *The Harvey Lectures, 1941–1942*. Series 37, pages 39–68; 1942.

RECEPTORS AND SENSORY PERCEPTION. R. Granit. Yale University Press, 1955.

SENSORY COMMUNICATION. Edited by Walter A. Rosenblith. John Wiley and Sons, Inc., 1961.

23. Eye and Camera

VISION AND THE EYE. M. H. Pirenne. The Pilot Press, Ltd., 1948.

THE RETINA. S. Polyak. University of Chicago Press, 1941.

THE PHOTOCHEMISTRY OF VISION. George Wald in *Documenta Ophthalmologica*, Vol. 3, page 94; 1949.

THE LIGHT REACTION IN THE BLEACHING OF RHODOPSIN. George Wald, Jack Durell and C. C. St. George in *Science*, Vol. 3, No. 2,877, pages 179–181; February 17, 1950.

24. Visual Pigments in Man

CHEMICAL BASIS OF HUMAN COLOUR VISION. W. A. H. Rushton in *Research*, Vol. 11, No. 12, pages 478–483; December, 1958.

THE VISUAL PIGMENTS. H. J. A. Dartnall. John Wiley and Sons, Inc., 1957.

VISUAL PIGMENTS IN MAN. W. A. H. Rushton. Liverpool University Press, 1962.

25. The Visual Cortex of the Brain

DISCHARGE PATTERNS AND FUNCTIONAL ORGANIZATION OF MAMMALIAN RETINA. Stephen W. Kuffler in *Journal of Neurophysiology*, Vol. 16, No. 1, pages 37–68; January, 1953.

INTEGRATIVE PROCESSES IN CENTRAL VISUAL PATHWAYS OF THE CAT. David M. Hubel in *Journal of the Optical Society of America*, Vol. 53, No. 1, pages 58–66; January, 1963.

RECEPTIVE FIELDS, BINOCULAR INTERACTION AND FUNCTIONAL ARCHITECTURE IN THE CAT'S VISUAL CORTEX. D. H. Hubel and T. N. Wiesel in *Journal of Physiology*, Vol. 160, No. 1, pages 106–154; January, 1962.

THE VISUAL PATHWAY. Ragnar Granit in *The Eye, Volume II: The Visual Process*, edited by Hugh Davson. Academic Press, 1962.

26. The Neurophysiology of Binocular Vision

THE NEURAL MECHANISM OF BINOCULAR DEPTH DISCRIMINATION. H. B. Barlow, C. Blakemore and J. D. Pettigrew in *Journal of Physiology*, Vol. 193, pages 327–342; 1967.

BINOCULAR INTERACTION ON SINGLE UNITS IN CAT STRIATE CORTEX: SIMULTANEOUS STIMULATION BY SINGLE MOVING SLIT WITH RECEPTIVE FIELDS IN CORRESPONDENCE. J. D. Pettigrew, T. Nikara and P. O. Bishop in *Experimental Brain Research*, Vol. 6, pages 391–410; 1968.

EYE DOMINANCE IN THE VISUAL CORTEX. Colin Blakemore and John D. Pettigrew in *Nature*, Vol. 225, No. 5231, pages 426–429; January 31, 1970.

FOUNDATIONS OF CYCLOPEAN PERCEPTION. Bela Julesz. University of Chicago Press, 1971.

VII   THE CONTROL OF BEHAVIOR

27. How We Control the Contraction of Our Muscles

TREATISE ON PHYSIOLOGICAL OPTICS: VOL. III. Hermann L. F. Helmholtz. English translation by James P. C. Southall. Dover Publications, Inc., 1962.

POSITION SENSE AND SENSE OF EFFORT. P. A. Merton in *Homeostasis and Feedback Mechanisms*. Symposia of the Society for Experimental Biology, Vol. 18, pages 387–400; 1964.

THE INNERVATION OF MAMMALIAN SKELETAL MUSCLE. D. Barker in *Ciba Foundation Symposium: Myotatic, Kinesthetic and Vestibular Mechanisms*, edited by A. V. S. de Reuck and Julie Knight. Little, Brown and Company, 1967.

SERVO ACTION AND STRETCH REFLEX IN HUMAN MUSCLE AND ITS APPARENT DEPENDENCE ON PERIPHERAL SENSATION. C. D. Marsden, P. A. Merton and H. B. Morton in *The Journal of Physiology*, Vol. 216, pages 21–22P; July, 1971.

PROJECTION FROM LOW THRESHOLD MUSCLE AFFERENTS OF HAND AND FOREARM TO AREA 3A OF BABOON'S CORTEX. C. G. Phillips, T. P. S. Powell and M. Wiesendanger in *The Journal of Physiology*, Vol. 217, pages 419–446; September, 1971.

28. Small Systems of Nerve Cells

HETEROSYNAPTIC FACILITATION IN NEURONS OF THE ABDOMINAL GANGLION OF APLYSIA DEPILANS. E. R. Kandel and L. Tauc in *Journal of Physiology* (London), Vol. 181, No. 1, pages 1–27; November, 1965.

ON THE FUNCTIONAL ANATOMY OF NEURONAL UNITS IN THE ABDOMINAL CORD OF THE CRAYFISH, PROCAMBARUS CLARKII (GIRARD). C. A. G. Wiersma and G. M. Hughes in *The Journal of Comparative Neurology*, Vol. 116, No. 2, pages 209–228; April, 1961.

RELEASE OF COORDINATED BEHAVIOR IN CRAYFISH BY SINGLE CENTRAL NEURONS. Donald Kennedy, W. H. Evoy and J. T. Hanawalt in *Science*, Vol. 154, No. 3751, pages 917–919; November 18, 1966.

TYPES OF INFORMATION STORED IN SINGLE NEURONS. Felix Strumwasser in *Invertebrate Nervous Systems: Their Significance for Mammalian Neurophysiology*, edited by Cornelius A. G. Wiersma. The University of Chicago Press, 1967.

29. The Flight-Control System of the Locust

THE CENTRAL NERVOUS CONTROL OF FLIGHT IN A LOCUST. Donald M. Wilson in *The Journal of Experimental Biology*, Vol. 38, No. 2, pages 471–490; June, 1961.

EXPLORATION OF NEURONAL MECHANISMS UNDERLYING BEHAVIOR IN INSECTS. Graham Hoyle in *Neural Theory and Modeling: Proceedings of the 1962 Ojai Symposium*, edited by Richard F. Reiss. Stanford University Press, 1964.

30. Moths and Ultrasound

THE DETECTION AND EVASION OF BATS BY MOTHS. Kenneth D. Roeder and Asher E. Treat in *American Scientist*, Vol. 49, No. 2, pages 135–148; June, 1961.

MOTH SOUNDS AND THE INSECT-CATCHING BEHAVIOR OF BATS. Dorothy C. Dunning and Kenneth D. Roeder in *Science*, Vol. 147, No. 3654, pages 173–174; January 8, 1965.

NERVE CELLS AND INSECT BEHAVIOR. Kenneth D. Roeder. Harvard University Press, 1963.

31. Language and the Brain

CEREBRAL DOMINANCE AND ITS RELATION TO PSYCHOLOGICAL FUNCTION. O. L. Zangwill. Oliver and Boyd, 1960.

DISCONNEXION SYNDROMES IN ANIMALS AND MAN: PART I. Norman Geschwind in *Brain*, Vol. 88, Part 2, pages 237–294; June, 1965.

DISCONNEXION SYNDROMES IN ANIMALS AND MAN: PART II. Norman Geschwind in *Brain*, Vol. 88, Part 3, pages 585–644; September, 1965.

HUMAN BRAIN: LEFT-RIGHT ASYMMETRIES IN TEMPORAL SPEECH REGION. Norman Geschwind and Walter Levitsky in *Science*, Vol. 161, No. 3837, pages 186–187; July 12, 1968.

TRAUMATIC APHASIA: ITS SYNDROMES, PSYCHOLOGY AND TREATMENT. A. R. Luria. Mouton and Co., 1970.

INDEX